Infosys Science Foundation Series

Infosys Science Foundation Series in Mathematical Sciences

The **Infosys Science Foundation Series in Mathematical Sciences,** a Scopus-indexed book series, focuses on high-quality content in the domain of pure and applied mathematics, biomathematics, financial mathematics, operations research and theoretical computer science. With this series, Springer hopes to provide readers with monographs, textbooks, edited volume, handbooks and professional books of the highest academic quality on current topics in relevant disciplines. Literature in this series will appeal to a wide audience of researchers, students, educators and professionals across the world.

Bang-Yen Chen • Majid Ali Choudhary
Editors

Geometric Inequalities and Applications

Springer

Editors
Bang-Yen Chen
Department of Mathematics
Michigan State University
Okemos, MI, USA

Majid Ali Choudhary
School of Sciences
Maulana Azad National Urdu University
Hyderabad, Telangana, India

ISSN 2363-6149 ISSN 2363-6157 (electronic)
Infosys Science Foundation Series
ISSN 2364-4036 ISSN 2364-4044 (electronic)
Infosys Science Foundation Series in Mathematical Sciences
ISBN 978-981-95-5147-7 ISBN 978-981-95-5148-4 (eBook)
https://doi.org/10.1007/978-981-95-5148-4

Mathematics Subject Classification: 53A04, 53A55, 53B15, 53B20, 53C40, 53E20, 58C40, 53Z05, 58D17, 53C20, 35Q51, 53B25, 53B50, 53B30, 53C25, 58J20, 70G45, 83C05, 53B05, 53B12, 53C15, 58J60, 53A45, 53C05, 53C21, 53C50, 53C12, 53C55, 58D1715, 53C80, 35C15

This Springer imprint is published by the registered company Springer Nature Singapore Pte Ltd.
The registered company address is: 152 Beach Road, #21-01/04 Gateway East, Singapore 189721, Singapore

Preface

Geometric inequalities play a crucial role in various branches of science and engineering, providing foundational tools for theoretical development and problem-solving. In mathematics, applications of geometric inequalities span across geometry, analysis, and optimization, among many others. For instance, the *priori* estimates for differential equations and the theory of functions of complex variables provide many such examples. In physics, particularly in general relativity and string theory, geometric inequalities help describe spacetime geometries and their properties under various physical conditions. For example, physicists can use geometric inequalities to derive constraints on energy distributions based on geometric configurations.

Geometric inequalities also play a crucial role in geometric analysis, particularly in studying elliptic partial differential equations related to geometric flows. They help establish regularity results for solutions to these equations by controlling various norms associated with functions defined on (Riemannian) manifolds. Also, the Sobolev inequalities provide bounds on functions defined on manifolds concerning their derivatives and integrals over certain domains. These inequalities are instrumental in studying partial differential equations on manifolds and have implications for embedding theorems.

In differential geometry, geometric inequalities play a crucial role, providing important tools for understanding the properties of geometric objects, such as Riemannian manifolds and their submanifolds. In addition, geometric inequalities help to analyze curvature properties of manifolds. For example, the Cauchy-Schwarz inequality has been applied to derive bounds on sectional curvature, which is crucial in studying Riemannian manifolds with constant curvature, such as spheres and hyperbolic spaces. Moreover, the relationship between curvature and geometric inequalities allows differential geometers to classify manifolds based on their curvature characteristics. In addition, geometric inequalities facilitate comparison theorems that relate different geometric structures. For example, Bonnet-Myers' theorem uses geometric inequalities to prove that if a Riemannian manifold has positive Ricci curvature, then it is always compact and it has a finite diameter. Such results are pivotal in understanding the global structure of Riemannian manifolds. In addition, embedding theorems often rely on geometric inequalities to establish conditions under which a manifold can be embedded into Euclidean space. For instance, the Whitney and Nash embedding theorems utilize concepts from geometric inequalities to demonstrate that any smooth manifold can be embedded into a Euclidean space of higher dimension.

This book is devoted to recent advances in a variety of geometric inequalities in differential geometry, as well as in the theory of solitons. As a result, this book consists of 15 chapters authored by leading mathematicians, encompassing a wide array of topics, including "Some Inequalities for Geometric Solitons" by A. M. Blaga, "Generalized Ricci-Yamabe Soliton on 3-Dimensional Lie Groups" by A. Delloum and G. Beldjilali, "Riemannian Invariants in Submanifold Theory" by A. Mihai, "Chen Inequalities for Submanifolds of Kenmotsu Space Forms" by I. Ünal, A. Barman and D. G. Prakasha, "Improved Chen-Ricci Inequalities for Semi-slant $\xi^{\perp}$-Riemannian Submersions from Sasakian Space Forms" by M. A. Akyol and N. Poyraz, "Characterizations of Perfect Fluid and Generalized Robertson-Walker Space-Time Admitting k Almost Ricci Yamabe Solitons" by K. De and U. C. De, "Riemannian Concircular Structure Manifold and Solitons" by S. K. Chaubey and A. Haseeb, "Statistical Maps and a Chen's First Inequality for These Maps" by S. Kazan and A. N. Siddiqui, "Hyperbolic Ricci-Yamabe Solitons and η-Hyperbolic Ricci-Yamabe Solitons" by M. D. Siddiqi, "A Survey on Hitchin-Thorpe Inequality and Its Extensions" by B.-Y. Chen, M. A. Choudhary, and M. Nisar, "The Principal Eigenvalue of a (p, q)-Biharmonic System Along the Ricci Flow" by S. Azami and Gh. Fasihi-Ramandi, "The Jacobi Geometry of Plane, Parametrized Curves and Associated Inequalities" by M. Crasmareanu, "B.-Y. Chen Inequalities for Submanifolds of a Conformally Flat Manifold" by C. Özür, "General Chen Inequalities for Statistical Submanifolds in Kenmotsu Statistical Manifolds of Constant ϕ-Sectional Curvature" by S. Decu and G.-E. Vilcu, and "B. Y. Chen Inequalities for Pointwise Quasi Hemi-Slant Submanifolds of a Kaehler Manifold" by N. Poyraz, M. A. Akyol, and Erol Yasar.

Both editors of this book hope that readers will find this book a valuable reference for geometrical inequalities, enabling them to perform their research more effectively, successfully, and creatively.

East Lansing, MI, USA — Bang-Yen Chen
Hyderabad, India — Majid Ali Choudhary
April, 2025

Contents

Some Inequalities for Geometric Solitons

Adara M. Blaga

Abstract For different types of geometric solitons with a certain kind of potential vector field, we provided some necessary and sufficient inequalities that must be satisfied by the Ricci and the scalar curvatures for the solitons to be trivial. By means of similar inequalities, we have also given new characterizations of a Euclidean sphere.

Keywords Trivial soliton · Ricci curvature · Euclidean sphere

1 Preliminaries

Let (M, g) be an n-dimensional Riemannian manifold ($n > 2$). We denote by $C^\infty(M)$ the set of smooth real functions on M, by Ric the Ricci curvature tensor field, by Q the Ricci operator, by r the scalar curvature, by ∇ the Levi-Civita connection of g, and by $\pounds_\zeta$ the Lie derivative operator in the direction of a smooth vector field ζ tangent to M. We will briefly recall the definitions of certain types of geometric solitons which we shall use in the sequel. Let ζ be a smooth vector field and η a 1-form on M. Then, (M, g, ζ) is said to be:

1. An *almost Ricci soliton* [23] if

$$\frac{1}{2}\pounds_\zeta g + \mathrm{Ric} = \lambda g, \quad \text{where } \lambda \in C^\infty(M) \tag{1.1}$$

2. An *almost η-Ricci soliton* [2, 3, 18] if

$$\frac{1}{2}\pounds_\zeta g + \mathrm{Ric} = \lambda g + \mu \eta \otimes \eta, \quad \text{where } \lambda, \mu \in C^\infty(M) \tag{1.2}$$

A. M. Blaga (✉)
Faculty of Physics and Mathematics, Department of Mathematics, West University of Timişoara, Timişoara, Romania

B.-Y. Chen, M. A. Choudhary (eds.), *Geometric Inequalities and Applications*, Infosys Science Foundation Series in Mathematical Sciences,
https://doi.org/10.1007/978-981-95-5148-4_1

3. An *almost Einstein soliton* [11, 17] if

$$\frac{1}{2}\pounds_\zeta g + \mathrm{Ric} = \left(\lambda + \frac{r}{2}\right) g, \quad \text{where } \lambda \in C^\infty(M) \tag{1.3}$$

4. An *almost Ricci–Bourguignon soliton* [19] if

$$\frac{1}{2}\pounds_\zeta g + \mathrm{Ric} = (\lambda + \rho r) g, \quad \text{where } \lambda \in C^\infty(M),\ \rho \in \mathbb{R} \tag{1.4}$$

5. A *generalized Ricci soliton* [24] if

$$\frac{1}{2}\pounds_\zeta g + \alpha\,\mathrm{Ric} = \beta g, \quad \text{where } \alpha, \beta \in C^\infty(M) \tag{1.5}$$

6. A *generalized soliton* [8] if

$$\frac{1}{2}\pounds_\zeta g + \alpha\,\mathrm{Ric} = \beta g + \gamma \eta \otimes \eta, \quad \text{where } \alpha, \beta, \gamma \in C^\infty(M) \tag{1.6}$$

7. A *hyperbolic Ricci soliton* [20] if

$$\pounds_\zeta \pounds_\zeta g + \lambda \pounds_\zeta g + \mathrm{Ric} = \mu g, \quad \text{where } \lambda, \mu \in \mathbb{R} \tag{1.7}$$

8. A *hyperbolic Yamabe soliton* [15] if

$$\pounds_\zeta \pounds_\zeta g + \lambda \pounds_\zeta g = (\mu - r) g, \quad \text{where } \lambda, \mu \in \mathbb{R} \tag{1.8}$$

For the abovementioned solitons, if the functions are constant, then we drop *almost*. Also, for all types of solitons, if the potential vector field is *of gradient type* [25], i.e., $\zeta = \nabla f$, where ∇f denotes the gradient of $f \in C^\infty(M)$, then we use the name of *gradient soliton*, and we call f a *potential function*. Any of these solitons is called a *trivial soliton* if its potential vector field is a *Killing vector field* [25], i.e., $\pounds_\zeta g = 0$.

Two essential problems in the theory of solitons are (i) finding conditions under which a soliton is trivial and (ii) finding characterizations of the Euclidean sphere. The present chapter aims to collect some characterization results for trivial solitons with different types of potential vector fields, as well as some characterizations of a Euclidean sphere, by means of (integral or not) inequalities satisfied by the Ricci curvature tensor field (see [2–16, 24]), completed by new results.

2 Some Ricci Inequalities

We have provided, in [4], a sufficient inequality satisfied by the Ricci and the scalar curvatures for a compact almost Ricci soliton to be a *Ricci-flat manifold*, i.e., $\mathrm{Ric} = 0$. We recall that a vector field ζ on (M, g) is called *concircular* [25] if $\nabla\zeta = fI$, where $f \in C^\infty(M)$ and I is the identity map on the smooth sections of M. As a particular case of Theorem 5 from [4], we have the following result.

Proposition 2.1 *Let (1.1) define a compact gradient almost Ricci soliton satisfying*

$$\int_M \mathrm{Ric}(\zeta, \zeta) \geq \int_M \left\{(r - (n-1)\lambda)^2 + (n-1)\lambda^2\right\}.$$

Then, ζ is a concircular vector field, and (M, g) is a Ricci-flat manifold.

Proof Since the soliton is of gradient type, the vector field ζ is closed, and (1.1) becomes $\nabla\zeta + Q = \lambda I$. Then, for $f = -1$, $\varphi = Q$, and $h = \lambda$ in Theorem 5 from [4], we get $Q = 0$ and $\nabla\zeta = \lambda I$. □

A similar inequality satisfied by the Ricci curvature and the potential vector field of a gradient almost Ricci soliton implies that the manifold is an *Einstein manifold* [1], i.e., $\mathrm{Ric} = \frac{r}{n}g$. We recall that a vector field ζ on (M, g) is called *conformal Killing* [25] if $\pounds_\zeta g = fg$, where $f \in C^\infty(M)$; *affine Killing* [25] if $\pounds_\zeta \nabla = 0$; and *affine conformal Killing* [25] if $\pounds_\zeta \nabla = df \otimes I + I \otimes df - g \otimes \nabla f$, where $f \in C^\infty(M)$. It is known [21] that the above condition from the definition of an affine conformal Killing vector field is equivalent to the following condition:

$$\pounds_\zeta g = 2fg + K, \quad \nabla K = 0, \tag{2.1}$$

where K is a symmetric $(0, 2)$-tensor field. In [9], we have proved the following result.

Proposition 2.2 *Let (1.1) define a compact and connected gradient almost Ricci soliton with affine conformal Killing potential vector field $\zeta = \nabla h$ satisfying*

$$\int_M \mathrm{Ric}(\nabla h, \nabla h) \geq n \int_M f\{(n-1)f + \mathrm{trace}(K)\} + \frac{n-1}{4n}\int_M \left(\mathrm{trace}(K)\right)^2.$$

Then, ζ is a conformal Killing vector field, and (M, g) is an Einstein manifold.

We recall that a vector field ζ on (M, g) is called 2-*Killing* [22] if $\pounds_\zeta \pounds_\zeta g = 0$.

A sufficient condition for a hyperbolic Ricci or a hyperbolic Yamabe soliton to be of constant scalar curvature is further provided (see [5]).

Proposition 2.3 *Let (1.7) (respectively, (1.8)) define a compact and connected gradient hyperbolic Ricci (respectively, a compact and connected gradient hyperbolic*

Yamabe) soliton with $\lambda \neq 0$ *such that* $\pounds_{\nabla f}\pounds_{\nabla f} g$ *is divergence-free and*

$$\lambda \int_M \mathrm{Ric}(\nabla f, \nabla r) \leq 0.$$

Then, (M, g) *is a manifold of constant scalar curvature.*

In particular, a compact and connected gradient Ricci (as well as a compact and connected gradient Yamabe) soliton with 2-Killing potential vector field $\zeta = \nabla f$ *satisfying*

$$\int_M \mathrm{Ric}(\nabla f, \nabla r) \leq 0$$

is a manifold of constant scalar curvature.

Proof The particular cases follow from Proposition 2.10 from [5], considering $\lambda = \frac{1}{2}$ in (1.7) (respectively, (1.8)) and taking into account that $\pounds_{\nabla f}\pounds_{\nabla f} g = 0$. □

In [7], we have obtained inequalities considering solitons defined by an affine connection associated with a 1-form. In [3, 10], and [16], respectively, we have found lower and upper bounds of the Ricci curvature tensor field's norm for gradient almost Ricci, gradient almost η-Ricci, and gradient almost Ricci–Bourguignon solitons. Similar inequalities can be determined for a gradient generalized soliton, recovering these as particular cases, as in the following.

Proposition 2.4 *If (1.6) defines a gradient generalized soliton,* $\zeta = \nabla f$*, and* $\eta = df$*, then*

$$\| \mathrm{Hess}(f)\|^2 + \gamma^2\|\nabla f\|^4 - \gamma\nabla f(\|\nabla f\|^2) - \frac{\left(\Delta(f) - \gamma\|\nabla f\|^2\right)^2}{n} \leq$$

$$\leq \alpha^2\| \mathrm{Ric} \|^2 \leq$$

$$\leq \| \mathrm{Hess}(f)\|^2 + \gamma^2\|\nabla f\|^4 - \gamma\nabla f(\|\nabla f\|^2) + \frac{\alpha^2 r^2}{n}.$$

In particular:

(i) *For a gradient almost* η*-Ricci soliton with* $\zeta = \nabla f$ *and* $\eta = df$*, we have*

$$\| \mathrm{Hess}(f)\|^2 + \mu^2\|\nabla f\|^4 - \mu\nabla f(\|\nabla f\|^2) - \frac{\left(\Delta(f) - \mu\|\nabla f\|^2\right)^2}{n} \leq$$

$$\leq \| \mathrm{Ric} \|^2 \leq$$

$$\leq \| \operatorname{Hess}(f)\|^2 + \mu^2\|\nabla f\|^4 - \mu\nabla f(\|\nabla f\|^2) + \frac{r^2}{n}.$$

(ii) For a gradient generalized Ricci soliton with $\zeta = \nabla f$*, we have*

$$\| \operatorname{Hess}(f)\|^2 - \frac{(\Delta(f))^2}{n} \leq \alpha^2\| \operatorname{Ric}\|^2 \leq \| \operatorname{Hess}(f)\|^2 + \frac{\alpha^2 r^2}{n}.$$

(iii) For a gradient almost Ricci soliton, a gradient almost Einstein soliton, as well as for a gradient almost Ricci–Bourguignon soliton with $\zeta = \nabla f$*, we have*

$$\| \operatorname{Hess}(f)\|^2 - \frac{(\Delta(f))^2}{n} \leq \| \operatorname{Ric}\|^2 \leq \| \operatorname{Hess}(f)\|^2 + \frac{r^2}{n}.$$

Proof By taking the trace in

$$\operatorname{Hess}(f) + \alpha \operatorname{Ric} = \beta g + \gamma df \otimes df,$$

we get

$$\Delta(f) + \alpha r = n\beta + \gamma\|\nabla f\|^2.$$

Now, taking successively the scalar product with $\operatorname{Hess}(f)$ and Ric, respectively, we obtain

$$\| \operatorname{Hess}(f)\|^2 + \alpha\langle \operatorname{Ric}, \operatorname{Hess}(f)\rangle = \beta\Delta(f) + \gamma \operatorname{Hess}(f)(\nabla f, \nabla f)$$

and

$$\langle \operatorname{Hess}(f), \operatorname{Ric}\rangle + \alpha\| \operatorname{Ric}\|^2 = \beta r + \gamma \operatorname{Ric}(\nabla f, \nabla f).$$

By multiplying the last relation with α and comparing it with the previous one, we infer

$$\| \operatorname{Hess}(f)\|^2 - \beta\Delta(f) - \gamma \operatorname{Hess}(f)(\nabla f, \nabla f) = \alpha^2\| \operatorname{Ric}\|^2 - \alpha\beta r \\ -\alpha\gamma \operatorname{Ric}(\nabla f, \nabla f).$$

Since we have

$$\alpha \operatorname{Ric}(\nabla f, \nabla f) = \beta\|\nabla f\|^2 + \gamma\|\nabla f\|^4 - \operatorname{Hess}(f)(\nabla f, \nabla f)$$

and

$$\operatorname{Hess}(f)(\nabla f, \nabla f) = g(\nabla_{\nabla f}\nabla f, \nabla f) = \frac{1}{2}\nabla f(\|\nabla f\|^2),$$

we obtain

$$\| \operatorname{Hess}(f)\|^2 + \gamma^2\|\nabla f\|^4 - \gamma\nabla f(\|\nabla f\|^2) - \frac{\left(\Delta(f) - \gamma\|\nabla f\|^2\right)^2}{n} + \frac{(\alpha r)^2}{n} = \alpha^2\| \operatorname{Ric} \|^2, \tag{2.2}$$

and we get the conclusion.

For $\alpha = 1, \beta = \lambda, \gamma = \mu$, we get (i); for $\gamma = 0$, we get (ii); for $\alpha = 1, \beta = \lambda, \gamma = 0$, for $\alpha = 1, \beta = \lambda + \frac{r}{2}, \gamma = 0$, and for $\alpha = 1, \beta = \lambda + \rho r, \gamma = 0$, we get (iii). The proof is complete. □

In [24], we have given a sufficient inequality for the potential function f of a gradient generalized Ricci soliton to be a harmonic function, i.e., $\Delta(f) = 0$, where Δ denotes the Laplacian operator. More precisely:

Proposition 2.5 *If (1.5) defines a gradient generalized Ricci soliton and*

$$\| \operatorname{Hess}(f)\|^2 \leq \alpha^2 \left(\| \operatorname{Ric} \|^2 - \frac{r^2}{n} \right),$$

then $\beta = \frac{r\alpha}{n}$, and f is a harmonic function.

In particular, for a gradient almost Ricci soliton satisfying

$$\| \operatorname{Hess}(f)\|^2 \leq \| \operatorname{Ric} \|^2 - \frac{r^2}{n},$$

f is a harmonic function.

Proof The particular case follows from Proposition 1 from [24], considering $\alpha = 1$ in (1.5). □

Under certain assumptions, if the potential function of an almost Einstein, of an almost Ricci–Bourguignon, or of a generalized soliton is a harmonic function, we obtained lower bounds for the Ricci curvature tensor field's norm in [6, 14, 16], respectively. For a gradient generalized soliton with harmonic potential function, we have

$$\| \operatorname{Hess}(f)\|^2 + \frac{(n-1)\left(\gamma\|\nabla f\|^2\right)^2}{n} - \gamma\nabla f(\|\nabla f\|^2) = \alpha^2 \left(\| \operatorname{Ric} \|^2 - \frac{r^2}{n} \right) \tag{2.3}$$

from (2.2), and we can state the following proposition:

Proposition 2.6 *A gradient generalized soliton defined by (1.6) with $\zeta = \nabla f$, $\eta = df$, such that f is a harmonic function, $\alpha(x) \neq 0$ for any $x \in M$, and*

$$\| \operatorname{Hess}(f)\|^2 + \frac{(n-1)\left(\gamma\|\nabla f\|^2\right)^2}{n} \leq \gamma \nabla f(\|\nabla f\|^2)$$

is an Einstein manifold.

Proof By means of Schwartz's inequality, $\| \operatorname{Ric} \|^2 - \frac{r^2}{n} \geq 0$, we conclude that $\operatorname{Ric} = \frac{r}{n} g$. □

From (2.3), we also deduce the following proposition:

Proposition 2.7 *A gradient generalized Ricci soliton defined by (1.5) with $\zeta = \nabla f$ such that f is a harmonic function satisfies*

$$\| \operatorname{Hess}(f)\|^2 = \alpha^2 \left(\| \operatorname{Ric} \|^2 - \frac{r^2}{n} \right).$$

In particular, for a gradient almost Ricci, for a gradient almost Einstein, and for a gradient almost Ricci–Bourguignon soliton with a harmonic potential function f, we have

$$\| \operatorname{Ric} \|^2 \geq \| \operatorname{Hess}(f)\|^2.$$

3 Trivial Solitons

We shall highlight conditions under which a soliton reduces to a trivial soliton. We remark that an Einstein manifold is a trivial Ricci soliton. Also, a manifold possessing a Ricci vector field satisfying $\nabla\zeta = -Q$ is a *steady* Ricci soliton (i.e., a Ricci soliton with $\lambda = 0$). We recall that a vector field ζ on (M, g) is called *parallel* [25] if $\nabla\zeta = 0$. Since any parallel vector field is a Killing vector field, the solitons with parallel potential vector fields are trivial solitons, too.

As particular cases of Theorems 8 and 9 from [4], we have the following results.

Proposition 3.1 *Let (1.1) define a compact steady Ricci soliton with Ricci potential vector field satisfying $\nabla\zeta = -Q$. If*

$$\int_M \operatorname{Ric}(\zeta,\zeta) \geq \frac{n-1}{n}\int_M r^2 \ \ or \ \int_M \operatorname{Ric}(\zeta,\zeta) \leq \int_M \left\{\frac{r^2}{n} + \zeta(r) - 2\| \operatorname{Ric} \|^2\right\},$$

then (M, g) is a Ricci-flat manifold and ζ is a parallel vector field (hence, the soliton is trivial).

Proof In this case, we have

$$\begin{aligned}(\pounds_\zeta \operatorname{Ric})(X, Y) &= \zeta(\operatorname{Ric}(X, Y)) - \operatorname{Ric}([\zeta, X], Y) - \operatorname{Ric}(X, [\zeta, Y]) \\ &= \zeta(g(X, QY)) - g(\nabla_\zeta X, QY) + g(\nabla_X \zeta, QY) \\ &\quad - g(QX, \nabla_\zeta Y) + g(QX, \nabla_Y \zeta) \\ &= g(X, \nabla_\zeta QY) - g(X, Q(\nabla_\zeta Y)) - 2g(QX, QY) \\ &= g(X, (\nabla_\zeta Q)Y) - 2g(QX, QY),\end{aligned}$$

for any vector fields X, Y tangent to M. Let $\{E_i\}_{1 \le i \le n}$ be a local orthonormal frame on (M, g). Then,

$$\begin{aligned}\operatorname{trace}(\pounds_\zeta \operatorname{Ric}) &= \sum_{i=1}^{n} (\pounds_\zeta \operatorname{Ric})(E_i, E_i) \\ &= \sum_{i=1}^{n} g(E_i, (\nabla_\zeta Q)E_i) - 2\sum_{i=1}^{n} g(QE_i, QE_i) \\ &= \zeta(r) - 2\| \operatorname{Ric} \|^2.\end{aligned}$$

Now we apply Theorems 8 and 9 from [4] for $a = -1$, and we get $Q = 0$ and $\nabla\zeta = 0$. □

Concerning the compact hyperbolic Ricci solitons, we have given in [5] the following sufficient condition for the soliton to be trivial.

Theorem 3.2 *Let (1.7) define a compact hyperbolic Ricci soliton. If* $\lambda \neq 0$, $\pounds_\zeta \pounds_\zeta g$ *is trace-free, and*

$$\int_M \operatorname{Ric}(\zeta, \zeta) \le 0,$$

then ζ *is a parallel vector field (hence, the soliton is trivial).*

Proof See Theorem 2.2 from [5]. □

Sufficient conditions for a compact hyperbolic Yamabe soliton to be trivial have been provided also in [15].

Theorem 3.3 *Let (1.8) define a compact hyperbolic Yamabe soliton.*

(i) If $\lambda \neq 0$, $\pounds_\zeta \pounds_\zeta g$ *is trace-free, and*

$$\int_M \operatorname{Ric}(\zeta, \zeta) \le 0,$$

then ζ *is a parallel vector field (hence, the soliton is trivial).*

(ii) If ζ is divergence-free, $\lambda \neq 0$, and

$$\int_M \|\pounds_\zeta \pounds_\zeta g\|^2 \leq n \int_M (r-\mu)^2 \text{ or } \int_M \mathrm{Ric}(\zeta,\zeta) \geq \int_M \left\{\|\nabla\zeta\|^2 + \frac{n(r-\mu)^2}{2\lambda^2}\right\},$$

then the soliton is trivial.

Proof For (i), see Theorem 2.2 from [5], and for (ii), see Proposition 2.3 from [15]. □

3.1 Solitons with Gradient Vector Fields

For compact gradient generalized, gradient hyperbolic Ricci and gradient hyperbolic Yamabe solitons, we have determined in [6] and [5] some triviality conditions.

Theorem 3.4 *If (1.6) defines a compact gradient generalized soliton with α, β, and γ constant, $\zeta = \nabla f$, $\eta = df$, and*

$$\alpha \int_M g(\nabla f, \nabla r) + \gamma \int_M \nabla f(\|\nabla f\|^2) \leq 0,$$

then the soliton is trivial.

In particular:

(i) A compact gradient η-Ricci soliton with $\zeta = \nabla f$, $\eta = df$, and

$$\int_M g(\nabla f, \nabla r) + \mu \int_M \nabla f(\|\nabla f\|^2) \leq 0$$

is a trivial soliton.

(ii) A compact gradient generalized Ricci soliton with α and β constant, $\zeta = \nabla f$, and

$$\alpha \int_M g(\nabla f, \nabla r) \leq 0$$

is a trivial soliton.

(iii) A compact gradient Ricci soliton with $\zeta = \nabla f$ and

$$\int_M g(\nabla f, \nabla r) \leq 0$$

is a trivial soliton.

Proof For the first statement, see Proposition 4 from [6].

For $\alpha = 1, \beta = \lambda, \gamma = \mu$, we get (i); for $\gamma = 0$, we get (ii); and for $\alpha = 1, \beta = \lambda, \gamma = 0$, we get (iii). The proof is complete. □

Theorem 3.5

(i) If (1.7) defines a compact gradient hyperbolic Ricci soliton with $\lambda \neq 0$ such that $£_{\nabla f}£_{\nabla f}g$ is trace-free and

$$\int_M \mathrm{Ric}(\nabla f, \nabla f) \geq \frac{1}{2\lambda}\int_M g(\nabla f, \nabla r) \quad or \quad \int_M \mathrm{Ric}(\nabla f, \nabla f) \geq \frac{1}{4\lambda^2}\int_M (r - n\mu)^2,$$

then the soliton is trivial.

(ii) If (1.8) defines a compact gradient hyperbolic Yamabe soliton with $\lambda \neq 0$ such that $£_{\nabla f}£_{\nabla f}g$ is trace-free and

$$\int_M \mathrm{Ric}(\nabla f, \nabla f) \geq \frac{n}{2\lambda}\int_M g(\nabla f, \nabla r) \quad or \quad \int_M \mathrm{Ric}(\nabla f, \nabla f) \geq \frac{n^2}{4\lambda^2}\int_M (r - \mu)^2,$$

then the soliton is trivial.

Proof For (i), see Theorems 2.5 and 2.7(i), and for (ii), see Theorems 2.4 and 2.7(ii) from [5]. □

Proposition 3.6 *If (1.7) (respectively, (1.8)) defines a compact gradient hyperbolic Ricci (respectively, a compact gradient hyperbolic Yamabe) soliton with $\lambda \neq 0$ such that $£_{\nabla f}£_{\nabla f}g$ is trace-free and*

$$\lambda \int_M g(\nabla f, \nabla r) \leq 0,$$

then the soliton is trivial.

Proof See Corollary 2.6 from [5]. □

3.2 *Solitons with Geodesic and Generalized Geodesic Vector Fields*

We recall that a vector field ζ on (M, g) is called *geodesic* [25] if $\nabla_\zeta \zeta = 0$, and we have from [11] and [24] the following characterizations of trivial almost Einstein and trivial generalized Ricci solitons.

Theorem 3.7 *If (1.3) defines a compact almost Einstein soliton with geodesic potential vector field and $r \neq 0$, then*

$$r\{(n-2)r + 2n\lambda\} \geq 0$$

if and only if the soliton is trivial.

Proof See Theorem 3.4 from [11]. □

Theorem 3.8 *If (1.3) defines a compact and connected almost Einstein soliton with geodesic potential vector field and* $r \neq 0$*, then* ζ *is an eigenvector of the Ricci operator with a constant eigenvalue* $\sigma \in \mathbb{R} \setminus \{0\}$ *and*

$$r(r - n\sigma) \leq 0$$

if and only if the soliton is trivial.

Proof See Theorem 3.5 from [11]. □

We shall further denote by $\theta := i_\zeta g$ the dual 1-form of ζ, and we define the $(1, 1)$-tensor field F by

$$g(FX, Y) := \frac{1}{2}(d\theta)(X, Y),$$

for any vector fields X, Y tangent to M. Then, in terms of F and Ric, we have from [11] the following characterizations of a trivial almost Einstein soliton with geodesic potential vector field.

Theorem 3.9 *If (1.3) defines a connected almost Einstein soliton with geodesic potential vector field, then*

$$\mathrm{Ric}(\zeta, \zeta) \geq \|F\|^2,$$

and the function $(n - 2)r + 2n\lambda$ *is constant on the integral curves of* ζ *if and only if the soliton is trivial.*

Proof See Theorem 3.7 from [11]. □

Theorem 3.10 *If (1.3) defines a compact and connected almost Einstein soliton with geodesic potential vector field, then*

$$\mathrm{Ric}(\zeta, \zeta) \geq \|F\|^2 + \frac{n-1}{4n}\lambda((n - 2)r + 2n\lambda))^2$$

if and only if the soliton is trivial.

Proof See Theorem 3.6 from [11]. □

Theorem 3.11 *If (1.3) defines a compact and connected almost Einstein soliton with geodesic potential vector field, then*

$$\mathrm{Ric}(F\zeta, F\zeta) \geq \frac{n-1}{n}(\mathrm{div}(F\zeta))^2 \text{ and } \lambda\{(n - 2)r + 2n\lambda\} \leq 0$$

if and only if the soliton is trivial.

Proof See Theorem 3.9 from [11]. □

Necessary and sufficient conditions for a generalized Ricci soliton with geodesic potential vector field to be trivial have been given in [24].

Theorem 3.12 *If (1.5) defines a compact and connected gradient generalized Ricci soliton with geodesic potential vector field, $r \neq 0$, and α and β are constant ($\alpha \neq 0$), then ∇f is an eigenvector of the Ricci operator with constant eigenvalue $\frac{\beta}{\alpha}$ and*

$$r\alpha(r\alpha - n\beta) \leq 0$$

if and only if the soliton is trivial.

Proof See Theorem 3 from [24]. □

Theorem 3.13 *If (1.5) defines a compact and connected gradient generalized Ricci soliton with geodesic potential vector field, then*

$$\mathrm{Ric}(\nabla f, \nabla f) \geq \frac{n-1}{n}(r\alpha - n\beta)^2$$

if and only if the soliton is trivial.

Proof See Theorem 4 from [24]. □

We recall that a vector field ζ on (M, g) is called *generalized geodesic* (see [13]) if $\nabla_{\zeta}\zeta = f\zeta$, where $f \in C^{\infty}(M)$, and we have proved in [13] that the following inequalities ensure that an almost Ricci or a generalized Ricci soliton with a generalized geodesic potential vector field is a trivial soliton.

Theorem 3.14 *If (1.5) defines a compact generalized Ricci soliton with generalized geodesic potential vector field and*

$$2\beta < r\alpha \leq n\beta \ \text{ and } \ \mathrm{Ric}\,(\nabla\alpha, \nabla\alpha) \geq r\,\|\nabla\alpha\|^2 + g\left(\frac{\alpha}{2}\nabla r - (n-1)\nabla\beta - \frac{r}{2}\zeta, \nabla\alpha\right),$$

then the soliton is trivial.

In particular, a compact almost Ricci soliton with generalized geodesic potential vector field and

$$2\lambda < r \leq n\lambda$$

is a trivial soliton.

Proof See Theorem 1 and Corollary 1 from [13]. □

3.3 Solitons with 2-Killing Vector Fields

In [13] we have shown that the following inequalities make a compact almost Ricci or generalized Ricci soliton with a 2-Killing potential vector field be a trivial soliton.

Theorem 3.15 *If (1.5) defines a compact generalized Ricci soliton with* 2*-Killing potential vector field and*

$$n\beta < r\alpha \leq 2n\beta \ \text{ and } \ \operatorname{Ric}(\nabla\alpha, \nabla\alpha) \leq r\,\|\nabla\alpha\|^2 + g\left(\frac{\alpha}{2}\nabla r - (n-1)\nabla\beta - \frac{r}{2}\zeta, \nabla\alpha\right),$$

then the soliton is trivial.

In particular, a compact almost Ricci soliton with 2*-Killing potential vector field and*

$$n\lambda < r \leq 2n\lambda$$

is a trivial soliton.

Proof See Theorem 2 and Corollary 2 from [13]. □

3.4 Solitons with Affine Killing and Affine Conformal Killing Vector Fields

Sufficient conditions for a compact and connected almost Ricci soliton with affine Killing or affine conformal Killing potential vector field to be a trivial soliton have been given in [12] and [9].

Theorem 3.16 *If (1.1) defines a compact and connected almost Ricci soliton with affine Killing potential vector field and*

$$r(r - n\lambda) \leq 0,$$

then the soliton is trivial.

Proof See Theorem 3 from [12]. □

Theorem 3.17 *If (1.1) defines a compact and connected gradient almost Ricci soliton with affine conformal Killing potential vector field and*

$$\int_M \operatorname{Ric}(\zeta, \zeta) \geq -n \int_M \zeta(\lambda),$$

then the soliton is trivial.

In particular, a compact and connected gradient Ricci soliton with

$$\int_M \mathrm{Ric}(\zeta, \zeta) \geq 0$$

is a trivial soliton.

Proof See Proposition 4.4 from [9]. □

4 New Characterizations of the Euclidean Spheres

Based on Obata's theorem, in [12] and [24], we have given new necessary and sufficient conditions for an almost Ricci or a generalized Ricci soliton to be isometric to a sphere.

Let (M, g) be an n-dimensional compact and connected Riemannian manifold $(n > 2)$.

Theorem 4.1 *Let (1.1) define an almost Ricci soliton with positive Ricci curvature. Then,*

$$\int_M \left\{ \mathrm{Ric}\,(c\zeta, c\zeta) + \frac{n-1}{n} (\Delta(\lambda))^2 \right\} \leq c \int_M \left\{ 2(n-1)\,\|\nabla\lambda\|^2 + r\Delta(\lambda) \right\},$$

for a nonzero constant c if and only if $c > 0$ and M is isometric to the sphere $\mathbf{S}^n(c)$.

Proof See Theorem 2 from [12]. □

Theorem 4.2 *Let (1.1) define a nontrivial almost Ricci soliton (i.e., λ is nonconstant) with Hodge decomposition of the potential vector field $\zeta = \bar{\zeta} + \nabla h$. Then,*

$$\int_M \mathrm{Ric}\,(\bar{\zeta}, \bar{\zeta}) \geq \int_M \|F\|^2 \quad \text{and} \quad \int_M r\,(r - n\lambda) \leq 0$$

if and only if M is isometric to the sphere $\mathbf{S}^n(c)$ with $c = \frac{r}{n(n-1)}$.

Proof See Theorem 4 from [12]. □

Theorem 4.3 *Let (1.5) define a gradient generalized Ricci soliton of constant scalar curvature, such that $\nabla(f + \alpha)$ is an eigenvector of the Ricci operator corresponding to the eigenvalue $\frac{r}{n}$. Then,*

$$\mathrm{Ric}\,(\nabla f, \nabla f) \geq \frac{r}{n}\,\|\nabla f\|^2$$

if and only if $r > 0$ and M is isometric to the sphere $\mathbf{S}^n(c)$ with $c = \frac{r}{n(n-1)}$.

Proof See Theorem 1 from [24]. □

Theorem 4.4 *Let (1.5) define a gradient generalized Ricci soliton. Then,*

$$\mathrm{Ric}\,(\nabla f, \nabla f) \geq (n-1)c\,\|\nabla f\|^2 \quad \textit{and} \quad \int_M (r\alpha - n\beta - cf)\,(r\alpha - n\beta - ncf) \leq 0$$

with $c > 0$ *(a constant) if and only if* M *is isometric to the sphere* $\mathbf{S}^n(c)$.

Proof See Theorem 2 from [24]. □

References

1. Besse, A.: Einstein Manifolds. Springer, Berlin (1987). https://doi.org/10.1007/978-3-540-74311-8
2. Blaga, A.M.: η-Ricci solitons on para-Kenmotsu manifolds. Balk. J. Geom. Appl. **20**(1), 1–13 (2015)
3. Blaga, A.M.: Almost η-Ricci solitons in $(LCS)_n$-manifolds. Bull. Belg. Math. Soc. Simon Stevin **25**(5), 641–653 (2018). https://doi.org/10.36045/bbms/1547780426
4. Blaga, A.M.: On the potential vector fields of soliton-type equations. Axioms **13**(7), 476 (2024). https://doi.org/10.3390/axioms13070476
5. Blaga, A.M.: On trivial gradient hyperbolic Ricci and gradient hyperbolic Yamabe solitons. J. Geom. **115**, 26 (2024). https://doi.org/10.1007/s00022-024-00725-6
6. Blaga, A.M.: Remarks on generalized solitons. Bol. Soc. Mat. Mex. **29**(2), 39 (2023). https://doi.org/10.1007/s40590-023-00511-5
7. Blaga, A.M., Chen, B.-Y.: Gradient solitons on statistical manifolds. J. Geom. Phys. **164**, 104195 (2021). https://doi.org/10.1016/j.geomphys.2021.104195
8. Blaga, A.M., Chen, B.-Y.: Harmonic forms and generalized solitons. Results Math. **79**(1), 16 (2024). https://doi.org/10.1007/s00025-023-02041-y
9. Blaga, A.M., Chen, B.-Y.: On conformal collineation and almost Ricci solitons. J. Geom. Phys. **207**, 105354 (2025). https://doi.org/10.1016/j.geomphys.2024.105354
10. Blaga, A.M., Crasmareanu, M.C.: Inequalities for gradient Einstein and Ricci solitons. Facta Universitatis. Ser. Math. Inform. **35**(2), 351–356 (2020). https://doi.org/10.22190/FUMI2002351B
11. Blaga, A.M., Deshmukh, S.: Einstein solitons with unit geodesic potential vector field. AIMS Math. **6**(8), 7961–7970 (2021). https://doi.org/10.3934/math.2021462
12. Blaga, A.M., Deshmukh, S.: Some properties of the potential field of an almost Ricci soliton. Mathematics **12**(19), 3049 (2024). https://doi.org/10.3390/math12193049
13. Blaga, A.M., Ishan, A., Deshmukh, S.: A note on solitons with generalized geodesic vector field. Symmetry **13**(7), 1104 (2021). https://doi.org/10.3390/sym13071104
14. Blaga, A.M., Laţcu, D.R.: Remarks on Einstein solitons with certain types of vector field. Int. Electron. J. Geom. **14**(1), 100–105 (2021). https://doi.org/10.36890/iejg.859572
15. Blaga, A.M., Özgür, C.: Some properties of hyperbolic Yamabe solitons. Phys. Scr. **100**(4), 045230 (2025). https://doi.org/10.1088/1402-4896/adb343
16. Blaga, A.M., Taştan, H.M.: Some results on almost η-Ricci–Bourguignon solitons. J. Geom. Phys. **168**, 104316 (2021). https://doi.org/10.1016/j.geomphys.2021.104316
17. Catino, G., Mazzieri, L.: Gradient Einstein solitons. Nonlinear Anal. **132**, 66–94 (2016). https://doi.org/10.1016/j.na.2015.10.021
18. Cho, J.T., Kimura, M.: Ricci solitons and real hypersurfaces in a complex space form. Tohoku Math. J. **61**(2), 205–212 (2009). https://doi.org/10.2748/tmj/1245849443
19. Dwivedi, S.: Some results on Ricci–Bourguignon solitons and almost solitons. Can. Math. Bull. 1–14 (2020). https://doi.org/10.4153/S0008439520000673
20. Faraji, H., Azami, S., Fasihi-Ramandi, G.: Three dimensional homogeneous hyperbolic Ricci solitons. J. Nonlinear Math. Phys. **30**, 135–155 (2023). https://doi.org/10.1007/s44198-022-00075-4

21. Katzin, G.H., Levine, J., Davis, W.R.: Curvature collineations: a fundamental symmetry property of the space-times of General Relativity defined by the vanishing Lie derivative of the Riemann curvature tensor. J. Math. Phys. **10**, 617–629 (1969). https://doi.org/10.1063/1.1664886
22. Oprea, T.: 2-Killing vector fields on Riemannian manifolds. Balk. J. Geom. Appl. **13**(1), 87–92 (2008)
23. Pigola, S., Rigoli, M., Rimoldi, M., Setti, A.: Ricci almost solitons. Ann. Sc. Norm. Super. Pisa Cl. Sci. Ser. 5 **10**(4), 757–799 (2011)
24. Turki, N.B., Blaga, A.M., Deshmukh, S.: Soliton-type equations on a Riemannian manifold. Mathematics **10**(4), 633 (2022). https://doi.org/10.3390/math10040633
25. Yano, K.: Integral Formulas in Riemannian Geometry. Marcel Dekker, New York (1970)

Generalized Ricci-Yamabe Soliton on Three-Dimensional Lie Groups

Adel Delloum and Gherici Beldjilali

Abstract We explore the presence of generalized Ricci-Yamabe solitons (briefly, **GRYS**) within the framework of three-dimensional left-invariant Lie groups.

Keywords Ricci soliton · Yamabe soliton · Lie groups

1 Introduction

Let (M^n, g) be an n-dimensional Riemannian manifold. The Riemannian curvature tensor R is defined by

$$R(X, Y)Z = \nabla_X \nabla_Y Z - \nabla_Y \nabla_X Z - \nabla_{[X,Y]} Z, \tag{1.1}$$

where ∇ is the Levi-Civita connection associated with g. The Ricci curvature tensor is formulated as

$$S(X, Y) = \sum_{i=1}^{n} g\big(R(X, e_i)e_i, Y\big), \tag{1.2}$$

where $\{e_i\}_{\{i=1,\ldots,n\}}$ is an orthonormal frame with respect to g. Herein, X, Y, Z are smooth vector fields on M.

The exploration of geometric properties on Riemannian manifolds constitutes a broad and dynamic area of research, drawing considerable attention in recent literature such as [3, 5, 6, 9]. Among these investigations, the elucidation of structures like the Ricci soliton holds particular significance, facilitating the development of indispensable geometric tools including specialized vector fields, metric deformations, and manifold products.

A. Delloum (✉) · G. Beldjilali
Faculty of Exact Sciences, Mascara University of Mustapha Stambouli, Mascara, Algeria
e-mail: adel.delloum@univ-mascara.dz; gherici.beldjilali@univ-mascara.dz

B.-Y. Chen, M. A. Choudhary (eds.), *Geometric Inequalities and Applications*,
Infosys Science Foundation Series in Mathematical Sciences,
https://doi.org/10.1007/978-981-95-5148-4_2

The seminal work of Hamilton [11] in 1982 introduced the concept of Ricci flow, aimed at deriving a canonical metric for smooth manifolds. Over time, Ricci flow has emerged as a potent analytical tool for examining Riemannian manifolds, particularly those exhibiting positive curvatures. Central to this framework is the notion of a Ricci soliton, which represents a distinct solution of the Ricci flow generated by a vector field V. Notably, a Ricci soliton assumes the form of a gradient Ricci soliton when the generating vector field V aligns with the gradient of a potential function. The generalized Ricci soliton equation (briefly, GRS), as formulated in a Riemannian manifold (M^n, g), is characterized by its definition (see [15])

$$\mathcal{L}_V g = -2c_1 V^\flat \otimes V^\flat + 2c_2 S + 2\lambda g, \tag{1.3}$$

where $\mathcal{L}_V g$ is the Lie derivative of the metric g along the vector field V

$$\big(\mathcal{L}_V g\big)(X, Y) = g\big(\nabla_X V, Y\big) + g\big(X, \nabla_Y V\big) \tag{1.4}$$

and $V^\flat$ is the g-dual of the vector V

$$V^\flat(X) = g(X, V).$$

Equation (1.3) is a generalization of

- Killing's equation $c_1 = c_2 = \lambda = 0$
- Equation for homotheties $c_1 = c_2 = 0$
- Ricci soliton $c_1 = 0$, $c_2 = -1$
- Cases of Einstein-Weyl $c_1 = 1$, $c_2 = \frac{-1}{n-2}$
- Metric projective structures with skew-symmetric Ricci tensor in projective class $c_1 = 1$, $c_2 = -\frac{1}{n-1}$, $\lambda = 0$
- Vacuum near-horizon geometry equation $c_1 = 1$, $c_2 = \frac{1}{2}$

Furthermore, when V is a Killing vector field (i.e., $\mathcal{L}_V g = 0$), Eq. (1.3) describes (M^n, g) as a perfect fluid space (briefly, **PFS**).

Recent research on generalized Ricci solitons has produced a substantial body of work, highlighting their significance in differential geometry and theoretical physics. These studies often explore the properties, classification, and applications of generalized Ricci solitons. For instance, researchers have investigated the classification of generalized Ricci solitons under various curvature conditions and symmetry constraints, providing insights into their geometric structures and potential applications in string theory and general relativity [4, 8]. Moreover, the stability and uniqueness of generalized Ricci solitons have been topics of considerable interest, with findings indicating conditions under which these solitons exhibit unique solutions and stability properties [13, 17]. Research on perfect fluid spaces has significantly advanced our understanding of their role in general relativity and cosmology. Perfect fluid spaces are essential in modeling astrophysical objects and cosmological scenarios, as they describe space-times filled with a fluid that has uniform properties at every point. Studies have extensively examined the properties and dynamics of these spaces, leading to new insights into their stability, evolution, and potential singularities [14, 18]. Furthermore, the interaction between perfect fluid spaces and other fields, such as electromagnetic fields, has been a focus of recent investigations, revealing complex behaviors and contributing to the broader understanding of gravitational interactions [1, 10]. These contributions are fundamental to both theoretical explorations and practical applications in astrophysics.

Conversely, a Ricci-Yamabe soliton (briefly, RYS) is defined as a semi-Riemannian manifold (M^n, g) equipped with a vector field V on M that satisfies

$$\mathcal{L}_V g = 2\alpha S + 2(\lambda + r\rho)g, \tag{1.5}$$

where $\rho \in \mathbb{R}$ is constant and r denotes the scalar curvature, defined as the trace of the Ricci tensor S with respect to the metric g

$$r = Tr_g S. \tag{1.6}$$

Likewise, Eq. (1.5) is a natural generalization of:

- Ricci soliton (briefly, **RS**) $\alpha = 1$, $\rho = 0$
- Ricci-Bourguignon soliton (briefly, **GBS**) $\alpha = 1$, $\rho \in \mathbb{R}$
- Yamabe soliton (briefly, **YS**) $\alpha = 0$, $\rho = -1$

Ricci-Yamabe solitons have been an active area of investigation in differential geometry. These solitons generalize both Ricci solitons and Yamabe solitons, serving as self-similar solutions to the Ricci flow and the Yamabe flow, respectively. Recent studies have explored various aspects of Ricci-Yamabe solitons, including their existence, uniqueness, and classification under different geometric conditions. Notable contributions include the work of Deshmukh and Alodan [7], which examined the geometric properties of Ricci-Yamabe solitons on warped product manifolds, and Blaga [2], who studied η-Ricci-Yamabe solitons in the context of almost contact metric manifolds. In their work, Traore et al. conducted a thorough investigation and provided detailed characterizations of the geometric properties of manifolds that admit almost η-Ricci-Bourguignon solitons, as documented in [19, 20]. These investigations provide valuable insights into the interplay between curvature and the underlying geometry of the manifolds.

Motivated by the work of [15], we define a generalized Ricci-Yamabe soliton (briefly, **GRYS**) as follows:

$$\mathcal{L}_V g = -2c_1 V^\flat \otimes V^\flat + 2c_2 S + 2(\lambda + r\rho)g. \tag{1.7}$$

Equation (1.7) is an immediate generalization of the following:

- **GRS** equation (1.3) for $\rho = 0$
- **RYS** (1.5) (briefly, RYS) Eq. (1.5) for $c_1 = 0$

In this chapter, we investigate and classify the existence of generalized Ricci-Bourguignon solitons (**GRYS**) (1.7) on left-invariant three-dimensional Lie groups (M^3, g).

This chapter is organized as follows: In the next section, we review the necessary prerequisites related to left-invariant three-dimensional Lie groups, their algebras, and curvatures. In the final section, we provide a complete classification of the **GRYS** associated with each algebra of three-dimensional left-invariant Lie groups.

2 Left-Invariant Three-Dimensional Lie Groups

A three-dimensional left-invariant Lie group G is a smooth manifold of dimension 3 equipped with a group structure such that left translations L_a defined by

$$\begin{aligned} L_a : G &\longrightarrow G \\ x &\longrightarrow L_a(x) = a.x \end{aligned}$$

for $a, x \in G$ are diffeomorphisms. This implies that the tangent space T_eG at the identity element $e \in G$, equipped with the Lie bracket operation derived from the group multiplication, forms a three-dimensional Lie algebra.

A Riemannian frame on a three-dimensional left-invariant Lie group (G, g) consists of three smooth vector fields $\{e_1, e_2, e_3\}$ on G, which are left-invariant and form an orthonormal basis with respect to the Riemannian metric g. Specifically, at each point $p \in G$,

$$g(e_i, e_j)_{|p} = \delta_{ij},$$

where δ_{ij} is the Kronecker delta.

The Riemannian metric g induces a natural Levi-Civita connection, which is torsion-free and compatible with g. This connection allows for the study of curvature properties within the framework of the Lie group structure.

In Jantzen's work [12], L. Bianchi compiled a catalog of three-dimensional real Lie algebras, accompanied by a demonstration that each three-dimensional Lie algebra finds isomorphism with a singular entry on his list. Given our focus on left-invariant structures, our analysis is confined to the Lie algebras associated with their respective Lie groups. The ensuing outcome elucidates the various categories of three-dimensional Lie algebras [16].

Proposition 2.1 *Let $\mathfrak{g}$ be a three-dimensional real Lie algebra. Then if $\mathfrak{g}$ is not abelian, it is isomorphic to one and only one of the Lie algebras listed below (Table 1):*

Table 1 Classification of three-dimensional real Lie algebras and their structure equations

Algebra	Structure equations
$\mathcal{A}_{3,1}$	$[e_2, e_3] = e_1$
$\mathcal{A}_{3,2}$	$[e_1, e_3] = e_1, [e_2, e_3] = e_1 + e_2$
$\mathcal{A}_{3,3}$	$[e_1, e_3] = e_1, [e_2, e_3] = e_2$
$\mathcal{A}_{3,4}$	$[e_1, e_3] = e_1, [e_2, e_3] = -e_2$
$\mathcal{A}^{\delta}_{3,5}$	$[e_1, e_3] = e_1, [e_2, e_3] = \delta e_2, (0 < \|\delta\| < 1)$
$\mathcal{A}_{3,6}$	$[e_1, e_3] = -e_2, [e_2, e_3] = e_1$
$\mathcal{A}^{\delta}_{3,7}$	$[e_1, e_3] = -\delta e_1 - e_2, [e_2, e_3] = e_1 + \delta e_2, (\delta > 0)$
$\mathcal{A}_{3,8}$	$[e_1, e_2] = e_1, [e_1, e_3] = -2e_2, [e_2, e_3] = e_3$
$\mathcal{A}_{3,9}$	$[e_1, e_2] = e_3, [e_1, e_3] = -e_2, [e_2, e_3] = e_1$

3 GRYS on Three-Dimensional Lie Groups

Our inquiry will delve into the presence of a **GRYS** (1.7) within the realm of three-dimensional left-invariant Lie groups, on each algebra $\mathcal{A}_{3,k}, \quad k \in \{1, ..., 9\}$, with the potential vector field V

$$V = ae_1 + be_2 + ce_3, \quad \text{and} \quad V^\flat \otimes V^\flat = \begin{pmatrix} a^2 & ab & ac \\ ab & b^2 & bc \\ ac & bc & c^2 \end{pmatrix}.$$

Obviously, formula (1.7) is symmetric, and we are lead to solve a system of six equations

$$(\mathcal{L}_V g)_{ij} = -2c_1 V_i V_j + 2c_2 S_{ij} + 2(\lambda + r\rho)\delta_{ij}. \tag{3.1}$$

3.1 The Algebra $\mathcal{A}_{3,1}$

The covariant derivatives of the basis elements are given by the following expressions:

$$\nabla_{e_1} e_1 = 0, \qquad \nabla_{e_1} e_2 = -\frac{1}{2} e_3, \qquad \nabla_{e_1} e_3 = \frac{1}{2} e_2,$$
$$\nabla_{e_2} e_1 = -\frac{1}{2} e_3, \qquad \nabla_{e_2} e_2 = 0, \qquad \nabla_{e_2} e_3 = \frac{1}{2} e_1,$$
$$\nabla_{e_3} e_1 = \frac{1}{2} e_2, \qquad \nabla_{e_3} e_2 = -\frac{1}{2} e_1, \qquad \nabla_{e_3} e_3 = 0.$$

Using formulas (1.2), (1.4), and (1.6), we obtain

$$(\mathcal{L}_V g)_{ij} = \begin{pmatrix} 0 & c & -b \\ c & 0 & 0 \\ -b & 0 & 0 \end{pmatrix}, \quad S_{ij} = \begin{pmatrix} \frac{1}{2} & 0 & 0 \\ 0 & -\frac{1}{2} & 0 \\ 0 & 0 & -\frac{1}{2} \end{pmatrix}, \tag{3.2}$$

and the scalar curvature (1.6) is

$$r = -\frac{1}{2}. \tag{3.3}$$

By directly substituting (3.2) into (3.1), we must address the challenge of solving

$$c_1 a^2 - \frac{c_2}{2} - \left(\lambda - \frac{\rho}{2}\right) = 0, \tag{3.4}$$

$$c_1 b^2 + \frac{c_2}{2} - \left(\lambda - \frac{\rho}{2}\right) = 0, \tag{3.5}$$

$$c_1 c^2 + \frac{c_2}{2} - \left(\lambda - \frac{\rho}{2}\right) = 0, \tag{3.6}$$

$$c + 2c_1 ab = 0, \tag{3.7}$$

$$b - 2c_1 ac = 0, \tag{3.8}$$

$$c_1 bc = 0. \tag{3.9}$$

By analyzing various cases related to Eq. (3.9), we obtain the following results:

- If $c_1 = 0$, then $b = c = 0$ is obtained from (3.7) and (3.8). The system becomes

$$-\frac{c_2}{2} - \left(\lambda - \frac{\rho}{2}\right) = 0, \tag{3.10}$$

$$\frac{c_2}{2} - \left(\lambda - \frac{\rho}{2}\right) = 0. \tag{3.11}$$

  From one hand, summing Eqs. (3.10) and (3.11) yields $\lambda = \frac{\rho}{2}$. On the other hand, subtracting Eqs. (3.10) and (3.11) results in $c_2 = 0$.
- If $c_1 \neq 0$, then $bc = 0$. Assume $b = 0$, then $c = 0$ by virtue of (3.7). Substituting in (3.5) yields $\lambda = \frac{c_2}{2} + \frac{\rho}{2}$, and this along with (3.4) yields $a = 0$. A similar result to this latter is obtained in the case where $c = 0$.

In summary, the solutions of Eq. (3.1) within the algebra $\mathcal{A}_{3,1}$ are as follows:

$$V = ae_1, \quad c_1 = \frac{c_2}{a^2}, \quad \lambda = \frac{\rho}{2} + \frac{c_2}{2}, \quad a \in \mathbb{R}^*, \quad \text{and} \quad c_2, \rho \in \mathbb{R}.$$

3.2 The Algebra $\mathcal{A}_{3,2}$

The derivatives with respect to covariant bases are delineated as follows:

$$\begin{aligned}
&\nabla_{e_1}e_1 = -e_3, && \nabla_{e_1}e_2 = -\frac{1}{2}e_3, && \nabla_{e_1}e_3 = e_1 + \frac{1}{2}e_2,\\
&\nabla_{e_2}e_1 = -\frac{1}{2}e_3, && \nabla_{e_2}e_2 = -e_3, && \nabla_{e_2}e_3 = \frac{1}{2}e_1 + e_2,\\
&\nabla_{e_3}e_1 = \frac{1}{2}e_2, && \nabla_{e_3}e_2 = -\frac{1}{2}e_1, && \nabla_{e_3}e_3 = 0.
\end{aligned}$$

Thus, from Eqs. (1.2), (1.4), and (1.6), we derive

$$(\mathcal{L}_V g)_{ij} = \begin{pmatrix} 2c & c & -a-b \\ c & 2c & -b \\ -a-b & -b & 0 \end{pmatrix}, \quad S_{ij} = \begin{pmatrix} -\frac{3}{2} & -1 & 0 \\ -1 & -\frac{5}{2} & 0 \\ 0 & 0 & -\frac{5}{2} \end{pmatrix}, \tag{3.12}$$

and the scalar curvature is

$$r = -\frac{13}{2}. \tag{3.13}$$

By employing Eqs. (3.12) and (3.13) within (1.7), we are prompted to undertake the challenge of resolving

$$c_1a^2 + \frac{3}{2}c_2 - \left(\lambda - \frac{13}{2}\rho\right) + c = 0, \tag{3.14}$$

$$c_1b^2 + \frac{5}{2}c_2 - \left(\lambda - \frac{13}{2}\rho\right) + c = 0, \tag{3.15}$$

$$c_1c^2 + \frac{5}{2}c_2 - \left(\lambda - \frac{13}{2}\rho\right) = 0, \tag{3.16}$$

$$c_1ab + c_2 + \frac{c}{2} = 0, \tag{3.17}$$

$$c_1ac - \frac{a+b}{2} = 0, \tag{3.18}$$

$$c_1bc - \frac{b}{2} = 0. \tag{3.19}$$

Exploring different scenarios informed by Eq. (3.19), we discover:

- If $b = 0$, then either $a = 0$ or $c_1c = \frac{1}{2}$, as indicated by (3.18):
  - If $a = 0$, the system of equations simplifies to

$$\frac{3}{2}c_2 - \left(\lambda - \frac{13}{2}\rho\right) + c = 0, \tag{3.20}$$

$$\frac{5}{2}c_2 - \left(\lambda - \frac{13}{2}\rho\right) + c = 0, \tag{3.21}$$

$$c_1c^2 + \frac{5}{2}c_2 - \left(\lambda - \frac{13}{2}\rho\right) = 0, \tag{3.22}$$

$$c_2 + \frac{c}{2} = 0. \tag{3.23}$$

Subtracting (3.20) from (3.21) yields $c_2 = 0$. Substituting this into (3.23) gives $c = 0$, and thus $\lambda = \frac{13}{2}\rho$.

– If $c_1c = \frac{1}{2}$, then substituting $c = \frac{1}{2c_1}$ results in

$$c_1a^2 + \frac{3}{2}c_2 - \left(\lambda - \frac{13}{2}\rho\right) + \frac{1}{2c_1} = 0, \tag{3.24}$$

$$\frac{5}{2}c_2 - \left(\lambda - \frac{13}{2}\rho\right) + \frac{1}{2c_1} = 0, \tag{3.25}$$

$$\frac{1}{4c_1} + \frac{5}{2}c_2 - \left(\lambda - \frac{13}{2}\rho\right) = 0, \tag{3.26}$$

$$c_2 + \frac{1}{4c_1} = 0. \tag{3.27}$$

Subtracting (3.25) from (3.26), we obtain $\frac{1}{c_1} = 0$, which has no solution.

- If $c_1c = \frac{1}{2}$, then from (3.18) necessarily $b = 0$ and from (3.17) $c_2 = -\frac{1}{4c_1}$. Direct substitution gives

$$\frac{1}{8c_1} - \left(\lambda - \frac{13}{2}\rho\right) = 0, \tag{3.28}$$

$$c_1b^2 - \left(\lambda - \frac{13}{2}\rho\right) - \frac{1}{8c_1} = 0, \tag{3.29}$$

$$\frac{3}{8c_1} + \left(\lambda - \frac{13}{2}\rho\right) = 0. \tag{3.30}$$

Summing Eqs. (3.28) and (3.30) gives $\frac{1}{c_1} = 0$, which has no solution.

Summarizing the above, Eq. (3.1) on the algebra $\mathcal{A}_{3,2}$ has no solution.

3.3 The Algebra $\mathcal{A}_{3,3}$

The covariant derivatives of the basis elements are as follows:

$$\begin{aligned} &\nabla_{e_1} e_1 = -e_3, && \nabla_{e_1} e_2 = 0, && \nabla_{e_1} e_3 = e_1, \\ &\nabla_{e_2} e_1 = 0, && \nabla_{e_2} e_2 = -e_3, && \nabla_{e_2} e_3 = e_2, \\ &\nabla_{e_3} e_1 = 0, && \nabla_{e_3} e_2 = 0, && \nabla_{e_3} e_3 = 0. \end{aligned}$$

Hence, from (1.2), (1.4), and (1.6), we have

$$(\mathcal{L}_V g)_{ij} = \begin{pmatrix} 2c & 0 & -a \\ 0 & 2c & -b \\ -a & -b & 0 \end{pmatrix}, \quad S_{ij} = \begin{pmatrix} -2 & 0 & 0 \\ 0 & -2 & 0 \\ 0 & 0 & -2 \end{pmatrix}, \tag{3.31}$$

and the scalar curvature (1.6) is

$$r = -6. \tag{3.32}$$

Upon substituting Eqs. (3.31) and (3.32) into (1.7), we find it necessary to address the task of resolving

$$c_1 a^2 + 2c_2 - \lambda + 6\rho + c = 0, \tag{3.33}$$

$$c_1 b^2 + 2c_2 - \lambda + 6\rho + c = 0, \tag{3.34}$$

$$c_1 c^2 + 2c_2 - \lambda + 6\rho = 0, \tag{3.35}$$

$$c_1 ab = 0, \tag{3.36}$$

$$2c_1 ac - a = 0, \tag{3.37}$$

$$2c_1 bc - b = 0. \tag{3.38}$$

Examining various scenarios based on Eq. (3.36), the outcomes are as follows:

- If $c_1 = 0$, then $a = b = 0$, and from (3.35) we get $\lambda = 2c_2 + 6\rho$. Substituting in either (3.33) or (3.34) yields $c = 0$.
- If $ab = 0$, we distinguish two particular cases:
  - If $a = 0$ and $b \neq 0$, then from (3.38) we have $c_1 c = \frac{1}{2}$. By direct substitution, we get

$$2c_2 - \lambda + 6\rho + \frac{1}{2c_1} = 0, \tag{3.39}$$

$$c_1 b^2 + 2c_2 - \lambda + 6\rho + \frac{1}{2c_1} = 0, \tag{3.40}$$

$$\frac{1}{4c_1} + 2c_2 - \lambda + 6\rho = 0. \tag{3.41}$$

Substituting (3.41) from (3.39) yields $\frac{1}{c_1} = 0$, which has no solution.

- The case where $a \neq 0$, $b = 0$, and $c_1 c = \frac{1}{2}$ yields an identical result as previously discussed.

- If $a = b = 0$, we obtain

$$2c_2 - \lambda + 6\rho + c = 0, \tag{3.42}$$

$$c_1 c^2 + 2c_2 - \lambda + 6\rho = 0. \tag{3.43}$$

Subtracting (3.41) from (3.43) provides $c(c_1 c - 1) = 0$. Therefore:

- Either $c = 0$ and $\lambda = 2c_2 + 6\rho$
- Or $c = \frac{1}{c_1}$ and $\lambda = 2c_2 + 6\rho + \frac{1}{c_1}$

In conclusion, the solutions of Eq. (3.1) on the algebra $\mathcal{A}_{3,3}$ are given by

$$V = ce_3, \quad c_1 = \frac{1}{c}, \quad \lambda = \frac{1}{c_1} + 2c_2 + 6\rho, \quad \text{where} \quad c \in \mathbb{R}^* \quad \text{and} \quad c_2, \rho \in \mathbb{R}.$$

3.4 The Algebra $\mathcal{A}_{3,4}$

The covariant derivatives of the basis elements are as follows:

$$\begin{aligned} \nabla_{e_1} e_1 &= -e_3, & \nabla_{e_1} e_2 &= 0, & \nabla_{e_1} e_3 &= e_1, \\ \nabla_{e_2} e_1 &= 0, & \nabla_{e_2} e_2 &= e_3, & \nabla_{e_2} e_3 &= -e_2, \\ \nabla_{e_3} e_1 &= 0, & \nabla_{e_3} e_2 &= 0, & \nabla_{e_3} e_3 &= 0. \end{aligned}$$

With the help of (1.2), (1.4), and (1.6), we have

$$(\mathcal{L}_V g)_{ij} = \begin{pmatrix} 2c & 0 & -a \\ 0 & -2c & b \\ -a & b & 0 \end{pmatrix}, \quad S_{ij} = \begin{pmatrix} 0 & 0 & 0 \\ 0 & 0 & 0 \\ 0 & 0 & -2 \end{pmatrix}, \tag{3.44}$$

and the scalar curvature is

$$r = -2. \tag{3.45}$$

After substituting Eqs. (3.44) and (3.45) into (3.1), we need to resolve

$$c_1a^2 - \lambda + 2\rho + c = 0, \tag{3.46}$$

$$c_1b^2 - \lambda + 2\rho - c = 0, \tag{3.47}$$

$$c_1c^2 + 2c_2 - \lambda + 2\rho = 0, \tag{3.48}$$

$$c_1ab = 0, \tag{3.49}$$

$$a - 2c_1ac = 0, \tag{3.50}$$

$$b + 2c_1bc = 0. \tag{3.51}$$

Considering different scenarios outlined in Eq. (3.49), the following observations arise:

- If $c_1 = 0$, then (3.50) and (3.51) yield $a = b = 0$. By direct substitution, we obtain

$$\lambda = 2\rho + c, \tag{3.52}$$

$$\lambda = 2\rho - c, \tag{3.53}$$

$$\lambda = 2c_2 + 2\rho. \tag{3.54}$$

  Subtracting Eq. (3.53) from (3.52) gives $c = 0$. Summing Eqs. (3.52) and (3.53) yields $\lambda = 2\rho$. Putting all of the above in (3.54), we obtain $c_2 = 0$.
- If $ab = 0$ and $c_1 \neq 0$, we consider three cases:
  - If $a = b = 0$, again using Eqs. (3.52), (3.53), and (3.54), we find $c = 0$, $\lambda = 2\rho$, and $c_2 = 0$.
  - If $a = 0$ and $b \neq 0$, then $c_1c = -\frac{1}{2}$. Substituting gives

$$\lambda + 2\rho - \frac{1}{2c_1} = 0, \tag{3.55}$$

$$c_1b^2 - \lambda + 2\rho + \frac{1}{2c_1} = 0, \tag{3.56}$$

$$\frac{1}{4c_1} + 2c_2 - \lambda + 2\rho = 0. \tag{3.57}$$

    From (3.55), we get $\lambda = 2\rho - \frac{1}{2c_1}$. Substituting in (3.56) yields $b^2 = -\frac{1}{2c_1^2}$, which has no real solutions.
  - The case where $b = 0$, $a \neq 0$, and $c_1c = -\frac{1}{2}$ is similar to the previous one.

Combining all the results, the solutions to Eq. (3.1) within the algebra $\mathcal{A}_{3,4}$ exhibit no solutions.

3.5 The Algebra $\mathcal{A}_{3,5}^{\delta}$

The covariant derivatives of the basis elements are as follows:

$$\nabla_{e_1}e_1 = -e_3, \qquad \nabla_{e_1}e_2 = 0, \qquad \nabla_{e_1}e_3 = e_1,$$
$$\nabla_{e_2}e_1 = 0, \qquad \nabla_{e_2}e_2 = -\delta e_3, \qquad \nabla_{e_2}e_3 = \delta e_2,$$
$$\nabla_{e_3}e_1 = 0, \qquad \nabla_{e_3}e_2 = 0, \qquad \nabla_{e_3}e_3 = 0,$$

where $0 < |\delta| < 1$. With direct computations, we have

$$(\mathcal{L}_V g)_{ij} = \begin{pmatrix} 2c & 0 & -a \\ 0 & 2c\delta & -b\delta \\ -a & -b\delta & 0 \end{pmatrix}, \quad S_{ij} = \begin{pmatrix} -1-\delta & 0 & 0 \\ 0 & -\delta^2-\delta & 0 \\ 0 & 0 & -\delta^2-1 \end{pmatrix}, \tag{3.58}$$

and the scalar curvature is

$$r = -2\delta^2 - 2\delta - 2. \tag{3.59}$$

Substituting Eq. (3.58) into (3.1), we are compelled to engage in the process of resolving

$$c_1a^2 + (1+\delta)c_2 - \lambda + 2(\delta^2+\delta+1)\rho + c = 0, \tag{3.60}$$
$$c_1b^2 + \delta(\delta+1)c_2 - \lambda + 2(\delta^2+\delta+1)\rho + \delta c = 0, \tag{3.61}$$
$$c_1c^2 + (\delta^2+1)c_2 - \lambda + 2(\delta^2+\delta+1)\rho = 0, \tag{3.62}$$
$$c_1ab = 0, \tag{3.63}$$
$$c_1ac - \frac{a}{2} = 0, \tag{3.64}$$
$$c_1bc - \delta\frac{b}{2} = 0. \tag{3.65}$$

Evaluating different possibilities with respect to Eq. (3.63), we conclude:

- If $c_1 = 0$, then Eqs. (3.64) and (3.65) give $a = b = 0$. Thus, we have

$$(1+\delta)c_2 - \lambda + 2(\delta^2+\delta+1)\rho + c = 0, \tag{3.66}$$
$$\delta(\delta+1)c_2 - \lambda + 2(\delta^2+\delta+1)\rho + \delta c = 0, \tag{3.67}$$
$$(\delta^2+1)c_2 - \lambda + 2(\delta^2+\delta+1)\rho = 0. \tag{3.68}$$

Subtracting Eqs. (3.66) from (3.67), we find $c = -(\delta + 1)c_2$. From Eq. (3.68), $\lambda = (\delta^2+1)c_2+2(\delta^2+\delta+1)\rho$. By substituting these results into either Eq. (3.66) or (3.67), we get $(\delta^2 + 1)c_2 = 0$, and hence $c_2 = 0$.

- If $a = b = 0$ and $c_1 \neq 0$, then Eqs. (3.66) and (3.67) result in $c = -(\delta + 1)c_2$ and $\lambda = 2(\delta^2 + \delta + 1)\rho$. Substituting into (3.62), we obtain $c_2 = -\frac{\delta^2+1}{(\delta^2+2\delta+1)c_1}$.
- If $a \neq 0$, $b = 0$, then from (3.64) we get $c = \frac{1}{2c_1}$. Thus

$$c_1a^2 + (1 + \delta)c_2 - \lambda + 2(\delta^2 + \delta + 1)\rho + \frac{1}{2c_1} = 0, \tag{3.69}$$

$$\delta(\delta + 1)c_2 - \lambda + 2(\delta^2 + \delta + 1)\rho + \frac{\delta}{2c_1} = 0, \tag{3.70}$$

$$\frac{1}{4c_1} + (\delta^2 + 1)c_2 - \lambda + 2(\delta^2 + \delta + 1)\rho = 0. \tag{3.71}$$

Subtracting Eq. (3.70) from (3.71), we find $c_1 = \frac{-2\delta+1}{4(\delta-1)c_2}$. Using these results along with Eq. (3.69), we find $a^2 = \frac{-2\delta^2+\delta-1}{4c_1^2}$, which leads to an impossibility due to $-2\delta^2 + \delta - 1 < 0$.

- If $a = 0$, $b \neq 0$, then from (3.65) we get $c = \frac{\delta}{2c_1}$. Using direct substitution, we have

$$(1 + \delta)c_2 - \lambda + 2(\delta^2 + \delta + 1)\rho + \frac{\delta}{2c_1} = 0, \tag{3.72}$$

$$c_1b^2 + \delta(\delta + 1)c_2 - \lambda + 2(\delta^2 + \delta + 1)\rho + \frac{\delta^2}{2c_1} = 0, \tag{3.73}$$

$$\frac{\delta^2}{4c_1} + (\delta^2 + 1)c_2 - \lambda + 2(\delta^2 + \delta + 1)\rho = 0. \tag{3.74}$$

Subtracting Eq. (3.73) from (3.74), we find $c_1 = \frac{\delta-2}{4(1-\delta)c_2}$. From (3.72) we pull

$$\lambda = -\frac{\delta^2 - \delta + 2}{4(1 - \delta)c_1}.$$

Finally, substituting in (3.73), we obtain

$$b^2 = -\frac{\delta^2 - \delta + 2}{4c_1^2},$$

which is absurd due to the fact $\delta^2 - \delta + 2 > 0$ for all $0 < |\delta| < 1$.

In summary, Eq. (3.1) within the algebra $\mathcal{A}^{\delta}_{3,5}$ admits the following solution:

$$V = ce_3, \quad c = -(\delta+1)c_2, \quad c_2 = -\frac{\delta^2+1}{(\delta^2+2\delta+1)c_1},$$
$$\lambda = 2(\delta^2+\delta+1)\rho,$$
$$c_1 \in \mathbb{R}^* \quad \text{and} \quad \rho \in \mathbb{R}.$$

3.6 *The Algebra* $\mathcal{A}_{3,6}$

The covariant derivatives of the basis elements are as follows:

$$\begin{aligned} &\nabla_{e_1}e_1 = 0, && \nabla_{e_1}e_2 = 0, && \nabla_{e_1}e_3 = 0,\\ &\nabla_{e_2}e_1 = 0, && \nabla_{e_2}e_2 = 0, && \nabla_{e_2}e_3 = 0,\\ &\nabla_{e_3}e_1 = e_2, && \nabla_{e_3}e_2 = -e_1, && \nabla_{e_3}e_3 = 0. \end{aligned}$$

With direct computations, we have

$$(\mathcal{L}_V g)_{ij} = \begin{pmatrix} 0 & 0 & -b \\ 0 & 0 & a \\ -b & a & 0 \end{pmatrix}, \quad S_{ij} = \begin{pmatrix} 0 & 0 & 0 \\ 0 & 0 & 0 \\ 0 & 0 & 0 \end{pmatrix}, \tag{3.75}$$

and the scalar curvature is

$$r = 0. \tag{3.76}$$

Substituting these into (3.1), we tackle the endeavor of resolving

$$c_1a^2 - \lambda = 0, \tag{3.77}$$
$$c_1b^2 - \lambda = 0, \tag{3.78}$$
$$c_1c^2 - \lambda = 0, \tag{3.79}$$
$$c_1ab = 0, \tag{3.80}$$
$$c_1ac - \frac{b}{2} = 0, \tag{3.81}$$
$$c_1bc + \frac{a}{2} = 0. \tag{3.82}$$

Reviewing several scenarios outlined by Eq. (3.80), the analysis indicates:

- If $c_1 = 0$, then from (3.77), (3.81), and (3.82) we get $\lambda = a = b = 0$.
- Consider $c_1 \neq 0$:
  - If $a = 0$, then from (3.81), $b = 0$, leading to $\lambda = 0$ and $c = 0$.
  - Similarly, if $b = 0$, from (3.82), $a = 0$, resulting in $\lambda = 0$ and $c = 0$.

Combining all the results, Eq. (3.1) within the algebra $\mathcal{A}_{3,6}$ satisfies only the Killing equation for $V = ce_3$ where $c \neq 0$.

3.7 The Algebra $\mathcal{A}_{3,7}^{\delta}$

The covariant derivatives of the basis elements are as follows:

$$\begin{array}{lll} \nabla_{e_1}e_1 = \delta e_3, & \nabla_{e_1}e_2 = 0, & \nabla_{e_1}e_3 = -\delta e_1, \\ \nabla_{e_2}e_1 = 0, & \nabla_{e_2}e_2 = -\delta e_3, & \nabla_{e_2}e_3 = \delta e_2, \\ \nabla_{e_3}e_1 = e_2, & \nabla_{e_3}e_2 = -e_1, & \nabla_{e_3}e_3 = 0, \end{array}$$

where $\delta > 0$.

With direct computations, we have

$$(\mathcal{L}_V g)_{ij} = \begin{pmatrix} -2c\delta & 0 & a\delta - b \\ 0 & 2c\delta & a - \delta b \\ a\delta - b & a - \delta b & 0 \end{pmatrix}, \quad S_{ij} = \begin{pmatrix} 0 & 2\delta & 0 \\ 2\delta & 0 & 0 \\ 0 & 0 & -2\delta^2 \end{pmatrix}, \tag{3.83}$$

and the scalar curvature is

$$r = -2\delta^2. \tag{3.84}$$

We need to solve the following equations from (3.1):

$$c_1a^2 - \lambda + 2\delta^2\rho - c\delta = 0, \tag{3.85}$$

$$c_1b^2 - \lambda + 2\delta^2\rho + c\delta = 0, \tag{3.86}$$

$$c_1c^2 + 2c_2\delta^2 - \lambda + 2\delta^2\rho = 0, \tag{3.87}$$

$$c_1ab - 2c_2\delta = 0, \tag{3.88}$$

$$c_1ac + \frac{a}{2}\delta - \frac{b}{2} = 0, \tag{3.89}$$

$$c_1bc + \frac{a}{2} - \frac{b}{2}\delta = 0. \tag{3.90}$$

Analyzing Eq. (3.89), we obtain $b = 2c_1ac + a\delta$. Substituting into (3.90) yields

$$a\left(c_1^2c^2 + \frac{1-\delta}{4}\right) = 0. \tag{3.91}$$

Reviewing several scenarios outlined by Eq. (3.91), the analysis indicates:

- If $a = 0$, substituting in (3.89) and using (3.88) give $b = 0$ and $c_2 = 0$. Substituting into (3.85) and (3.86) gives $c = 0$ and $\lambda = 2\delta^2\rho$.
- If $c_1c = \frac{\sqrt{\delta-1}}{2}$, which is valid only for $\delta \geq 1$, then by direct substitution we get:
  - If $\delta > 1$, then $a = 0$, and similar results are obtained as discussed previously.
  - If $\delta = 1$, then $c_1 = 0$. In the first case, from (3.88) we get $c_2 = 0$ and from (3.89) $a = b$. Hence, we are left with

$$-\lambda + 2\rho - c = 0, \tag{3.92}$$

$$-\lambda + 2\rho + c = 0, \tag{3.93}$$

$$-\lambda + 2\rho = 0, \tag{3.94}$$

which clearly gives $\lambda = 2\rho$ and $c = 0$.

In the second case, where $c = 0$, again from (3.89) we have $a = b$, and using (3.88) we get $a = \sqrt{\frac{2c_2}{c_1}}$. Finally, from (3.87) we obtain $\lambda = 2c_2 + 2\rho$.

Combining all the results, the solutions to Eq. (3.1) within the algebra $\mathcal{A}_{3,7}^{\delta}$ are

$$V = a(e_1 + e_2), \quad c_1 = \frac{2c_2}{a^2}, \quad \lambda = 2c_2 + 2\rho, \quad a \in \mathbb{R}^*, \quad \text{and} \quad c_2, \rho \in \mathbb{R}.$$

3.8 The Algebra $\mathcal{A}_{3,8}$

The covariant derivatives of the basis elements are

$$\begin{aligned}
&\nabla_{e_1}e_1 = -e_2, && \nabla_{e_1}e_2 = e_1 + e_3, && \nabla_{e_1}e_3 = -e_2,\\
&\nabla_{e_2}e_1 = e_3, && \nabla_{e_2}e_2 = 0, && \nabla_{e_2}e_3 = -e_1,\\
&\nabla_{e_3}e_1 = e_2, && \nabla_{e_3}e_2 = -e_1 - e_3, && \nabla_{e_3}e_3 = e_2.
\end{aligned}$$

From direct computations, we obtain

$$(\mathcal{L}_V g)_{ij} = \begin{pmatrix} 2b & -a-2c & 0\\ -a-2c & 0 & 2a+c\\ 0 & 2a+c & -2b \end{pmatrix}, \quad S_{ij} = \begin{pmatrix} -2 & 0 & -2\\ 0 & 0 & 0\\ -2 & 0 & -2 \end{pmatrix}. \tag{3.95}$$

The scalar curvature is given by

$$r = -4. \tag{3.96}$$

Therefore, upon substituting (3.95) and (3.96) into (3.1), the following equations must be satisfied:

$$c_1a^2 + 2c_2 - \lambda + 4\rho + b = 0, \tag{3.97}$$

$$c_1b^2 - \lambda + 4\rho = 0, \tag{3.98}$$

$$c_1c^2 + 2c_2 - \lambda + 4\rho - b = 0, \tag{3.99}$$

$$c_1ab - \frac{a}{2} - c = 0, \tag{3.100}$$

$$c_1ac + 2c_2 = 0, \tag{3.101}$$

$$c_1bc + a + \frac{c}{2} = 0. \tag{3.102}$$

From (3.100), we deduce $c = c_1ab - \frac{a}{2}$. Substituting this into (3.102) gives

$$a\left(c_1^2b^2 + \frac{3}{4}\right) = 0.$$

Hence, $a = 0$ and $c = 0$. This implies $c_2 = 0$. Substituting $c_2 = 0$ into (3.98) yields $\lambda = 4\rho$. Finally, substituting $\lambda = 4\rho$ into either (3.97) or (3.99) provides $b = 0$.

In conclusion, Eq. (3.1) in the algebra $\mathcal{A}_{3,8}$ has no solution.

3.9 The Algebra $\mathcal{A}_{3,9}$

The covariant derivatives of the basis elements are

$$\begin{aligned}
&\nabla_{e_1}e_1 = 0, && \nabla_{e_1}e_2 = \frac{1}{2}e_3, && \nabla_{e_1}e_3 = -\frac{1}{2}e_2,\\
&\nabla_{e_2}e_1 = -\frac{1}{2}e_3, && \nabla_{e_2}e_2 = 0, && \nabla_{e_2}e_3 = \frac{1}{2}e_1,\\
&\nabla_{e_3}e_1 = \frac{1}{2}e_2, && \nabla_{e_3}e_2 = -\frac{1}{2}e_1, && \nabla_{e_3}e_3 = 0.
\end{aligned}$$

From direct computations, we obtain

$$(\mathcal{L}_V g)_{ij} = \begin{pmatrix} 0 & 0 & 0 \\ 0 & 0 & 0 \\ 0 & 0 & 0 \end{pmatrix}, \quad S_{ij} = \begin{pmatrix} \frac{1}{2} & 0 & 0 \\ 0 & \frac{1}{2} & 0 \\ 0 & 0 & \frac{1}{2} \end{pmatrix}, \tag{3.103}$$

and the scalar curvature is

$$r = \frac{3}{2}. \tag{3.104}$$

Thus, the following equations must be satisfied:

$$2c_1a^2 - c_2 - 2\lambda - 3\rho = 0, \tag{3.105}$$

$$2c_1b^2 - c_2 - 2\lambda - 3\rho = 0, \tag{3.106}$$

$$2c_1c^2 - c_2 - 2\lambda - 3\rho = 0, \tag{3.107}$$

$$c_1ab = 0, \tag{3.108}$$

$$c_1ac = 0, \tag{3.109}$$

$$c_1bc = 0. \tag{3.110}$$

Analyzing Eqs. (3.108)–(3.110) yields:

- If $c_1 = 0$, then $\lambda = -\frac{1}{2}c_2 - \frac{3}{2}\rho$ and $c_2, \rho \in \mathbb{R}$.
- If $c_1 \neq 0$, then $a = b = c = 0$, $\lambda = -\frac{1}{2}c_2 - \frac{3}{2}\rho$, and $c_2, \rho \in \mathbb{R}$.

Therefore, the solution set is

$$V = ae_1 + be_2 + ce_3, \quad c_1 = 0, \quad \lambda = -\frac{1}{2}c_2 - \frac{3}{2}\rho, \quad \text{and} \quad a, b, c, c_2, \rho \in \mathbb{R}.$$

Theorem 3.1 *The generalized Ricci-Yamabe soliton equation*

$$\mathcal{L}_V g = -2c_1 V^\flat \otimes V^\flat + 2c_2 S + 2(\lambda + r\rho)g$$

admits the following solutions on three-dimensional left-invariant Lie algebras:

- ***Algebra*** $\mathcal{A}_{3,1}$:

$$V = ae_1, \quad c_1 = \frac{c_2}{a^2}, \quad \lambda = \frac{\rho}{2} + \frac{c_2}{2}, \quad a \in \mathbb{R}^* \quad \textit{and} \quad c_2, \rho \in \mathbb{R}.$$

- ***Algebra*** $\mathcal{A}_{3,3}$:

$$V = ce_3, \quad c_1 = \frac{1}{c}, \quad \lambda = \frac{1}{c_1} + 2c_2 + 6\rho, \quad \textit{where} \quad c \in \mathbb{R}^* \quad \textit{and} \quad c_2, \rho \in \mathbb{R}.$$

- ***Algebra*** $\mathcal{A}_{3,5}^{\delta}$:

$$V = ce_3, \quad c = -(\delta + 1)c_2, \quad c_2 = -\frac{\delta^2 + 1}{(\delta^2 + 2\delta + 1)c_1},$$

$$\lambda = 2(\delta^2 + \delta + 1)\rho, \quad where \quad c_1 \in \mathbb{R}^* \quad and \quad \rho \in \mathbb{R}.$$

- ***Algebra*** $\mathcal{A}_{3,7}^{\delta}$:

$$V = a(e_1 + e_2), \quad c_1 = \frac{2c_2}{a^2}, \quad \lambda = 2c_2 + 2\rho, \quad a \in \mathbb{R}^* \quad and \quad c_2, \rho \in \mathbb{R}.$$

- ***Algebra*** $\mathcal{A}_{3,9}$:

$$V = ae_1 + be_2 + ce_3, \quad c_1 = 0, \quad \lambda = -\frac{1}{2}c_2 - \frac{3}{2}\rho, \quad and \quad a, b, c, c_2, \rho \in \mathbb{R}.$$

4 Conclusion

In this chapter, we have extended the concept of the Ricci-Yamabe soliton through Eq. (1.7) and explored the presence of this structure on left-invariant three-dimensional Lie groups. The findings provide concrete examples that substantiate the existence of this structure, thus demonstrating its viability. This work opens a wide range of possibilities for future research in this area. We can summarize the existence of various solitonic structures on left-invariant three-dimensional Lie algebras in the following (Table 2):

Table 2 Possible solitonic structure on left-invariant three-dimensional Lie algebras

Algebra	GRYS	GRS	RBS	RS	YS	PFS
$\mathcal{A}_{3,1}$	✓	✓	✓	✓	✓	✓
$\mathcal{A}_{3,2}$	✗	✗	✗	✗	✗	✗
$\mathcal{A}_{3,3}$	✓	✓	✓	✓	✓	✗
$\mathcal{A}_{3,4}$	✗	✗	✗	✗	✗	✗
$\mathcal{A}_{3,5}^{\delta}$	✓	✓	✓	✓	✓	✗
$\mathcal{A}_{3,6}$	✗	✗	✗	✗	✗	✗
$\mathcal{A}_{3,7}^{\delta}$	✓	✓	✓	✓	✓	✗
$\mathcal{A}_{3,8}$	✗	✗	✗	✗	✗	✗
$\mathcal{A}_{3,9}$	✓	✓	✓	✓	✓	✗

References

1. Barnes, A., Rowlingson, R.R.: Irrotational perfect fluids with a purely electric Weyl tensor. Class. Quantum Grav. **6**(7), 949–960 (1989). https://doi.org/10.1088/0264-9381/6/7/010
2. Blaga, A.M.: On η-Ricci-Yamabe solitons in almost contact metric manifol. Symmetry **13**(4), 643 (2021). https://doi.org/10.3390/sym13040643
3. Blaga, A.M., Crasmareanu, M.C.: Torse-forming η-Ricci solitons in almost paracontact. Filomat. https://doi.org/10.2298/FIL1702499B
4. Catino, G., Mazzieri, L., Roncoroni, A.: Rigidity of gradient Einstein shrinkers. Commun. Contemp. Math. **18**(06), 1650033 (2016). https://doi.org/10.1142/S0219199716500336
5. Chen, B.-Y., Deshmukh, S.: Geometry of compact shrinking Ricci solitons. Balkan J. Geom. Appl. https://api.semanticscholar.org/CorpusID:12020344
6. Deshmukh, S.: Jacobi-type vector fields on Ricci solitons. Bull. Math. Soc. Sci. Roumanie (N.S.) **55**(1), 41–50 (2012)
7. Deshmukh, S., Alodan, A.: Ricci-Yamabe solitons on warped product manifolds. Mediterr. J. Math. **17**(5), 1–11 (2020). https://doi.org/10.1007/s00009-020-01552-4
8. Fernández-López, M., García-Río, E.: Rigidity of shrinking Ricci solitons. Math. Z. **259**(3), 611–628 (2008). https://doi.org/10.1007/s00209-007-0296-7
9. Freitas Filho, A.A.: Solitons de Ricci com Estrutura de Produto Deformado. Tese (Doutorado em Matemática), Universidade Federal do Amazonas, Manaus (2017). http://tede.ufam.edu.br/handle/tede/5872
10. Glass, E.N.: Shear-free irrotational perfect fluids. J. Math. Phys. **26**(12), 3141–3143 (1985). https://doi.org/10.1063/1.526864
11. Hamilton, R.S.: Three-manifolds with positive Ricci curvature. J. Differ. Geom. **17**, 255–306 (1982)
12. Jantzen, R.: Editor's note: On the three-dimensional spaces which admit a continuous group of motions by Luigi Bianchi. Gen. Relativ. Gravit. **33**, 2157–2170 (2001). https://doi.org/10.1023/A:1015326128022
13. Lafuente, R., Lauret, J.: Structure of homogeneous Ricci solitons and the Alekseevskii conjecture. J. Differ. Geom. **98**(2), 315–347 (2011). https://doi.org/10.4310/jdg/1314666228
14. Mars, M.: Spacetime Ehlers group: transformation law for the Weyl tensor. Class. Quantum Grav. **17**(17), 3353–3372 (2000). https://doi.org/10.1088/0264-9381/17/17/307
15. Nurowski, P., Randall, M.: Generalized Ricci solitons. J. Geom. Anal. **26**, 1280–1345 (2016) https://doi.org/10.1007/s12220-015-9592-8
16. Patera, J., Sharp, R.T., Winternitz, P., Zassenhaus, H.: Invariants of real low dimension Lie algebras. J. Math. Phys. **17**, 986–994 (1976). https://doi.org/10.1063/1.522992
17. Petersen, P., Wylie, W.: On the classification of gradient Ricci solitons. Geom. Topol. **14**(4), 2277–2300 (2014). https://doi.org/10.2140/gt.2010.14.2277
18. Stephani, H., Kramer, D., MacCallum, M., Hoenselaers, C., Herlt, E.: Exact Solutions of Einstein's Field Equations. Cambridge University Press, Cambridge (2003). https://doi.org/10.1017/CBO9780511535185
19. Traore, M., Taçstan, H.M., Gerdan Aydn, S.: Some characterizations on gradient almost η-Ricci-Bourguignon solitons. Preprint (2022). https://doi.org/10.21203/rs.3.rs-2397382/v1
20. Traore, M., Taçstan, H.M., Gerdan Aydn, S.: On almost η-Ricci-Bourguignon solitons. Miskolc Math. Notes **25**(1), 493–508 (2024). https://doi.org/10.18514/MMN.2024.4378

Riemannian Invariants in Submanifold Theory

Adela Mihai

Abstract The basics of submanifolds in complex space forms and Sasakian space forms are recalled, and then Chen-type inequalities for different submanifolds in complex and Sasakian space forms are presented.

The most important Chen inequalities in real space forms are stated. We give a general construction method for purely real submanifolds and present geometric inequalities for purely submanifolds in complex space forms. We obtain an improved Chen-Ricci inequality for Kaehlerian slant submanifolds in complex space forms. Works on DDVV conjecture are also presented. Next, results on submanifolds in Sasakian manifolds are presented. We prove Chen's first inequality for contact slant submanifolds in Sasakian space forms. We define Chen-type Sasakian invariants, obtain sharp inequalities for these invariants, and derive characterizations of the equality case in terms of the shape operator. We generalize a result of Chen and obtain a Chen-Ricci inequality for purely real submanifolds with T parallel with respect to the Levi-Civita connection. Another subsection presents certain results for submanifolds in space forms with semi-symmetric metric (respectively, nonmetric) connections. We study statistical submanifolds and their behavior in statistical manifolds of constant curvature.

Next, we present results on warped product submanifolds in complex space forms, generalized complex space forms, and quaternion space forms.

After that, a new characterization of Einstein spaces by using their curvatures symmetries is given.

This chapter represents a collection of results from the author's papers on this topic; the proofs are given in detail, so the reader can follow the techniques.

Keywords Kaehler manifolds · Sasakian manifolds · Einstein manifolds · Submanifolds · Riemannian invariants · Chen invariants · Chen inequalities

---

A. Mihai (✉)
Technical University of Civil Engineering Bucharest, Bucharest, Romania

Transilvania University of Braşov, Braşov, Romania
e-mail: adela.mihai@utcb.ro; adela.mihai@unitbv.ro

B.-Y. Chen, M. A. Choudhary (eds.), *Geometric Inequalities and Applications*,
Infosys Science Foundation Series in Mathematical Sciences,
https://doi.org/10.1007/978-981-95-5148-4_3

1 Preliminaries

In Riemannian Geometry the manifolds endowed with certain endomorphisms of their tangent bundles play an important role.

Among these, the most important ones are the almost complex structures (on even-dimensional manifolds) and almost contact structures (on odd-dimensional manifolds). In particular the Kaehler manifolds and the Sasakian manifolds, respectively, are the most studied such manifolds, because they have the most interesting properties and applications.

In order to have the highest degree of homogeneity (i.e. their group of isometries has the maximum dimension), the spaces of constant sectional curvatures are the most investigated. It is known that a Kaehler manifold with constant sectional curvature is flat. For this reason the notion of complex space form (a Kaehler manifold with constant holomorphic sectional curvature) was introduced. Analogously, the Sasakian space forms were defined.

On the other hand, starting from the classical theory of curves and surfaces in Euclidean spaces, the theory of submanifolds is an important field of research in Riemannian Geometry.

There are certain important specific classes of submanifolds in Kaehler manifolds and Sasakian manifolds, respectively, for example complex and Lagrangian submanifolds in Kaehler manifolds and invariant and Legendrian submanifolds in Sasakian manifolds. For a comprehensive study on submanifolds see [21].

Let $\tilde{M}$ be a complex manifold of dimension m and J its standard almost complex structure. A *Hermitian metric* on $\tilde{M}$ is a Riemannian metric g invariant with respect to J, i.e., $g(JX, JY) = g(X, Y)$, $\forall X, Y \in \Gamma(T\tilde{M})$.

The pairing $(\tilde{M}, g)$ is called a *Hermitian* manifold.

Any complex manifold admits a Hermitian metric.

A Hermitian metric g on a complex manifold $\tilde{M}$ defines a nondegenerate 2-form $\omega(X, Y) = g(JX, Y)$, $X, Y \in \Gamma(T\tilde{M})$, which is called the *fundamental 2-form*. Clearly, $\omega(JX, JY) = \omega(X, Y)$.

A Hermitian manifold is called a *Kaehler* manifold if the fundamental 2-form ω is closed.

Necessary and sufficient conditions for a Hermitian manifold to be a Kaehler manifold are given by the following:

Theorem ([94]) *Let $(\tilde{M}, g)$ be an m-dimensional Hermitian manifold and $\tilde{\nabla}$ the Levi-Civita connection associated with g. The following statements are equivalent to each other:*

- *(i) M is a Kaehler manifold.*
- *(ii) The standard almost complex structure J on $\tilde{M}$ is parallel with respect to $\tilde{\nabla}$, i.e., $\tilde{\nabla} J = 0$.*
- *(iii) For any $z_0 \in \tilde{M}$, there exists a holomorphic coordinate system in a neighborhood of z_0 such that*

$$g = (\delta_{kj} + h_{kj})dz^k d\bar{z}^j,$$

where $h_{kj}(z_0) = \frac{\partial h_{kj}}{\partial z^l}(z_0) = 0$, *for any* $k, j, l = 1, \ldots, m$.

(iv) *Locally, there exists a real differentiable function* F *such that the fundamental 2-form is given by* $\omega = i\partial\bar{\partial}F$, *where the exterior differentiation* d *is decomposed in* $d\alpha = \partial\alpha + \bar{\partial}\alpha$.

Examples of Kaehler manifolds are:

1. $\mathbf{C}^n$ with the Euclidean metric $g = \sum_{k=1}^{n} dz^k d\bar{z}^k$.
2. The *complex torus* $T^n = \mathbf{C}^n/_G$ with the Hermitian structure induced by the Euclidean metric of $\mathbf{C}^n$.
3. The *complex projective space* $P^n(\mathbf{C})$ endowed with the Fubini-Study metric, which, in local coordinates, is given by

$$g_{j\bar{k}} = \frac{(1 + z^s\bar{z}^s)\delta_{jk} - z^k\bar{z}^j}{(1 + z^s\bar{z}^s)^2}.$$

4. The *complex Grassmann manifold* $G_p(\mathbf{C}^{p+q})$ with a generalized Fubini-Study metric.
5. Let $D^n = \text{Int } S^{2n-1}$ be the unit disk in $\mathbf{C}^n$, i.e., $D^n = \{z \in \mathbf{C}^n \mid \sum_{j=1}^{n} |z^j|^2 < 1\}$, endowed with the Bergman metric

$$g_{j\bar{k}} = \frac{(1 - z^s\bar{z}^s)\delta_{jk} + \bar{z}^j z^k}{(1 - z^s\bar{z}^s)^2}.$$

6. Any orientable surface is a Kaehler manifold.

There are obstructions to the existence of Kaehlerian metrics on a compact complex manifold.

Theorem **([69])** *On a compact Kaehler manifold the Betti numbers of even order are nonzero.*

As an application, the Calabi manifolds $S^{2m+1} \times S^{2n+1}$ do not admit any Kaehler metric if $(m, n) \neq (0, 0)$. In particular, Hopf manifolds are not Kaehler manifolds.

Theorem **([69])** *On a compact Kaehler manifold the Betti numbers of odd order are even.*

A sectional curvature of a Kaehler $\tilde{M}$ in direction of an invariant 2-plane section by J is called a *holomorphic sectional curvature* of $\tilde{M}$.

For the 2-plane section π invariant by J, one considers an orthonormal basis $\{X, JX\}$, with unit X. Then the holomorphic sectional curvature is given by $\tilde{K}(\pi) = \tilde{R}(X, JX, X, JX)$.

Let $\tilde{M}$ be a Kaehler manifold. If the function holomorphic sectional curvature K is constant for all 2-plane sections π of $T_p\tilde{M}$ invariant by J for any $p \in \tilde{M}$, then $\tilde{M}$

is called a *space with constant holomorphic sectional curvature* (or *complex space form*).

The curvature tensor of a complex space form of constant holomorphic sectional curvature $4c$ has the expression

$$\begin{aligned}\tilde{R}(X, Y, Z, W) = c[&g(X, Z)g(Y, W) - g(X, W)g(Y, Z)\\ &+ g(X, JZ)g(Y, JW) - g(X, JW)g(Y, JZ)\\ &+ 2g(X, JY)g(Z, JW)],\end{aligned}$$

for any tangent vector fields X, Y, Z, W.

Recall that a Riemannian manifold (M, g) is an *Einstein manifold* if the Ricci tensor S is proportional to the Riemannian metric g, i.e., $S = \lambda g$, where λ is a real number.

Each complex space form is an Einstein manifold.

Examples of complex space forms:

1. $\mathbf{C}^n$ with the Euclidean metric is a flat complex space form.
2. $P^n(\mathbf{C})$ with the Fubini-Study metric has holomorphic sectional curvature equal to 4.
3. D^n with the Bergman metric has holomorphic sectional curvature equal to -4.

Conversely, the following result holds good.

Theorem **([69])** *Let M be a connected, simply connected, and complete complex space form. Then M is isometric to either $\mathbf{C}^n$, $P^n(\mathbf{C})$, or D^n.*

Let $(\tilde{M}, J, g)$ be an m-dimensional Kaehler manifold and M an n-dimensional submanifold of $\tilde{M}$. The induced Riemannian metric on M is also denoted by g. We denote by $\tilde{\nabla}$ and ∇ the Levi-Civita connections on $\tilde{M}$ and M, respectively. The fundamental formulae and equations for a submanifold are recalled below.

Let h be the second fundamental form of the submanifold M. Then the *Gauss formula* is written as

$$\tilde{\nabla}_X Y = \nabla_X Y + h(X, Y),$$

for any $X, Y \in \Gamma(TM)$.

Denoting by $\nabla^{\perp}$ the connection in the normal bundle and by A the shape operator, one has the *Weingarten formula:*

$$\tilde{\nabla}_X \xi = -A_{\xi} X + \nabla^{\perp}_X \xi,$$

for any $X \in \Gamma(TM)$ and $\xi \in \Gamma(T^{\perp}M)$.

Let $\tilde{R}$, R, and $R^{\perp}$ be the curvature tensors with respect to $\tilde{\nabla}$, ∇, and $\nabla^{\perp}$, respectively.

For any $X, Y, Z, W \in \Gamma(TM)$, the *Gauss equation* is expressed by

$$\tilde{R}(X, Y, Z, W) = R(X, Y, Z, W) - g(h(X, Z), h(Y, W)) + g(h(X, W), h(Y, Z)).$$

One denotes

$$(\nabla_X h)(Y, Z) = \nabla_X^{\perp} h(Y, Z) - h(\nabla_X Y, Z) - h(X, \nabla_Y Z);$$

then the normal component of $\tilde{R}(X, Y)Z$ is given by

$$(\widetilde{R}(X, Y)Z)^{\perp} = (\nabla_X h)(Y, Z) - (\nabla_Y h)(X, Z).$$

The above relation represents the *Codazzi equation.*

Using the Weingarten formula, one obtains the *Ricci equation*:

$$\widetilde{R}(X, Y, \xi, \eta) = R^{\perp}(X, Y, \xi, \eta) - g(A\eta A_{\xi} X, Y) + g(A_{\xi} A\eta X, Y)$$

$$= R^{\perp}(X, Y, \xi, \eta) + g([A_{\xi}, A\eta]X, Y),$$

for any $X, Y \in \Gamma(TM)$ and $\xi, \eta \in \Gamma(T^{\perp}M)$.

If the second fundamental form h vanishes identically, M is a *totally geodesic* submanifold.

Let $\{e_1, \ldots, e_n\}$ be an orthonormal basis of the tangent space T_pM, $p \in M$, and H be the *mean curvature vector*, i.e.,

$$H(p) = \frac{1}{n} \sum_{i=1}^{n} h(e_i, e_i).$$

The submanifold M is said to be *minimal* if $H(p) = 0, \forall p \in M$.

There are no compact minimal submanifolds of $\mathbf{R}^m$.

For a normal section V on M, if A_V is everywhere proportional to the identity transformation I, i.e., $A_V = aI$, for some function a, then V is called an *umbilical section* on M, or M is said to be *umbilical with respect to* V. If the submanifold M is umbilical with respect to every local normal section of M, then M is said to be *totally umbilical.*

An equivalent definition is the following: M is *totally umbilical* if $h(X, Y) = g(X, Y)H$, for any vector fields X, Y tangent to M.

If the second fundamental form and the mean curvature of M in $\widetilde{M}$ satisfy $g(h(X, Y), H) = fg(X, Y)$ for some function f on M, then M is called *pseudo-umbilical.*

The submanifold M is a *parallel* submanifold if the second fundamental form h is parallel, that is, $\nabla h = 0$, identically.

According to the behavior of the tangent spaces of a submanifold M under the action of the almost complex structure J of the ambient space $\tilde{M}$, we distinguish two special classes of submanifolds:

(i) *Complex submanifolds*, if $J(T_pM) = T_pM, \forall p \in M$
(ii) *Totally real submanifolds*, if $J(T_pM) \subset T_p^{\perp}M, \forall p \in M$

Any complex submanifold of a Kaehler manifold is a Kaehler manifold and a minimal submanifold.

If the real dimension of the totally real submanifold M is equal to the complex dimension of the Kaehler manifold $\tilde{M}$, then M is called a *Lagrangian submanifold.* In other words, a *Lagrangian submanifold* is a totally real submanifold of maximum dimension.

Other classes of submanifolds in Kaehler manifolds are of interest in submanifold theory.

A *slant submanifold* [23, 84] is a submanifold M of a Kaehler manifold $(\tilde{M}, J, g)$ such that, for any nonzero vector X $\in T_pM$, the angle $\theta(X)$ between JX and the tangent space T_pM is a constant (which is independent of the choice of the point $p \in M$ and the choice of the tangent vector X in the tangent plane T_pM).

It is obvious that complex submanifolds and totally real submanifolds are special classes of slant submanifolds. A slant submanifold is called *proper* if it is neither a complex submanifold nor a totally real submanifold.

A submanifold M of a Kaehler manifold $\tilde{M}$ is said to be a *CR-submanifold* if it admits a holomorphic differentiable distribution $\mathcal{D}$, i.e., $J(\mathcal{D}_p) = \mathcal{D}_P$, $\forall p \in M$, such that its complementary orthogonal distribution $\mathcal{D}^{\perp}$ is totally real, i.e., $J(\mathcal{D}_p^{\perp}) \subset T_p^{\perp}M$, $p \in M$.

The CR-submanifolds were studied by B.Y. Chen [22], A. Bejancu [11], K. Yano and M. Kon [123], etc.

Both complex and totally real submanifolds are improper CR-submanifolds.

It is easily seen that a real hypersurface of a Kaehler manifold is a proper CR-submanifold.

Roughly speaking, a Sasakian manifold is the odd-dimensional correspondent of a Kaehler manifold.

A $(2m + 1)$-dimensional Riemannian manifold $(\widetilde{M}, g)$ is said to be a *Sasakian manifold* if it admits an endomorphism ϕ of its tangent bundle $T\widetilde{M}$, a vector field ξ, and a 1-form η, satisfying

$$\begin{cases} \phi^2 = -Id + \eta \otimes \xi, \ \eta(\xi) = 1, \ \phi\xi = 0, \ \eta \circ \phi = 0, \\ g(\phi X, \phi Y) = g(X, Y) - \eta(X)\eta(Y), \ \eta(X) = g(X, \xi), \\ (\tilde{\nabla}_X \phi)Y = -g(X, Y)\xi + \eta(Y)X, \ \tilde{\nabla}_X \xi = \phi X, \end{cases}$$

for any vector fields X, Y on $\widetilde{M}$, where $\widetilde{\nabla}$ denotes the Levi-Civita connection with respect to g.

A plane section π in $T_p\tilde{M}$ is called a *ϕ-section* if it is spanned by X and ϕX, where X is a unit tangent vector orthogonal to ξ. The sectional curvature of a ϕ-

section is called a *ϕ-sectional curvature*. A Sasakian manifold with constant ϕ-sectional curvature c is said to be a *Sasakian space form* and is denoted by $\widetilde{M}(c)$.

The curvature tensor of $\widetilde{R}$ of a Sasakian space form $\widetilde{M}(c)$ is given by [13]

$$\begin{aligned}\widetilde{R}(X,Y)Z &= \frac{c+3}{4}\{g(Y,Z)X - g(X,Z)Y\}\\ &+ \frac{c-1}{4}\{\eta(X)\eta(Z)Y - \eta(Y)\eta(Z)X\\ &+ g(X,Z)\eta(Y)\xi - g(Y,Z)\eta(X)\xi\\ &+ g(\phi Y,Z)\phi X - g(\phi X,Z)\phi Y - 2g(\phi X,Y)\phi Z\},\end{aligned}$$

for any tangent vector fields X, Y, Z on $\widetilde{M}(c)$.

As examples of Sasakian space forms we mention $\mathbf{R}^{2m+1}$ and S^{2m+1}, with standard Sasakian structures (see [13, 14, 124]).

Let M be an n-dimensional submanifold in a Sasakian manifold $\widetilde{M}$.

By analogy with the submanifolds of a Kaehler manifold, we distinguish special classes of submanifolds of Sasakian manifolds.

A submanifold M normal to ξ in a Sasakian manifold $\widetilde{M}$ is said to be a *C-totally real* submanifold. In this case, it follows that ϕ maps any tangent space of M into the normal space, that is, $\phi(T_pM) \subset T_p^{\perp}M$, for every $p \in M$.

In particular, if $\dim \widetilde{M} = 2\dim M + 1$, then M is called a *Legendrian* submanifold.

For submanifolds tangent to the structure vector field ξ, there are different classes of submanifolds. We mention the following:

(i) A submanifold M tangent to ξ is called an *invariant* submanifold if ϕ preserves any tangent space of M, that is, $\phi(T_pM) \subset T_pM$, for every $p \in M$.
An invariant submanifold of a Sasakian manifold is a Sasakian manifold and a minimal submanifold.

(ii) A submanifold M tangent to ξ is called an *anti-invariant* submanifold if ϕ maps any tangent space of M into the normal space, that is, $\phi(T_pM) \subset T_p^{\perp}M$, for every $p \in M$.

(iii) A *contact slant submanifold* is a submanifold M tangent to ξ of a Sasakian manifold $(\widetilde{M}, \phi, \xi, \eta, g)$ such that, for any vector X $\in T_pM$ linearly independent with ξ_p, the angle $\theta(X)$ between ϕX and the tangent space T_pM is a constant (which is independent of the choice of the point $p \in M$ and the choice of the tangent vector X in the tangent plane T_pM).

(iv) A submanifold M tangent to ξ is called a *contact CR-submanifold* if it admits an invariant differentiable distribution $\mathcal{D}$ with respect to ϕ whose orthogonal complementary orthogonal distribution $\mathcal{D}^{\perp}$ is anti-invariant, that is, $TM = \mathcal{D} \oplus \mathcal{D}^{\perp}$, with $\phi(\mathcal{D}_p) \subset \mathcal{D}_p$ and $\phi(\mathcal{D}_p^{\perp}) \subset T_p^{\perp}M$, for every $p \in M$.

2 Chen Invariants and Chen-Type Inequalities

The Riemannian invariants of a Riemannian manifold are the intrinsic characteristics of the Riemannian manifold. Among the Riemannian invariants, the most studied were sectional, scalar, and Ricci curvatures. We recall a string of Riemannian invariants on a Riemannian manifold, which are known as *Chen invariants* [30].

The *Chen first invariant* of a Riemannian manifold M is given by $\delta_M(p) = \tau(p) - (\inf K)(p)$, $p \in M$, where K and τ are the sectional curvature and the scalar curvature of M, respectively.

For an integer $k \geq 0$, we denote by $S(n,k)$ the finite set that consists of k-tuples $(n_1, \ldots, n_k)$ of integers ≥ 2 satisfying $n_1 < n$, $n_1 + \ldots + n_k \leq n$. Denote by $S(n)$ the set of k-tuples with $k \geq 0$ for a fixed n.

For each k-tuple $(n_1, \ldots, n_k) \in S(n)$, Chen introduced a Riemannian invariant defined by

$$\delta(n_1, \ldots, n_k)(p) = \tau(p) - \inf\{\tau(L_1) + \ldots + \tau(L_k)\},$$

where $L_1, \ldots, L_k$ run over all k mutually orthogonal subspaces of T_pM such that $\dim L_j = n_j$, $j = 1, \ldots, k$.

In the introduction of the article [29], B.Y. Chen recalled one of the basic problems in submanifold theory:

Find simple relationship between the main extrinsic invariants and the main intrinsic invariants of a submanifold.

We recall the most important Chen inequalities obtained by B.Y. Chen for submanifolds in real space forms.

Theorem 2.1 ([24]) *Let M^n be an n-dimensional ($n \geq 3$) submanifold of a real space form $\widetilde{M}^m(c)$ of constant sectional curvature c. Then*

$$\delta_M \leq \frac{n-2}{2}\left\{\frac{n^2}{n-1}\|H\|^2 + (n+1)c\right\}. \tag{2.1}$$

Equality holds if and only if, with respect to suitable frame fields $\{e_1, \ldots, e_n, e_{n+1}, \ldots, e_m\}$, the shape operators take the following forms:

$$A_{n+1} = \begin{pmatrix} a & 0 & 0 \ldots 0 \\ 0 & \mu - a & 0 \ldots 0 \\ 0 & 0 & \mu \ldots 0 \\ \vdots & \vdots & \vdots \ddots \vdots \\ 0 & 0 & 0 \ldots \mu \end{pmatrix},$$

$$A_r = \begin{pmatrix} h_{11}^r & h_{12}^r & 0 \dots 0 \\ h_{12}^r & -h_{11}^r & 0 \dots 0 \\ 0 & 0 & 0 \dots 0 \\ \vdots & \vdots & \vdots \ddots \vdots \\ 0 & 0 & 0 \dots 0 \end{pmatrix}, \quad r = n+2, \dots, m.$$

Furthermore, when the equality sign of (2.1) *holds at a point* $p \in M^n$, *we also have* $K(e_1 \wedge e_2) = \inf K$ *at point* p.

For each $(n_1, \dots, n_k) \in S(n)$, one defines

$$d(n_1, \dots, n_k) = \frac{n^2 \left(n + k - 1 - \sum_{j=1}^{k} n_j \right)}{2 \left(n + k - \sum_{j=1}^{k} n_j \right)},$$

$$b(n_1, \dots, n_k) = \frac{1}{2} \left[n(n-1) - \sum_{j=1}^{k} n_j(n_j - 1) \right].$$

For proving the above inequality, B.Y. Chen uses the following algebraic lemma (which we call from now as Chen's lemma).

Lemma ([24]) *Let* $n \geq 3$ *be an integer and* $a_1, a_2, \dots, a_n, b$ *real numbers such that*

$$\left(\sum_{i=1}^{n} a_i \right)^2 = (n-1) \left(\sum_{i=1}^{n} a_i^2 + b \right).$$

Then $2a_1a_2 \geq b$.
The equality holds if and only if $a_1 + a_2 = a_3 = \dots = a_n$.

The following sharp inequality involving the Chen invariants and the squared mean curvature obtained in [30] plays the most fundamental role in this topic.

Theorem 2.2 ([30]) *For each* $(n_1, \dots, n_k) \in S(n)$ *and each* n*-dimensional submanifold* M *in a Riemannian space form* $\widetilde{M}^m(c)$ *of constant sectional curvature* c*, we have*

$$\delta(n_1, \dots, n_k) \leq d(n_1, \dots, n_k) \, \|H\|^2 + b(n_1, \dots, n_k)c. \tag{2.2}$$

The equality case of inequality (2.2) *holds at a point* $p \in M$ *if and only if there exists an orthonormal basis* $\{e_1, \dots, e_m\}$ *at* p *such that the shape operators of* M *in*

$\widetilde{M}^m(c)$ at p take the following forms:

$$A_{n+1} = \begin{pmatrix} a_1 & 0 & 0 & \dots & 0 \\ 0 & a_2 & 0 & \dots & 0 \\ 0 & 0 & a_3 & \dots & 0 \\ \vdots & \vdots & \vdots & \ddots & \vdots \\ 0 & 0 & 0 & \dots & a_n \end{pmatrix},$$

$$A_r = \begin{pmatrix} A_1^r & \dots & 0 & 0 & \dots & 0 \\ \vdots & \ddots & \vdots & \vdots & \ddots & \vdots \\ 0 & \dots & A_k^r & 0 & \dots & 0 \\ 0 & \dots & 0 & \mu_r & \dots & 0 \\ \vdots & \ddots & \vdots & \vdots & \ddots & \vdots \\ 0 & \dots & 0 & 0 & \dots & \mu_r \end{pmatrix}, \quad r = n+2, \dots, m,$$

where $a_1, \dots, a_n$ satisfy

$$a_1 + \dots + a_{n_1} = \dots = a_{n_1+\dots+n_{k-1}+1} + \dots + a_{n_1+\dots+n_k} = a_{n_1+\dots+n_k+1} = \dots = a_n,$$

and each A_j^r is a symmetric $n_j \times n_j$ submatrix satisfying

$$trace(A_1^r) = \dots = trace(A_k^r) = \mu_r.$$

2.1 Purely Real Submanifolds

A proper slant submanifold M of a Kaehler manifold is said to be *Kaehlerian slant* if the canonical endomorphism P, i.e., the restriction of J to TM, is parallel; more precisely, $\nabla P = 0$, where ∇ is the Levi-Civita connection on M.

A Kaehlerian slant submanifold is a Kaehler manifold with respect to the induced metric and the almost complex structure $\widetilde{J} = (\sec\theta)J$, where θ is the slant angle.

On a Kaehlerian slant submanifold the coefficients of the second fundamental form have the symmetry property: $h_{ij}^k = h_{jk}^i = h_{ki}^j$. Examples of proper slant submanifolds and Kaehlerian slant submanifolds are given in [23].

We recall now few properties of P. Denoting $Q = P^2$, then Q is a self-adjoint endomorphism of TM. Each tangent space T_pM admits an orthogonal decomposition of eigenspaces of Q:

$$T_pM = D_p^1 \oplus \dots \oplus D_p^{k(p)}.$$

Moreover, each eigenvalue λ_i of Q lies in $[-1, 0]$.

If $\lambda_i \neq 0$, then the corresponding eigenspace D^i_p is of even dimension and invariant under P, $P(D^i_p) = D^i_p$; if $\lambda_i \neq -1$, then $\dim F(D^i_p) = \dim D^i_p$, and the normal subspaces $F(D^i_p)$ are mutually perpendicular, where FX is the normal component of JX.

Definition ([23]) A submanifold M is called a *purely real submanifold* if every eigenvalue of $Q = P^2$ lies in $(-1, 0]$, i.e., $FX \neq 0$, for any nonzero vector X tangent to M.

Thus, by definition, the class of purely real submanifolds contains both slant submanifolds and totally real submanifolds (in particular Lagrangian submanifolds, i.e., totally real submanifolds of maximum dimension).

To generalize Kaehlerian slant submanifolds we will consider purely real submanifolds with $\nabla P = 0$.

Submanifolds with $\nabla P = 0$ are characterized by the following proposition:

Proposition ([23]) *Let M be a submanifold of an almost Hermitian manifold $\widetilde{M}$. Then $\nabla P = 0$ if and only if M is locally the Riemannian product $M_1 \times \ldots \times M_k$, where each M_i is either a complex submanifold, a totally real submanifold, or a Kaehlerian slant submanifold of $\widetilde{M}$.*

Also, the following result holds.

Proposition ([23]) *Let M be an irreducible submanifold of an almost Hermitian manifold $\widetilde{M}$. If M is neither invariant nor totally real, then M is a Kaehlerian slant submanifold if and only if the endomorphism P is parallel, i.e., $\nabla P = 0$.*

Lemma ([23]) *For submanifolds M of a Kaehler manifold $\widetilde{M}$, the condition $\nabla P = 0$ is equivalent to $A_{FX}Y = A_{FY}X$, for any vectors X and Y tangent to M, where A denotes the shape operator.*

The following proposition gives a characterization of submanifolds with $\nabla Q = 0$.

Proposition ([23]) *Let M be a submanifold of an almost Hermitian manifold $\widetilde{M}$. Then the self-adjoint endomorphism $Q = P^2$ is parallel (i.e., $\nabla Q = 0$) if and only if:*

(i) Each eigenvalue λ_i of Q is constant on M.
(ii) Each distribution D^i associated with the eigenvalue λ_i is completely integrable.
(iii) M is locally the Riemannian product $M_1 \times \ldots \times M_k$ of the leaves of the distributions.

B.Y. Chen [32] proved the following sharp estimate of the squared mean curvature in terms of the scalar curvature for Kaehlerian slant submanifolds in complex space forms.

Theorem 2.3 ([32]) *Let M be an n-dimensional ($n \geq 2$) Kaehlerian slant submanifold of an n-dimensional complex space form $\widetilde{M}(4c)$ of constant holomorphic sectional curvature $4c$. Then*

$$\|H\|^2 \geq \frac{2(n+2)}{n^2(n-1)}\tau - \frac{n+2}{n}\left(1 + 3\frac{\cos^2\theta}{n-1}\right)c, \tag{2.3}$$

where θ is the slant angle of M.

In particular, for Lagrangian submanifolds, one derives the following:

Corollary ([32]) *Let M be a Lagrangian submanifold of an n-dimensional ($n > 1$) complex space form $\widetilde{M}(4c)$ of constant holomorphic sectional curvature $4c$. Then*

$$\|H\|^2 \geq \frac{2(n+2)}{n^2(n-1)}\tau - \frac{n+2}{n}c. \tag{2.4}$$

The inequality (2.4) was first obtained in [25].

On the other hand, it is known that any proper slant surface is Kaehlerian slant. Thus, the previous theorem implies the following.

Corollary ([32]) *Let M be a proper slant surface in a complex space form $\widetilde{M}(4c)$ of complex dimension 2. Then the squared mean curvature $\|H\|^2$ and the Gaussian curvature G of M satisfy*

$$\|H\|^2 \geq 2\big[G - (1 + 3\cos^2\theta)c\big], \tag{2.5}$$

at each point $p \in M$, where θ denotes the slant angle of the slant surface.

The above inequality was obtained by B.Y. Chen in [27] and as a corollary of a result from [39].

Theorem 2.4 ([27]) *Let M be a purely real surface in a complex space form $\widetilde{M}(4c)$ of complex dimension 2. Then*

$$\|H\|^2 \geq 2\big[G - \|\nabla\alpha\|^2 - (1 + 3\cos^2\theta)c\big] + 4g(\nabla\alpha, Jh(e_1, e_2))\csc\alpha,$$

with respect to any orthonormal frame $\{e_1, e_2\}$ satisfying $g(\nabla\alpha, e_2) = 0$ (α is the Wirtinger angle, i.e., $\cos\alpha = g(Je_1, e_2)$, and $\nabla\alpha$ is the gradient of α).

We first present the results obtained in [80]. More precisely, we generalized Theorem 2.3 for purely real submanifolds with P parallel with respect to the Levi-Civita connection.

Theorem 2.5 ([80]) *Let M be a purely real n-dimensional ($n \geq 2$) submanifold with $\nabla P = 0$ of an n-dimensional complex space form $\widetilde{M}(4c)$ of constant holomorphic sectional curvature $4c$. Then*

$$\|H\|^2 \geq \frac{2(n+2)}{n^2(n-1)}\tau - \frac{n+2}{n}\left[1 + 3\frac{\|P\|^2}{n(n-1)}\right]c. \tag{2.6}$$

Proof Let $p \in M$ and $\{e_1, e_2, \ldots, e_n\}$ be an orthonormal basis of the tangent space T_pM such that all e_j's are eigenvectors of P^2. An orthonormal basis $\{e_1^*, e_2^*, \ldots, e_n^*\}$ of the normal space $T_p^{\perp}M$ is defined by $e_i^* = \frac{Fe_i}{\|Fe_i\|}$, $i = \overline{1, n}$.

For a purely real submanifold with $\nabla P = 0$, one has

$$A_{FX}Y = A_{FY}X, \quad \forall X, Y \in \Gamma TM,$$

or equivalently,

$$h_{ij}^k = h_{ik}^j = h_{kj}^i,$$

where A means the shape operator and $h_{ij}^k = g(h(e_i, e_j), e_k^*)$, $i, j, k = 1, \ldots, n$.

From the Gauss equation, it follows that

$$2\tau = n^2\|H\|^2 - \|h\|^2 + c\big[n(n-1) + 3\|P\|^2\big].$$

By the definition, the squared mean curvature is given by

$$n^2\|H\|^2 = \sum_i \left[\sum_j (h_{jj}^i)^2 + 2\sum_{j<k} h_{jj}^i h_{kk}^i\right].$$

We derive

$$\tau = \frac{n(n-1) + 3\|P\|^2}{2}c + \sum_i \sum_{j<k} h_{jj}^i h_{kk}^i - \sum_{i\neq j} (h_{jj}^i)^2 - 3\sum_{i<j<k} (h_{ij}^k)^2.$$

If we denote $m = \frac{n+2}{n-1}$, we get

$$n^2\|H\|^2 - m\big[2\tau - n(n-1)c - 3\|P\|^2 c\big]$$

$$= \sum_i (h_{ii}^i)^2 + (1+2m)\sum_{i\neq j} (h_{jj}^i)^2 + 6m\sum_{i<j<k} (h_{ij}^k)^2 - 2(m-1)\sum_i \sum_{j<k} h_{jj}^i h_{kk}^i$$

$$= \sum_i (h_{ii}^i)^2 + 6m\sum_{i<j<k} (h_{ij}^k)^2 + (m-1)\sum_i \sum_{j<k} (h_{jj}^i - h_{kk}^i)^2$$

$$+ [1 + 2m - (n-2)(m-1)]\sum_{i\neq j} (h_{jj}^i)^2 - 2(m-1)\sum_{i\neq j} h_{ii}^i h_{jj}^i$$

$$= 6m \sum_{i<j<k} \left(h_{ij}^k\right)^2 + (m-1) \sum_{i \neq j,k} \sum_{j<k} \left(h_{jj}^i - h_{kk}^i\right)^2$$

$$+ \frac{1}{n-1} \sum_{i \neq j} \left[h_{ii}^i - (n-1)(m-1)h_{jj}^i\right]^2 \geq 0.$$

It follows that

$$n^2 \|H\|^2 - m\left[2\tau - n(n-1)c - 3\|P\|^2 c\right] \geq 0.$$

Using the definition of the real number m, the previous relation becomes

$$n^2 \|H\|^2 - \frac{n+2}{n-1}\left[2\tau - n(n-1)c - 3\|P\|^2 c\right] \geq 0,$$

which is equivalent to the inequality to prove.

In [104] we proved Chen inequalities for slant submanifolds M in complex space forms $\tilde{M}(c)$ of constant holomorphic sectional curvature c.

We considered 2-plane sections π invariant by P and defined

$$\delta'_M(p) = \tau(p) - \inf\{K(\pi) | \pi \subset T_pM, \ \dim \pi = 2, \text{invariant by } P\},$$

where, as usual, we denoted by $K(\pi)$ the sectional curvature associated with the 2-plane section π and by $\tau(p)$ the scalar curvature at $p \in M$, $\tau(p) = \sum_{1 \leq i<j \leq n} K(e_i \wedge e_j)$, with $\{e_1, e_2, \ldots, e_n\}$ an orthonormal basis of T_pM.

The Chen first inequality has the following form.

Theorem 2.6 ([104]) *Given an m-dimensional complex space form $\tilde{M}(4c)$ and a θ-slant submanifold M,* $\dim M = n$, $n \geq 3$, *we have*

$$\delta'_M(p) \leq \frac{n-2}{2} \left\{\frac{n^2}{n-1} \|H\|^2 + (n+1+3\cos^2\theta)c\right\}. \tag{2.7}$$

The equality case of the inequality holds at a point $p \in M$ if and only if there exist an orthonormal basis $\{e_1, e_2, \ldots, e_n\}$ of T_pM and an orthonormal basis $\{e_{n+1}, \ldots, e_{2m}\}$ of $T_p^{\perp}M$ such that the shape operators of M in $\tilde{M}(4c)$ at p have the following forms:

$$A_{n+1} = \begin{pmatrix} a & 0 & 0 & \ldots & 0 \\ 0 & b & 0 & \ldots & 0 \\ 0 & 0 & \mu & \ldots & 0 \\ \vdots & \vdots & \vdots & \ddots & \vdots \\ 0 & 0 & 0 & \ldots & \mu \end{pmatrix}, \quad a+b=\mu,$$

$$A_r = \begin{pmatrix} h_{11}^r & h_{12}^r & 0 \dots 0 \\ h_{12}^r & -h_{11}^r & 0 \dots 0 \\ 0 & 0 & 0 \dots 0 \\ \vdots & \vdots & \vdots \ddots \vdots \\ 0 & 0 & 0 \dots 0 \end{pmatrix},$$

where one denotes $A_r = A_{e_r}$, $r = n+1, \dots, 2m$, *and* $h_{ij}^r = g(h(e_i, e_j), e_r)$, $i, j = 1, \dots, n$, $r = n+1, \dots, 2m$.

Remark The equality case implies the minimality for Kaehlerian slant submanifolds with $n = m$ [95].

In [80] we extended the above inequality to purely real submanifolds M in complex space forms $\tilde{M}(4c)$.

For a 2-plane section $\pi \subset T_pM$, $p \in M$, we denoted

$$\Phi^2(\pi) = g^2(Je_1, e_2),$$

where $\{e_1, e_2\}$ is an orthonormal basis of π (see [20]). Then $\Phi^2(\pi)$ is a real number in [0, 1], which is independent of the choice of the orthonormal basis $\{e_1, e_2\}$ of π.

We proved the following optimal inequality.

Theorem 2.7 ([80]) *Let M be an n-dimensional ($n \geq 3$) purely real submanifold of an m-dimensional complex space form $\tilde{M}(4c)$, $p \in M$, and $\pi \subset T_pM$ a 2-plane section. Then*

$$\tau(p) - K(\pi) \leq \frac{n^2(n-2)}{2(n-1)} \|H\|^2 + \left[(n+1)(n-2) + 3\|P\|^2 - 6\Phi^2(\pi)\right]\frac{c}{2}. \tag{2.8}$$

Moreover, the equality case of the inequality holds at a point $p \in M$ if and only if there exist an orthonormal basis $\{e_1, e_2, \dots, e_n\}$ of T_pM and an orthonormal basis $\{e_{n+1}, \dots, e_{2m}\}$ of $T_p^\perp M$ such that the shape operators take the following forms:

$$A_{n+1} = \begin{pmatrix} a & 0 & 0 & \dots & 0 \\ 0 & b & 0 & \dots & 0 \\ 0 & 0 & \mu & \dots & 0 \\ \vdots & \vdots & \vdots & \ddots & \vdots \\ 0 & 0 & 0 & \dots & \mu \end{pmatrix}, \quad a + b = \mu,$$

$$A_r = \begin{pmatrix} h_{11}^r & h_{12}^r & 0 \dots 0 \\ h_{12}^r & -h_{11}^r & 0 \dots 0 \\ 0 & 0 & 0 \dots 0 \\ \vdots & \vdots & \vdots \ddots \vdots \\ 0 & 0 & 0 \dots 0 \end{pmatrix}, \quad r \in \{n+1, \dots, 2m\}.$$

Proof Let $p \in M$ and $\pi \subset T_pM$ be a 2-plane section and $\{e_1, e_2\}$ an orthonormal basis of π. We construct $\{e_1, e_2, e_3, \ldots, e_n\}$ an orthonormal basis of T_pM.

The Gauss equation implies

$$2\tau = n^2 \|H\|^2 - \|h\|^2 + \left[n(n-1) + 3\|P\|^2\right]c.$$

We put

$$\varepsilon = 2\tau - \frac{n^2(n-2)}{n-1}\|H\|^2 - \left[n(n-1) + 3\|P\|^2\right]c.$$

From the above two equations, we get

$$n^2 \|H\|^2 = (n-1)\left(\varepsilon + \|h\|^2\right). \tag{2.9}$$

We take e_{n+1} parallel with H and construct $\{e_{n+1}, \ldots, e_{2m}\}$ an orthonormal basis of $T_p^\perp M$. Equation (2.9) becomes

$$\left(\sum_{i=1}^{n} h_{ii}^{n+1}\right)^2 = (n-1)\left(\sum_{r=n+1}^{2m}\sum_{i,j=1}^{n}(h_{ij}^r)^2 + \varepsilon\right),$$

or equivalently,

$$\left(\sum_{i=1}^{n} h_{ii}^{n+1}\right)^2 = (n-1)\left[\sum_{i=1}^{n}(h_{ii}^{n+1})^2 + \sum_{i\neq j}(h_{ij}^{n+1})^2 + \sum_{r=n+2}^{2m}\sum_{i,j=1}^{n}(h_{ij}^r)^2 + \varepsilon\right].$$

By applying Chen's lemma, we obtain

$$2h_{11}^{n+1}h_{22}^{n+1} \geq \sum_{i\neq j}\left(h_{ij}^{n+1}\right)^2 + \sum_{r=n+2}^{2m}\sum_{i,j=1}^{n}\left(h_{ij}^r\right)^2 + \varepsilon.$$

The Gauss equation gives

$$K(\pi) = \left[1 + 3\Phi^2(\pi)\right]c + \sum_{r=n+1}^{2m}\left[h_{11}^r h_{22}^r - \left(h_{12}^r\right)^2\right]$$

$$\geq [1 + 3\Phi^2(\pi)]c + \frac{1}{2}\sum_{i\neq j}\left(h_{ij}^{n+1}\right)^2 + \frac{1}{2}\sum_{r=n+2}^{2m}\sum_{i,j=1}^{n}\left(h_{ij}^r\right)^2 + \frac{\varepsilon}{2}$$

$$+\sum_{r=n+2}^{2m} h_{11}^r h_{22}^r - \sum_{r=n+1}^{2m} (h_{12}^r)^2$$

$$= \left[1 + 3\Phi^2(\pi)\right]c + \frac{1}{2}\sum_{i\neq j>2} \left(h_{ij}^{n+1}\right)^2 + \frac{1}{2}\sum_{r=n+2}^{2m}\sum_{3\leq i<j\leq n} (h_{ij}^r)^2$$

$$+\frac{1}{2}\sum_{r=n+2}^{2m} (h_{11}^r + h_{22}^r)^2 + \sum_{r=n+1}^{2m}\sum_{j=3}^{n}\left[(h_{1j}^r)^2 + (h_{2j}^r)^2\right] + \frac{\varepsilon}{2}$$

$$\geq [1 + 3\Phi^2(\pi)]c + \frac{\varepsilon}{2},$$

which implies the inequality to prove.

We have an equality at a point $p \in M$ if and only if all the above inequalities become equalities and the equality case of Chen's lemma holds. Thus the shape operators take the desired forms.

For n-dimensional Kaehlerian slant submanifolds a particular case of purely real submanifolds in n-dimensional complex space form $\widetilde{M}(4c)$, we proved an improved Chen first inequality.

Theorem 2.8 ([80]) *Let M be an n-dimensional ($n \geq 3$) Kaehlerian slant submanifold in the complex space form $\tilde{M}(4c)$,* $\dim_{\mathbb{C}} \tilde{M}(4c) = n$*, and $p \in M$, $\pi \subset T_pM$ a 2-plane section. Then*

$$\tau(p) - K(\pi) \leq \frac{n^2(2n-3)}{2(2n+3)}\|H\|^2 + \left[(n+1)(n-2) + 3n\cos^2\theta - 6\Phi^2(\pi)\right]\frac{c}{2}. \tag{2.10}$$

Moreover, the equality case of the inequality (2.10) *holds at a point $p \in M$ if and only if there exists an orthonormal basis $\{e_1, e_2, \ldots, e_n\}$ at p such that with respect to this basis the second fundamental form takes the following form:*

$$h(e_1, e_1) = ae_1^* + 3be_3^* \qquad h(e_1, e_3) = 3be_1^* \qquad h(e_3, e_j) = 4be_j^*$$

$$h(e_2, e_2) = -ae_1^* + 3be_3^* \quad h(e_2, e_3) = 3be_2^* \qquad h(e_j, e_k) = 4be_3^*\delta_{jk}$$

$$h(e_1, e_2) = -ae_2^* \qquad h(e_3, e_3) = 12be_3^* \qquad h(e_1, e_j) = h(e_2, e_j) = 0,$$

for some numbers a, b and $j, k = 4, \ldots, n$, where $e_i^* = \frac{Fe_i}{\|Fe_i\|}$, $i = 1, \ldots, n$.

Proof Let $p \in M$ and $\pi \subset T_pM$ be a 2-plane section and $\{e_1, e_2, \ldots, e_n\}$ an orthonormal basis of the tangent space T_pM such that $e_1, e_2 \in \pi$. An orthonormal basis $\{e_1^*, e_2^*, \dots, e_n^*\}$ of the normal space $T_p^{\perp}M$ is defined by $e_i^* = \frac{Fe_i}{\sin\theta}, i = \overline{1,n}$. We denote $h_{ij}^k = g(h(e_i, e_j), e_k^*)$.

The Gauss equation implies

$$\tau(p) = \sum_{r=1}^{n} \sum_{1\le i<j\le n} \left[h_{ii}^r h_{jj}^r - \left(h_{ij}^r\right)^2\right] + \left[n(n-1) + 3n\cos^2\theta\right]\frac{c}{2}, \tag{2.11}$$

and

$$K(\pi) = \sum_{r=1}^{n} \left[h_{11}^r h_{22}^r - \left(h_{12}^r\right)^2\right] + \left[1 + 3\Phi^2(\pi)\right]c, \tag{2.12}$$

respectively. Since M is a Kaehlerian slant submanifold, we have $h_{ij}^k = h_{jk}^i = h_{ki}^j$.

From formulas (2.11) and (2.12), we obtain

$$\begin{aligned} \tau(p) - K(\pi) = \sum_{r=1}^{n} \Bigg\{ & \sum_{j=3}^{n} \left(h_{11}^r + h_{22}^r\right) h_{jj}^r + \sum_{3\le i<j\le n} h_{ii}^r h_{jj}^r \\ & - \sum_{j=3}^{n} \left[\left(h_{1j}^r\right)^2 + \left(h_{2j}^r\right)^2\right] \Bigg\} \\ & + \left[(n+1)(n-2) + 3n\cos^2\theta - 6\Phi^2(\pi)\right]\frac{c}{2}. \end{aligned} \tag{2.13}$$

It follows that

$$\tau(p) - K(\pi) \le$$

$$\begin{aligned} \le & \sum_{r=1}^{n} \left[\sum_{j=3}^{n} \left(h_{11}^r + h_{22}^r\right) h_{jj}^r + \sum_{3\le i<j\le n} h_{ii}^r h_{jj}^r \right] \\ & - \sum_{j=3}^{n} \left(h_{11}^j\right)^2 - \sum_{j=3}^{n} \left(h_{jj}^1\right)^2 - \sum_{2\le i\ne j\le n} \left(h_{jj}^i\right)^2 \\ & + \left[(n+1)(n-2) + 3n\cos^2\theta - 6\Phi^2(\pi)\right]\frac{c}{2}. \end{aligned}$$

In order to achieve the proof, we will use some ideas and results from [17]. We point out the following inequalities (see [17]):

$$\sum_{j=3}^{n} \left(h_{11}^r + h_{22}^r\right)^2 h_{jj}^r + \sum_{3\le i<j\le n} h_{ii}^r h_{jj}^r - \sum_{j=3}^{n} \left(h_{jj}^r\right)^2 \tag{2.14}$$

$$\leq \frac{n-2}{2(n+1)}\left(h_{11}^r + \ldots + h_{nn}^r\right)^2 \leq \frac{2n-3}{2(2n+3)}\left(h_{11}^r + \ldots + h_{nn}^r\right)^2,$$

for $r = 1, 2$. The first inequality is equivalent to

$$\sum_{j=3}^{n}\left(h_{11}^r + h_{22}^r - 3h_{jj}^r\right)^2 + 3\sum_{3\leq i<j\leq n}\left(h_{ii}^r - h_{jj}^r\right)^2 \geq 0.$$

The equality holds if and only if $3h_{jj}^r = h_{11}^r + h_{22}^r$, $\forall j = 3, \ldots, n$.
The equality holds in the second inequality if and only if $h_{11}^r + h_{22}^r + \ldots + h_{nn}^r = 0$.
Also, we have

$$\sum_{j=3}^{n}\left(h_{11}^r + h_{22}^r\right)^2 h_{jj}^r + \sum_{3\leq i<j\leq n} h_{ii}^r h_{jj}^r - \sum_{j=3}^{n}\left(h_{jj}^r\right)^2 \leq \frac{2n-3}{2(2n+3)}\left(h_{11}^r + \ldots + h_{nn}^r\right)^2,$$

for $r = 3, \ldots, n$, which is equivalent to (see [17])

$$\sum_{3\leq j\leq n, j\neq r}\left[2(h_{11}^r + h_{22}^r) - 3h_{jj}^r\right]^2 + (2n+3)(h_{11}^r - h_{22}^r)^2 \tag{2.15}$$

$$+ 6\sum_{3\leq j\leq n, j\neq r}(h_{ii}^r - h_{jj}^r)^2 + 2\sum_{j=3}^{n}(h_{rr}^r - h_{jj}^r)^2 + 3\left[h_{rr}^r - 2(h_{11}^r + h_{22}^r)\right]^2 \geq 0.$$

The equality holds if and only if

$$\begin{cases} h_{11}^r = h_{22}^r = 3\lambda^r, \\ h_{jj}^r = 4\lambda^r, \ \forall j = 3, \ldots, n, \ j \neq r, \\ h_{rr}^r = 12\lambda^r, \end{cases} \qquad \lambda^r \in \mathbf{R}.$$

By summing the inequalities (2.14) and (2.15) we obtained the inequality (2.10).
Combining the above equality cases, we get the desired forms of the second fundamental form.

In particular, we derive the following.

Theorem 2.9 ([80]) *Let M be an n-dimensional ($n \geq 3$) Kaehlerian slant submanifold in the complex space form $\tilde{M}(4c)$, $\dim_{\mathbf{C}} \tilde{M}(4c) = n$, $p \in M$, $\pi \subset T_pM$ a 2-plane section. Then*

$$\delta_M'(p) \leq \frac{n^2(2n-3)}{2(2n+3)}\|H\|^2 + (n-2)\left[n + 1 + 3\cos^2\theta\right]\frac{c}{2}. \tag{2.16}$$

The equality case of the inequality holds at a point $p \in M$ if and only if, with respect to a suitable orthonormal basis $\{e_1, e_2, \ldots, e_n\}$ of T_pM the second fundamental form h takes the same form as in Theorem 2.8.

Proof If M is Kaehlerian slant and π is invariant by P, one has $\Phi^2(\pi) = \cos^2\theta$.

In contrast with the last remark, the equality case does not imply the minimality of the submanifold. However we stated the following result.

Theorem 2.10 ([80]) *Let M be an n-dimensional Kaehlerian slant submanifold in the complex space form $\tilde{M}(4c)$, $\dim_{\mathbf{C}} \tilde{M}(4c) = n$, and $n \geq 4$. If the equality case holds identically in* (2.16)*, then M is a minimal submanifold.*

The *proof* is similar to that of Theorem 3 from [17].

In the case $n = 3$ there is an example of non-minimal Lagrangian submanifold in $\mathbf{C}P^3$ satisfying the equality case of (2.10) [18].

In [47] we presented ***a general construction method to obtain the explicit expression of a purely real submanifold in the complex hyperbolic plane*** $CH^n(-4)$ via Hopf's fibration. For the details, see [45] (the same method applies to $\mathbf{C}P^n(4)$ with minor modification).

Let $\mathbf{C}_1^{n+1}$ denote the complex $(n+1)$-space together with the Hermitian inner product:

$$F(z, w) = -z_0\bar{w}_0 + \sum_{j=1}^{n} z_j\bar{w}_j,$$

for $z = (z_0, \ldots, z_n)$ and $w = (w_0, \ldots, w_n)$ in $\mathbf{C}_1^{n+1}$. We denote

$$H_1^{2n+1} = \{z \in \mathbf{C}_1^{n+1} : F(z, z) = -1\}. \tag{2.17}$$

Then H_1^{2n+1} is a real hypersurface of $\mathbf{C}_1^{n+1}$ whose tangent space at $z \in H_1^{2n+1}$ is

$$T_zH^{2n+1} = \{w \in \mathbf{C}_1^{n+1} : Re\, F(z, w) = 0\}.$$

The restriction of the real part of F, Re F, on H_1^{2n+1} gives rise to a pseudo-Riemannian metric g on H_1^{2n+1}. It is well known that H_1^{2n+1} together with g is a Lorentzian manifold of constant sectional curvature -1, which is known as the *anti-de Sitter space.*

We put

$$T_z' = \{u \in \mathbf{C}_1^{n+1} : \mathrm{Re}\, F(u, v) = \mathrm{Re}\, F(u, iz) = 0\}$$

and $H_1^1 = \{\eta \in \mathbf{C} : \eta\bar{\eta} = 1\}$. Then we have an H_1^1-action on H_1^{2n+1}, $z \mapsto \eta z$. At $z \in H_1^{2n+1}(-1)$, the vector iz is tangent to the flow of the action. Since F is Hermitian, we have Re $F(iz, iz) = -1$. Notice that the orbit is given by $z_t = e^{it}z$

with $dz_t/dt = iz_t$ which lies in the negative-definite plane spanned by z and iz. The quotient $H_1^{2n+1}/\sim$ under the action is the complex hyperbolic n-space $CH^n(-4)$ of constant holomorphic sectional curvature -4.

The complex structure J on $CH^n(-4)$ is induced from the complex structure J on $\mathbf{C}_1^{n+1}$ via the *Hopf fibration*:

$$\pi : H_1^{2n+1} \to CH^n(-4), \tag{2.18}$$

which is a Riemannian submersion with totally geodesic fibers. If $z \in H_1^{2n+1}$, we put

$$V = Jz. \tag{2.19}$$

Then V is a time-like unit vector tangent to the fiber of the submersion at z.

Denote by $\hat{\nabla}$ and $\tilde{\nabla}$ the Levi-Civita connections of H_1^{2n+1} and $\mathbf{C}H^n(-4)$, respectively. Let X^* denote the horizontal lift of a vector X on M. For vector fields X, Y tangent to $CH^n(-4)$ and V normal to $CH^n(-4)$, we have

$$\hat{\nabla}_{X^*}Y^* = (\tilde{\nabla}_X Y)^* + \langle JX, Y\rangle V, \tag{2.20}$$

$$\hat{\nabla}_{X^*}V = \hat{\nabla}_V X^* = (JX)^*. \tag{2.21}$$

Let $\phi : M \to CH^n(-4)$ be an isometric immersion from a Riemannian m-manifold M into $CH^n(-4)$. Then the pre-image $\hat{M} = \pi^{-1}(M)$ is a principal circle bundle over M with totally geodesic fibers. The lift $\hat{\phi} : \hat{M} \to H_1^{2n+1}$ of ϕ is an isometric immersion such that the diagram

$$\begin{array}{ccc} \hat{M} & \xrightarrow{\hat{\phi}} & H_1^{2n+1}(-1) \\ \pi \downarrow & & \downarrow \pi \\ M & \xrightarrow{\phi} & CH^n(-4) \end{array} \tag{2.22}$$

commutes.

Conversely, if $\psi : \hat{M} \to H_1^{2n+1}(-1)$ is an isometric immersion invariant under the action, there is a unique isometric immersion $\psi_\pi : \pi(\hat{M}) \to CH^n(-4)$, called the *projection of* ψ, such that the associated diagram commutes.

Since V generates the vertical subspaces of the Riemannian submersion (2.18), we have the following orthogonal decomposition:

$$T_z\hat{M} = (T_{\pi(z)}M)^* \oplus \mathrm{Span}\{V\}.$$

Let ∇ be the Levi-Civita connection of M. Then we have

$$\hat{\nabla}_{X^*}Y^* = (\nabla_X Y)^* + (h(X, Y))^* + \langle JX, Y\rangle V, \tag{2.23}$$

for X, Y tangent to M, where h is the second fundamental form of M in $CH^n(-4)$.

If ξ is a normal vector field of M in $CH^n(-4)$, then (2.20) yields

$$\hat{\nabla}_{X^*}\xi^* = (\tilde{\nabla}_X\xi)^* + \langle JX, \xi\rangle V. \tag{2.24}$$

Hence, by the Weingarten formula, we have

$$\hat{A}_{\xi^*}X^* = (A_\xi X)^* - \langle FX, \xi\rangle V, \tag{2.25}$$

$$\hat{D}_{X^*}\xi^* = (D_X\xi)^*, \tag{2.26}$$

where A and $\hat{A}$ are the shape operators of M in $CH^n(-4)$ and $\hat{M}$ in H_1^{2n+1}, respectively, and D and $\hat{D}$ are the corresponding normal connections.

From (2.20) we have

$$\hat{h}(X^*, Y^*) = (h(X, Y))^*, \tag{2.27}$$

where $\hat{h}$ denotes the second fundamental form of $\hat{M}$ in H_1^{2n+1}. By using (2.27) we also have

$$\nabla'_{X^*}Y^* = (\nabla_X Y)^* + \langle JX, Y\rangle V, \tag{2.28}$$

where ∇' is the Levi-Civita connection of $\hat{M}$.

Also, it follows from (2.21) that

$$\hat{h}(X^*, V) = (FX)^*, \quad \nabla'_{X^*}V = \nabla'_V X^* = (FX)^* \tag{2.29}$$

for X tangent to M.

Let $z : H_1^5 \to \mathbf{C}_1^3$ denote the standard inclusion and $\breve{\nabla}$ be the Levi-Civita connection of $\mathbf{C}_1^3$. If M is a purely real surface of $CH^2(-4)$, then it follows from (2.17), (2.19), (2.21), (2.23), (2.27), (2.28), and (2.29) that

$$\begin{cases} \breve{\nabla}_{X^*}Y^* = (\nabla_X Y)^* + (h(X, Y))^* + \langle JX, Y\rangle V + \langle X, Y\rangle z, \\ \breve{\nabla}_{X^*}V = \breve{\nabla}_V X^* = (JX)^*, \\ \breve{\nabla}_V V = -z, \end{cases} \tag{2.30}$$

for X, Y tangent to M.

On $H_1^{2n+1}(-1) \subset \mathbf{C}_1^{n+1}$ we consider the induced Sasakian structure (g, ϕ, ξ), where the (1,1)-tensor ϕ is obtained from the projection of the canonical complex structure J of $\mathbf{C}_1^{n+1}$ onto the tangent bundle of H_1^{2n+1} and $\xi = V$.

Now, in [47] we defined the notion of *contact purely real submanifolds* as follows.

Let $(\tilde{M}^{2m+1}, g, \phi, \xi)$ be an almost contact metric $(2m + 1)$-manifold endowed with a Riemannian (or pseudo-Riemannian) metric g, an almost contact (1, 1)-

tensor ϕ, and the structure vector field ξ. Then an immersion $f : N \to \tilde{M}^{2m+1}$ of a manifold N into $\tilde{M}^{2m+1}$ is called *contact purely real* if it satisfies:

(i) The structure vector field ξ of $\tilde{M}^{2m+1}$ is tangent to $f_*(TN)$.
(ii) For any nonzero vector X tangent to $f_*(T_pN)$ and perpendicular to ξ, $\phi(X)$ is transversal to $f_*(T_pN)$.

The following lemma is easy to verify:

Lemma 2.11 *The immersion $\psi : M \to CH^n(-4)$ is purely real if and only if the lift $\hat{\psi} : \pi^{-1}(M) \to H_1^{2n+1}$ of ψ is contact purely real, where $\pi : H_a^{2n+1} \to CH^n(-4)$ is the Hopf fibration.*

In principle, the **method to obtain the representation of a purely real surface in** $CH^2(-4)$ is by solving the PDE system (2.30). This procedure goes as follows:

First, we determine both the intrinsic and extrinsic structures of the purely real surface. Next, we construct a coordinate system on the associated contact purely real surface $\hat{M} = \pi^{-1}(M)$. After that we solve the PDE system via the coordinate system on $\hat{M}$ to obtain the explicit solution of the system. Such a solution gives rise to the desired explicit expression of the contact purely real surface $\hat{M}$ of H_1^5 via π.

The same method also applies to purely surfaces in $CP^2(4)$.

The following general optimal inequality was given in [39].

Theorem 2.12 *Let M be a purely real surface in a complex space form $\tilde{M}^2(4\varepsilon)$. Then we have*

$$H^2 \geq 2\{K - ||\nabla\alpha||^2 - (1 + 3\cos^2\alpha)\varepsilon\} + 4\langle\nabla\alpha, Jh(e_1, e_2)\rangle \csc\alpha \tag{2.31}$$

with respect to any orthonormal frame $\{e_1, e_2\}$ satisfying $\langle\nabla\alpha, e_2\rangle = 0$, where $\nabla\alpha$ is the gradient of the Wirtinger angle α and H^2 and K are the squared mean curvature and the Gauss curvature of M, respectively.

The equality case of (2.31) *holds identically if and only if the shape operators take the following forms:*

$$A_{e_3} = \begin{pmatrix} 3\varphi & \delta \\ \delta & \varphi \end{pmatrix}, \quad A_{e_4} = \begin{pmatrix} \delta + e_1\alpha & \varphi \\ \varphi & 3\delta + 3e_1\alpha \end{pmatrix}, \tag{2.32}$$

with respect to some suitable adapted orthonormal frame $\{e_1, e_2, e_3, e_4\}$.

A purely real surface in a complex space form $\tilde{M}^2(4\varepsilon)$ is said to *satisfy the basic equality* if it satisfies the equality case of inequality (2.31) identically.

We classified the minimal surfaces in $\tilde{M}^2(4\varepsilon)$ satisfying the basic equality.

Theorem 2.13 ([47]) *Let M be a purely real minimal surface of a complex space form $\tilde{M}^2(4\varepsilon)$. If M satisfies the basic equality, then we have either:*

(a) $\varepsilon > 0$ *and M is a totally geodesic Lagrangian surface or*
(b) $\varepsilon \leq 0$.

Proof Assume that M is a purely real minimal surface in a complex space form $\tilde{M}^2(4\varepsilon)$. We choose an adapted orthonormal frame $\{e_1, e_2, e_3, e_4\}$ such that the gradient of α is parallel to e_1 at p. So, we have $\nabla\alpha = (e_1\alpha)e_1$. Let us put

$$h(e_1, e_1) = \beta e_3 + \gamma e_4, \ h(e_1, e_2) = \delta e_3 + \varphi e_4, \ \ h(e_2, e_2) = \lambda e_3 + \mu e_4. \quad (2.33)$$

If M satisfies the equality case of (2.31), then Theorem 2.12 implies that the second fundamental form satisfies

$$h(e_1, e_1) = 0, \ h(e_1, e_2) = -(e_1\alpha)e_3, \ \ h(e_2, e_2) = 0. \quad (2.34)$$

If M is a slant surface, then α is constant. So, it follows from (2.32) that M is a totally geodesic purely real surface. In this case, M is a totally geodesic Lagrangian surface of constant curvature ε (cf. [50, Theorem 3.1]).

Next, assume that M is a non-slant minimal surface, i.e., $\nabla\alpha \neq 0$ holds. Then M contains only isolated totally geodesic points since M is minimal. Therefore, $U = \{p \in M : \nabla\alpha(p) \neq 0\}$ is a dense open subset of M.

Because Span $\{e_1\}$ and Span $\{e_2\}$ are one-dimensional distributions, there exists a local coordinate system $\{x, y\}$ on U such that $\partial/\partial x$ and $\partial/\partial y$ are parallel to e_1 and e_2, respectively. Thus, the metric tensor g on U takes the following form:

$$g = E^2dx^2 + G^2dy^2, \quad (2.35)$$

where E and G are positive functions.

The Levi-Civita connection on M satisfies

$$\begin{cases} \nabla_{\frac{\partial}{\partial x}} \frac{\partial}{\partial x} = \frac{E_x}{E}\frac{\partial}{\partial x} - \frac{EE_y}{G^2}\frac{\partial}{\partial y}, \\ \nabla_{\frac{\partial}{\partial x}} \frac{\partial}{\partial y} = \frac{E_y}{E}\frac{\partial}{\partial x} + \frac{G_x}{G}\frac{\partial}{\partial y}, \\ \nabla_{\frac{\partial}{\partial y}} \frac{\partial}{\partial y} = -\frac{GG_x}{E^2}\frac{\partial}{\partial x} + \frac{G_y}{G}\frac{\partial}{\partial y}. \end{cases} \quad (2.36)$$

Let us put

$$e_1 = \frac{1}{E}\frac{\partial}{\partial x}, \quad e_2 = \frac{1}{G}\frac{\partial}{\partial y}. \quad (2.37)$$

From $e_2\alpha = 0$, we have $\alpha = \alpha(x)$. It follows from (2.37) that

$$\begin{cases} h\left(\frac{\partial}{\partial x}, \frac{\partial}{\partial x}\right) = h\left(\frac{\partial}{\partial y}, \frac{\partial}{\partial y}\right) = 0, \\ h\left(\frac{\partial}{\partial x}, \frac{\partial}{\partial y}\right) = -\frac{\alpha'(x)G}{E}e_3. \end{cases} \quad (2.38)$$

Recall the following lemma from [40].

Lemma *Let M be a purely real surface in a Kaehler surface. Then, with respect to an adapted orthonormal frame* $\{e_1, e_2, e_3, e_4\}$, *we have*

$$e_1\alpha = h^4_{11} - h^3_{12}, \quad e_2\alpha = h^4_{12} - h^3_{22},$$

$$\begin{cases} \Phi_1 = \omega_1 - \left(h^3_{11} + h^4_{12}\right)\cot\alpha, \\ \Phi_2 = \omega_2 - \left(h^3_{12} + h^4_{22}\right)\cot\alpha, \end{cases}$$

where $\nabla_X e_1 = \omega(X)e_2$, $\nabla^\perp_X e_3 = \Phi(X)e_4$, $\omega_j = \omega(e_j)$, *and* $\Phi_j = \Phi(e_j)$ *for* $j = 1, 2$.

It follows from (2.34), (2.36), (2.37), and the above lemma that

$$\Phi\left(\frac{\partial}{\partial x}\right) = -\frac{E_y}{G}, \quad \Phi\left(\frac{\partial}{\partial y}\right) = \frac{G_x + \alpha'(x)G\cot\alpha}{E}. \tag{2.39}$$

Thus, we find from (2.38) and (2.39) that

$$\begin{cases} \left(\bar{\nabla}_{\frac{\partial}{\partial y}} h\right)\left(\frac{\partial}{\partial x}, \frac{\partial}{\partial x}\right) = \frac{2\alpha'(x)G_x}{E} e_3, \\ \left(\bar{\nabla}_{\frac{\partial}{\partial x}} h\right)\left(\frac{\partial}{\partial x}, \frac{\partial}{\partial y}\right) = \frac{2\alpha'(x)E_xG - \alpha''(x)EG}{E^2} e_3 + \frac{\alpha'(x)E_y}{E} e_4, \\ \left(\bar{\nabla}_{\frac{\partial}{\partial x}} h\right)\left(\frac{\partial}{\partial y}, \frac{\partial}{\partial y}\right) = \frac{2\alpha'(x)E_yG}{E^2} e_3, \\ \left(\bar{\nabla}_{\frac{\partial}{\partial y}} h\right)\left(\frac{\partial}{\partial x}, \frac{\partial}{\partial y}\right) = \frac{2\alpha'(x)E_yG}{E^2} e_3 - \frac{\alpha'(x)G(G_x + \alpha'(x)G\cot\alpha)}{E^2} e_4. \end{cases} \tag{2.40}$$

It follows that

$$\begin{cases} \left(\frac{\partial}{\partial x}\right)^\perp = 3\varepsilon E^2 G \sin\alpha\cos\alpha e_3, \\ \left(\tilde{R}\left(\frac{\partial}{\partial x}, \frac{\partial}{\partial y}\right)\frac{\partial}{\partial y}\right)^\perp = -3\varepsilon E G^2 \sin\alpha\cos\alpha e_4, \\ \tilde{g}\left(\tilde{R}\left(\frac{\partial}{\partial x}, \frac{\partial}{\partial y}\right)\frac{\partial}{\partial y}, \frac{\partial}{\partial x}\right) = (1 + 3\cos^2\alpha)\varepsilon E^2 G^2. \end{cases} \tag{2.41}$$

Hence, we discover from (2.40), (2.41), and the equation of Codazzi that

$$E_y = 0, \tag{2.42}$$

$$2\alpha'(x)(EG_x - E_xG) + \alpha''(x)EG = 3\varepsilon E^4 G\sin\alpha\cos\alpha, \tag{2.43}$$

$$\alpha'(x)G_x + \alpha'^2(x)G\cot\alpha = -3\varepsilon E^3 G\sin\alpha\cos\alpha. \tag{2.44}$$

It follows from (2.42) that $E = E(x)$. Thus, we may choose x, y such that

$$g = dx^2 + G^2dy^2. \tag{2.45}$$

From this we see that the Gauss curvature is given by

$$K = -\frac{G_{xx}}{G}. \tag{2.46}$$

On the other hand, from (2.32), the last equation in (2.41), and (2.46), we have

$$G_{xx} = G(\alpha_x^2 - 3\varepsilon\cos^2\alpha - \varepsilon). \tag{2.47}$$

Since E is assumed to be one, (2.43) and (2.44) reduce to

$$2\alpha'(x)G_x + G\alpha''(x) = 3\varepsilon G\sin\alpha\cos\alpha, \tag{2.48}$$

$$\alpha'(x)G_x + \alpha'^2(x)G\cot\alpha = -3\varepsilon G\sin\alpha\cos\alpha. \tag{2.49}$$

Summing up the last two relations, we get

$$\frac{\alpha''(x)}{\alpha'(x)} = -3\frac{G_x}{G} - \alpha'(x)\cot\alpha. \tag{2.50}$$

After solving the last equation for G, we obtain

$$G = \frac{f(y)\csc^{\frac{1}{3}}\alpha}{\alpha'^{\frac{1}{3}}(x)} \tag{2.51}$$

for some nonzero function $f(y)$. Substituting this into (2.49) gives

$$\alpha'' = 2\alpha'^2\cot\alpha + 9\varepsilon\sin\alpha\cos\alpha. \tag{2.52}$$

Now, by substituting (2.51) into (2.47) we discover

$$3\alpha'\alpha''' = 4\alpha''^2 + \alpha'^2(9\varepsilon(1+3\cos^2\alpha) - \alpha''\cot\alpha) + \alpha'^4(4\cot^2\alpha - 6). \tag{2.53}$$

From the last two relations we obtain $\varepsilon(4\alpha'^2 + 9\varepsilon\sin^2 2\alpha) = 0$, which is impossible unless $\varepsilon \leq 0$. This proves the theorem.

In view of Theorem 2.13, we classified purely real minimal surfaces in $CH^2(-4)$ satisfying the basic equality.

Theorem 2.14 ([47]) *Let M be a purely real minimal surface of the complex hyperbolic plane $CH^2(-4)$. Then M satisfies the basic equality if and only if it is congruent to an open portion of one of the following two surfaces:*

1. *A real hyperbolic plane H^2 of constant curvature -1 embedded in $CH^2(-4)$ as a totally geodesic Lagrangian surface.*

2. *A surface in* $CH^2(-4)$ *given by* $\pi \circ z$, *where* $\pi : H_1^5(-1) \to CH^2(-4)$ *is the Hopf fibration and* $z : \mathbf{R}^3 \to H_1^5 \subset \mathbf{C}_1^3$ *is given by*

$$z(x,u,t) = \frac{e^{i(t+u)-x}}{3\sqrt[3]{4}}\Big(3i\sqrt[3]{2}\sinh(\sqrt{3}u) + \sqrt{3}\big(\sqrt[3]{2} + 2e^{2x}\big)\cosh(\sqrt{3}u),$$

$$3e^{2x}\cosh(\sqrt{3}u) + i\sqrt{3}\big(2\sqrt[3]{2} + e^{2x}\big)\sinh(\sqrt{3}u), \frac{\sqrt{3}}{e^{3iu}}\big(\sqrt[3]{2} - e^{2x}\big)\Big)$$

This purely real surface has the nonconstant Wirtinger angle $\alpha = \arctan(e^{3x})$.

Next, we classified purely real surfaces with circular ellipse of curvature in $CP^2(4)$ and in $CH^2(-4)$ which satisfy the basic equality.

Theorem 2.15 ([47]) *Let* M *be a purely real surface with circular ellipse of curvature in a complex space form* $\tilde{M}^2(4\varepsilon)$, $\varepsilon = \pm 1$. *If* M *satisfies the basic equality, then we have either:*

1. M *is a Lagrangian surface satisfying the equality*

$$H^2 = 2K - 2\varepsilon$$

identically or

2. $\varepsilon = -1$, *and* M *is congruent to an open portion of a proper slant surface in* $CH^2(-4)$ *given by* $\pi \circ z$, *where* $\pi : H_1^5 \to CH^2(-4)$ *is the Hopf fibration and* $z : M \to H_1^5 \subset \mathbf{C}_1^3$ *is defined by*

$$z(u,v,t) = e^{it}\left(\frac{3}{2}\cosh av + \frac{1}{6}u^2 e^{-av} - \frac{i}{6}\sqrt{6}u(1+e^{-av}) - \frac{1}{2},\right.$$

$$\frac{1}{3}(1+2e^{-av})u + i\sqrt{6}\left(-\frac{1}{3} + \frac{1}{4}e^{av} + e^{-av}\left(\frac{1}{12} + \frac{1}{18}u^2\right)\right),$$

$$\left.\frac{\sqrt{2}}{6}(1-e^{-av})u + i\sqrt{3}\left(\frac{1}{6} + \frac{1}{4}e^{av} + e^{-av}\left(-\frac{5}{12} + \frac{1}{18}u^2\right)\right)\right), \quad a = \sqrt{\frac{2}{3}}.$$

Remark Lagrangian surfaces in a complex space form $\tilde{M}^2(4\varepsilon)$ with $\epsilon = 1$ or -1 satisfying the equality have been completely classified in [25, 51]. Such surfaces have circular ellipse of curvature.

A surface M in a Kaehler surface is said to have *full second fundamental form* if its first normal space $\operatorname{Im} h$ satisfies $\dim(\operatorname{Im} h) = 2$ identically. It is said to have *degenerate second fundamental form* if $\dim(\operatorname{Im} h) < 2$ holds at each point in M.

Theorem 2.16 ([47]) *Let M be a purely real surface satisfying the basic equality in a complex space form $\tilde{M}^2(4\varepsilon)$ with $\varepsilon = \pm 1$. If M has the degenerate second fundamental form, then either:*

(i) M is a totally geodesic Lagrangian surface or
(ii) $\epsilon = -1$, and M is locally congruent to the surface given by $\pi \circ z$, where $\pi : H_1^5(-1) \to CH^2(-4)$ is the Hopf fibration and $z : \mathbf{R}^3 \to H_1^5 \subset \mathbf{C}_1^3$ is

$$z(x, u, t) = \frac{e^{i(t+u)-x}}{3\sqrt[3]{4}} \Big(3i\sqrt[3]{2}\sinh(\sqrt{3}u) + \sqrt{3}(\sqrt[3]{2} + 2e^{2x}) \cosh(\sqrt{3}u),$$

$$3e^{2x}\cosh(\sqrt{3}u) + i\sqrt{3}(2\sqrt[3]{2} + e^{2x}) \sinh(\sqrt{3}u), \frac{\sqrt{3}}{e^{3iu}}(\sqrt[3]{2} - e^{2x})\Big).$$

By continuing this idea, in [46] we considered purely real surfaces with harmonic Wirtinger function and purely real surfaces with closed canonical form. In order to do so, first we proved a general formula for the Laplacian of the Wirtinger function involving the canonical form. Then we provide a necessary and sufficient condition for non-minimal proper purely real surfaces to have closed canonical form. As applications, we obtained several classification results for purely real surfaces to have harmonic Wirtinger function or with closed canonical form.

Let M be a purely real surface in a Kaehler surface $\tilde{M}$. We recall a 1-form Ψ_H on M defined by Chen [23]

$$\Psi_H(X) = (\csc^2 \alpha)\langle H, JX\rangle, \tag{2.54}$$

for $X \in TM$, where α is the Wirtinger function of M. Since M is purely real, Ψ_H is well defined on M. We call this 1-form Ψ_H the *canonical form*.

For a Lagrangian surface in $\mathbf{C}^2$, Ψ_H is a closed form. Moreover, up constants, it represents the Maslov class of the Lagrangian surface (cf. [101]). However, when the purely real surface is non-Lagrangian, this form is non-closed in general.

We established the following general formula for purely real surfaces in a complex space form $\tilde{M}^2(4\varepsilon)$.

Theorem 2.17 ([46]) *Let M be a purely real surface in a complex space form $\tilde{M}^2(4\varepsilon)$ of constant holomorphic sectional curvature 4ε. Then we have*

$$\Delta\alpha = \Big\{||\nabla\alpha||^2 + 6\varepsilon \sin^2 \alpha\Big\} \cot \alpha + 2(\sin \alpha) * d\Psi_H, \tag{2.55}$$

where $\nabla\alpha$ is the gradient of α, $\Delta\alpha := div(\nabla\alpha)$ is the Laplacian of α, and $*$ is the Hodge star operator.

An immediate consequence of Theorem 2.17 is the following.

Corollary 2.18 ([48]) *Every slant surface M in $\mathbf{C}^2$ satisfies $d\Psi_H = 0$.*

Remark This result is false if the ambient space $\mathbf{C}^2$ is replaced by a non-flat complex space form $\tilde{M}^2(4\varepsilon)$.

Besides minimal surfaces and slant surfaces, there exist many non-minimal, non-slant purely real surfaces in $\mathbf{C}^2$ which satisfy $\Delta\alpha = 0$ or $d\Psi_H = 0$. For instance, a large family of such surfaces can be obtained from surfaces of revolution.

Proposition 2.19 ([46]) *Let $\phi(s) = (r(s), z(s))$ be a unit space curve in $\mathbb{E}^2$ with $r(s) > 0$ and $r'(s) \neq 1$. Consider the surface of revolution in $\mathbf{C} \times \mathbf{R} \subset \mathbf{C}^2$ defined by*

$$L(s,\theta) = \big(r(s)e^{i\theta}, z(s)\big). \tag{2.56}$$

Then we have

(i) The surface of revolution in $\mathbf{C}^2$ has harmonic Wintinger function if and only if $r(s)$ satisfies

$$r'''(s) = \frac{r'(s)r''(s)[r(s)r''(s) - r'(s)^2 + 1]}{r(s)(r'(s)^2 - 1)}. \tag{2.57}$$

(ii) The surface of revolution in $\mathbf{C}^2$ is a non-slant, non-minimal surface with $d\Psi_H = 0$ if and only if $r(s)$ satisfies

$$r'''(s) = \frac{r'(s)r''(s)[2r(s)r''(s) - r'(s)^2 + 1]}{r(s)[r'(s)^2 - 1]}, \tag{2.58}$$

$$r''(s) \neq 0, \quad rr'' + r'^2 \neq 1. \tag{2.59}$$

An immediate consequence of Theorem 2.17 is the following.

Corollary 2.20 ([40]) *Let M be a purely real minimal surface in a complex space form $\tilde{M}^2(4\varepsilon)$. Then we have*

$$\Delta\alpha = \Big\{||\nabla\alpha||^2 + 6\varepsilon\sin^2\alpha\Big\}\cot\alpha.$$

By applying the above corollary one obtains the following.

Corollary 2.21 ([50]) *Every slant surface in a complex space form $\tilde{M}^2(4\varepsilon)$ with $\varepsilon \neq 0$ is non-minimal unless it is either Lagrangian or complex.*

Corollary 2.22 ([40]) *Let M be a purely real minimal surface in the complex Euclidean plane $\mathbf{C}^2$. If the Wirtinger function α is a harmonic function, then M is slant.*

Corollary 2.23 ([50]) *Let M be a purely real minimal surface in complex projective plane CP^2. If the Wirtinger function α is harmonic, then M is Lagrangian.*

For minimal surfaces in CH^2, we have the following.

Theorem 2.24 ([46]) *A minimal surface in the complex hyperbolic plane CH^2 has the harmonic Wirtinger function if and only if it is either a complex surface or a minimal Lagrangian surface.*

Remark Obviously, there exist many complex surfaces in $CH^2(-4)$. In fact, every holomorphic curve in $CH^2(-4)$ is such an example.

Remark There exist many minimal Lagrangian surfaces in $CH^2(-4)$. The simplest one is the hyperbolic plane $H^2(-1)$ of curvature -1 isometrically immersed into $CH^2(-4)$ as a totally geodesic Lagrangian surface.

For non-totally geodesic Lagrangian minimal surfaces in $CH^2(-4)$, we have the following existence and uniqueness result from Corollary 3.6 in [28, p. 667] (for $\varphi = \mu^{2/3}$).

Proposition 2.25 ([28]) *If M is a minimal Lagrangian surface without totally geodesic points in $CH^2(-4)$, then, with respect to suitable coordinates x, y, we have:*

(*a*) *The metric tensor of M takes the form of $g = \varphi^{-1}\left(dx^2 + dy^2\right)$ for some positive function φ satisfying Poisson's equation:*

$$\frac{\partial^2(\ln\varphi)}{\partial x^2} + \frac{\partial^2(\ln\varphi)}{\partial y^2} = -\frac{2\left(1+2\varphi^3\right)}{\varphi}. \tag{2.60}$$

(*b*) *The second fundamental form h is given by*

$$\begin{aligned} &h\left(\tfrac{\partial}{\partial x}, \tfrac{\partial}{\partial x}\right) = -\varphi J\left(\tfrac{\partial}{\partial x}\right), \quad h\left(\tfrac{\partial}{\partial x}, \tfrac{\partial}{\partial y}\right) = \varphi J\left(\tfrac{\partial}{\partial y}\right), \\ &h\left(\tfrac{\partial}{\partial y}, \tfrac{\partial}{\partial y}\right) = \varphi J\left(\tfrac{\partial}{\partial x}\right). \end{aligned} \tag{2.61}$$

Conversely, if φ is a positive function defined on a simply connected domain U of $\mathbf{R}^2$ satisfying (2.60) *and if $g = \varphi^{-1}\left(dx^2 + dy^2\right)$ is the metric tensor on U, then, up to rigid motions of $CH^2(-4)$, there is a unique minimal Lagrangian isometric immersion of U into $CH^2(-4)$ whose second fundamental form is given by* (2.61).

Because (2.60) admits infinitely many solutions, there are abundant examples of minimal Lagrangian surfaces in $CH^2(-4)$.

It follows that each solution of this Poisson equation is nonconstant. Thus, an immediate consequence of Proposition 2.25 is the following (see also [49, 57]).

Corollary 2.26 *Every non-totally geodesic Lagrangian minimal surface in $CH^2(-4)$ has nonconstant Gauss curvature $K = -(1+2\varphi^3) \leq -1$ (associated with some solution φ of* (2.60)*).*

Trivially, we have $\Psi_H = 0$ for minimal purely real surfaces. Moreover, it was proved in [23] that $d\Psi_H = 0$ holds for every slant surface in $\mathbf{C}^2$.

We studied purely real surfaces in a Kaehler surface with closed canonical form.

For a non-minimal purely real surface M in a Kaehler surface $\tilde{M}^2$, we may choose an adapted frame e_1, e_2, e_3, e_4 such that H is parallel to e_3, $H = me_3$, $\mu = -\gamma$, where $m = (\beta + \lambda)/2$ is the mean curvature of M. We call such a frame on M a *H-adapted frame*.

Proposition 2.27 ([46]) *Let M be a non-minimal proper purely real surface in a Kaehler surface $\tilde{M}^2$. Then the canonical form Ψ_H is closed if and only if, with respect to a H-adapted frame, there exists an orthogonal coordinate system $\{x, y\}$ such that $e_1 = p^{-1}\frac{\partial}{\partial x}$ and $e_2 = q^{-1}\frac{\partial}{\partial y}$ with $p = (f(x)\sin\alpha)/m$ for some function $f(x)$.*

Obviously, every purely real minimal surface in a complex space form $M^2(4\varepsilon)$ satisfies $d\Psi_H = DH = 0$. The next theorem determines purely real surfaces in complex space forms $M^2(4\varepsilon)$ satisfying $d\Psi_H = DH = 0$.

Theorem 2.28 ([46]) *Every purely real surface in a complex space form $\tilde{M}^2(4\varepsilon)$ satisfying $d\Psi_H = DH = 0$ is either a minimal surface or a Lagrangian surface.*

Recall that Lagrangian surfaces in complex space forms $\tilde{M}^2(4\varepsilon)$ satisfy $d\Psi_H = 0$ (see, for instance, [118]).

It is known that a Lagrangian surface in $\mathbf{C}^2$ with a nonzero parallel mean curvature vector is either an open part of the product two circles $S^1(r_1) \times S^1(r_2) \subset \mathbf{C} \times \mathbf{C}$ or an open part of a circular cylinder $S^1 \times \mathbb{E}^1 \subset \mathbf{C} \times \mathbf{C}$ (cf. [23], p. 50, Theorem 1.1).

Lagrangian surfaces with a nonzero parallel mean curvature vector in $CP^2(4)$ and in $CH^2(-4)$ are parallel surfaces. Such surfaces have already been classified (cf. [43, Appendix]). In fact, we have the following:

A Lagrangian surface with a nonzero parallel mean curvature vector in $CP^2(4)$ is a flat surface whose immersion is congruent to $\pi \circ L$, where $\pi : S^5(1) \to CP^2(4)$ is the Hopf fibration and $L : M^2 \to S^5(1) \subseteq \mathbf{C}^3$ is given by

$$L(x, y) = \left(\frac{a\, e^{-ix/a}}{\sqrt{1+a^2}}, \frac{e^{i(ax+by)}}{\sqrt{1+a^2+b^2}} \sin\left(\sqrt{1+a^2+b^2}\, y\right), \right.$$

$$\left. \frac{e^{i(ax+by)}}{\sqrt{1+a^2}} \left(\cos\left(\sqrt{1+a^2+b^2}\, y\right) - \frac{ib}{\sqrt{1+a^2+b^2}} \sin\left(\sqrt{1+a^2+b^2}\, y\right)\right)\right),$$

where a and b are real numbers with $a \neq 0$.

A Lagrangian surface with nonzero parallel mean curvature vector in $CH^2(-4)$ is a flat surface whose immersion is congruent to $\pi \circ L$, where $\pi : H_1^5(-1) \to CH^2(-4)$ is the Hopf fibration and $L : M^2 \to H_1^5(-1) \subseteq \mathbf{C}_1^3$ is one of the following six maps:

1.

$$L(x,y)=\left(\frac{e^{i(ax+by)}}{\sqrt{1-a^2}}\left(\cosh\left(\sqrt{1-a^2-b^2}\,y\right)-\frac{ib\,\sinh\left(\sqrt{1-a^2-b^2}\,y\right)}{\sqrt{1-a^2-b^2}}\right),\right.$$

$$\left.\frac{e^{i(ax+by)}}{\sqrt{1-a^2-b^2}}\sinh\left(\sqrt{1-a^2-b^2}\,y\right),\frac{a\,e^{ix/a}}{\sqrt{1-a^2}}\right),$$

with $a, b \in \mathbf{R},\ a \neq 0$ and $a^2 + b^2 < 1$;

2.

$$L(x,y)=\left(\left(\frac{i}{b}+y\right)e^{i(\sqrt{1-b^2}x+by)},\,ye^{i(\sqrt{1-b^2}x+by)},\,\frac{\sqrt{1-b^2}}{b}e^{ix/\sqrt{1-b^2}}\right),$$

with $b \in \mathbf{R},\ 0 < b^2 < 1$;

3.

$$L(x,y)=\left(\frac{e^{i(ax+by)}}{\sqrt{1-a^2}}\left(\cos\left(\sqrt{a^2+b^2-1}\,y\right)-\frac{ib\,\sin\left(\sqrt{a^2+b^2-1}\,y\right)}{\sqrt{a^2+b^2-1}}\right),\right.$$

$$\left.\frac{e^{i(ax+by)}}{\sqrt{a^2+b^2-1}}\sin\left(\sqrt{a^2+b^2-1}\,y\right),\frac{a\,e^{ix/a}}{\sqrt{1-a^2}}\right),$$

with $a, b \in \mathbf{R},\ 0 < a^2 < 1$ and $a^2 + b^2 > 1$;

4.

$$L(x,y)=\left(\frac{a\,e^{ix/a}}{\sqrt{a^2-1}},\frac{e^{i(ax+by)}}{\sqrt{a^2+b^2-1}}\sin\left(\sqrt{a^2+b^2-1}\,y\right),\right.$$

$$\left.\frac{e^{i(ax+by)}}{\sqrt{a^2-1}}\left(\cos\left(\sqrt{a^2+b^2-1}\,y\right)-\frac{ib\,\sin\left(\sqrt{a^2+b^2-1}\,y\right)}{\sqrt{a^2+b^2-1}}\right)\right),$$

with $a, b \in \mathbf{R},\ a^2 > 1$;

5.

$$L(x,y)=\frac{e^{ix}}{8b^2}\left(i+8b^2(i+x)-4by,\,i+8b^2x-4by,\,4be^{2iby}\right),$$

with $b \in \mathbf{R},\ b \neq 0$;

6.

$$L(x, y) = e^{ix}\left(1 + \frac{y^2}{2} - ix, y, \frac{y^2}{2} - ix\right).$$

The following two propositions follow easily from Theorem 2.17.

Proposition 2.29 ([46]) *Let M be a purely real surface in the complex Euclidean plane $\mathbf{C}^2$. Then M is a slant surface if and only if we have $\Delta\alpha = d\Psi_H = 0$.*

Proposition 2.30 ([46]) *Let M be a purely real surface in the complex projective plane CP^2. Then M is a Lagrangian surface if and only if we have $\Delta\alpha = d\Psi_H = 0$.*

In contrast to Corollary 2.18, we have the following consequence of Proposition 2.30.

Corollary 2.31 ([46]) *Every proper slant surface M in CP^2 satisfies $d\Psi_H \neq 0$.*

For purely real surfaces in CH^2, we also have the following.

Theorem 2.32 ([46]) *A purely real surface M in a complex hyperbolic plane CH^2 satisfies $\Delta\alpha = d\Psi_H = 0$ if and only if M is a Lagrangian surface.*

Next, we consider again Kaehlerian slant submanifolds in complex space forms.

B.Y. Chen proved in [31] an optimal inequality for Lagrangian submanifolds (a particular case of purely real submanifolds) in complex space forms in terms of the Ricci curvature and the squared mean curvature, well known as the Chen-Ricci inequality. Recently, the Chen-Ricci inequality was improved in [55] for Lagrangian submanifolds in complex space forms.

We extended the improved Chen-Ricci inequality to Kaehlerian slant submanifolds in complex space forms. We also investigated the equality case of the inequality.

Definition A slant *H-umbilical submanifold* of a Kaehler manifold $\widetilde{M}^n$ is a slant submanifold for which the second fundamental form takes the following forms:

$$h(e_1, e_1) = \lambda e_1^*, \quad h(e_2, e_2) = \cdots = h(e_n, e_n) = \mu e_1^*,$$

$$h(e_1, e_j) = \mu e_j^*, \quad h(e_j, e_k) = 0, \quad 2 \leq j \neq k \leq n,$$

where $e_1^*, \ldots, e_n^*$ are defined as before.

Recall some known results.

In [29], B.Y. Chen established a sharp relationship between the Ricci curvature and the squared mean curvature for any n-dimensional Riemannian submanifold of a real space form $\widetilde{M}(c)$ of constant sectional curvature c, namely

$$\mathrm{Ric}(X) \leq (n-1)c + \frac{n^2}{4}\,\|H\|^2,$$

which is well known as the Chen-Ricci inequality. The same inequality holds for Lagrangian submanifolds in a complex space form $\widetilde{M}(4c)$ as well (see [31]).

I. Mihai proved a similar inequality in [93] for certain submanifolds of Sasakian space forms.

In [73], Matsumoto, Mihai, and Oiagă extended the Chen-Ricci inequality to the following inequality for submanifolds in complex space forms.

Theorem ([73]) *Let M be an n-dimensional submanifold of a complex m-dimensional complex space form $\widetilde{M}(4c)$. Denote by J the complex structure of $\widetilde{M}(4c)$. Then:*

(i) For each vector $X \in T_pM$, we have

$$Ric(X) \leq (n-1)c + \frac{n^2}{4} \|H\|^2 + 3c \|PX\|^2,$$

where PX is the tangential component of JX.

(ii) If $H(p) = 0$, then a unit tangent vector X at p satisfies the equality case if and only if $X \in \ker h_p$.

(iii) The equality case holds identically for all unit tangent vectors at p if and only if p is a totally geodesic point or $n = 2$ and p is a totally umbilical point.

In particular, for θ-slant submanifolds, the following result holds.

Corollary ([73]) *Let M be an n-dimensional θ-slant submanifold of a complex space form $\widetilde{M}(4c)$. Then,*

(i) For each vector $X \in T_pM$, we have

$$Ric(X) \leq (n-1)c + \frac{n^2}{4} \|H\|^2 + 3c\cos^2\theta.$$

(ii) If $H(p) = 0$, then a unit tangent vector X at p satisfies the equality case if and only if $X \in \ker h_p$.

(iii) The equality case holds identically for all unit tangent vectors at p if and only if p is a totally geodesic point or $n = 2$ and p is a totally umbilical point.

The Chen-Ricci inequality was further improved to the following for Lagrangian submanifolds (cf. [55]).

Theorem ([55]) *Let M be a Lagrangian submanifold of dimension $n \geq 2$ in a complex space form $\widetilde{M}(4c)$ of constant holomorphic sectional curvature $4c$ and X a unit tangent vector in T_pM, $p \in M$. Then, we have*

$$Ric(X) \leq (n-1)\left(c + \frac{n}{4} \|H\|^2\right).$$

The equality sign holds for any unit tangent vector at p if and only if either:

(i) *p is a totally geodesic point or*
(ii) $n = 2$ *and* p *is a H-umbilical point with* $\lambda = 3\,\mu$.

Lagrangian submanifolds in complex space forms satisfying the equality case of the inequality were determined by Deng in [55]. More precisely, he proved the following.

Corollary **([55])** *Let* M *be a Lagrangian submanifold of real dimension* $n \geq 2$ *in a complex space form* $\widetilde{M}(4c)$. *If*

$$Ric(X) = (n-1)\left(c + \frac{n}{4}\,\|H\|^2\right),$$

for any unit tangent vector X *of* M, *then either:*

(i) M *is a totally geodesic submanifold in* $\widetilde{M}(4c)$ *or*
(ii) $n = 2$, *and* M *is a Lagrangian H-umbilical submanifold of* $\widetilde{M}(4c)$ *with* $\lambda = 3\mu$.

We extended the last theorem to Kaehlerian slant submanifolds in complex space forms, by applying the following two lemmas from [55].

Lemma 2.33 *Let* $f_1(x_1, x_2, \ldots, x_n)$ *be a function in* $\mathbf{R}^n$ *defined by*

$$f_1(x_1, x_2, \ldots, x_n) = x_1 \sum_{j=2}^{n} x_j - \sum_{j=2}^{n} x_j^2.$$

If $x_1 + x_2 + \ldots + x_n = 2na$, *then we have*

$$f_1(x_1, x_2, \ldots, x_n) \leq \frac{n-1}{4n}\,(x_1 + x_2 + \ldots + x_n)^2,$$

with the equality sign holding if and only if $\frac{1}{n+1} x_1 = x_2 = \ldots = x_n = a$.

Lemma 2.34 *Let* $f_2(x_1, x_2, \ldots, x_n)$ *be a function in* $\mathbf{R}^n$ *defined by*

$$f_2(x_1, x_2, \ldots, x_n) = x_1 \sum_{j=2}^{n} x_j - x_1^2.$$

If $x_1 + x_2 + \ldots + x_n = 4a$, *then we have*

$$f_2(x_1, x_2, \ldots, x_n) \leq \frac{1}{8}\,(x_1 + x_2 + \ldots + x_n)^2,$$

with the equality sign holding if and only if $x_1 = a$ *and* $x_2 + \ldots + x_n = 3a$.

Our main result is the following theorem.

Theorem 2.35 ([89]) *Let* M *be an* n*-dimensional Kaehlerian* θ*-slant submanifold in a complex* n*-dimensional complex space form* $\widetilde{M}(4c)$ *of constant holomorphic* *sectional curvature 4c. Then for any unit tangent vector X to M we have*

$$Ric(X) \leq (n-1)\left(c + \frac{n}{4}\|H\|^2\right) + 3c\cos^2\theta. \tag{2.62}$$

The equality sign of (2.62) holds identically if and only if either:

(i) $c = 0$ *and M is totally geodesic or*
(ii) $n = 2$, $c < 0$, *and M is a slant H-umbilical surface with* $\lambda = 3\mu$.

Proof For a given point $p \in M$ and a given unit vector $X \in T_pM$, we choose an orthonormal basis $\{e_1 = X, e_2, \dots, e_n\} \subset T_pM$ and

$$\left\{e_1^* = \frac{Fe_1}{\sin\theta}, \dots, e_n^* = \frac{Fe_n}{\sin\theta}\right\} \subset T_p^{\perp}M.$$

Now we put in the Gauss equation $X = Z = e_1$ and $Y = W = e_j$, for $j = 2, \dots, n$. Then the Gauss equation gives

$$\begin{aligned}\widetilde{R}(e_1, e_j, e_1, e_j) = R(e_1, e_j, e_1, e_j) - g(h(e_1, e_1), h(e_j, e_j))\\ + g(h(e_1, e_j), h(e_1, e_j)),\end{aligned}$$

or equivalently,

$$\widetilde{R}(e_1, e_j, e_1, e_j) = R(e_1, e_j, e_1, e_j) - \sum_{r=1}^{n}\left(h_{11}^r h_{jj}^r - (h_{1j}^r)^2\right), \ \forall j \in \overline{2, n}.$$

Since the Riemannian curvature tensor of M is given by

$$\begin{aligned}\widetilde{R}(X, Y, Z, W) = c\{g(X, Z)g(Y, W) - g(X, W)g(Y, Z)\\ + g(JX, Z)g(JY, W) - g(JX, W)g(JY, Z) + 2g(JX, Y)g(JZ, W)\},\end{aligned}$$

we find

$$\widetilde{R}(e_1, e_j, e_1, e_j) = c[1 + 3g^2(Je_1, e_j)]. \tag{2.63}$$

By summing after $j = \overline{2, n}$, we get

$$(n - 1 + 3\|PX\|^2)c = \text{Ric}(X) - \sum_{r=1}^{n}\sum_{j=2}^{n}\left[h_{11}^r h_{jj}^r - (h_{1j}^r)^2\right],$$

or,

$$(n-1+3\cos^2\theta)c = \mathrm{Ric}(X) - \sum_{r=1}^{n}\sum_{j=2}^{n}\left[h_{11}^r h_{jj}^r - (h_{1j}^r)^2\right].$$

It follows that

$$\mathrm{Ric}(X) - (n-1+3\cos^2\theta)c = \sum_{r=1}^{n}\sum_{j=2}^{n}\left[h_{11}^r h_{jj}^r - (h_{1j}^r)^2\right] \tag{2.64}$$

$$\leq \sum_{r=1}^{n}\sum_{j=2}^{n} h_{11}^r h_{jj}^r - \sum_{j=2}^{n}(h_{1j}^1)^2 - \sum_{j=2}^{n}(h_{1j}^j)^2.$$

Since M is a Kaehlerian slant submanifold, we have $h_{1j}^1 = h_{11}^j$ and $h_{1j}^j = h_{jj}^1$, and then

$$\mathrm{Ric}(X) - (n-1+3\cos^2\theta)c \leq \sum_{r=1}^{n}\sum_{j=2}^{n} h_{11}^r h_{jj}^r - \sum_{j=2}^{n}(h_{11}^j)^2 - \sum_{j=2}^{n}(h_{jj}^1)^2. \tag{2.65}$$

Now we put

$$f_1\left(h_{11}^1, h_{22}^1, \ldots, h_{nn}^1\right) = h_{11}^1 \sum_{j=2}^{n} h_{jj}^1 - \sum_{j=2}^{n}(h_{jj}^1)^2$$

and

$$f_r\left(h_{11}^r, h_{22}^r, \ldots, h_{nn}^r\right) = h_{11}^r \sum_{j=2}^{n} h_{jj}^r - (h_{11}^r)^2, \ \forall r \in \overline{2, n}.$$

Since $nH^1 = h_{11}^1 + h_{22}^1 + \ldots + h_{nn}^1$, we obtain by using Lemma 2.33 that

$$f_1\left(h_{11}^1, h_{22}^1, \ldots, h_{nn}^1\right) \leq \frac{n-1}{4n}\left(nH^1\right)^2 = \frac{n(n-1)}{4}\left(H^1\right)^2. \tag{2.66}$$

By applying Lemma 2.34 for $2 \leq r \leq n$, we get

$$f_r\left(h_{11}^r, h_{22}^r, \ldots, h_{nn}^r\right) \leq \frac{1}{8}(nH^r)^2 = \frac{n^2}{8}(H^r)^2 \leq \frac{n(n-1)}{4}(H^r)^2. \tag{2.67}$$

From (2.65), (2.66) and (2.67), we obtain

$$\mathrm{Ric}(X) - (n-1+3\cos^2\theta)c \leq \frac{n(n-1)}{4}\sum_{r=1}^{n}(H^r)^2 = \frac{n(n-1)}{4}\|H\|^2.$$

Thus we have

$$\mathrm{Ric}(X) \leq (n-1+3\cos^2\theta)c + \frac{n(n-1)}{4}\,\|H\|^2,$$

which implies (2.62).

Next, we shall study the equality case. For $n \geq 3$, we choose Fe_1 parallel to H. Then we have $H^r = 0$, for $r \geq 2$. Thus, by Lemma 2.34, we get

$$h^1_{1j} = h^j_{11} = \frac{nH^j}{4} = 0, \ \forall j \geq 2,$$

and

$$h^1_{jk} = 0, \ \forall j,k \geq 2, \ j \neq k.$$

From Lemma 2.33, we have $h^1_{11} = (n+1)a$ and $h^1_{jj} = a, \forall j \geq 2$, with $a = \frac{H^1}{2}$.

In (2.64) we compute $\mathrm{Ric}(X) = \mathrm{Ric}(e_1)$. Similarly, by computing $\mathrm{Ric}(e_2)$ and using the equality, we get

$$h^r_{2j} = h^2_{jr} = 0, \quad \forall r \neq 2, \quad j \neq 2, \quad r \neq j.$$

Then we obtain

$$\frac{h^2_{11}}{n+1} = h^2_{22} = \ldots = h^2_{nn} = \frac{H^2}{2} = 0.$$

The argument is also true for matrices $\left(h^r_{jk}\right)$ because the equality holds for all unit tangent vectors, so $h^2_{2j} = h^j_{22} = \frac{H^j}{2} = 0, \forall j \geq 3$.

The matrix $\left(h^2_{jk}\right)$ (respectively, the matrix $\left(h^r_{jk}\right)$) has only two possible nonzero entries $h^2_{12} = h^2_{21} = h^1_{22} = \frac{H^1}{2}$ (respectively, $h^r_{1r} = h^r_{r1} = h^1_{rr} = \frac{H^1}{2}\ \forall r \geq 3$). Now, after putting $X = Z = e_2$ and $Y = W = e_j$, $j = 2, \ldots, n$, in the Guss equation, we obtain

$$\widetilde{R}(e_2, e_j, e_2, e_j) = R(e_2, e_j, e_2, e_j) - \left(\frac{H^1}{2}\right)^2, \ \forall j \geq 3.$$

If we put $X = Z = e_2$ and $Y = W = e_1$ in the Guss equation, we get

$$\widetilde{R}(e_2, e_1, e_2, e_1) = R(e_2, e_1, e_2, e_1) - (n+1)\left(\frac{H^1}{2}\right)^2 + \left(\frac{H^1}{2}\right)^2.$$

After combining the last two relations, we find

$$\text{Ric}(e_2) - (n - 1 + 3\cos^2\theta)c = 2(n-1)\left(\frac{H^1}{2}\right)^2.$$

On the other hand, the equality case of (2.62) implies that

$$\text{Ric}(e_2) - (n - 1 + 3\cos^2\theta)c = \frac{n(n-1)}{4}\|H\|^2 = n(n-1)\left(\frac{H^1}{2}\right)^2.$$

Since $n \neq 1, 2$, by equating the last two equations we find $H^1 = 0$. Thus, $\left(h^r_{jk}\right)$ are all zero, i.e., M is a totally geodesic submanifold in $\widetilde{M}(4c)$. In particular, M is a curvature-invariant submanifold of $\widetilde{M}(4c)$. Therefore, when $c \neq 0$, it follows from a result of Chen and Ogiue [49] that M is either a complex submanifold or a Lagrangian submanifold of $\widetilde{M}(4c)$. Hence, M is a non-proper θ-slant, which is a contradiction. Consequently, we have either:

1. $c = 0$ and M is totally geodesic or
2. $n = 2$.

If (1) occurs, we obtain (i) of the theorem. Now, let us assume that $n = 2$. A result of Chen from [27] states that if M is a proper slant surface in a complex two-dimensional complex space form $\widetilde{M}^2(4c)$ satisfying the equality case of (2.62) identically, then M is either totally geodesic or $c < 0$. In particular, when M is not totally geodesic, one has

$$h(e_1, e_1) = \lambda e_1^*, h(e_2, e_2) = \mu e_1^*, h(e_1, e_2) = \mu e_2^*,$$

with $\lambda = 3\mu = \frac{3H^1}{2}$, i.e., M is H-umbilical. This gives case (ii) of the theorem.

Since a proper slant surface is Kaehlerian slant automatically (cf. [23]), we rediscovered the following result of Chen (see [27]).

Theorem 2.36 *If M is a proper slant surface in a complex space form $\widetilde{M}(4c)$ of complex dimension 2, then the squared mean curvature and the Gaussian curvature of M satisfy*

$$\|H\|^2 \geq 2[G - (1 + 3\cos^2\theta)c]$$

at each point $p \in M$, where θ is the slant angle of the slant surface.

Example 2.37 The explicit representation of the slant surface in $CH^2(-4)$ satisfying the equality case of inequality (2.62) was determined by Chen and Tazawa in [50, Theorem 5.2] as follows:

Let z be the immersion $z : \mathbf{R}^3 \to \mathbf{C}^3_1$ defined by

$$z(u, v, t) = e^{it}\left(1 + \frac{3}{2}\left(\cosh\left(\frac{\sqrt{2}}{\sqrt{3}}v\right) - 1\right) + \frac{u^2}{6}e^{-\sqrt{\frac{2}{3}}v} - i\frac{u}{\sqrt{6}}\left(1 + e^{-\sqrt{\frac{2}{3}}v}\right),\right.$$

$$\frac{u}{3}\left(1+2e^{-\sqrt{\frac{2}{3}}v}\right)+\frac{i}{6\sqrt{6}}e^{-\sqrt{\frac{2}{3}}v}\left(\left(e^{\sqrt{\frac{2}{3}}v}-1\right)\left(9e^{\sqrt{\frac{2}{3}}v}-3\right)+2u^2\right),$$

$$\frac{u}{3\sqrt{2}}\left(1-e^{-\sqrt{\frac{2}{3}}v}\right)+\frac{i}{12\sqrt{3}}\left(6-15e^{-\sqrt{\frac{2}{3}}v}+9e^{\sqrt{\frac{2}{3}}v}+2e^{-\sqrt{\frac{2}{3}}v}u^2\right)\Big).$$

It was proved that $\langle z, z\rangle = -1$. Hence, z defines an immersion from $\mathbf{R}^3$ into the anti-de Sitter space-time $H_1^5(-1)$. Moreover, the image $z(\mathbf{R}^3)$ in $H_1^5(-1)$ is invariant under the action of $\mathbf{C}^* = \mathbf{C} - \{0\}$. Let $\pi : H_1^4(-1) \to CH^2(-4)$ denote the Hopf fibration. It was also shown that the composition

$$\pi \circ z : \mathbf{R}^3 \to CH_1^2(-4)$$

defines a slant surface with slant angle $\theta = \cos^{-1}(\frac{1}{3})$. Also, the authors proved that $\pi \circ z$ is a H-umbilical immersion satisfying $\lambda = 3\mu$. Consequently, this example of slant H-umbilical surface satisfies the equality case of inequality (2.62) identically.

The next part of this section presents the work we have done on (generalized) Wintgen inequality.

In [54] the authors obtained a pointwise inequality for submanifolds M^n in space forms $N^{n+2}(c)$ of dimension $n \geq 2$ and codimension 2 relating the main scalar invariants, namely the scalar curvature (intrinsic invariant) and the squared mean curvature and scalar normal curvature (extrinsic invariant). A corresponding inequality for dimension $n = 3$ was proved in [52]. On the other hand, in [74] we gave a sharp estimate of the normal curvature for totally real surfaces in complex space forms. In [79] we extended this inequality to three-dimensional Lagrangian submanifolds in complex space forms. We mention that a corresponding inequality for Kaehler hypersurfaces was obtained in [112].

Let M^n be an n-dimensional submanifold isometrically immersed into the real space form $N^m(c)$.

The *normalized scalar curvature of the tangent bundle* (intrinsic invariant) is defined by

$$\rho = \frac{2}{n(n-1)} \sum_{1\leq i<j\leq n} g(R(e_i, e_j)e_j, e_i) = \frac{2}{n(n-1)}\tau.$$

The *normalized scalar curvature of the normal bundle* (extrinsic invariant) is given by

$$\rho^{\perp} = \frac{2}{n(n-1)}\sqrt{\sum_{1\leq i<j\leq n}\ \sum_{n+1\leq r<s\leq m} [g(R^{\perp}(e_i, e_j)\xi_r, \xi_s)]^2},$$

where $\{e_1, \ldots, e_n\}$ is an orthonormal basis of the tangent space, $\{\xi_{n+1}, \ldots, \xi_m\}$ is an orthonormal basis of the normal space at a point $p \in M$, R and $R^{\perp}$ are the curvature tensors of the tangent bundle and normal bundle, respectively, and τ is the scalar curvature.

In [26] B.Y. Chen proved a pointwise inequality for an n-dimensional submanifold M^n isometrically immersed into the real space form $N^m(c)$

$$\|H\|^2 \geq \rho - c. \tag{2.68}$$

B. Suceavă [115] gave another proof of the same inequality.

In [73] we found that for an n-dimensional submanifold M^n isometrically immersed into the complex space form $\tilde{M}(4c)$ one has

$$\|H\|^2 \geq \rho - c - \frac{3\|P\|^2}{n(n-1)}c,$$

where J is the complex structure of the ambient space, and $JX = PX + FX$, where PX is the tangential component and FX is the normal component.

Also, in [78], for M^n a purely real n-dimensional submanifold with $\nabla P = 0$ isometrically immersed into the complex space form $\tilde{M}(4c)$, the following inequality was proved

$$\|H\|^2 \geq \frac{n+2}{n}(\rho - c) - \frac{3\|P\|^2}{n(n-1)}c.$$

We recall the following result from [54].

Theorem 2.38 *Let $\phi : M^n \to N^{n+2}(c)$ be an isometric immersion. Then at every point p*

$$\|H\|^2 \geq \rho + \rho^{\perp} - c,$$

Remark This inequality generalized the Wintgen inequality; it is valid for all submanifolds M^n of all real space forms $N^{n+2}(c)$ with $n \geq 2$ and codimension 2.

In the same paper, the authors stated the following conjecture, afterward named the *DDVV conjecture*.

Conjecture *Let $\phi : M^n \to N^m(c)$ be an isometric immersion. Then at every point p the inequality occurs:*

$$\|H\|^2 \geq \rho + \rho^{\perp} - c. \tag{DDVV}$$

The DDVV conjecture was proved first in the following cases:

(a) $n = 2, m = 4, c = 0$ by P. Wintgen [120].
(b) $n = 2, m \geq 4$ by I.V. Guadalupe and L. Rodriguez [63].

Remark In both cases, equality is realized in (DDVV) inequality at a point p if and only if the ellipse of curvature is a circle. In the case of trivial normal connection, (DDVV) reduces to Chen's inequality.

(c) $n = 3, m \geq 5$ by T. Choi and Z. Lu [52]

For a totally real surface M of a complex space form $\tilde{M}(4c)$ of arbitrary codimension, we obtained an inequality relating the squared mean curvature $\|H\|^2$, the holomorphic sectional curvature c, the Gauss curvature K, and the elliptic curvature K^E of the surface. Using the notion of ellipse of curvature, we proved a characterization of the equality and gave an example of a Lagrangian surface of $\mathbf{C}^2$ satisfying the equality case [74].

Let M be a totally real surface of the complex space form $\widetilde{M}(c)$ of constant holomorphic sectional curvature c and of complex dimension n.

For a point $p \in M$, let $\{e_1, e_2\}$ be an orthonormal basis of the tangent plane T_pM and $\{e_3, \ldots, e_{2n}\}$ an orthonormal basis of the normal space $T_p^{\perp}M$.

The *ellipse of curvature* at a point $p \in M$ is the subspace E_p of the normal space given by

$$E_p = \{h_p(X, X) \mid X \in T_pM, \|X\| = 1\}.$$

For any vector $X = \cos\theta \cdot e_1 + \sin\theta \cdot e_2, \theta \in [0, 2\pi]$, we have

$$h_p(X, X) = H(p) + \cos 2\theta \cdot \frac{h_{11} - h_{22}}{2} + \sin 2\theta \cdot h_{12}.$$

The following result [63] holds good:

Proposition *If the ellipse of curvature is nondegenerated, then the vectors $h_{11}-h_{22}$ and h_{12} are linearly independent.*

Using a similar method as in [62, 63] and the previous proposition, we can define a 2-plane subbundle P of the normal bundle, with the induced connection.

We will define then the *elliptic curvature* by the formula

$$K^E = g([A_{e_3}, A_{e_4}]e_1, e_2),$$

where $\{e_1, e_2\}, \{e_3, e_4\}$ are orthonormal basis of T_pM, respectively of P_p, and A is the Weingarten operator.

Remark This definition of the elliptic curvature coincides with the definition of the *normal curvature* (given by Wintgen [120] and also Guadalupe and Rodriguez [63] by the formula $K^N = g(R^{\perp}(e_1, e_2)e_3, e_4))$ if the ambient space $\tilde{M}(c)$ is a real space form.

We can choose $\{e_1, e_2\}$ such that the vectors $u = \frac{h_{11}-h_{22}}{2}$ and $v = h_{12}$ are perpendicular, in which case they coincide with the half-axis of the ellipse. Then we will take $e_3 = \frac{u}{\|u\|}$ and $e_4 = \frac{v}{\|v\|}$.

From the equation of Ricci and the definition of K^E, we have

$$K^E = -\|h_{11} - h_{22}\| \cdot \|h_{12}\| . \tag{2.69}$$

Also, from the Gauss equation we obtain the formula of the Gauss curvature K of the totally real surface M of the complex space form $\widetilde{M}(c)$:

$$K = g(h_{11}, h_{22}) - \|h_{12}\|^2 + \frac{c}{4}. \tag{2.70}$$

By the definition of the mean curvature vector, the above equation, and the relation $\|h\|^2 = \|h_{11}\|^2 + \|h_{22}\|^2 + 2\|h_{12}\|^2$, we have

$$4\|H\|^2 = \|h\|^2 + 2\Big(K - \frac{c}{4}\Big). \tag{2.71}$$

Then,

$$\begin{aligned} 0 \le (\|h_{11} - h_{22}\| - 2\|h_{12}\|)^2 &= \|h\|^2 - 2\Big(K - \frac{c}{4}\Big) + 4K^E \qquad (2.72)\\ &= 4\|H\|^2 - 4\Big(K - \frac{c}{4}\Big) + 4K^E, \end{aligned}$$

which is equivalent to

$$\|H\|^2 \ge K - K^E - \frac{c}{4}. \tag{2.73}$$

The equality sign is realized if and only if $\|h_{11} - h_{22}\| = 2\|h_{12}\|$, i.e., $\|u\| = \|v\|$, so the ellipse of curvature is a circle.

Thus, we proved the following:

Theorem 2.39 ([74]) *Let M be a totally real surface of the complex space form $\widetilde{M}(c)$ of constant holomorphic sectional curvature c and of complex dimension n. Then, at any point $p \in M$, we have*

$$\|H\|^2 \ge K - K^E - \frac{c}{4}.$$

Moreover, the equality sign is realized if and only if the ellipse of curvature is a circle.

We will give one **example** of a Lagrangian surface in $\mathbf{C}^2$ with the standard almost complex structure J_0, for which the equality sign is realized (which we call an *ideal surface*).

Let M be the *rotation surface* of Vrănceanu, given by

$$X(u, v) = r(u)(\cos u \cos v, \cos u \sin v, \sin u \cos v, \sin u \sin v),$$

where r is a C^∞-differentiable function with positive values.

Let $\{e_1, e_2\}$ be an orthonormal basis of the tangent plane and $\{e_3, e_4\}$ an orthonormal basis of the normal plane.

We have the following expressions for $e_i, i \in \{1, \dots, 4\}$ (see also [109]):

$$e_1 = (-\cos u \sin v, \cos u \cos v, -\sin u \sin v, \sin u \cos v),$$

$$e_2 = \frac{1}{A}(B \cos v, B \sin v, C \cos v, C \sin v),$$

$$e_3 = \frac{1}{A}(-C \cos v, -C \sin v, B \cos v, B \sin v),$$

$$e_4 = (-\sin u \sin v, \sin u \cos v, \cos u \sin v, -\cos u \cos v),$$

where $A = [r^2 + (r')^2]^{\frac{1}{2}}$, $B = r' \cos u - r \sin u$, $C = r' \sin u + r \cos u$.

M is a totally real surface of maximum dimension, so is a *Lagrangian surface* of $\mathbf{C}^2$. Also, M verifies the equality sign of the inequality proved in the previous theorem (it is an *ideal surface*) if and only if

$$r(u) = \frac{1}{(|\cos 2u|)^{\frac{1}{2}}}$$

(the ellipse of curvature at every point of M is a circle).

Moreover, M is a minimal surface (see [100]), and $X = c_1 \otimes c_2$ is the tensor product immersion of $c_1(u) = \dfrac{1}{(|\cos 2u|)^{\frac{1}{2}}}(\cos u, \cos v)$ (an orthogonal hyperbola) and $c_2(u) = (\cos v, \sin v)$ (a circle).

Later, in [79] we established a generalized Wintgen inequality for three-dimensional Lagrangian submanifolds in complex space forms.

Let M^n be an n-dimensional Lagrangian submanifold isometrically immersed into the complex space form $\tilde{M}(4c)$. Let $\{e_1, \dots, e_n\}$ be an orthonormal frame on M. Then a normal frame is given by $\xi_1 = Je_1, \dots, \xi_n = Je_n$.

Following [122], the *scalar normal curvature* K_N is defined by

$$K_N = \frac{1}{2} \sum_{1 \le r < s \le n} (\mathrm{Tr}\,[A_r, A_s])^2 = \sum_{1 \le r < s \le n} \sum_{1 \le i < j \le n} (g([A_r, A_s]e_i, e_j))^2.$$

The *normalized scalar normal curvature* is given by

$$\rho_N = \frac{2}{n(n-1)} \sqrt{K_N}.$$

In particular, for $n = 2$ the above definition agrees with the definition of the squared elliptic curvature which we considered in [74].

For $n = 3$ we proved a sharp estimate of ρ_N, in terms of ρ and $\|H\|^2$.

Theorem 2.40 ([79]) *Let M be a three-dimensional Lagrangian submanifold isometrically immersed into the complex space form $\tilde{M}(4c)$. Then we have*

$$\|H\|^2 \geq \rho + \rho_N - c.$$

Proof Let $\{e_1, e_2, e_3\}$ be an orthonormal frame on M and $\{\xi_1 = Je_1, \xi_2 = Je_2, \xi_3 = Je_3\}$ the corresponding normal frame. With respect to these frames, we have

$$9\,\|H\|^2 = \sum_{r=1}^{3}\left(\sum_{i=1}^{3} h_{ii}^r\right)^2 = \frac{1}{2}\sum_{r=1}^{3}\sum_{1\leq i<j\leq 3}^{3}(h_{ii}^r - h_{jj}^r)^2 + 3\sum_{r=1}^{3}\sum_{1\leq i<j\leq 3}^{3} h_{ii}^r h_{jj}^r. \tag{2.74}$$

From the Gauss equation it follows that

$$2\tau = 9\,\|H\|^2 - \|h\|^2 + 6c. \tag{2.75}$$

Then

$$18\,\|H\|^2 = \sum_{r=1}^{3}\sum_{1\leq i<j\leq 3}^{3}(h_{ii}^r - h_{jj}^r)^2 + 6(\tau - 3c) + 6\sum_{r=1}^{3}\sum_{1\leq i<j\leq 3}^{3}(h_{ij}^r)^2. \tag{2.76}$$

Choi and Lu [52] proved that

$$\begin{aligned}&\sum_{r=1}^{3}\sum_{1\leq i<j\leq 3}^{3}(h_{ii}^r - h_{jj}^r)^2 + 6\sum_{r=1}^{3}\sum_{1\leq i<j\leq 3}^{3}(h_{ij}^r)^2\\ &\geqslant 6\left\{\sum_{1\leq r<s\leq 3}\sum_{1\leq i<j\leq 3}\left[\sum_{k=1}^{3}(h_{ik}^r h_{jk}^s - h_{ik}^s h_{jk}^r)\right]\right\}^{\frac{1}{2}}.\end{aligned} \tag{2.77}$$

Obviously,

$$K_N = \sum_{1\leq r<s\leq 3}\sum_{1\leq i<j\leq 3}\left[\sum (h_{ik}^r h_{jk}^s - h_{ik}^s h_{jk}^r)\right]^2. \tag{2.78}$$

Combining the last three equations, we find

$$18(\|H\|^2 - \rho + c) \geqslant 18\rho_N,$$

i.e.,

$$\rho + \rho_N \leq \|H\|^2 + c.$$

Remark The same inequality can be obtained for three-dimensional totally real submanifold with arbitrary codimension (with the corresponding definition of K_N).

F. Dillen, J. Fastenakels, and J. Van der Veken obtained in [56] an estimate of $\rho^\perp$ for invariant (Kaehler) submanifolds in complex space forms of arbitrary dimension and codimension.

We have obtained the following estimation.

Proposition 2.41 ([79]) *For an n-dimensional Lagrangian submanifold in a complex space form, we have*

$$\frac{n^2(n-1)^2}{4}(\rho^\perp)^2 = \frac{n^2(n-1)^2}{4}\rho_N^2 + \frac{n(n-1)}{2}c^2 + \frac{c}{2}\|h\|^2. \tag{2.79}$$

Proof From the Ricci equation it follows that

$$\begin{aligned} g(R^\perp(e_i, e_j)\xi_r, \xi_s) &= c[g(Je_i, \xi_r), g(Jej, \xi_s) - g(Je_i, \xi_s)g(Je_j, \xi_r)] \\ &\quad + g([A_r, A_s], e_i, e_j) \\ &= c(\delta_{ir}\delta_{js} - \delta_{is}\delta_{jr}) + g([A_r, A_s], e_i, e_j). \end{aligned}$$

Thus $\tau^\perp = \frac{n(n-1)}{2}\rho^\perp$ is given by

$$(\tau^\perp)^2 = \sum_{1\le r<s\le n}\sum_{1\le i<j\le n} g^2(R^\perp(e_i, e_j)\xi_r, \xi_s)$$

$$= \sum_{1\le r<s\le n}\sum_{1\le i<j\le n} [c(\delta_{ir}\delta_{js} - \delta_{is}\delta_{jr}) + g([A_r, A_s], e_i, e_j)]^2$$

$$= \frac{n^2(n-1)^2}{4}\rho_N^2 + c^2 \sum_{1\le r<s\le n}\sum_{1\le i<j\le n} (\delta_{ir}\delta_{js} - \delta_{is}\delta_{jr})^2$$

$$+ 2c \sum_{1\le r<s\le n}\sum_{1\le i<j\le n} (\delta_{ir}\delta_{js} - \delta_{is}\delta_{jr})g([A_r, A_s], e_i, e_j)$$

$$= \frac{n^2(n-1)^2}{4}\rho_N^2 + \frac{n(n-1)}{2}c^2 + \frac{c}{2}\|h\|^2.$$

Remark In particular, in the case of three-dimensional Lagrangian submanifold in complex space form:

$$9(\rho^\perp)^2 = 9\rho_N^2 + 3c^2 + \frac{c}{2}\|h\|^2.$$

Corollary 2.42 ([79]) *For a three-dimensional Lagrangian submanifold of a complex space form,*

$$(\rho^{\perp})^2 \leq (\|H\|^2 - \rho + c)^2 + \frac{c}{2}\left(\|H\|^2 - \frac{2}{3}\rho\right) + \frac{2}{3}c^2.$$

Proof It follows from Theorem 2.40, the above remark, and the Gauss equation.

Remark 2.43 We want to mention that the above result was discovered before the DDVV conjecture was solved in the most general setting by Lu [72] and Ge and Tang [60], independently, for submanifolds in Riemannian space forms.

Remark 2.44 Recently, the generalized Wintgen inequality for Lagrangian submanifolds of arbitrary dimension in complex space forms was established by I. Mihai [96].

2.2 *Submanifolds in Sasakian Manifolds*

In [53] we established Chen inequalities for contact slant submanifolds in Sasakian space forms, by using subspaces orthogonal to the Reeb vector field ξ.

We proved the Chen first inequality for contact slant submanifolds in a Sasakian space form. We give the whole proof for illustrating the techniques and the particular choice of the orthonormal basis of the tangent space. We pointed out that we considered plane sections π orthogonal to ξ. It is known that the sectional curvature of a plane section tangent to ξ is 1.

Theorem 2.45 ([53]) *Let M be an $(n = 2k + 1)$-dimensional contact θ-slant submanifold in a $(2m + 1)$-dimensional Sasakian space form $\tilde{M}(c)$. Then we have*

$$\delta_M \leq \frac{n-2}{2}\left\{\frac{n^2}{n-1}\|H\|^2 + \frac{(c+3)(n+1)}{4}\right\} + \frac{(c-1)}{8}\big[3(n-3)\cos^2\theta - 2(n-1)\big]. \tag{2.80}$$

The equality case of the inequality (2.80) *holds at a point $p \in M$ if and only if there exists an orthonormal basis $\{e_1, \ldots, e_n = \xi\}$ of T_pM and an orthonormal basis $\{e_{n+1}, \ldots, e_{2m}, e_{2m+1}\}$ of $T_p^{\perp}M$ such that the shape operators of M in $\tilde{M}(c)$ at p have the following forms:*

$$A_{n+1} = \begin{pmatrix} a & 0 & 0 & . & . & & . & 0 \\ 0 & b & 0 & . & . & & . & 0 \\ 0 & 0 & & & \mu I_{n-2} & & & \end{pmatrix}, \quad a + b = \mu, \tag{2.81}$$

$$A_r = \begin{pmatrix} h_{11}^r & h_{12}^r & 0 \, . \, . & . \, 0 \\ h_{12}^r & -h_{11}^r & 0 \, . \, . & . \, 0 \\ 0 & 0 & \mathbf{0}_{n-2} & \end{pmatrix}, \quad r \in \{n+2, \ldots, 2m+1\}. \tag{2.82}$$

Proof Since $\tilde{M}(c)$ is a Sasakian space form, then we have

$$\tilde{R}(X, Y, Z, W) = \frac{c+3}{4}\{-g(Y, Z)g(X, W) + g(X, Z)g(Y, W)\} \tag{2.83}$$

$$+ \frac{c-1}{4}\{-\eta(X)\eta(Z)g(Y, W) + \eta(Y)\eta(Z)g(X, W) - g(X, Z)\eta(Y)g(\xi, W)$$

$$+ g(Y, Z)\eta(X)g(\xi, W) - g(\phi Y, Z)g(\phi X, W) + g(\phi X, Z)g(\phi Y, W)$$

$$+ 2g(\phi X, Y)g(\phi Z, W)\},$$

for any $X, Y, Z, W \in \Gamma(TM)$.

Let $p \in M$ and $\{e_1, \ldots, e_n = \xi\}$ be an orthonormal basis of T_pM and $\{e_{n+1}, \ldots, e_{2m}, e_{2m+1}\}$ an orthonormal basis of $T_p^{\perp}M$. For $X = Z = e_i, Y = W = e_j, \forall i, j \in \{1, \ldots, n\}$. From Eq. (2.83), it follows that

$$\tilde{R}(e_i, e_j, e_i, e_j) = \frac{c+3}{4}(-n + n^2) + \frac{c-1}{4}\left\{-2(n-1) + 3\sum_{i,j=1}^{n} g^2(\phi e_i, e_j)\right\}. \tag{2.84}$$

Let $M \subset \tilde{M}(c)$ be a contact θ-slant submanifold, $\dim M = n = 2k+1$.
For $X \in \Gamma(TM)$, we put

$$\phi X = PX + FX, \quad PX \in \Gamma(TM), \ FX \in \Gamma(T^{\perp}M).$$

Let $p \in M$ and $\{e_1, \ldots, e_n = \xi\}$ be an orthonormal basis of T_pM, with

$$e_1, e_2 = \frac{1}{\cos\theta}Pe_1, \ldots, e_{2k} = \frac{1}{\cos\theta}Pe_{2k-1}, e_{2k+1} = \xi.$$

We have

$$g(\phi e_1, e_2) = g(\phi e_1, \frac{1}{\cos\theta}Pe_1) = \frac{1}{\cos\theta}g(\phi e_1, Pe_1)$$

$$= \frac{1}{\cos\theta}g(Pe_1, Pe_1) = \cos\theta$$

and, in the same way,

$$g^2(\phi e_i, e_{i+1}) = \cos^2\theta;$$

then

$$\sum_{i,j=1}^{n} g^2(\phi e_i, e_j) = (n-1)\cos^2\theta.$$

The relation (2.84) implies that

$$\tilde{R}(e_i, e_j, e_i, e_j) = \frac{c+3}{4}(n^2-n) + \frac{c-1}{4}\big[3(n-1)\cos^2\theta - 2(n-1)\big]. \tag{2.85}$$

Denoting

$$\|h\|^2 = \sum_{i,j=1}^{n} g(h(e_i, e_j), h(e_i, e_j)), \tag{2.86}$$

the relation (2.85) implies that

$$\frac{c+3}{4}n(n-1)+\frac{c-1}{4}[3(n-1)\cos^2\theta-2n+2] = 2\tau - n^2\|H\|^2+\|h\|^2, \tag{2.87}$$

or equivalently,

$$2\tau = n^2\|H\|^2 - \|h\|^2 + \frac{c+3}{4}n(n-1)+\frac{c-1}{4}[3(n-1)\cos^2\theta-2n+2]. \tag{2.88}$$

If we put

$$\varepsilon = 2\tau - \frac{n^2}{n-1}(n-2)\|H\|^2 - \frac{c+3}{4}n(n-1) - \frac{c-1}{4}[3(n-1)\cos^2\theta - 2n+2], \tag{2.89}$$

we obtain

$$n^2\|H\|^2 = (n-1)(\varepsilon + \|h\|^2). \tag{2.90}$$

Let $p \in M$, $\pi \subset T_pM$, $\dim\pi = 2$, $\pi = sp\{e_1, e_2\}$. We take $e_{n+1} = \frac{H}{\|H\|}$. The relation (2.90) becomes

$$\left(\sum_{i=1}^{n} h_{ii}^{n+1}\right)^2 = (n-1)\left\{\sum_{i,j=1}^{n}\sum_{r=n+1}^{2m+1}(h_{ij}^r)^2 + \varepsilon\right\},$$

or equivalently,

$$\left(\sum_{i=1}^{n} h_{ii}^{n+1}\right)^2 = (n-1)\left\{\sum_{i=1}^{n}\big[(h_{ii}^{n+1})^2\big] + \sum_{i\neq j}(h_{ij}^{n+1})^2 + \sum_{r=n+2}^{2m+1}\sum_{i,j=1}^{n}(h_{ij}^r)^2 + \varepsilon\right\}. \tag{2.91}$$

By using the algebraic Chen's lemma, we derive from (2.91)

$$2h_{11}^{n+1}h_{22}^{n+1} \geq \sum_{i\neq j}(h_{ij}^{n+1})^2 + \sum_{i,j=1}^{n}\sum_{r=n+2}^{2m+1}(h_{ij}^r)^2 + \varepsilon. \tag{2.92}$$

From the Gauss equation for $X = Z = e_1$, $Y = W = e_2$, we obtain

$$K(\pi) = \frac{c+3}{4} + 3\cos^2\theta \cdot \frac{c-1}{4} + \sum_{r=n+1}^{2m+1}\left[h_{11}^r h_{22}^r - (h_{12}^r)^2\right]$$

$$\geq \frac{c+3}{4} + 3\cos^2\theta \cdot \frac{c-1}{4} + \frac{1}{2}\sum_{i\neq j}(h_{ij}^{n+1})^2 + \frac{1}{2}\sum_{i,j=1}^{n}\sum_{r=n+2}^{2m+1}(h_{ij}^r)^2 + \frac{\varepsilon}{2}$$

$$+ \sum_{r=n+2}^{2m+1} h_{11}^r h_{22}^r - \sum_{r=n+1}^{2m+1}(h_{12}^r)^2$$

$$= \frac{c+3}{4} + 3\cos^2\theta \cdot \frac{c-1}{4} + \frac{1}{2}\sum_{i\neq j}(h_{ij}^{n+1})^2 + \frac{1}{2}\sum_{r=n+2}^{2m+1}\sum_{i,j>2}(h_{ij}^r)^2$$

$$+ \frac{1}{2}\sum_{r=n+2}^{2m+1}(h_{11}^r + h_{22}^r)^2 + \sum_{j>2}\left[(h_{1j}^{n+1})^2 + (h_{2j}^{n+1})^2\right] + \frac{\varepsilon}{2}$$

$$\geq \frac{c+3}{4} + 3\cos^2\theta \cdot \frac{c-1}{4} + \frac{\varepsilon}{2},$$

or equivalently,

$$K(\pi) \geq \frac{c+3}{4} + 3\cos^2\theta \cdot \frac{c-1}{4} + \frac{\varepsilon}{2}. \tag{2.93}$$

Then

$$\inf K \geq \frac{c+3}{4} + 3\cos^2\theta \cdot \frac{c-1}{4} + \tau \tag{2.94}$$

$$- \left\{\frac{c+3}{8}(n^2-n) + \frac{c-1}{8}[3(n-1)\cos^2\theta - 2n + 2]\right\} - \frac{n^2(n-2)}{2(n-1)}\|H\|^2.$$

The last relation implies that

$$\delta_M \leq \frac{n-2}{2}\left\{\frac{n^2}{n-1}\|H\|^2 + \frac{(c+3)(n+1)}{4}\right\}$$
$$+\frac{(c-1)}{8}\left[3(n-3)\cos^2\theta - 2(n-1)\right],$$

where δ_M is the Chen first invariant.

This relation represents the inequality to prove.

The case of equality at a point $p \in M$ holds if and only if it achieves the equality in the previous inequality, and we have the equality in the lemma, i.e.,

$$\begin{cases} h_{ij}^{n+1} = 0, \quad \forall i \neq j, i, j > 2, \\ h_{ij}^r = 0, \quad \forall i \neq j, i, j > 2, r = n+1, \ldots, 2m+1, \\ h_{11}^r + h_{22}^r = 0, \quad \forall r = n+2, \ldots, 2m+1, \\ h_{1j}^{n+1} = h_{2j}^{n+1} = 0, \quad \forall j > 2, \\ h_{11}^{n+1} + h_{22}^{n+1} = h_{33}^{n+1} = \ldots = h_{nn}^{n+1}. \end{cases}$$

We may choose $\{e_1, e_2\}$ such that $h_{12}^{n+1} = 0$, and we denote $a = h_{11}^r, b = h_{22}^r, \mu = h_{33}^{n+1} = \ldots = h_{nn}^{n+1}$. It follows that the shape operators take the desired forms.

Corollary 2.46 ([53]) *Let M be an $(n = 2k+1)$-dimensional invariant submanifold in a $(2m+1)$-dimensional Sasakian space form $\tilde{M}(c)$. Then we have:*

$$\delta_M \leq \frac{(c+3)(n-2)(n+1)}{8} + \frac{(c-1)(n-7)}{8}.$$

Corollary 2.47 ([53]) *Let M be an n-dimensional anti-invariant submanifold in a $(2m+1)$-dimensional Sasakian space form $\tilde{M}(c)$. Then we have*

$$\delta_M \leq \frac{n-2}{2}\left\{\frac{n^2}{n-1}\|H\|^2 + \frac{(c+3)(n+1)}{4}\right\} - \frac{(c-1)(n-1)}{4}.$$

We generalized Theorem 2.45, using Chen invariants $\delta(n_1, \ldots, n_k)$.

We notice that we consider only subspaces orthogonal to ξ.

Theorem 2.48 ([53]) *Let M be an $(n = 2k+1)$-dimensional contact θ-slant submanifold in a $(2m+1)$-dimensional Sasakian space form $\tilde{M}(c)$. Then we have*

$$\delta(n_1, \ldots, n_k) \leq d(n_1, \ldots, n_k)\,\|H\|^2 + b(n_1, \ldots, n_k)\frac{c+3}{8} \tag{2.95}$$
$$+\frac{c-1}{8}\left\{3(n-1)\cos^2\theta - 6\sum_{j=1}^{k} m_j \cos^2\theta\right\},$$

where $m_j = \left[\frac{n_j}{2}\right]$, $\forall j \in \{1, \ldots, k\}$.

The proof is based on the following:

Lemma 2.49 ([53]) *Let* M *be an* $(n = 2k+1)$*-dimensional contact* θ*-slant submanifold in a* $(2m+1)$*-dimensional Sasakian space form* $\tilde{M}(c)$*. Let* $n_1, \ldots, n_k$ *be integers* ≥ 2 *satisfying* $n_1 < n$ *and* $n_1 + \ldots + n_k \leq n$*. For* $p \in M$*, let* $L_j \subset T_pM$ *be a subspace of* T_pM*,* $\dim L_j = n_j$*,* $\forall j \in \{1, \ldots, k\}$*. Then we have*

$$\begin{aligned}\tau - \sum_{j=1}^{k} \tau(L_j) &\leq d(n_1, \ldots, n_k)\, \|H\|^2 \\ &+ \left\{\frac{c+3}{8} n(n-1) + \frac{c-1}{8}(3\,\|P\|^2 - 2n + 2)\right\} \\ &- \sum_{j=1}^{k} \left\{\frac{c+3}{8} n_j(n_j - 1) + \frac{c-1}{4} 3\Psi(L_j)\right\},\end{aligned} \tag{2.96}$$

where $\Psi(L) = \sum_{1 \leq i < j \leq r} g^2(Pu_i, u_j)$ *and* $\{u_1, \ldots, u_r\}$ *is an orthonormal basis of the* r*-dimensional subspace* L *of* T_pM*.*

This lemma is a contact version of a lemma from [30].

Corollary 2.50 ([53]) *Let* M *be an* $(n = 2k+1)$*-dimensional invariant submanifold in a* $(2m+1)$*-dimensional Sasakian space form* $\tilde{M}(c)$*. Then we have*

$$\delta(n_1, \ldots, n_k) \leq b(n_1, \ldots, n_k)\frac{c+3}{8} + \frac{c-1}{8}\left\{3(n-1) - 6\sum_{j=1}^{k} m_j\right\},$$

where $n_j = 2m_j + \varphi_j$, $\varphi_j \in \{0, 1\}$, $\forall j \in \{1, \ldots, k\}$.

Corollary 2.51 ([53]) *Let* M *be an* $(n = 2k+1)$*-dimensional anti-invariant submanifold in a* $(2m+1)$*-dimensional Sasakian space form* $\tilde{M}(c)$*. Then we have*

$$\delta(n_1, \ldots, n_k) \leq d(n_1, \ldots, n_k)\, \|H\|^2 + b(n_1, \ldots, n_k)\frac{c+3}{8}.$$

Next we present the inequalities obtained for the new defined Chen-type invariants for invariant submanifolds in Sasakian space forms.

In [34], B.Y. Chen introduced a series of Riemannian invariants on Kaehler manifolds and proved sharp estimates of these invariants for Kaehler submanifolds in complex space forms in terms of the main extrinsic invariant, namely the squared mean curvature.

It is well known that the Sasakian manifolds are the odd version of Kaehler manifolds, and the geometry studying Sasakian manifolds, i.e., *contact geometry*, is an important field of differential geometry.

In [90] we defined analogous Chen invariants for Sasakian manifolds and obtained inequalities involving these invariants for invariant submanifolds in Sasakian space forms.

It is known that any invariant submanifold of a Sasakian manifold is Sasakian. In this respect, we considered that is interesting to study the behavior of invariant submanifolds of Sasakian manifolds from this point of view, of Riemannian invariants, and, more precisely, corresponding Chen-like invariants to those introduced by B.Y. Chen in [34].

In this study of such submanifolds (we must observe that the dimension of the submanifold should be ≥ 5) in Sasakian space forms, we dealt with the notion of totally real plane section (similar to that defined by Chen in Kaehler geometry); we need to impose the condition that the plane section must be orthogonal to the Reeb vector field ξ.

We estimated the sectional curvature of totally real plane sections of an invariant submanifold in terms of the ϕ-sectional curvature of the embedding Sasakian space form; the characterization of the equality case is given.

We defined a series of Chen-like invariants δ_k^r on any Sasakian manifold. By using the above estimate of the sectional curvature of totally real plane sections, we obtained sharp inequalities for these invariants for invariant submanifolds of a Sasakian space form.

Also, we derived characterizations of the equality cases in terms of the shape operators and give one example which shows that the inequality fails for $k \geq 4$.

We recall important results about invariant submanifolds in Sasakian manifolds [122].

Proposition 2.52 *Every invariant submanifold of a Sasakian manifold is a Sasakian manifold.*

Proposition 2.53 *Every invariant submanifold of a Sasakian manifold is minimal.*

Proposition 2.54 *If the second fundamental form of an invariant submanifold M^n of a Sasakian space form $\widetilde{M}^{2m+1}(c)$ is parallel, then M^n is totally geodesic.*

Proposition 2.55 *Let M^n be an invariant submanifold of a Sasakian space form $\widetilde{M}^{2m+1}(c)$with ϕ-sectional curvature c. Then M^n is totally geodesic if and only if M^n has constant ϕ-sectional curvature c.*

We put $2q = 2m + 1 - n$ and choose $\{e_{n+1}, \dots, e_{n+q}, e_{n+q+1} = \phi e_{n+1}, \dots, e_{2m+1} = \phi e_{n+q}\}$ an orthonormal normal frame. Then the shape operators $A_\alpha = A_{e_{n+\alpha}}$ and $A_{\alpha^*} = A_{e_{n+q+\alpha}}, \alpha, \alpha^* = \overline{1, q}$, of an invariant submanifold M^n in a Sasakian manifold $\widetilde{M}^{2m+1}$ take the forms:

$$A_\alpha = \begin{pmatrix} A'_\alpha & A''_\alpha & 0 \\ A''_\alpha & -A'_\alpha & 0 \\ 0 & 0 & 0 \end{pmatrix}, \quad A_{\alpha^*} = \begin{pmatrix} -A''_\alpha & A'_\alpha & 0 \\ A'_\alpha & A''_\alpha & 0 \\ 0 & 0 & 0 \end{pmatrix}, \tag{2.97}$$

where A'_α and A''_α are $n \times n$ matrices.

We recall now two important examples of invariant submanifolds.

Example 2.56 Let S^{2m+1} be a unit sphere with a standard Sasakian structure. An odd-dimensional unit sphere $S^{2n+1}(n < m)$ with induced structure is a totally geodesic Sasakian submanifold of S^{2m+1}. Obviously the Sasakian space form $\mathbf{R}^{2n+1}(-3)$ in $\mathbf{R}^{2m+1}(-3)$ is a totally geodesic Sasakian submanifold.

Example 2.57 The circle bundle (Q^n, S^1) over a hyperquadric in $\mathbf{C}P^{n+1}$ is a Sasakian submanifold of S^{2n+3} which is an η-Einstein manifold.

Let M^{2n+1} be a Sasakian manifold. For each real number k and $p \in M^{2n+1}$, we define an invariant δ_k^r by $\delta_k^r(p) = \tau(p) - k \inf K^r(p)$, where $\inf K^r(p) = \inf_{\pi^r}\{K(\pi^r)\}$ and π^r runs over all totally real plane sections in T_pM^{2n+1} (i.e., $\phi(\pi^r)$ is perpendicular to π^r).

The next theorem gives an inequality between the infimum of K^r (intrinsic invariant) of an invariant submanifold and the ϕ-sectional curvature of the Sasakian space form (extrinsic invariant), i.e., the embedding space; the characterization of the equality case is given.

Theorem 2.58 ([90]) *For any invariant submanifold M^n in a Sasakian space form $\widetilde{M}^{2m+1}(c)$, we have*

$$\inf K^r \leq \frac{c+3}{4}. \tag{2.98}$$

The equality case holds if and only if M^n is a totally geodesic submanifold.

Proof By a ϕ-sectional curvature $H(X)$ of M^n with respect to a unit tangent vector X orthogonal to ξ, we mean the sectional curvature $K(X, \phi X)$ spanned by the vectors X and ϕX. Let $K(X, Y)$ be the sectional curvature determined by orthonormal vectors X and Y, with X,Y orthogonal to ξ, $g(X, \phi Y) = 0$. Then we have (see [13], p. 111)

$$K(X, Y) + K(X, \phi Y) = \frac{1}{4}[(H(X + \phi Y) + H(X - \phi Y)$$

$$+ H(X + Y) + H(X - Y) - H(X) - H(Y) + 6].$$

Let T^1M^n denote the unit sphere bundle of M^n consisting of all unit tangent vectors on M^n. For each $x \in M^n$, we put

$$W_x = \{(X,Y);\ X,Y \in T_x^1 M^n,\ g(X,\xi) = g(Y,\xi) = g(X,Y) = g(X,\phi Y) = 0\}.$$

Then W_x is a closed subset of $T_x^1 M^n \times T_x^1 M^n$, and it is easy to verify that if $\{X, Y\}$ spans a totally real plane section, then both $\{X + \phi Y, X - \phi Y\}$ and $\{X + Y, X - Y\}$ also span totally real plane sections. We define a function $\hat{H} : W_x \to \mathbf{R}$ by

$$\hat{H}(X,Y) = H(X) + H(Y),\ (X,Y) \in W_x.$$

Suppose that (X_m, Y_m) is a point in W_x such that $\hat{H}$ attains an absolute maximum value, say m_x, at (X_m, Y_m). It follows that

$$K(X_m, Y_m) + K(X_m, \phi Y_m) \leq \frac{1}{4}[\hat{H}(X_m, Y_m) + 6].$$

On the other hand, it is known that $H(X) \leq c$ (as in the Kaehler case, see [38]). Thus, from the previous relation, we obtain

$$K(X_m, Y_m) + K(X_m, \phi Y_m) \leq \frac{c+3}{2},$$

which implies the inequality (2.98).

For the equality case the proof is similar to the proof of Theorem 1 from [34].

Remark It is well known that the sectional curvature of a plane section which contains the vector ξ is equal to 1, i.e., $K(X,\xi) = 1$; thus we have considered only the case when X and Y are both orthogonal to ξ.

Also we obtained an inequality for δ_k^r of an invariant submanifold of a Sasakian space form and characterize the equality case for $k < 4$ (the submanifold is then totally geodesic) and $k = 4$ (in terms of the shape operator). For $k > 4$ the inequality fails.

Theorem 2.59 ([90]) *For any invariant submanifold M^n in a Sasakian space form $\widetilde{M}^{2m+1}(c)$, the following statements hold:*

1. *For each $k \in (-\infty, 4]$, δ_k^r satisfies*

$$\delta_k^r \leq [n(n-1) - 2k]\frac{c+3}{8} + (n-1)\frac{c-1}{8}. \tag{2.99}$$

2. *Inequality (2.99) fails for every $k > 4$.*
3. *$\delta_k^r = [n(n-1) - 2k]\frac{c+3}{8} + (n-1)\frac{c-1}{8}$ holds for some $k \in (-\infty, 4)$ if and only if M^n is a totally geodesic submanifold of $\widetilde{M}^{2m+1}(c)$.*
4. *The invariant submanifold M^n satisfies*

$$\delta_4^r = [n(n-1) - 8]\frac{c+3}{8} + (n-1)\frac{c-1}{8}$$

at a point $p \in M^n$ *if and only if there exists an orthonormal basis*

$$\big\{e_1, e_2 = \phi e_1, e_3, e_4 = \phi e_3, \ldots, e_{2k-1}, e_{2k} = \phi e_{2k-1},$$
$$e_{2k+1} = \xi, e_{n+1}, \ldots, e_{2m+1}\big\}$$

of $\widetilde{M}^{2m+1}(c)$ *such that, with respect to this basis, the shape operator of* M^n *takes the forms* (2.97), *with*

$$A'_\alpha = \begin{pmatrix} a_\alpha & b_\alpha & 0 \\ b_\alpha & -a_\alpha & 0 \\ 0 & 0 & 0 \end{pmatrix}, \quad A''_\alpha = \begin{pmatrix} a^*_\alpha & b^*_\alpha & 0 \\ b^*_\alpha & -a^*_\alpha & 0 \\ 0 & 0 & 0 \end{pmatrix},$$

where $a_\alpha, b_\alpha, a_{\alpha^*}, b_{\alpha^*}$ *are real numbers.*

In [91] we proved a Chen inequality involving the scalar curvature and a Chen-Ricci inequality for special contact slant submanifolds of Sasakian space forms, as the contact versions of the inequalities obtained by the first author in [80] and by both authors in [89], respectively.

The class of slant submanifolds of almost contact metric manifolds was introduced by A. Lotta [71] and studied by many authors [19]. In [98] the authors defined special contact slant submanifolds of Sasakian space forms and proved the minimality of such submanifolds satisfying the equality case of a Chen-Ricci inequality, identically.

A submanifold M tangent to ξ in a Sasakian manifold is called a *contact* θ*-slant submanifold* [19] if for any $p \in M$ and any $X \in T_pM$ linearly independent on ξ, the angle between φX and T_pM is a constant θ, called the *slant angle* of M.

A *proper contact* θ*-slant submanifold* is a contact slant submanifold which is neither invariant nor anti-invariant, i.e., $\theta \neq 0$ and $\theta \neq \frac{\pi}{2}$.

A proper contact θ-slant submanifold is a *special contact* θ*-slant submanifold* [98] if

$$(\nabla_X T)Y = \cos^2\theta[g(X,Y)\xi - \eta(Y)\xi], \; \forall X, Y \in \Gamma TM,$$

where TX is the tangential component of φX for any vector field X tangent to M.

We denote $\|T\|^2 = \sum_{i,j=1}^n g^2(Te_i, e_j)$, where $\{e_1, \ldots, e_n\}$ is an orthonormal basis of T_pM, $p \in M$.

We remark that any three-dimensional proper contact slant submanifold of a Sasakian manifold is a special contact slant submanifold [19]. Other examples can be found in [19].

B.Y. Chen [32] proved a sharp estimate of the squared mean curvature in terms of the scalar curvature for Kaehlerian slant submanifolds in complex space forms.

In [80] we generalized the abovementioned inequality for purely real submanifolds with P parallel with respect to the Levi-Civita connection (as usual, we denote by J the standard complex structure on the ambient complex space form and by PX

the tangential component of JX, for any tangent vector field X to M) (see Sect. 2.1, Theorem 2.5).

On the other hand, B.Y. Chen [29] proved an estimate of the squared mean curvature of an n-dimensional submanifold M in a real space form $\tilde{M}(c)$ of constant sectional curvature c in terms of its Ricci curvature. For any unit tangent vector X at $p \in M$, one has

$$\mathrm{Ric}(X) \leq (n-1)c + \frac{n^2}{4} \|H\|^2 .$$

The above inequality is known as the *Chen-Ricci* inequality.

The same inequality holds for Lagrangian submanifolds in a complex space form $\tilde{M}(4c)$ as well (see [31]).

S. Deng [55] improved the Chen-Ricci inequality for Lagrangian submanifolds in complex space forms (see Sect. 2.1).

The Whitney 2-sphere in $\mathbf{C}^2$ is a nontrivial example of a Lagrangian submanifold which satisfies the equality case of the improved Chen-Ricci inequality identically.

Recall that in [89] we extended Deng's inequality to Kaehlerian slant submanifolds in complex space forms (see Sect. 2.1, Theorem 2.35).

A nontrivial example of a slant surface satisfying the equality case identically is given in the same paper [89].

We obtained corresponding results for special contact slant submanifolds in Sasakian space forms, more precisely the following Chen inequality for the scalar curvature.

Theorem 2.60 ([91]) *Let M be an $(n+1)$-dimensional special contact slant submanifold of a $(2n+1)$-dimensional Sasakian space form $\tilde{M}(c)$. Then*

$$\|H\|^2 \geq \frac{2(n+2)}{(n-1)\,(n+1)}\tau - \frac{n(n+2)}{(n-1)(n+1)} \cdot \frac{c+3}{4} \tag{2.100}$$
$$- \frac{n(n+2)}{(n-1)\,(n+1)^2}\left(3\cos^2\theta - 2\right)\frac{c-1}{4} + \frac{n}{(n+1)^2}\sin^2\theta.$$

The equality holds at any point $p \in M$ if and only if there exists a real function μ on M such that the second fundamental form satisfies the relations

$$h(e_1, e_1) = 3\mu e_1^*,\ h(e_2, e_2) = \ldots, = h(e_n, e_n) = \mu e_1^*,$$

$$h(e_i, e_j) = \mu e_i^*,\ h(e_j, e_k) = 0\,(2 \leq j \neq k \neq n),$$

with respect to a suitable orthonormal frame $\{e_0 = \xi, e_1, \ldots, e_n\}$ on M, where $e_k^* = \frac{1}{\sin\theta} N e_k$, $k \in \{1, \ldots, n\}$.

A submanifold M whose second fundamental form satisfies the above relations is called a *H-umbilical* submanifold.

Remark In particular, for $c = -3$ and $\theta = \frac{\pi}{2}$ (i.e., M is anti-invariant submanifold) we find a result from [15].

Corollary ([15]) *Let M be an $(n+1)$-dimensional anti-invariant submanifold of the Sasakian space form $\mathbf{R}^{2n+1}$. Then, at any point $p \in M$, the squared mean curvature and the scalar curvature satisfy the inequality*

$$\|H\|^2 \geq \frac{2(n+2)}{(n-1)(n+1)}\tau.$$

Moreover, the equality holds at any point $p \in M$ if and only if there exists a real function μ on M such that the second fundamental form satisfies the relations

$$h(e_1, e_1) = 3\mu\varphi e_1,\ h(e_2, e_2) = \ldots, = h(e_n, e_n) = \mu\varphi e_1,$$

$$h(e_i, e_j) = \mu\varphi e_i,\ h(e_j, e_k) = 0\,(2 \leq j \neq k \neq n),$$

with respect to a suitable orthonormal frame $\{e_0 = \xi, e_1, \ldots, e_n\}$ of T_pM.

A nontrivial example of an anti-invariant submanifold in the Sasakian space form $\mathbf{R}^{2n+1}$ which satisfies the equality case of the above inequality identically is the Riemannian product of the Whitney n-sphere and the real line $\mathbf{R}$.

On the other hand, I. Mihai [93] proved Chen-Ricci inequalities for submanifolds in Sasakian space forms.

Theorem 2.61 ([93]) *Let M be an n-dimensional C-totally real submanifold of a $(2m+1)$-dimensional Sasakian space form $\tilde{M}(c)$. Then, for each unit vector $X \in T_pM$, we have*

$$Ric(X) \leq \frac{1}{4}[(n-1)(c+3) + n^2\|H\|^2].$$

Theorem 2.62 ([93]) *Let $\tilde{M}(c)$ be a $(2m+1)$-dimensional Sasakian space form and M an n-dimensional submanifold tangent to ξ. Then, for each unit vector $X \in T_pM$ orthogonal to ξ, we have*

$$Ric(X) \leq \frac{1}{4}[(n-1)(c+3) + 3(||TX||^2 - 2)(c-1) + n^2\|H\|^2].$$

I. Mihai and I.N. Rădulescu [99] improved the inequality from Theorem 2.61 for Legendrian submanifolds in Sasakian space forms.

Theorem 2.63 ([99]) *Let M^n be an n-dimensional Legendrian submanifold in a Sasakian space form $\tilde{M}^{2n+1}(c)$ of constant φ-sectional curvature c. Then, for any unit tangent vector X to M^n, we have*

$$Ric(X) \le \frac{n-1}{4}(c+3+n||H||^2).$$

The equality sign holds identically if and only if either:

(i) M^n *is totally geodesic or*
(ii) $n=2$ *and* M^2 *is a H-umbilical Legendrian surface with* $\lambda = 3\mu$.

In the following we improved the inequality from Theorem 2.62 for special contact slant submanifolds in Sasakian space forms.

Let M be an $(n+1)$-dimensional special contact slant submanifold of a $(2n+1)$-dimensional Sasakian space form $\widetilde{M}(c)$. We will take an orthonormal basis of T_pM, respectively $T_p^{\perp}M$, in the same manner as in the previous section. For a contact θ-slant submanifold $\sum_{j=2}^{n} g^2(Te_1, e_j) = cos^2\theta$.

By taking $X = Z = e_1$, $Y = W = e_j$, $j = 2, \ldots, n$, in the expression of the curvature tensor $\widetilde{R}$ of the Sasakian space form $\widetilde{M}(c)$, we obtain

$$\tilde{R}(e_1, e_j, e_1, e_j) = \frac{c+3}{4}\left[g(e_1, e_1)g(e_j, e_j) - g(e_j, e_1)g(e_1, e_j)\right] \tag{2.101}$$

$$+\frac{c-1}{4}\left[-\eta(e_1)\eta(e_1)g(e_j, e_j) + \eta(e_j)\eta(e_1)g(e_1, e_j)\right]$$

$$-g(e_1, e_1)\eta(e_j)\eta(e_j) + g(e_j, e_1)\eta(e_1)\eta(e_j)$$

$$-g(\phi e_j, e_1)g(\phi e_1, e_j) + g(\phi e_1, e_1)g(\phi e_j, e_j) + 2g(\phi e_1, e_j)g(\phi e_1, e_j)$$

$$= \frac{c+3}{4} + \frac{3}{4}g^2(\phi e_1, e_j)(c-1).$$

Then

$$\sum_{j=2}^{n} \tilde{R}\left(e_1, e_j, e_1, e_j\right) + \tilde{R}\left(e_1, e_0, e_1, e_0\right)$$

$$= n\frac{c+3}{4} + \frac{3}{4}\sum_{j=2}^{n} g^2(\phi e_1, e_j)(c-1) + 1 = n\frac{c+3}{4} + \frac{3}{4}(c-1)\cos^2\theta + 1.$$

We consider $e_1 = X$. From the Gauss equation

$$\mathrm{Ric}(X) = n\frac{c+3}{4} + \frac{3}{4}(c-1)\cos^2\theta + 1 +$$

$$+\sum_{r=1}^{n}\left[h_{11}^r h_{00}^r - (h_{10}^r)^2\right] + \sum_{j=2}^{n}\left[h_{11}^r h_{jj}^r - \left(h_{1j}^r\right)^2\right].$$

But $e_0 = \xi$ and

$$h^r_{00} = g\left(h\left(e_0, e_0\right), e^*_r\right) = g\left(h\left(\xi, \xi\right), e^*_r\right) = g\left(\tilde{\nabla}_\xi \xi, e^*_r\right) = 0,$$

because $\tilde{\nabla}_\xi \xi = -\phi\xi = 0$.

Also,

$$\sum_{r=1}^{n} \left(h^r_{10}\right)^2 = \sum_{r=1}^{n} g^2\left(h\left(e_1, e_0\right), e^*_r\right) = \sum_{r=1}^{n} g^2\left(\tilde{\nabla}_{e_1} e_0, e^*_r\right)$$
$$= \sum_{r=1}^{n} g^2\left(\phi e_1, e^*_r\right) = \sin^2\theta.$$

The relation (2.101) implies

$$\mathrm{Ric}(X) = n\frac{c+3}{4} + \frac{3}{4}(c-1)\cos^2\theta + 1 - \sin^2\theta \tag{2.102}$$
$$+ \sum_{r=1}^{n}\sum_{j=1}^{n}\left[h^r_{11}h^r_{jj} - (h^r_{1j})^2\right].$$

Using the same arguments as in the proof of Theorem 3.3 from [89], we obtain from (2.102)

$$\mathrm{Ric}(X) - n\frac{c+3}{4} - \frac{3}{4}(c-1)\cos^2\theta - \cos^2\theta \le \frac{(n-1)(n+1)}{4}\,\|H\|^2.$$

Therefore we proved the following improved Chen-Ricci inequality.

Theorem 2.64 ([91]) *Let M be an $(n+1)$-dimensional special contact slant submanifold of a $(2n+1)$-dimensional Sasakian space form $\widetilde{M}(c)$. Then, for any unit tangent vector X to M, we have*

$$Ric(X) \le \frac{(n-1)(n+1)}{4}\,\|H\|^2 + n\frac{c+3}{4} + \frac{3}{4}(c-1)\cos^2\theta + \cos^2\theta. \tag{2.103}$$

The equality holds at every point $p \in M$ if and only if either:

(i) M is a totally contact geodesic submanifold, i.e.,

$$h(X, Y) = \eta(X)h(Y, \xi) + \eta(Y)h(X, \xi),$$

for any $X, Y \in \Gamma TM$ or

(ii) $n = 2$ and M is a three-dimensional H-umbilical contact slant submanifold, i.e.,

$$h(e_1, e_1) = 3\mu e_1^*, \ h(e_2, e_2) = \mu e_1^*, \ h(e_1, e_2) = \mu e_2^*,$$

with respect to an orthonormal frame $\{e_0 = \xi, e_1, e_2\}$.

Remark The inequality (2.103) is also true for anti-invariant submanifolds in Sasakian space forms.

2.3 Submanifolds with Semi-symmetric Metric (Nonmetric) Connections

In [64], H.A. Hayden introduced the notion of a semi-symmetric metric connection on a Riemannian manifold. K. Yano studied in [121] some properties of a Riemannian manifold endowed with a semi-symmetric metric connection. In [65] and [66], T. Imai found some properties of a Riemannian manifold and a hypersurface of a Riemannian manifold with a semi-symmetric metric connection. Z. Nakao [102] studied submanifolds of a Riemannian manifold with semi-symmetric connections. In [86, 87] we proved Chen inequalities for submanifolds of real space forms endowed with a semi-symmetric metric connection, i.e., relations between the mean curvature associated with the semi-symmetric metric connection, scalar and sectional curvatures, Ricci curvatures, and the sectional curvature of the ambient space. The equality cases are considered.

Let N^{n+p} be an $(n+p)$-dimensional Riemannian manifold and $\widetilde{\nabla}$ a linear connection on N^{n+p}. If the torsion tensor $\widetilde{T}$ of $\widetilde{\nabla}$, defined by

$$\widetilde{T}\left(\widetilde{X}, \widetilde{Y}\right) = \widetilde{\nabla}_{\widetilde{X}}\widetilde{Y} - \widetilde{\nabla}_{\widetilde{Y}}\widetilde{X} - [\widetilde{X}, \widetilde{Y}],$$

for any vector fields $\widetilde{X}$ and $\widetilde{Y}$ on N^{n+p} satisfies

$$\widetilde{T}\left(\widetilde{X}, \widetilde{Y}\right) = \phi(\widetilde{Y})\widetilde{X} - \phi(\widetilde{X})\widetilde{Y},$$

for a 1-form ϕ, then the connection $\widetilde{\nabla}$ is called a *semi-symmetric connection.*

Let g be a Riemannian metric on N^{n+p}. If $\widetilde{\nabla} g = 0$, then $\widetilde{\nabla}$ is called a *semi-symmetric metric connection* on N^{n+p}.

Following [121], a semi-symmetric metric connection $\widetilde{\nabla}$ on N^{n+p} is given by

$$\widetilde{\nabla}_{\widetilde{X}}\widetilde{Y} = \overset{\circ}{\widetilde{\nabla}}_{\widetilde{X}} \widetilde{Y} + \phi(\widetilde{Y})\widetilde{X} - g(\widetilde{X}, \widetilde{Y})P,$$

for any vector fields $\widetilde{X}$ and $\widetilde{Y}$ on N^{n+p}, where $\overset{\circ}{\widetilde{\nabla}}$ denotes the Levi-Civita connection with respect to the Riemannian metric g and P is a vector field defined by $g(P, \widetilde{X}) = \phi(\widetilde{X})$, for any vector field $\widetilde{X}$.

We will consider a Riemannian manifold N^{n+p} endowed with a semi-symmetric metric connection $\widetilde{\nabla}$ and the Levi-Civita connection denoted by $\overset{\circ}{\widetilde{\nabla}}$.

Let M^n be an n-dimensional submanifold of an $(n+p)$-dimensional Riemannian manifold N^{n+p}. On the submanifold M^n we consider the induced semi-symmetric metric connection denoted by ∇ and the induced Levi-Civita connection denoted by $\overset{\circ}{\nabla}$.

Let $\widetilde{R}$ be the curvature tensor of N^{n+p} with respect to $\widetilde{\nabla}$ and $\overset{\circ}{\widetilde{R}}$ the curvature tensor of N^{n+p} with respect to $\overset{\circ}{\widetilde{\nabla}}$. We also denote by R and $\overset{\circ}{R}$ the curvature tensors of ∇ and $\overset{\circ}{\nabla}$, respectively, on M^n.

The Gauss formulas with respect to ∇ and, respectively, $\overset{\circ}{\nabla}$ can be written as

$$\widetilde{\nabla}_X Y = \nabla_X Y + h(X, Y), \quad X, Y \in \chi(M),$$

$$\overset{\circ}{\widetilde{\nabla}}_X Y = \overset{\circ}{\nabla}_X Y + \overset{\circ}{h} (X, Y), \quad X, Y \in \chi(M),$$

where $\overset{\circ}{h}$ is the second fundamental form of M^n in N^{n+p} and h is a $(0, 2)$-tensor on M^n. According to the formula (7) from [102] h is also symmetric.

One denotes by $\overset{\circ}{H}$ the mean curvature vector of M^n in N^{n+p}.

Let $N^{n+p}(c)$ be a real space form of constant sectional curvature c endowed with a semi-symmetric metric connection $\widetilde{\nabla}$.

The curvature tensor $\overset{\circ}{\widetilde{R}}$ with respect to the Levi-Civita connection $\overset{\circ}{\widetilde{\nabla}}$ on $N^{n+p}(c)$ is expressed by

$$\overset{\circ}{\widetilde{R}} (X, Y, Z, W) = c\,\{g(X, W)g(Y, Z) - g(X, Z)g(Y, W)\}.$$

Then the curvature tensor $\widetilde{R}$ with respect to the semi-symmetric metric connection $\widetilde{\nabla}$ on $N^{n+p}(c)$ can be written as [66]

$$\begin{aligned}\widetilde{R}(X, Y, Z, W) = \overset{\circ}{\widetilde{R}} (X, Y, Z, W) - \alpha(Y, Z)g(X, W) \\ +\alpha(X, Z)g(Y, W) - \alpha(X, W)g(Y, Z) + \alpha(Y, W)g(X, Z),\end{aligned}$$

for any vector fields $X, Y, Z, W \in \chi(M^n)$, where α is a $(0, 2)$-tensor field defined by

$$\alpha(X, Y) = \left(\overset{\circ}{\widetilde{\nabla}}_X \phi\right) Y - \phi(X)\phi(Y) + \frac{1}{2}\phi(P)g(X, Y), \quad \forall X, Y \in \chi(M).$$

From the last relations it follows that the curvature tensor $\widetilde{R}$ can be expressed as

$$\begin{aligned}\widetilde{R}(X,Y,Z,W) = c\,\{g(X,W)g(Y,Z) - g(X,Z)g(Y,W)\}\\ -\alpha(Y,Z)g(X,W) + \alpha(X,Z)g(Y,W)\\ -\alpha(X,W)g(Y,Z) + \alpha(Y,W)g(X,Z).\end{aligned}$$

Denote by λ the trace of α.

For submanifolds of real space forms endowed with a semi-symmetric metric connection, we established the following optimal inequality, which we will call the *Chen first inequality*:

Theorem 2.65 ([86]) *Let $M^n, n \geq 3$, be an n-dimensional submanifold of an $(n+p)$-dimensional real space form $N^{n+p}(c)$ of constant sectional curvature c, endowed with a semi-symmetric metric connection $\widetilde{\nabla}$. We have*

$$\tau(x) - K(\pi) \leq (n-2)\left[\frac{n^2}{2(n-1)}\,\|H\|^2 + (n+1)\frac{c}{2} - \lambda\right] - trace\left(\alpha_{|_{\pi^\perp}}\right), \tag{2.104}$$

where π is a 2-plane section of T_xM^n, $x \in M^n$.

Recall the following important result (Proposition 1.2) from [65].

Proposition *The mean curvature H of M^n with respect to the semi-symmetric metric connection coincides with the mean curvature $\overset{\circ}{H}$ of M^n with respect to the Levi-Civita connection if and only if the vector field P is tangent to M^n.*

Remark According to the formula (7) from [102] it follows that $h = \overset{\circ}{h}$ if P is tangent to M^n.

For P tangent to M^n the inequality (2.104) is written as in the following

Corollary 2.66 ([86]) *Under the same assumptions as in Theorem 2.72, if the vector field P is tangent to M^n, then we have*

$$\tau(x) - K(\pi) \leq (n-2)\left[\frac{n^2}{2(n-1)}\left\|\overset{\circ}{H}\right\|^2 + (n+1)\frac{c}{2} - \lambda\right] - trace\left(\alpha_{|_{\pi^\perp}}\right). \tag{2.105}$$

Theorem 2.67 ([86]) *If the vector field P is tangent to M^n, then the equality case of inequality* (2.104) *holds at a point $x \in M^n$ if and only if there exist an orthonormal basis $\{e_1, e_2, \ldots, e_n\}$ of T_xM^n and an orthonormal basis $\{e_{n+1}, \ldots, e_{n+p}\}$ of*

$T_x^{\perp}M^n$ *such that the shape operators of* M^n *in* $N^{n+p}(c)$ *at* x *have the following forms:*

$$A_{e_{n+1}} = \begin{pmatrix} a & 0 & 0 & \cdots & 0 \\ 0 & b & 0 & \cdots & 0 \\ 0 & 0 & \mu & \cdots & 0 \\ \vdots & \vdots & \vdots & \ddots & \vdots \\ 0 & 0 & 0 & \cdots & \mu \end{pmatrix}, \quad a+b=\mu,$$

$$A_{e_{n+i}} = \begin{pmatrix} h_{11}^r & h_{12}^r & 0 & \cdots & 0 \\ h_{12}^r & -h_{11}^r & 0 & \cdots & 0 \\ 0 & 0 & 0 & \cdots & 0 \\ \vdots & \vdots & \vdots & \cdots & \vdots \\ 0 & 0 & 0 & \cdots & 0 \end{pmatrix}, \quad 2 \le i \le p,$$

where $h_{ij}^r = g(h(e_i, e_j), e_r)$, $1 \le i, j \le n$, *and* $n+1 \le r \le n+p$.

We also established a sharp relation between the Ricci curvature in the direction of a unit tangent vector X and the mean curvature H with respect to the semi-symmetric metric connection $\widetilde{\nabla}$.

We denote

$$N(x) = \left\{X \in T_xM^n \mid h(X,Y) = 0,\ \forall Y \in T_xM^n\right\}.$$

Theorem 2.68 ([87]) *Let* M^n, $n \ge 3$, *be an* n*-dimensional submanifold of an* $(n+p)$*-dimensional real space form* $N^{n+p}(c)$ *of constant sectional curvature* c *endowed with a semi-symmetric metric connection* $\widetilde{\nabla}$. *Then:*

(i) For each unit vector X *in* T_xM *we have*

$$\|H\|^2 \ge \frac{4}{n^2}\left[Ric(X) - (n-1)c + \lambda + (n-2)\alpha(X,X)\right]. \tag{2.106}$$

(ii) If $H(x) = 0$, *then a unit tangent vector* X *at* x *satisfies the equality case of* (2.106) *if and only if* $X \in N(x)$.

Corollary 2.69 ([86]) *If the vector field* P *is tangent to* M^n, *then the equality case of inequality* (2.106) *holds identically for all unit tangent vectors at* x *if and only if either* x *is a totally geodesic point or* $n = 2$ *and* x *is a totally umbilical point.*

We stated a relationship between the sectional curvature of a submanifold M^n of a real space form $N^{n+p}(c)$ of constant sectional curvature c endowed with a semi-symmetric metric connection $\widetilde{\nabla}$ and the associated squared mean curvature $\|H\|^2$. Using this inequality, we proved a relationship between the k-Ricci curvature of M^n

(intrinsic invariant) and the squared mean curvature $\|H\|^2$ (extrinsic invariant). We assume that the vector field P is tangent to M^n.

Theorem 2.70 ([86]) *Let M^n, $n \geq 3$, be an n-dimensional submanifold of an $(n+p)$-dimensional real space form $N^{n+p}(c)$ of constant sectional curvature c endowed with a semi-symmetric metric connection $\widetilde{\nabla}$ such that the vector field P is tangent to M^n. Then we have*

$$\|H\|^2 \geq \frac{2\tau}{n(n-1)} - c + \frac{2}{n}\lambda. \tag{2.107}$$

Using Theorem 2.70, we obtain the following:

Theorem 2.71 ([86]) *Let M^n, $n \geq 3$, be an n-dimensional submanifold of an $(n+p)$-dimensional real space form $N^{n+p}(c)$ of constant sectional curvature c endowed with a semi-symmetric metric connection $\widetilde{\nabla}$, such that the vector field P is tangent to M^n. Then, for any integer k, $2 \leq k \leq n$, and any point $p \in M^n$, we have*

$$\|H\|^2(p) \geq \Theta_k(p) - c + \frac{2}{n}\lambda, \tag{2.108}$$

where $\Theta_k(p) = \frac{1}{k-1}\inf_{X,L^k} Ric_{L^k}(X)$, L^k runs over all k-plane sections in T_pM, and X runs over all unit vectors in L^k.

In [88] we continued the study of Chen inequalities for submanifolds in space forms with semi-symmetric metric connections, more precisely Chen inequalities for submanifolds in complex, respectively, Sasakian space forms endowed with semi-symmetric metric connection.

Let N^{2m} be a Kaehler manifold and J the canonical almost complex structure. The sectional curvature of N^{2m} in the direction of an invariant 2-plane section by J is called the *holomorphic sectional curvature*. If the holomorphic sectional curvature is constant $4c$ for all plane sections π of T_xN^{2m} invariant by J for any $x \in N^{2m}$, then N^{2m} is called a *complex space form* and is denoted by $N^{2m}(4c)$. The curvature tensor $\overset{\circ}{\widetilde{R}}$ with respect to the Levi-Civita connection $\overset{\circ}{\widetilde{\nabla}}$ on $N^{2m}(4c)$ is given by

$$\begin{aligned}\overset{\circ}{\widetilde{R}}(X, Y, Z, W) = c[&g(X, W)g(Y, Z) - g(X, Z)g(Y, W) - g(JX, Z)g(JY, W) \\ &+ g(JX, W)g(JY, Z) - 2g(X, JY)g(Z, JW)].\end{aligned} \tag{2.109}$$

If $N^{2m}(4c)$ is a complex space form of constant holomorphic sectional curvature $4c$ with a semi-symmetric metric connection $\widetilde{\nabla}$, then the curvature tensor $\widetilde{R}$ of $N^{2m}(4c)$ can be expressed as

$$\widetilde{R}(X, Y, Z, W) = c[g(X, W)g(Y, Z) - g(X, Z)g(Y, W) - g(JX, Z)g(JY, W) \tag{2.110}$$

$$+ g(JX,W)g(JY,Z) - 2g(X,JY)g(Z,JW)] - \alpha(Y,Z)g(X,W)$$

$$+ \alpha(X,Z)g(Y,W) - \alpha(X,W)g(Y,Z) + \alpha(Y,W)g(X,Z).$$

Let $M^n, n \geq 3$, be an n-dimensional submanifold of a $2m$-dimensional complex space form $N^{2m}(4c)$ of constant holomorphic sectional curvature $4c$. For any tangent vector field X to M^n, we put $JX = PX + FX$, where PX and FX are the tangential and normal components of JX, respectively. We define

$$\|P\|^2 = \sum_{i,j=1}^{n} g^2(Je_i, e_j).$$

Following [3], we denoted $\Theta^2(\pi) = g^2(Pe_1, e_2) = g^2(Je_1, e_2)$, where $\{e_1, e_2\}$ is an orthonormal basis of a 2-plane section π. $\Theta^2(\pi)$ is a real number in $[0, 1]$, independent of the choice of e_1, e_2.

We proved the following:

Theorem 2.72 ([88]) *Let $M^n, n \geq 3$, be an n-dimensional submanifold of a $2m$-dimensional complex space form $N^{2m}(4c)$ of constant holomorphic sectional curvature $4c$, endowed with a semi-symmetric metric connection $\widetilde{\nabla}$. We have*

$$\tau(x) - K(\pi) \leq \frac{n-2}{2}\left[\frac{n^2}{n-1}\|H\|^2 + (n+1)c - 2\lambda\right] - \left[6\Theta^2(\pi) - 3\|P\|^2\right]\frac{c}{2}$$
$$- \mathit{trace}\left(\alpha_{|_{\pi^\perp}}\right),$$

where π is a 2-plane section of T_xM^n, $x \in M^n$.

Remark Because $h = \overset{\circ}{h}$ if U is tangent to M^n (according to the formula (7) from [102]; see also Proposition 1.2 from [65]), the inequality proved in Theorem 2.72 becomes

$$\tau(x) - K(\pi) \leq \frac{n-2}{2}\left[\frac{n^2}{n-1}\left\|\overset{\circ}{H}\right\|^2 + (n+1)c - 2\lambda\right] - \left[6\Theta^2(\pi) - 3\|P\|^2\right]\frac{c}{2}$$
$$- \operatorname{trace}\left(\alpha_{|_{\pi^\perp}}\right).$$

Theorem 2.73 ([88]) *Under the same assumptions as in Theorem 2.72, if the vector field U is tangent to M^n, then the equality case of inequality from Theorem 2.72 holds at a point $x \in M^n$ if and only if there exist an orthonormal basis $\{e_1, e_2, \ldots, e_n\}$ of T_xM^n and an orthonormal basis $\{e_{n+1}, \ldots, e_{2m}\}$ of $T_x^\perp M^n$ such that the shape operators of M^n in $N^{2m}(4c)$ at x have the following forms:*

$$A_{e_{n+1}} = \begin{pmatrix} a & 0 & 0 & \cdots & 0 \\ 0 & b & 0 & \cdots & 0 \\ 0 & 0 & \mu & \cdots & 0 \\ \vdots & \vdots & \vdots & \ddots & \vdots \\ 0 & 0 & 0 & \cdots & \mu \end{pmatrix}, \quad a+b=\mu,$$

$$A_{e_r} = \begin{pmatrix} h^r_{11} & h^r_{12} & 0 & \cdots & 0 \\ h^r_{12} & -h^r_{11} & 0 & \cdots & 0 \\ 0 & 0 & 0 & \cdots & 0 \\ \vdots & \vdots & \vdots & \cdots & \vdots \\ 0 & 0 & 0 & \cdots & 0 \end{pmatrix}, \quad n+2 \le i \le 2m,$$

where we denote $h^r_{ij} = g(h(e_i, e_j), e_r)$, $1 \le i, j \le n$, *and* $n+2 \le r \le 2m$.

We proved relationships between the Ricci curvature of a submanifold M^n of a complex space form $N^{2m}(4c)$ of constant holomorphic sectional curvature, endowed with a semi-symmetric metric connection, and the squared mean curvature $\|H\|^2$, under the assumption that the vector field U is tangent to M^n.

Theorem 2.74 ([88]) *Let* $M^n, n \ge 3$, *be an n-dimensional submanifold of a 2m-dimensional complex space form* $N^{2m}(4c)$ *of constant holomorphic sectional curvature* $4c$ *endowed with a semi-symmetric metric connection* $\widetilde{\nabla}$ *such that the vector field* U *is tangent to* M^n. *Then we have*

$$\|H\|^2 \ge \frac{2\tau}{n(n-1)} + \frac{2}{n}\lambda - c - \frac{3c}{n(n-1)} \|P\|^2. \tag{2.111}$$

Using the above theorem, we obtain the following:

Theorem 2.75 ([88]) *Let* $M^n, n \ge 3$, *be an n-dimensional submanifold of a 2m-dimensional complex space form* $N^{2m}(4c)$ *of constant holomorphic sectional curvature* $4c$ *endowed with a semi-symmetric metric connection* $\widetilde{\nabla}$, *such that the vector field* U *is tangent to* M^n. *Then, for any integer* k, $2 \le k \le n$, *and any point* $x \in M^n$, *we have*

$$\|H\|^2(x) \ge \Theta_k(p) + \frac{2}{n}\lambda - c - \frac{3c}{n(n-1)} \|P\|^2.$$

We considered also submanifolds of Sasakian space forms.

Recall that a $(2m+1)$-dimensional Riemannian manifold (N^{2m+1}, g) is a *Sasakian manifold* if it admits a $(1,1)$-tensor field φ, a vector field ξ, and a 1-form η satisfying

$$\varphi^2 X = -X + \eta(X)\xi, \ \eta(\xi) = 1,$$

$$g(\varphi X, \varphi Y) = g(X, Y) - \eta(X)\eta(Y),\ g(X, \xi) = \eta(X),$$

$$g(X, \phi Y) = d\eta(X, Y),$$

for any vector fields X, Y on TN, and

$$[\phi, \phi] + 2d\eta \otimes \xi = 0,$$

where $[\phi, \phi]$ is the Nijenhuis torsion of ϕ.

A Sasakian manifold with constant φ-sectional curvature c is said to be a *Sasakian space form* and is denoted by $N^{2m+1}(c)$. The curvature tensor $\overset{\circ}{\widetilde{R}}$ with respect to the Levi-Civita connection $\overset{\circ}{\widetilde{\nabla}}$ on $N^{2m+1}(c)$ is expressed by

$$\begin{aligned}\overset{\circ}{\widetilde{R}}(X, Y, Z, W) &= \frac{c+3}{4}[g(X, W)g(Y, Z) - g(X, Z)g(Y, W)] \\ &+ \frac{c-1}{4}[\eta(X)\eta(Z)g(Y, W) - \eta(Y)\eta(Z)g(X, W) \\ &+ \eta(Y)\eta(W)g(X, Z) - \eta(X)\eta(W)g(Y, Z) \\ &+ g(X, \varphi Z)g(\varphi Y, W) - g(Y, \varphi Z)g(\varphi X, W) + 2g(X, \varphi Y)g(\varphi Z, W)],\end{aligned} \tag{2.112}$$

for vector fields X, Y, Z, W on $N^{2m+1}(c)$.

If $N^{2m+1}(c)$ is a $(2m+1)$-dimensional Sasakian space form of constant φ-sectional curvature c endowed with a semi-symmetric metric connection $\widetilde{\nabla}$, it follows that the curvature tensor $\widetilde{R}$ of $N^{2m+1}(c)$ can be expressed as

$$\begin{aligned}\widetilde{R}(X, Y, Z, W) &= \frac{c+3}{4}[g(X, W)g(Y, Z) - g(X, Z)g(Y, W)] \\ &+ \frac{c-1}{4}[\eta(X)\eta(Z)g(Y, W) - \eta(Y)\eta(Z)g(X, W) \\ &+ \eta(Y)\eta(W)g(X, Z) - \eta(X)\eta(W)g(Y, Z) \\ &+ g(X, \varphi Z)g(\varphi Y, W) - g(Y, \varphi Z)g(\varphi X, W) + 2g(X, \varphi Y)g(\varphi Z, W)] \\ &- \alpha(Y, Z)g(X, W) + \alpha(X, Z)g(Y, W) - \alpha(X, W)g(Y, Z) + \alpha(Y, W)g(X, Z).\end{aligned} \tag{2.113}$$

Let $M^n, n \geq 3$, be an n-dimensional submanifold of a $(2m+1)$-dimensional Sasakian space form of constant φ-sectional curvature $N^{n+p}(c)$ of constant sectional curvature c. For any tangent vector field X to M^n, we put $\varphi X = PX + FX$, where PX and FX are tangential and normal components of φX, respectively, and we decompose $\xi = \xi^\top + \xi^\perp$, where $\xi^\top$ and $\xi^\perp$ denote the tangential and normal parts of ξ.

Recall that $\Theta^2(\pi) = g^2(Pe_1, e_2) = g^2(Je_1, e_2)$, where $\{e_1, e_2\}$ is an orthonormal basis of a 2-plane section π, is a real number in $[0, 1]$, independent of the choice of e_1, e_2.

For submanifolds of Sasakian space forms endowed with a semi-symmetric metric connection, we established the following optimal inequality.

Theorem 2.76 ([88]) *Let $M^n, n \geq 3$, be an n-dimensional submanifold of a $(2m+1)$-dimensional Sasakian space form $N^{2m+1}(c)$ of constant ϕ-sectional curvature endowed with a semi-symmetric metric connection $\widetilde{\nabla}$. We have*

$$\tau(x) - K(\pi) \leq (n-2)\left[\frac{n^2}{2(n-1)}\|H\|^2 + (n+1)\frac{c+3}{8} - \lambda\right] + \tag{2.114}$$

$$+\frac{c-1}{8}\left[3\|P\|^2 - 6\Theta^2(\pi) - 2(n-1)\|\xi^\top\|^2 + 2\|\xi_\pi\|^2\right] - trace\left(\alpha_{|_{\pi^\perp}}\right),$$

where π is a 2-plane section of T_xM^n, $x \in M^n$.

Corollary 2.77 ([88]) *Under the same assumptions as in Theorem 2.76, if ξ is tangent to M^n, we have*

$$\tau(x) - K(\pi) \leq (n-2)\left[\frac{n^2}{2(n-1)}\|H\|^2 + (n+1)\frac{c+3}{8} - \lambda\right]$$

$$+\frac{c-1}{8}\left[3\|P\|^2 - 6\Theta^2(\pi) - 2(n-1) + 2\|\xi_\pi\|^2\right] - trace\left(\alpha_{|_{\pi^\perp}}\right).$$

If ξ is normal to M^n, we have

$$\tau(x) - K(\pi) \leq (n-2)\left[\frac{n^2}{2(n-1)}\|H\|^2 + (n+1)\frac{c+3}{8} - \lambda\right] - trace\left(\alpha_{|_{\pi^\perp}}\right).$$

Remark Because $h = \overset{\circ}{h}$, if U is tangent to M^n [102], the inequality (2.114) becomes

$$\tau(x) - K(\pi) \leq (n-2)\left[\frac{n^2}{2(n-1)}\left\|\overset{\circ}{H}\right\|^2 + (n+1)\frac{c+3}{8} - \lambda\right]$$

$$+\frac{c-1}{8}\left[3\|P\|^2 - 6\Theta^2(\pi) - 2(n-1)\|\xi^\top\|^2 + 2\|\xi_\pi\|^2\right] - \mathrm{trace}\left(\alpha_{|_{\pi^\perp}}\right).$$

Theorem 2.78 ([88]) *If the vector field U is tangent to M^n, then the equality case of inequality* (2.114) *holds at a point $x \in M^n$ if and only if there exist an orthonormal basis $\{e_1, e_2, \ldots, e_n\}$ of T_xM^n and an orthonormal basis $\{e_{n+1}, \ldots, e_{n+p}\}$ of $T_x^\perp M^n$ such that the shape operators of M^n in $N^{2m+1}(c)$ at x have the following* *forms:*

$$A_{e_{n+1}} = \begin{pmatrix} a & 0 & 0 & \cdots & 0 \\ 0 & b & 0 & \cdots & 0 \\ 0 & 0 & \mu & \cdots & 0 \\ \vdots & \vdots & \vdots & \ddots & \vdots \\ 0 & 0 & 0 & \cdots & \mu \end{pmatrix}, \quad a + b = \mu,$$

$$A_{e_r} = \begin{pmatrix} h^r_{11} & h^r_{12} & 0 \cdots 0 \\ h^r_{12} & -h^r_{11} & 0 \cdots 0 \\ 0 & 0 & 0 \cdots 0 \\ \vdots & \vdots & \vdots \cdots \vdots \\ 0 & 0 & 0 \cdots 0 \end{pmatrix}, \quad n + 2 \leq i \leq 2m + 1,$$

where we denote $h^r_{ij} = g(h(e_i, e_j), e_r)$, $1 \leq i, j \leq n$, *and* $n + 2 \leq r \leq 2m + 1$.

We also stated a relationship between the sectional curvature of a submanifold M^n of a Sasakian space form $N^{2m+1}(c)$ of constant ϕ-sectional curvature c endowed with a semi-symmetric metric connection $\widetilde{\nabla}$ and the squared mean curvature $\|H\|^2$. Using this inequality, we prove a relationship between the k-Ricci curvature of M^n (intrinsic invariant) and the squared mean curvature $\|H\|^2$ (extrinsic invariant).

Theorem 2.79 ([88]) *Let* $M^n, n \geq 3$, *be an* n*-dimensional submanifold of a* $(2m + 1)$*-dimensional real space form* $N^{2m+1}(c)$ *of constant* ϕ*-sectional curvature* c *endowed with a semi-symmetric metric connection* $\widetilde{\nabla}$ *such that the vector field* U *is tangent to* M^n*. Then we have*

$$\|H\|^2 \geq \frac{2\tau}{n(n-1)} + \frac{2}{n}\lambda - \frac{c+3}{4} - \frac{c-1}{4n(n-1)}\left[-2(n-1)\|\xi^\top\|^2 + \|P\|^2\right]. \tag{2.115}$$

From the above theorem we derived the following:

Theorem 2.80 ([88]) *Let* $M^n, n \geq 3$, *be an* n*-dimensional submanifold of a* $(2m + 1)$*-dimensional Sasakian space form* $N^{2m+1}(c)$ *of constant* φ*-sectional curvature* c *endowed with a semi-symmetric metric connection* $\widetilde{\nabla}$*, such that the vector field* U *is tangent to* M^n*. Then, for any integer* k*,* $2 \leq k \leq n$*, and any point* $x \in M^n$*, we have*

$$\|H\|^2(x) \geq \Theta_k(x) + \frac{2}{n}\lambda - \frac{c+3}{4} - \frac{c-1}{4n(n-1)}\left[-2(n-1)\|\xi^\top\|^2 + \|P\|^2\right]. \tag{2.116}$$

Motivated by the above studies, in [81] we improved Chen-Ricci inequalities for a Lagrangian submanifold M^n of dimension n $(n \geq 2)$ in a complex space form $\widetilde{M}^{2n}(4c)$ of constant holomorphic sectional curvature c with a semi-symmetric metric connection and a Legendrian submanifold M^n in a Sasakian space form $\widetilde{M}^{2n+1}(c)$ of constant φ-sectional curvature c with a semi-symmetric metric connection, respectively.

Let M^n, $n \geq 2$, be an n-dimensional submanifold of a $2m$-dimensional complex space form $\widetilde{M}^{2m}(4c)$ of constant holomorphic sectional curvature $4c$. If $J(T_pM^n) \subset T_p^\perp M^n$, then M^n is called an *anti-invariant submanifold* of $\widetilde{M}^{2m}$. For an anti-invariant submanifold of a Kaehlerian manifold, it is known that (see [122])

$$\overset{\circ}{A}_{JX} Y = \overset{\circ}{A}_{JY} X, \quad X, Y \in T_pM,$$

or equivalently, $\overset{\circ}{h}{}^k_{ij} = \overset{\circ}{h}{}^j_{ik} = \overset{\circ}{h}{}^i_{jk}$, $\forall i, j, k = 1, \ldots, n$, where $\overset{\circ}{A}$ is the shape operator with respect to $\overset{\circ}{\tilde{\nabla}}$ and

$$\overset{\circ}{h}{}^k_{ij} = g(\overset{\circ}{h}(e_i, e_j), Je_k), \; i, j, k = 1, \ldots, n.$$

Recall that a *Lagrangian submanifold* is a totally real submanifold of maximum dimension.

Theorem 2.81 ([81]) *Let M^n be a Lagrangian submanifold of dimension n ($n \geq 2$) in a $2n$-dimensional complex space form $\widetilde{M}^{2n}(4c)$ of constant holomorphic sectional curvature c with a semi-symmetric metric connection such that the vector field P is tangent to M^n. Then for any unit tangent vector X to M^n we have*

$$Ric(X) + (n-2)\alpha(X, X) + trace\,\alpha \leq (n-1)\left(c + \frac{n}{4}\|H\|^2\right). \tag{2.117}$$

The equality sign holds identically if and only if either:

(i) M^n is totally geodesic or
(ii) $n = 2$, and M^2 is a H-umbilical Lagrangian surface with $\lambda = 3\mu$.

We improved Chen-Ricci inequality for submanifolds of Sasakian space forms with a semi-symmetric metric connection.

A submanifold M^n of a Sasakian manifold $\widetilde{M}^{2m+1}$ normal to ξ is called a *C-totally real submanifold*. On such a submanifold, φ maps any tangent vector to M^n at $p \in M^n$ into the normal space $T_p^\perp M^n$. In particular, if $n = m$, i.e., M^n has a maximum dimension, then it is a *Legendrian submanifold*. For a Legendrian submanifold M^n we may choose an orthonormal basis of $T_p^\perp M^n$ of the form $\{e_{n+1} = \varphi e_1, \ldots, e_{2n} = \varphi e_n, e_{2n+1} = \xi\}$. One has (see [122])

$$\overset{\circ}{A}_{\varphi X} Y = \overset{\circ}{A}_{\varphi Y} X, \quad X, Y \in T_pM^n,$$

or equivalently, $\overset{\circ}{h}{}^k_{ij} = \overset{\circ}{h}{}^j_{ik} = \overset{\circ}{h}{}^i_{jk}$, $\forall i, j, k = 1, \ldots, n$, where $\overset{\circ}{A}$ is the corresponding shape operator and $\overset{\circ}{h}{}^k_{ij} = g(\overset{\circ}{h}(e_i, e_j), \varphi e_k)$, $i, j, k = 1, \ldots, n$.

Theorem 2.82 ([81]) *Let M^n be an n-dimensional Legendrian submanifold in a Sasakian space form $\widetilde{M}^{2n+1}(c)$ of constant φ-sectional curvature c with a semi-symmetric metric connection such that the vector field P is tangent to M^n. Then, for any unit tangent vector X to M^n, we have*

$$Ric(X) + (n-2)\alpha(X, X) + trace\alpha \le \frac{n-1}{4}\left(c + 3 + n\,\|H\|^2\right). \qquad (2.118)$$

The equality sign holds identically if and only if either:

(i) M^n is totally geodesic or
(ii) $n = 2$, and M^2 is a H-umbilical Legendrian surface with $\lambda = 3\mu$.

The notion of a *connection* is one of the most important in Geometry. Its history is long and interesting, being written by Christoffel, Ricci, Levi-Civita, Cartan, Darboux, and Koszul (see, e.g., [70]).

There are various physical problems involving the semi-symmetric metric connection. In [111] the following two examples are given.

If a man is moving on the surface of the earth always facing one definite point, say Jerusalem or Mekka or the North Pole, then this displacement is semi-symmetric and metric.

During the mathematical congress in Moscow in 1934 *one evening mathematicians invented the Moscow displacement. The streets of Moscow are approximately straight lines through the Kremlin and concentric circles around it. If a person walks in the street always facing the Kremlin, then this displacement is semi-symmetric and metric.*

In [82], we constructed examples of different types of connections starting from a semi-symmetric metric connection $\widetilde{\nabla}$ on a Riemannian manifold, for example, a connection which is a symmetric metric connection with respect to a conformally related metric g^*, but symmetric nonmetric with respect to the initial metric g.

We formulated an open problem: find a parallel complex structure on a Kaehler manifold with respect to such a new connection.

We recall that K. Yano [121] showed that a semi-symmetric metric connection $\widetilde{\nabla}$ is given by

$$\widetilde{\nabla}_X Y = \nabla^\circ_X Y + \Phi(Y)X - g(X, Y)P,$$

where ∇° is the Levi-Civita connection on $\widetilde{N}$ with respect to g and P is a vector field defined by $P = \Phi^\#$, equivalent to $g(P, X) = \Phi(X)$, for any vector field X. So, the above relation can be written as

$$\widetilde{\nabla}_X Y = \nabla^\circ_X Y + g(P, Y)X - g(X, Y)P. \qquad (2.119)$$

Let us consider only a part of formula (2.119) and define

$$\nabla'_X Y = \nabla^\circ_X Y + \Phi(Y)X,$$

with Φ a 1-form.

We proved that ∇' is a semi-symmetric connection, but it is not metric.

More precisely we have the following proposition:

Proposition 2.83 ([82]) *Let $(\tilde{N}, g)$ be an n-dimensional Riemannian manifold and ∇° its Levi-Civita connection with respect to g. Then the connection ∇' defined by*

$$\nabla'_X Y = \nabla^\circ_X Y + \Phi(Y)X$$

with Φ a 1-form on $\tilde{N}$ is a semi-symmetric nonmetric connection on $\tilde{N}$.

On the other hand, on the Riemannian manifold $(\tilde{N}, g)$ denote by $\Omega^1(\tilde{N})$ the space of 1-forms on $\tilde{N}$.

Following Yano [121], to any 1-form Φ corresponds to a metric semi-symmetric connection

$$\widetilde{\nabla}_X Y = \nabla^\circ_X Y + \Phi(Y)X - g(X, Y)\Phi^\natural.$$

We shall consider two cases:

(i) Φ is closed.
(ii) Φ is exact.

By direct calculation we have

$$d\Phi(X, Y) = X\Phi(Y) - Y\Phi(X) - \Phi([X, Y]) = Xg(P, Y) - Yg(P, X) - g(P, [X, Y])$$

$$= g(\nabla^\circ_X P, Y) + g(P, \nabla^\circ_X Y) - g(\nabla^\circ_Y P, X) - g(P, \nabla^\circ_Y X) - g(P, [X, Y])$$

$$= g(\nabla^\circ_X P, Y) - g(\nabla^\circ_Y P, X) + [g(P, \nabla^\circ_X Y) - g(P, \nabla^\circ_Y X) - g(P, [X, Y]]$$

$$= g(\nabla^\circ_X P, Y) - g(\nabla^\circ_Y P, X).$$

Then Φ is closed if and only if $g(\nabla^\circ_X P, Y) - g(\nabla^\circ_Y P, X) = 0$.

In the case (ii), Φ exactly implies that $\exists\, f \in C^\infty(\tilde{N})$ such that $\Phi = df$. Then $g(P, X) = \Phi(X) = df(X) = Xf$, $P = \operatorname{grad} f$, and, Φ being closed, we have

$$g(\nabla^\circ_X \operatorname{grad} f, Y) = g(\nabla^\circ_Y \operatorname{grad} f, X).$$

For an exact 1-form Φ, i.e., $\exists f \in C^\infty(\tilde{N})$ such that $\Phi = df$, we define a *conformally related metric* g^* by $g^* = e^{2f} g$ (which remains Riemannian metric) and denote by ∇^* its Levi-Civita connection (on $(\tilde{N}, g^*)$).

Proposition 2.84 ([82]) *Let $(\tilde{N}, g)$ be an n-dimensional Riemannian manifold and $g^* = e^{2f} g$ a conformally related metric to g, with $f \in C^\infty(\tilde{N})$. Let ∇^* be the Levi-Civita connection with respect to g^*. Then:*

(i) The connection ∇^* is given by

$$\nabla^*_X Y = \nabla^\circ_X Y + \Phi(Y)X - g(X, Y)\Phi^\natural + \Phi(X)Y,$$

i.e.,

$$\nabla^*_X Y = \widetilde{\nabla}_X Y + \Phi(X)Y,$$

where $\widetilde{\nabla}$ is the semi-symmetric metric connection with respect to g.

(ii) The connection ∇^* is a symmetric nonmetric connection with respect to g.

A Kaehler manifold is one of the most interesting manifolds from the class of complex manifolds and is well determined by its metric g and its almost complex structure J and then is usually denoted by (M, g, J). It is known that a Hermitian manifold (M, g, J) is Kaehler if and only if its almost complex structure J is parallel with respect to the Levi-Civita connection associated with the Riemannian metric g, i.e., $\nabla^\circ J = 0$.

Let $\widetilde{\nabla}$ be the semi-symmetric metric connection with respect to g on a Kaehler manifold (M, g, J) and ∇° be the Levi-Civita connection associated with g.

We calculate

$$(\widetilde{\nabla}_X J)Y = \widetilde{\nabla}_X JY - J\widetilde{\nabla}_X Y$$

$$= \nabla^\circ_X JY + \Phi(JY)X - g(X, JY)\Phi^\# - J(\nabla^\circ_X Y + \Phi(Y)X - g(X, Y)\Phi^\#)$$

$$= \nabla^\circ_X JY - J(\nabla^\circ_X Y) + \Phi(JY)X - J\Phi(Y)X - g(X, JY)\Phi^\# + Jg(X, Y)\Phi^\#$$

$$= \Phi(JY)X - J\Phi(Y)X - g(X, JY)\Phi^\# + Jg(X, Y)\Phi^\#.$$

Remark $\widetilde{\nabla} J \neq 0$, so J cannot be parallel with respect to the semi-symmetric metric connection $\widetilde{\nabla}$.

Indeed, if X is orthogonal to P and JP, then P, JP, X, and JX are linearly independent; therefore $(\widetilde{\nabla}_X J)Y \neq 0$.

Starting from the semi-symmetric connection $\widetilde{\nabla}$ on the Kaehler manifold (M, g, J), we can derive another connection ∇^*.

We formulated the following **open problem**:

Find another almost complex structure J^* on the Kaehler manifold (M, g, J) such that J^* is parallel with respect to ∇^* (i.e., $\nabla^* J^* = 0$).

On the other hand, when the real space form is endowed with a semi-symmetric nonmetric connection, in [107, 108] we proved Chen inequalities for its submanifolds, more precisely relations between the mean curvature associated with a semi-symmetric nonmetric connection, scalar and sectional curvatures, Ricci curvatures, and the sectional curvature of the ambient space. The equality cases were considered.

Let g be a Riemannian metric on N^{n+p}. If $\widetilde{\nabla} g \neq 0$, where $\widetilde{\nabla}$ is a semi-symmetric connection, then $\widetilde{\nabla}$ is called a *semi-symmetric nonmetric connection* on N^{n+p}.

Following [1], a semi-symmetric nonmetric connection $\widetilde{\nabla}$ on N^{n+p} is given by

$$\widetilde{\nabla}_{\widetilde{X}}\widetilde{Y} = \overset{\circ}{\widetilde{\nabla}}_{\widetilde{X}} \widetilde{Y} + \phi(\widetilde{Y})\widetilde{X},$$

for any vector fields $\widetilde{X}$ and $\widetilde{Y}$ on N^{n+p}, where $\overset{\circ}{\widetilde{\nabla}}$ denotes the Levi-Civita connection with respect to the Riemannian metric g and ϕ is a 1-form. Denote $P = \phi^{\sharp}$, i.e., the vector field P is defined by $g(P, \widetilde{X}) = \phi(\widetilde{X})$, for any vector field $\widetilde{X}$ on N^{n+p}.

We will consider a Riemannian manifold N^{n+p} endowed with a semi-symmetric nonmetric connection $\widetilde{\nabla}$ and the Levi-Civita connection denoted by $\overset{\circ}{\widetilde{\nabla}}$.

Let M^n be an n-dimensional submanifold of an $(n+p)$-dimensional Riemannian manifold N^{n+p}. On the submanifold M^n we consider the induced semi-symmetric nonmetric connection denoted by ∇ and the induced Levi-Civita connection denoted by $\overset{\circ}{\nabla}$.

Let $\widetilde{R}$ be the curvature tensor of N^{n+p} with respect to $\widetilde{\nabla}$ and $\overset{\circ}{\widetilde{R}}$ the curvature tensor of N^{n+p} with respect to $\overset{\circ}{\widetilde{\nabla}}$. We also denote by R and $\overset{\circ}{R}$ the curvature tensors of ∇ and $\overset{\circ}{\nabla}$, respectively, on M^n.

The Gauss formulas with respect to ∇ and, respectively, $\overset{\circ}{\nabla}$ can be written as

$$\widetilde{\nabla}_X Y = \nabla_X Y + h(X, Y), \ X, Y \in \Gamma(TM^n),$$

$$\overset{\circ}{\widetilde{\nabla}}_X Y = \overset{\circ}{\nabla}_X Y + \overset{\circ}{h}(X, Y), \ X, Y \in \Gamma(TM^n),$$

where $\overset{\circ}{h}$ is the second fundamental form of M^n in N^{n+p} and h is a (0, 2)-tensor on M^n. According to the formula (3.4) in [2],

$$h = \overset{\circ}{h}. \tag{2.120}$$

One denotes by H the mean curvature vector of M^n in N^{n+p}.

Let $N^{n+p}(c)$ be a real space form of constant sectional curvature c endowed with a semi-symmetric nonmetric connection $\widetilde{\nabla}$.

The curvature tensor $\overset{\circ}{\widetilde{R}}$ with respect to the Levi-Civita connection $\overset{\circ}{\widetilde{\nabla}}$ on $N^{n+p}(c)$ is expressed by

$$\overset{\circ}{\widetilde{R}}(X, Y, Z, W) = c\{g(X, W)g(Y, Z) - g(X, Z)g(Y, W)\}.$$

Then the curvature tensor $\widetilde{R}$ with respect to the semi-symmetric nonmetric connection $\widetilde{\nabla}$ on $N^{n+p}(c)$ can be written as (see [1])

$$\widetilde{R}(X, Y, Z, W) = \overset{\circ}{\widetilde{R}}(X, Y, Z, W) + s(X, Z)g(Y, W) - s(Y, Z)g(X, W),$$

for any vector fields $X, Y, Z, W \in \chi(M^n)$, where s is a $(0, 2)$-tensor field defined by

$$s(X, Y) = \left(\overset{\circ}{\widetilde{\nabla}}_X \phi\right) Y - \phi(X)\phi(Y), \quad \forall X, Y \in \chi(M^n).$$

It follows that the curvature tensor $\widetilde{R}$ can be expressed as

$$\widetilde{R}(X, Y, Z, W) = c\{g(X, W)g(Y, Z) - g(X, Z)g(Y, W)\} \tag{2.121}$$
$$+ s(X, Z)g(Y, W) - s(Y, Z)g(X, W).$$

Denote by λ the trace of s.

Using (2.120), the Gauss equation for the submanifold M^n into the real space form $N^{n+p}(c)$ is

$$\overset{\circ}{\widetilde{R}}(X, Y, Z, W) = \overset{\circ}{R}(X, Y, Z, W) + g(h(X, Z), h(Y, W)) - g(h(X, W), h(Y, Z)). \tag{2.122}$$

Decomposing the vector field P on M uniquely into its tangent and normal components P^T and $P^\perp$, respectively, we have $P = P^T + P^\perp$.

Denote

$$\Omega(X) = s(X, X) + g(P^\perp, h(X, X)), \tag{2.123}$$

for a unit vector X tangent to M^n at a point x.

In general for submanifolds M^n of a real space form endowed with a semi-symmetric nonmetric connection, the sectional curvature $K(\pi)$ of a plane section (and consequently the Chen invariants) cannot be defined by the standard definition because it depends on the choice of the orthonormal basis of π. For this reason we put the condition $\Omega(X) =$ constant for all unit vectors tangent to M^n.

For submanifolds of real space forms endowed with a semi-symmetric nonmetric connection, we establish the following optimal inequality, which will call *Chen first inequality*:

Theorem 2.85 ([107]) *Let $M^n, n \geq 3$, be an n-dimensional submanifold of an $(n+p)$-dimensional real space form $N^{n+p}(c)$ of constant sectional curvature c, endowed with a semi-symmetric nonmetric connection $\widetilde{\nabla}$. We have*

$$\delta_{M^n}(x) \leq \Omega + (n-2)\left[\frac{n^2}{2(n-1)}\|H\|^2 + (n+1)\frac{c}{2}\right] - \frac{1}{2}(n-1)\lambda - \frac{1}{2}n^2\phi(H),$$

where π is a 2-plane section of T_xM^n, $x \in M^n$. Equality holds at a point $x \in M^n$ if and only if there exist an orthonormal basis $\{e_1, e_2, \ldots, e_n\}$ of T_xM^n and an orthonormal basis $\{e_{n+1}, \ldots, e_{n+p}\}$ of $T_x^\perp M^n$ such that the shape operators of M^n in $N^{n+p}(c)$ at x have the following forms:

$$A_{e_{n+1}} = \begin{pmatrix} a & 0 & 0 & \cdots & 0 \\ 0 & b & 0 & \cdots & 0 \\ 0 & 0 & \mu & \cdots & 0 \\ \vdots & \vdots & \vdots & \ddots & \vdots \\ 0 & 0 & 0 & \cdots & \mu \end{pmatrix}, \quad a + b = \mu,$$

$$A_{e_{n+i}} = \begin{pmatrix} h_{11}^{n+i} & h_{12}^{n+i} & 0 & \cdots & 0 \\ h_{12}^{n+i} & -h_{11}^{n+i} & 0 & \cdots & 0 \\ 0 & 0 & 0 & \cdots & 0 \\ \vdots & \vdots & \vdots & \ddots & \vdots \\ 0 & 0 & 0 & \cdots & 0 \end{pmatrix}, \quad 2 \leq i \leq p.$$

Proof From [2], the Gauss equation with respect to the semi-symmetric nonmetric connection is

$$\begin{aligned} \widetilde{R}(X, Y, Z, W) = R(X, Y, Z, W) + g(h(X, Z), h(Y, W)) - g(h(Y, Z), h(X, W)) \\ + g(P^\perp, h(Y, Z))g(X, W) - g(P^\perp, h(X, Z))g(Y, W). \end{aligned} \tag{2.124}$$

Let $x \in M^n$ and $\{e_1, e_2, \ldots, e_n\}$ and $\{e_{n+1}, \ldots, e_{n+p}\}$ be orthonormal basis of T_xM^n and $T_x^\perp M^n$, respectively. For $X = W = e_i, Y = Z = e_j, i \neq j$, from Eq. (2.121) it follows that

$$\tilde{R}(e_i, e_j, e_j, e_i) = c - s(e_j, e_j). \tag{2.125}$$

From the last two equations we get

$$\begin{aligned} c - s(e_j, e_j) = R(e_i, e_j, e_j, e_i) + g(h(e_i, e_j), h(e_i, e_j)) \\ -g(h(e_i, e_i), h(e_j, e_j)) + \phi(h(e_j, e_j)). \end{aligned}$$

By summation after $1 \leq i, j \leq n$, it follows from the above relation that

$$(n^2 - n)c - (n-1)\lambda = 2\tau + \|h\|^2 - n^2\|H\|^2 + n^2\phi(H), \tag{2.126}$$

where we recall that λ is the trace of s and denote

$$\|h\|^2 = \sum_{i,j=1}^{n} g(h(e_i, e_j), h(e_i, e_j)),$$

$$H = \frac{1}{n}\text{trace}\, h, \quad \phi(H) = \frac{1}{n}\sum_{j=1}^{n} \phi(h(e_j, e_j)) = g(P^{\perp}, H).$$

One takes

$$\varepsilon = 2\tau - \frac{n^2(n-2)}{n-1}\|H\|^2 + (n-1)\lambda - (n^2 - n)c + n^2\phi(H). \tag{2.127}$$

Then (2.123) becomes

$$n^2\|H\|^2 = (n-1)\left(\|h\|^2 + \varepsilon\right). \tag{2.128}$$

Let $x \in M^n$, $\pi \subset T_xM^n$, $\dim \pi = 2$, $\pi = sp\{e_1, e_2\}$. We define $e_{n+1} = \frac{H}{\|H\|}$, and from the relation (2.128) we obtain

$$\left(\sum_{i=1}^{n} h_{ii}^{n+1}\right)^2 = (n-1)\left(\sum_{i,j=1}^{n}\sum_{r=n+1}^{n+p}(h_{ij}^r)^2 + \varepsilon\right),$$

or equivalently,

$$\left(\sum_{i=1}^{n} h_{ii}^{n+1}\right)^2 = (n-1)\left\{\sum_{i=1}^{n}(h_{ii}^{n+1})^2 + \sum_{i\neq j}(h_{ij}^{n+1})^2 + \sum_{i,j=1}^{n}\sum_{r=n+2}^{n+p}(h_{ij}^r)^2 + \varepsilon\right\}. \tag{2.129}$$

By using the Chen lemma we derive

$$2h_{11}^{n+1}h_{22}^{n+1} \geq \sum_{i\neq j}(h_{ij}^{n+1})^2 + \sum_{i,j=1}^{n}\sum_{r=n+2}^{n+p}(h_{ij}^r)^2 + \varepsilon. \tag{2.130}$$

The Gauss equation for $X = W = e_1$, $Y = Z = e_2$ gives

$$K(\pi) = R(e_1, e_2, e_2, e_1) = c - s(e_2, e_2)$$

$$- g(P^{\perp}, h(e_2, e_2)) + \sum_{r=n+1}^{p}\left[h_{11}^r h_{22}^r - (h_{12}^r)^2\right]$$

$$\geq c - s(e_2, e_2) - \phi(h(e_2, e_2)) + \frac{1}{2}\left[\sum_{i\neq j}(h_{ij}^{n+1})^2 + \sum_{i,j=1}^{n}\sum_{r=n+2}^{n+p}(h_{ij}^r)^2 + \varepsilon\right]$$

$$+ \sum_{r=n+2}^{n+p} h_{11}^r h_{22}^r - \sum_{r=n+1}^{n+p}(h_{12}^r)^2 = c - s(e_2, e_2) - \phi(h(e_2, e_2))$$

$$+ \frac{1}{2}\sum_{i\neq j}(h_{ij}^{n+1})^2 + \frac{1}{2}\sum_{i,j=1}^{n}\sum_{r=n+2}^{n+p}(h_{ij}^r)^2 + \frac{1}{2}\varepsilon + \sum_{r=n+2}^{n+p} h_{11}^r h_{22}^r - \sum_{r=n+1}^{n+p}(h_{12}^r)^2$$

$$= c - s(e_2, e_2) - g(P^{\perp}, h(e_2, e_2)) + \frac{1}{2}\sum_{i\neq j}(h_{ij}^{n+1})^2 + \frac{1}{2}\sum_{r=n+2}^{n+p}\sum_{i,j>2}(h_{ij}^r)^2$$

$$+ \frac{1}{2}\sum_{r=n+2}^{n+p}(h_{11}^r + h_{22}^r)^2 + \sum_{j>2}\left[(h_{1j}^{n+1})^2 + (h_{2j}^{n+1})^2\right] + \frac{1}{2}\varepsilon$$

$$\geq c - s(e_2, e_2) - g(P^{\perp}, h(e_2, e_2)) + \frac{\varepsilon}{2},$$

which implies

$$K(\pi) \geq c - s(e_2, e_2) - g(P^{\perp}, h(e_2, e_2)) + \frac{\varepsilon}{2}.$$

Let $\{e_1, e_2, \ldots, e_n\}$ be an orthonormal basis of T_xM^n with $\pi = sp\,\{e_1, e_2\}$. The formula (2.123) implies that

$$\Omega(e_1) = \Omega(e_2) = \ldots = \Omega(e_n).$$

Denote it simply by Ω. By using (2.127) we get

$$K(\pi) \geq \tau - \Omega - (n-2)\left[\frac{n^2}{2(n-1)}\|H\|^2 + (n+1)\frac{c}{2}\right] + \frac{1}{2}(n-1)\lambda + \frac{1}{2}n^2\phi(H),$$

which represents the inequality to prove.

The equality case holds at a point $x \in M^n$ if and only if it achieves the equality in all the previous inequalities and, we have the equality in the lemma.

$$h_{ij}^{n+1} = 0,\ \forall i \neq j, i, j > 2,$$

$$h_{ij}^r = 0, \forall i \neq j, i, j > 2, r = n+1, \ldots, n+p,$$

$$h_{11}^r + h_{22}^r = 0, \ \forall r = n+2, \ldots, n+p,$$

$$h_{1j}^{n+1} = h_{2j}^{n+1} = 0, \ \forall j > 2,$$

$$h_{11}^{n+1} + h_{22}^{n+1} = h_{33}^{n+1} = \ldots = h_{nn}^{n+1}.$$

We may choose $\{e_1, e_2\}$ such that $h_{12}^{n+1} = 0$, and we denote $a = h_{11}^r, b = h_{22}^r, \mu = h_{33}^{n+1} = \ldots = h_{nn}^{n+1}$.

It follows that the shape operators take the desired forms.

We also established a sharp relation between the Ricci curvature in the direction of a unit tangent vector X and the mean curvature H with respect to the semi-symmetric nonmetric connection $\widetilde{\nabla}$.

Denote the relative null subspace of the tangent space by

$$N(x) = \left\{X \in T_x M^n \mid h(X, Y) = 0, \ \forall Y \in T_x M^n\right\}.$$

Theorem 2.86 ([108]) *Let $M^n, n \geq 3$, be an n-dimensional submanifold of an $(n+p)$-dimensional real space form $N^{n+p}(c)$ of constant sectional curvature c endowed with a semi-symmetric nonmetric connection $\widetilde{\nabla}$. Then:*

(i) For each unit vector X in $T_x M$ we have

$$\|H\|^2 \geq \frac{4}{n^2}[Ric(X) - (n-1)(c - \Omega)]. \tag{2.131}$$

(ii) If $H(x) = 0$, then a unit tangent vector X at x satisfies the equality case of (2.131) *if and only if $X \in N(x)$.*
(iii) The equality case of inequality (2.131) *holds identically for all unit tangent vectors at x if and only if either x is a totally geodesic point or $n = 2$ and x is a totally umbilical point.*

A relationship between the sectional curvature of a submanifold M^n of a real space form $N^{n+p}(c)$ of constant sectional curvature c endowed with a semi-symmetric nonmetric connection $\widetilde{\nabla}$ and the associated squared mean curvature $\|H\|^2$ was established in [107]. Using this inequality, we proved a relationship between the k-Ricci curvature of M^n (intrinsic invariant) and the squared mean curvature $\|H\|^2$ (extrinsic invariant).

Theorem 2.87 ([107]) *Let $M^n, n \geq 3$, be an n-dimensional submanifold of an $(n+p)$-dimensional real space form $N^{n+p}(c)$ of constant sectional curvature c endowed with a semi-symmetric nonmetric connection $\widetilde{\nabla}$. Then we have*

$$\|H\|^2 \geq \frac{2\tau}{n(n-1)} - c + \frac{1}{n}\lambda + \frac{n}{n-1}\phi(H). \tag{2.132}$$

Using the above theorem, we obtain the following:

Corollary 2.88 ([107]) *Let $M^n, n \geq 3$, be an n-dimensional submanifold of an $(n+p)$-dimensional real space form $N^{n+p}(c)$ of constant sectional curvature c endowed with a semi-symmetric nonmetric connection $\widetilde{\nabla}$. Then, for any integer k, $2 \leq k \leq n$, and any point $x \in M^n$, we have*

$$\|H\|^2(p) \geq \Theta_k(p) - c + \frac{1}{n}\lambda + \frac{n}{n-1}\phi(H). \tag{2.133}$$

Recently, in [85] we have proposed a definition of the sectional curvature of the semi-symmetric nonmetric connection, which does not depend on the orthonormal basis of a 2-plane section, i.e., is well defined.

More precisely, let (M, g) be a Riemannian manifold endowed with a semi-symmetric nonmetric connection ∇. Recall that

$$\nabla_X Y = \nabla^0_X Y + \omega(Y)X,$$

where ∇^0 is the Levi-Civita connection on (M, g).

We remarked in the previous section that one cannot define the sectional curvature of a plane section $\pi = \text{span}\,\{e_1, e_2\} \subset T_pM$, $p \in M$, by $g(R(e_1, e_2)e_2, e_1)$.

This is the reason for which a well-defined sectional curvature is necessary; the steps to get there are below (see [85]).

First we consider the linear connection

$$\nabla'_X Y = \nabla^0_X Y - g(X, Y)P.$$

Then we prove the first result.

Proposition 2.89 ([85]) *Let (M, g) be a Riemannian manifold, ∇ a semi-symmetric nonmetric connection given by*

$$\nabla_X Y = \nabla^0_X Y + \omega(Y)X,$$

and ∇' a linear connection defined by

$$\nabla'_X Y = \nabla^0_X Y - g(X, Y)P.$$

Then ∇ and ∇' are conjugate connections, i.e.,

$$Zg(X, Y) = g(\nabla_Z X, Y) + g(X, \nabla'_Z Y), \ \forall X, Y, Z \in \Gamma(TM).$$

Proof Let $X, Y, Z \in \Gamma(TM)$. Then

$$g(\nabla_Z X, Y) + g(X, \nabla'_Z Y) = g(\nabla^0_Z X + \omega(X)Z, Y) + g(X, \nabla^0_Z Y - g(Z, Y)P)$$

$$= Zg(X, Y) + \omega(X)g(Z, Y) - g(Z, Y)g(X, P) = Zg(X, Y).$$

The basic properties of the connection ∇' are given in the following.

Proposition 2.90 ([85]) *Let* (M, g) *be a Riemannian manifold and* ∇' *the connection defined by* $\nabla'_X Y = \nabla^0_X Y - g(X, Y)P$, *where* ∇^0 *is the Levi-Civita connection. Then:*

(i) ∇' *is symmetric, i.e., its torsion* $T' = 0$.
(ii) ∇' *is nonmetric.*

Proof Let $X, Y, Z \in \Gamma(TM)$. We have:

(i)

$$T'(X, Y) = \nabla'_X Y - \nabla'_Y X - [X, Y]$$

$$= \nabla^0_X Y - g(X, Y)P - \nabla^0_Y X + g(X, Y)P - [X, Y] = 0.$$

(ii)

$$(\nabla'_X g)(Y, Z) = Xg(Y, Z) - g(\nabla^0_X Y - g(X, Y)P, Z) - g(Y, \nabla^0_X Z - g(X, Z)P)$$

$$= -g(X, Y)\omega(Z) + g(X, Z)\omega(Y) \neq 0.$$

Next, we prove an important relation between the curvatures of the conjugate connections ∇ and ∇'.

Theorem 2.91 ([85]) *Let* (M, g) *be a Riemannian manifold,* ∇ *a semi-symmetric nonmetric connection, and* ∇' *its conjugate connection defined by*

$$\nabla_X Y = \nabla^0_X Y + \omega(Y)X,$$

$$\nabla'_X Y = \nabla^0_X Y - g(X, Y)P.$$

Then

$$g(R'(X, Y)Z, W) = -g(R(X, Y)W, Z).$$

Proof Let $X, Y, Z, W \in \Gamma(TM)$. Then,

$$g(R(X, Y)Z, W) = g(\nabla_X \nabla_Y Z - \nabla_Y \nabla_X Z - \nabla_{[X,Y]} Z, W)$$

$$= Xg(\nabla_Y Z, W) - g(\nabla_Y Z, \nabla'_X W) - Yg(\nabla_X Z, W)$$

$$+ g(\nabla_X Z, \nabla'_Y W) - [X, Y]g(Z, W) + g(Z, \nabla'_{[X,Y]} W)$$

$$= XYg(Z, W) - Xg(Z, \nabla'_Y W) - Yg(Z, \nabla'_X W) + g(Z, \nabla'_Y \nabla'_X W)$$

$$- YXg(Z, W) + Yg(Z, \nabla'_X W) + Xg(Z, \nabla'_Y W) - g(Z, \nabla'_X \nabla'_Y W)$$

$$- [X, Y]g(Z, W) + g(Z, \nabla'_{[X,Y]} W) = -g(Z, R'(X, Y)W).$$

Inspired by an idea of B. Opozda [106], we then define a (0, 4)-tensor field S by

$$S(X, Y, Z, W) = \frac{1}{2}[g(R(X, Y)W, Z) + g(R'(X, Y)W, Z)].$$

Theorem 2.91 implies

$$S(X, Y, Z, W) = -S(X, Y, W, Z)$$

and

$$S(X, Y, Z, W) = \frac{1}{2}[g(R(X, Y)W, Z) - g(R(X, Y)Z, W)].$$

Let $p \in M$ and $\pi \subset T_pM$ a plane section. For an orthonormal basis $\{e_1, e_2\}$ of π, we derive

$$S(e_1, e_2, e_1, e_2) = \frac{1}{2}[g(R(e_1, e_2)e_2, e_1) - g(R(e_1, e_2)e_1, e_2)].$$

By the formula of the curvature tensor of a semi-symmetric nonmetric connection, it follows that

$$S(e_1, e_2, e_1, e_2) = R^0(e_1, e_2, e_1, e_2) - \frac{1}{2}[s(e_2, e_2) + s(e_1, e_1)],$$

which does not depend on the orthonormal basis $\{e_1, e_2\}$ of π.

Therefore, we are now able to introduce the following definition of a sectional curvature of the semi-symmetric nonmetric connection ∇.

Definition The *sectional curvature* of the plane section $\pi \subset T_pM$ spanned by the orthonormal basis $\{e_1, e_2\}$ is defined by

$$K(\pi) = \frac{1}{2}[g(R(e_1, e_2)e_2, e_1) + g(R(e_2, e_1)e_1, e_2)].$$

Using the above definition, we can compute the scalar curvature and the Ricci curvature of a Riemannian space form admitting a semi-symmetric nonmetric connection.

Let $M(c)$ be an n-dimensional Riemannian space form (the sectional curvature associated with the Levi-Civita connection is a constant c) admitting a semi-symmetric nonmetric connection ∇. Let $p \in M(c)$ and $\{e_1, \ldots, e_n\}$ be an orthonormal basis of T_pM.

The *scalar curvature* with respect to ∇ is

$$\tau = \sum_{1 \leq i < j \leq n} K(e_i \wedge e_j),$$

where $e_i \wedge e_j$ is the plane section spanned by e_i and e_j.

By using the definition of the sectional curvature K, we have

$$\tau = \frac{1}{2} \sum_{1 \leq i < j \leq n} [g(R(e_i, e_j)e_j, e_i) + g(R(e_j, e_i)e_i, e_j)] = \sum_{1 \leq i, j \leq n} g(R(e_i, e_j)e_j, e_i).$$

By the formula of the curvature tensor of a semi-symmetric nonmetric connection, it follows that

$$\tau = \frac{1}{2}n(n-1)c + \frac{1}{2}(n-1)\operatorname{trace} s.$$

Let $p \in M(c)$, $X \in T_pM$ unit, and $\{e_1 = X, e_2, \ldots, e_n\}$ be an orthonormal basis of T_pM. It is known that

$$\operatorname{Ric}(X) = \sum_{j=2}^{n} K(X \wedge e_j) = \frac{1}{2} \sum_{j=2}^{n} [g(R(X, e_j)e_j, X) + g(R(e_j, X)X, e_j)]$$

$$= (n-1)c + \frac{1}{2}[(n-2)s(X, X) + \operatorname{trace} s].$$

On the other hand, recall that B.Y. Chen [29] established an estimate of the mean curvature in terms of the Ricci curvature for any Riemannian submanifold of dimension n in a Riemannian space form $\tilde{M}(c)$ of constant sectional curvature c:

$$\operatorname{Ric}(X) \leq (n-1)c + \frac{n^2}{4}\|H\|^2,$$

known as the *Chen-Ricci inequality*.

As an application of the new definition, we established in [85] the Chen-Ricci inequality for submanifolds in a Riemannian space form admitting a semi-symmetric nonmetric connection by using the sectional curvature defined in the previous section.

Let $\tilde{M}(c)$ be an m-dimensional Riemannian space form, $\tilde{\nabla}$ a semi-symmetric nonmetric connection on $\tilde{M}(c)$, and M an n-dimensional ($n \geq 2$) submanifold of $\tilde{M}(c)$.

The Gauss formulae for the semi-symmetric connection $\tilde{\nabla}$ and the Levi-Civita connection $\tilde{\nabla}^0$, respectively, are written as

$$\tilde{\nabla}_X Y = \nabla_X Y + h(X, Y),$$

$$\tilde{\nabla}^0_X Y = \nabla^0_X Y + h^0(X, Y),$$

for all vector fields X, Y on the submanifold M. In the above formulae, h^0 is the second fundamental form of M, and h is a (0, 2)-tensor on M. In [2], it is proven that $h^0 = h$.

We decompose the vector field P on M uniquely into its tangent and normal components $P^\top$ and $P^\perp$, respectively; we have $P = P^\top + P^\perp$.

The Gauss equation with respect to the semi-symmetric nonmetric connection is given by (see also [2])

$$\begin{aligned} g(\tilde{R}(X, Y)Z, W) &= g(R(X, Y)Z, W) + g(h(X, Z), h(Y, W)) \\ &\quad - g(h(X, W), h(Y, Z)) \\ &\quad + g(P^\perp, h(Y, Z))g(X, W) - g(P^\perp, h(X, Z))g(Y, W), \end{aligned}$$

for any vector fields X, Y, Z, and W on M.

Theorem 2.92 ([85]) *Let $\tilde{M}(c)$ be an m-dimensional Riemannian space form, $\tilde{\nabla}$ a semi-symmetric nonmetric connection on it, and M an n-dimensional ($n \geq 2$) submanifold of $\tilde{M}(c)$. Then we have the following:*

1. *For each unit vector $X \in T_pM$,*

$$Ric(X) \leq \frac{n^2}{4}\|H\|^2 + (n-1)c - \frac{1}{2}[trace\, s + (n-2)s(X, X)] \qquad (2.134)$$
$$- \frac{1}{2}[n\omega(H) + (n-2)g(P^\perp, h(X, X))].$$

2. *If $H(p) = 0$, then a unit tangent vector X at p satisfies the equality case of the inequality* (2.134) *if and only if $X \in N_p$, where $N_p = \{X \in T_pM | h(X, Y) = 0, \forall Y \in T_pM\}$.*
3. *The equality case of the inequality* (2.134) *holds identically for all unit tangent vectors at p if and only if either:*

   (i) p is a totally geodesic point or
   (ii) $n = 2$ and p is a totally umbilical point.

Proof 1. Let $p \in M$ and $X \in T_pM$ be a unit tangent vector. Consider an orthonormal basis $\{e_1, \ldots, e_n, e_{n+1}, \ldots, e_m\}$ in $T_p\tilde{M}(c)$, with $e_1 = X, e_2, \ldots, e_n$ tangent to M at p.

As usual, one denotes $h^r_{ij} = g(h(e_i, e_j), e_r)$, $i, j \in \{1, \ldots, n\}$, $r \in \{n+1, \ldots, m\}$. We have

$$\mathrm{Ric}(X) = \sum_{j=2}^{n} K(e_1 \wedge e_j).$$

If we take $X = W = e_1$ and $Y = Z = e_j$ in the Gauss equation, we have

$$g(R(e_1, e_j)e_j, e_1) = c - s(e_j, e_j) + \sum_{r=n+1}^{m} \left[h^r_{11}h^r_{jj} - (h^r_{1j})^2\right] - g(P^{\perp}, h(e_j, e_j)),$$

respectively, and from the Gauss equation if we put $X = Z = e_1$, $Y = W = e_j$, we obtain

$$g(R(e_j, e_1)e_1, e_j) = c - s(e_1, e_1) + \sum_{r=n+1}^{m} \left[h^r_{11}h^r_{jj} - (h^r_{1j})^2\right] - g(P^{\perp}, h(e_1, e_1)).$$

Because

$$K(e_1 \wedge e_j) = \frac{1}{2}[g(R(e_1, e_j)e_j, e_1) + g(R(e_j, e_1)e_1, e_j)],$$

from the previous two relations, we have

$$K(e_1 \wedge e_j) = c - \frac{1}{2}[s(e_j, e_j) + s(e_1, e_1)] + \sum_{r=n+1}^{m} \left[h^r_{11}h^r_{jj} - (h^r_{1j})^2\right]$$

$$- \frac{1}{2}\left[g(P^{\perp}, h(e_j, e_j)) + g(P^{\perp}, h(e_1, e_1))\right].$$

By substitution we find

$$\mathrm{Ric}(X) = (n-1)c - \frac{1}{2}[\mathrm{trace}\, s + (n-2)s(X, X)]$$

$$- \frac{1}{2}[n\omega(H) + (n-2)g(P^{\perp}, h(X, X))]$$

$$+ \sum_{j=2}^{n} \sum_{r=n+1}^{m} \left[h^r_{11}h^r_{jj} - (h^r_{1j})^2\right].$$

The last equation implies

$$\mathrm{Ric}(X) \leq (n-1)c - \frac{1}{2}[\mathrm{trace}\, s + (n-2)s(X,X)]$$

$$-\frac{1}{2}[n\omega(H) + (n-2)g(P^{\perp}, h(X,X))] + \sum_{j=2}^{n}\sum_{r=n+1}^{m} h_{11}^{r} h_{jj}^{r}.$$

Obviously one has

$$h_{11}^{r}\left(\sum_{j=2}^{n} h_{jj}^{r}\right) \leq \frac{1}{4}\left(\sum_{i=1}^{n} h_{ii}^{r}\right)^{2},$$

with equality if and only if

$$h_{11}^{r} = h_{22}^{r} + \cdots + h_{nn}^{r}.$$

It follows that

$$\mathrm{Ric}(X) \leq \frac{n^2}{4}\|H\|^2 + (n-1)c - \frac{1}{2}[\mathrm{trace}\, s + (n-2)s(X,X)]$$

$$-\frac{1}{2}[n\omega(H) + (n-2)g(P^{\perp}, h(X,X))].$$

2. If a unit vector X at p satisfies the equality case in (2.134), we get

$$h_{1i}^{r} = 0,\ 2 \leq i \leq n,\ \forall r \in \{n+1,\ldots,m\},$$

$$h_{11}^{r} = h_{22}^{r} + \ldots + h_{nn}^{r},\ \forall r \in \{n+1,\ldots,m\}.$$

Therefore, because $H(p) = 0$, we have $h_{1j}^{r} = 0$, for all $j \in \{1,\ldots,n\}$, $r \in \{n+1,\ldots,m\}$, that is, $X \in N_p$.

3. The equality case of the inequality (2.134) holds for all unit tangent vectors at p if and only if

$$h_{ij}^{r} = 0, \quad 1 \leq i \neq j \leq n, \quad r \in \{n+1,\ldots,m\},$$

$$h_{11}^{r} + \ldots + h_{nn}^{r} - 2h_{ii}^{r} = 0, \quad i \in \{1,\ldots,n\}, \quad r \in \{n+1,\ldots,m\},$$

which imply $h(e_i, e_j) = 0,\ 1 \leq i \neq j \leq n$, and $(n-2)H(p) = 0$.

We distinguish two cases:

(i) $n \neq 2$; $h(e_i, e_j) = 0, \forall i, j \in \{1,\ldots,n\}$, i.e., h_p vanishes on T_pM.

(ii) $n = 2$; then $h(e_i, e_j) = g(e_i, e_j)H(p)$, for any $i, j \in \{1, 2\}$, i.e., p is a totally umbilical point.

Remark This definition of the sectional curvature of the semi-symmetric nonmetric connection was used in the very recent work of M.E. Aydin, R. Lopez, and A. Mihai (see [6, 9]) for the study of constant sectional curvature surfaces with a semi-symmetric nonmetric connection, respectively, in the classification of translation surfaces in $\mathbf{R}^3$ with constant sectional curvature.

2.4 Statistical Submanifolds

In this subsection, we study the behavior of submanifolds in statistical manifolds of constant curvature. We investigate curvature properties of such submanifolds. Some inequalities for submanifolds with any codimension and hypersurfaces of statistical manifolds of constant curvature are also established.

Statistical manifolds introduced, in 1985, by Amari have been studied in terms of information geometry. Since the geometry of such manifolds includes the notion of dual connections, also called conjugate connections in affine geometry, it is closely related to affine differential geometry. Also, a statistical structure is a generalization of a Hessian structure [114].

Let $\left(\tilde{M}, \tilde{g}\right)$ be a Riemannian manifold and M a submanifold of $\tilde{M}$. If (M, ∇, g) is a statistical manifold, then we call (M, ∇, g) a statistical submanifold of $\left(\tilde{M}, \tilde{g}\right)$, where ∇ is an affine connection on M and g is the metric tensor on M induced from the Riemannian metric $\tilde{g}$ on $\tilde{M}$. Let $\tilde{\nabla}$ be an affine connection on $\tilde{M}$. If $\left(\tilde{M}, \tilde{g}, \tilde{\nabla}\right)$ is a statistical manifold and M a submanifold of $\tilde{M}$, then (M, ∇, g) is also a statistical manifold by induced connection ∇ and metric g. In the case that $\left(\tilde{M}, \tilde{g}\right)$ is a semi-Riemannian manifold, the induced metric g has to be nondegenerate. For details, see [116, 119].

In the geometry of submanifolds, Gauss formula, Weingarten formula, and the equations of Gauss, Codazzi, and Ricci are known as fundamental equations. Corresponding fundamental equations on statistical submanifolds were obtained in [119]. A condition for the curvature of a statistical manifold to admit a kind of standard hypersurface was given by H. Furuhata [58, 59], and he introduced a complex version of the notion of statistical structures as well.

On the other hand, B.Y. Chen [24] established basic inequalities for submanifolds in real space forms, well known as Chen inequalities. In particular, a sharp relationship between the Ricci curvature and the squared mean curvature for any n-dimensional Riemannian submanifold of a real space form was proved in [29], which is known as the Chen-Ricci inequality. Moreover, Chen's inequalities for submanifolds of real space forms endowed with a semi-symmetric metric connection were obtained in [86, 87].

In [7] we obtained some inequalities for submanifolds with any codimension and hypersurfaces of statistical manifolds.

We first introduce the statistical submanifolds.

Let $\left(\tilde{M}, \tilde{g}\right)$ be a Riemannian manifold of dimension $(n+k)$ and $\tilde{\nabla}$ an affine connection on $\tilde{M}$. One denotes the set of sections of a vector bundle $E \longrightarrow \tilde{M}$ by $\Gamma(E)$. Thus, the set of tensor fields of type (p, q) on $\tilde{M}$ is denoted by $\Gamma\left(T\tilde{M}^{(p,q)}\right)$.

Definition ([58]) Let $\tilde{T} \in \Gamma\left(T\tilde{M}^{(1,2)}\right)$ be the torsion tensor field of $\tilde{\nabla}$. Then a pair $\left(\tilde{\nabla}, \tilde{g}\right)$ is called a *statistical structure* on $\tilde{M}$ if

$$\left(\tilde{\nabla}_X\tilde{g}\right)(Y, Z) - \left(\tilde{\nabla}_Y\tilde{g}\right)(X, Z) = \tilde{g}\left(\tilde{T}(X, Y), Z\right) \tag{2.135}$$

holds for $X, Y, Z \in \Gamma\left(T\tilde{M}\right)$, and

$$\tilde{T} = 0. \tag{2.136}$$

A *statistical manifold* is a Riemannian manifold $\left(\tilde{M}, \tilde{g}\right)$ of dimension $(n+k)$, endowed with a pair of torsion-free affine connections $\tilde{\nabla}$ and $\tilde{\nabla}^*$ satisfying

$$Z\tilde{g}(X, Y) = \tilde{g}\left(\tilde{\nabla}_Z X, Y\right) + \tilde{g}\left(X, \tilde{\nabla}^*_Z Y\right), \tag{2.137}$$

for any X, Y, and $Z \in \Gamma\left(T\tilde{M}\right)$. One denotes a statistical manifold by $\left(\tilde{M}, \tilde{g}, \tilde{\nabla}\right)$. The connections $\tilde{\nabla}$ and $\tilde{\nabla}^*$ are called *dual connections*, and it is easily shown that $\left(\tilde{\nabla}^*\right)^* = \tilde{\nabla}$. If $\left(\tilde{\nabla}, \tilde{g}\right)$ is a statistical structure on $\tilde{M}$, then $\left(\tilde{\nabla}^*, \tilde{g}\right)$ is also a statistical structure [4, 119].

On the other hand, any torsion-free affine connection $\tilde{\nabla}$ always has a dual connection given by

$$\tilde{\nabla} + \tilde{\nabla}^* = 2\tilde{\nabla}^0, \tag{2.138}$$

where $\tilde{\nabla}^0$ is Levi-Civita connection on $\tilde{M}$.

Denote by $\tilde{R}$ and $\tilde{R}^*$ the curvature tensor fields of $\tilde{\nabla}$ and $\tilde{\nabla}^*$, respectively.

A statistical structure $\left(\tilde{\nabla}, \tilde{g}\right)$ is said to be *of constant curvature* $c \in \mathbb{R}$ if

$$\tilde{R}(X, Y)Z = c\left\{\tilde{g}(Y, Z)X - \tilde{g}(X, Z)Y\right\}. \tag{2.139}$$

A statistical structure $\left(\tilde{\nabla}, \tilde{g}\right)$ of constant curvature 0 is called *a Hessian structure.*

The curvature tensor fields $\tilde{R}$ and $\tilde{R}^*$ of dual connections satisfy

$$\tilde{g}\left(\tilde{R}^*(X,Y)Z,W\right) = -\tilde{g}\left(Z,\tilde{R}(X,Y)W\right), \tag{2.140}$$

from which it follows immediately that if $\left(\tilde{\nabla},\tilde{g}\right)$ is a statistical structure of constant structure c, then $\left(\tilde{\nabla}^*,\tilde{g}\right)$ is also a statistical structure of constant c. In particular, if $\left(\tilde{\nabla},\tilde{g}\right)$ is Hessian, so is $\left(\tilde{\nabla}^*,\tilde{g}\right)$.

Let M be an n-dimensional submanifold of $\tilde{M}$. Then, for any $X, Y \in \Gamma(TM)$, according to [119], the corresponding Gauss formulas are

$$\tilde{\nabla}_X Y = \nabla_X Y + h(X,Y), \tag{2.141}$$

$$\tilde{\nabla}^*_X Y = \nabla^*_X Y + h^*(X,Y), \tag{2.142}$$

where h and $h^*$are symmetric and bilinear, called the *imbedding curvature tensor* of M in $\tilde{M}$ for $\tilde{\nabla}$ and the *imbedding curvature tensor* of M in $\tilde{M}$ for $\tilde{\nabla}^*$, respectively.

In [119], it is also proved that (∇, g) and (∇^*, g) are dual statistical structures on M, where g is induced metric on $\Gamma(TM)$ from the Riemannian metric $\tilde{g}$ on $\tilde{M}$.

Since h and $h^*$are bilinear, we have the linear transformations A_ξ and A^*_ξ defined by

$$g\left(A_\xi X, Y\right) = \tilde{g}\left(h(X,Y),\xi\right), \tag{2.143}$$

$$g\left(A^*_\xi X, Y\right) = \tilde{g}\left(h^*(X,Y),\xi\right), \tag{2.144}$$

for any $\xi \in \Gamma\left(TM^\perp\right)$ and $X, Y \in \Gamma(TM)$. Further, in [119], the corresponding Weingarten formulas are as follows:

$$\tilde{\nabla}_X \xi = -A^*_\xi X + \nabla^\perp_X \xi, \tag{2.145}$$

$$\tilde{\nabla}^*_X \xi = -A_\xi X + \nabla^{*\perp}_X \xi, \tag{2.146}$$

for any $\xi \in \Gamma\left(TM^\perp\right)$ and $X \in \Gamma(TM)$. The connections $\nabla^\perp_X$ and $\nabla^{*\perp}_X$ given by (2.145) and (2.146) are Riemannian dual connections with respect to induced metric on $\Gamma\left(TM^\perp\right)$.

The corresponding Gauss, Codazzi, and Ricci equations are given by the following:

Proposition 2.93 ([119]) *Let $\tilde{\nabla}$ be a dual connection on $\tilde{M}$ and ∇ the induced connection on M. Let $\tilde{R}$ and R be the Riemannian curvature tensors of $\tilde{\nabla}$ and ∇, respectively. Then*

$$\tilde{g}\left(\tilde{R}(X,Y)Z,W\right)=g(R(X,Y)Z,W)+\tilde{g}\left(h(X,Z),h^*(Y,W)\right) \tag{2.147}$$
$$-\tilde{g}\left(h^*(X,W),h(Y,Z)\right),$$

$$\left(\tilde{R}(X,Y)Z\right)^{\perp}=\nabla_X^{\perp}h(Y,Z)-h(\nabla_XY,Z)-h(Y,\nabla_XZ) \tag{2.148}$$
$$-\nabla_Y^{\perp}h(Y,Z)+h(\nabla_YX,Z)+h(X,\nabla_YZ),$$

$$\tilde{g}\left(R^{\perp}(X,Y)\xi,\eta\right)=\tilde{g}\left(\tilde{R}(X,Y)\xi,\eta\right)+g\left(\left[A_\xi^*,A_\eta\right]X,Y\right), \tag{2.149}$$

where $R^{\perp}$ *is the Riemannian curvature tensor on* $TM^{\perp}$, $\xi,\eta\in\Gamma\left(TM^{\perp}\right)$ *and* $\left[A_\xi^*,A_\eta\right]=A_\xi^*A_\eta-A_\eta A_\xi^*$.

For the equations of Gauss, Codazzi, and Ricci with respect to the dual connection $\tilde{\nabla}^*$ on $\tilde{M}$, we have the following proposition:

Proposition 2.94 ([119]) *Let* $\tilde{\nabla}^*$ *be a dual connection on* $\tilde{M}$ *and* ∇^* *the induced connection on* M. *Let* $\tilde{R}^*$ *and* R^* *be the Riemannian curvature tensors for* $\tilde{\nabla}^*$ *and* ∇^*, *respectively. Then*

$$\tilde{g}\left(\tilde{R}^*(X,Y)Z,W\right)=g\left(R^*(X,Y)Z,W\right)+\tilde{g}\left(h^*(X,Z),h(Y,W)\right) \tag{2.150}$$
$$-\tilde{g}\left(h(X,W),h^*(Y,Z)\right),$$

$$\left(\tilde{R}^*(X,Y)Z\right)^{\perp}=\nabla_X^{*\perp}h^*(Y,Z)-h^*\left(\nabla_X^*Y,Z\right)-h^*\left(Y,\nabla_X^*Z\right) \tag{2.151}$$
$$-\nabla_Y^{*\perp}h^*(Y,Z)+h^*\left(\nabla_Y^*X,Z\right)+h^*\left(X,\nabla_Y^*Z\right),$$

$$\tilde{g}\left(R^{*\perp}(X,Y)\xi,\eta\right)=\tilde{g}\left(\tilde{R}^*(X,Y)\xi,\eta\right)+g\left(\left[A_\xi,A_\eta^*\right]X,Y\right), \tag{2.152}$$

where $R^{*\perp}$ *is the Riemannian curvature tensor for* $\nabla^{\perp *}$ *on* $TM^{\perp}$, $\xi,\eta\in\Gamma\left(TM^{\perp}\right)$ *and* $\left[A_\xi,A_\eta^*\right]=A_\xi A_\eta^*-A_\eta^*A_\xi$.

Let $\left(\tilde{M},\tilde{g},\tilde{\nabla}\right)$ be a statistical manifold and $f:M\longrightarrow\tilde{M}$ be an immersion. We define a pair g and ∇ on M by

$$g=f^*\tilde{g},\quad g(\nabla_XY,Z)=\tilde{g}\left(\tilde{\nabla}_Xf_*Y,f_*Z\right), \tag{2.153}$$

for any $X,Y,Z\in\Gamma(TM)$, where the connection induced from $\tilde{\nabla}$ by f on the induced bundle $f^*T\tilde{M}\longrightarrow M$ is denoted by the same symbol $\tilde{\nabla}$. Then the pair

(∇, g) is a *statistical structure* on M, which is called the statistical structure *induced* by f from $\left(\tilde{\nabla}, \tilde{g}\right)$ (cf. [58]).

Definition **([58])** Let (M, g, ∇) and $\left(\tilde{M}, \tilde{g}, \tilde{\nabla}\right)$ be two statistical manifolds. An immersion $f : M \longrightarrow \tilde{M}$ is called a *statistical immersion* if (∇, g) coincides with the induced statistical structure, i.e., if (2.153) holds.

Let $f : (M, g, \nabla) \longrightarrow \left(\tilde{M}, \tilde{g}, \tilde{\nabla}\right)$ be a statistical immersion of codimension one, and $\xi \in \Gamma\left(f^* T\tilde{M}\right)$ be a unit normal vector field of f. Also we denote the dual connection of $\tilde{\nabla}$ with respect to $\tilde{g}$ by $\tilde{\nabla}^*$. Thus, from [58], we have the following Gauss and Weingarten formulas:

$$\tilde{\nabla}_X f_* Y = f_* \nabla_X Y + h(X, Y)\xi, \tag{2.154}$$

$$\tilde{\nabla}_X \xi = -f_* A^* X + \tau^*(X)\xi, \tag{2.155}$$

$$\tilde{\nabla}^*_X f_* Y = f_* \nabla^*_X Y + h^*(X, Y)\xi, \tag{2.156}$$

$$\tilde{\nabla}^*_X \xi = -f_* A X + \tau(X)\xi, \tag{2.157}$$

where $h, h^* \in \Gamma\left(TM^{(0,2)}\right)$, $A, A^* \in \Gamma\left(TM^{(1,1)}\right)$, and $\tau, \tau^* \in \Gamma(TM^*)$ satisfy

$$h(X, Y) = g(AX, Y); \quad h^*(X, Y) = g\left(A^* X, Y\right), \tag{2.158}$$

$$\tau(X) + \tau^*(X) = 0, \tag{2.159}$$

for any $X, Y \in \Gamma(TM)$.

Denote by $\tilde{R}, \tilde{R}^*, R$, and R^* the curvature tensor fields of the connections $\tilde{\nabla}, \tilde{\nabla}^*, \nabla$, and ∇^*, respectively. Then, for the Gauss equation of a statistical hypersurface, we calculate

$$\begin{aligned}\tilde{R}(X, Y)Z = {} & R(X, Y)Z - h(Y, Z)A^* X + h(X, Z)A^* Y + (\nabla_X h)(Y, Z)\xi \\ & - (\nabla_Y h)(X, Z)\xi + \tau^*(X)h(Y, Z)\xi - \tau^*(Y)h(X, Z)\xi.\end{aligned} \tag{2.160}$$

The normal component of $\tilde{R}(X, Y)Z$ is

$$\left(\tilde{R}(X, Y)Z\right)^{\perp} = (\nabla_X h)(Y, Z)\xi \tag{2.161}$$

$$- (\nabla_Y h)(X, Z)\xi + \tau^*(X)h(Y, Z)\xi - \tau^*(Y)h(X, Z)\xi,$$

which is known as the Codazzi equation. Similarly we get the Ricci equation of a statistical hypersurface as follows:

$$\tilde{R}(X, Y)\xi = -\left(\nabla_X A^*\right)Y + \left(\nabla_Y A^*\right)X - \tau^*(Y)A^*X + \tau^*(X)A^*Y \\ - h\left(X, A^*Y\right)\xi + h\left(A^*X, Y\right)\xi + d\tau^*(X, Y)\xi. \tag{2.162}$$

The equations of Gauss, Codazzi, and Ricci with respect to the dual connection $\tilde{\nabla}^*$ on $\tilde{M}$ are

$$\tilde{R}^*(X, Y)Z = R^*(X, Y)Z - h^*(Y, Z)AX + h^*(X, Z)AY + \left(\nabla^*_X h^*\right)(Y, Z)\xi \\ - \left(\nabla^*_Y h^*\right)(X, Z)\xi + \tau(X)h^*(Y, Z)\xi - \tau(Y)h^*(X, Z)\xi, \tag{2.163}$$

$$\left(\tilde{R}^*(X, Y)Z\right)^{\perp} = \left(\nabla^*_X h^*\right)(Y, Z)\xi \tag{2.164}$$

$$- \left(\nabla^*_Y h^*\right)(X, Z)\xi + \tau(X)h^*(Y, Z)\xi - \tau(Y)h^*(X, Z)\xi,$$

$$\tilde{R}^*(X, Y)\xi = -\left(\nabla^*_X A\right)Y + \left(\nabla^*_Y A\right)X - \tau(Y)AX + \tau(X)AY \\ - h^*(X, AY)\xi + h^*(AX, Y)\xi + d\tau(X, Y)\xi. \tag{2.165}$$

In the case when the ambient space is of constant curvature c, the equations of Gauss, Codazzi, and Ricci reduce to

$$R(X, Y)Z = c\{g(Y, Z)X - g(X, Z)Y\} + \left\{h(Y, Z)A^*X - h(X, Z)A^*Y\right\}, \tag{2.166}$$

$$(\nabla_X h)(Y, Z) + \tau^*(X)h(Y, Z) = (\nabla_Y h)(X, Z) + \tau^*(Y)h(X, Z), \tag{2.167}$$

$$\left(\nabla_X A^*\right)Y - \tau^*(X)A^*Y = \left(\nabla_Y A^*\right)X - \tau^*(Y)A^*X, \tag{2.168}$$

$$h\left(X, A^*Y\right) - h\left(A^*X, Y\right) = d\tau^*(X, Y), \tag{2.169}$$

and the dual ones reduce to

$$R^*(X, Y)Z = c\{g(Y, Z)X - g(X, Z)Y\} + \left\{h^*(Y, Z)AX - h^*(X, Z)AY\right\}, \tag{2.170}$$

$$\left(\nabla^*_X h^*\right)(Y, Z) + \tau(X)h^*(Y, Z) = \left(\nabla^*_Y h^*\right)(X, Z) + \tau(Y)h^*(X, Z), \tag{2.171}$$

$$\left(\nabla^*_X A\right)Y - \tau(X)AY = \left(\nabla^*_Y A\right)X - \tau(Y)AX, \tag{2.172}$$

$$h^*(X, AY) - h^*(AX, Y) = d\tau(X, Y). \tag{2.173}$$

In [7] we obtained general inequalities for statistical submanifolds.

Let $\tilde{M}$ be an $(n + k)$-dimensional statistical manifold of constant curvature $c \in \mathbb{R}$, denoted by $\tilde{M}(c)$, and M be an n-dimensional statistical submanifold of $\tilde{M}(c)$.

We use the notation

$$R(X, Y, Z, W) = g(R(X, Y)W, Z)$$

and, similarly,

$$R^*(X, Y, Z, W) = g\left(R^*(X, Y)W, Z\right),$$

where R and R^* are the curvature tensor fields of ∇ and ∇^*.

Let $\{e_1, \ldots, e_n\}$ and $\{e_{n+1}, \ldots, e_{n+k}\}$ be orthonormal tangent and normal frames, respectively, on M.

The mean curvature vector fields are given by

$$H = \frac{1}{n}\sum_{i=1}^{n} h(e_i, e_i) = \frac{1}{n}\sum_{\alpha=1}^{k}\left(\sum_{i=1}^{n} h_{ii}^{\alpha}\right) e_{n+\alpha},\ h_{ij}^{\alpha} = \tilde{g}\left(h\left(e_i, e_j\right), e_{n+\alpha}\right) \tag{2.174}$$

and

$$H^* = \frac{1}{n}\sum_{i=1}^{n} h^*(e_i, e_i) = \frac{1}{n}\sum_{\alpha=1}^{k}\left(\sum_{i=1}^{n} h_{ii}^{*\alpha}\right) e_{n+\alpha},\ h_{ij}^{*\alpha} = \tilde{g}\left(h^*\left(e_i, e_j\right), e_{n+\alpha}\right). \tag{2.175}$$

Then we have the following.

Proposition 2.95 ([7]) *Let M be an n-dimensional submanifold of an $(n+k)$-dimensional statistical manifold $\tilde{M}(c)$ of constant curvature $c \in \mathbb{R}$. Assume that the imbedding curvature tensors h and h^* satisfy*

$$h(X, Y) = g(X, Y)H, \quad h^*(X, Y) = g(X, Y)H^*,$$

for any $X, Y \in \Gamma(TM)$. Then M is also a statistical manifold of constant curvature $c + g(H, H^*)$, whenever $g(H, H^*)$ is constant.

Definition ([103]) Let $\tilde{M}$ be an $(n+k)$-dimensional statistical manifold. Then the *Ricci tensor* $\tilde{S}$ (of type $(0, 2)$) is defined by

$$\tilde{S}(Y, Z) = \text{trace}\left\{X \longrightarrow \tilde{R}(X, Y)Z\right\},$$

where $\tilde{R}$ is the curvature tensor field of the affine connection $\tilde{\nabla}$ on $\tilde{M}$.

Thus we have the following result.

Theorem 2.96 ([7]) *Let $\tilde{M}(c)$ be an $(n+k)$-dimensional statistical manifold of constant curvature $c \in \mathbb{R}$ and M an n-dimensional statistical submanifold of $\tilde{M}(c)$. Also let $\{e_1, \ldots, e_n\}$ and $\{n_1, \ldots, n_k\}$ be orthonormal tangent and normal frames, respectively, on M. Then Ricci tensor S and dual Ricci tensor S^* of M satisfy*

$$S(X,Y)=c(n-1)g(X,Y)+\sum_{i=1}^{k}\left[g\left(A_{n_i}X,Y\right)tr\,A_{n_i}^{*}-g\left(A_{n_i}Y,A_{n_i}^{*}X\right)\right], \tag{2.176}$$

$$S^{*}(X,Y)=c(n-1)g(X,Y)+\sum_{i=1}^{k}g\left(A_{n_i}^{*}X,Y\right)tr\,A_{n_i}-g\left(A_{n_i}X,A_{n_i}^{*}Y\right). \tag{2.177}$$

The proof is technical.

Definition **([103])** Let ∇ be a torsion-free affine connection on a Riemannian manifold M that admits a parallel volume element ω. If ω is a volume element on M such that $\nabla\omega=0$, then (∇,ω) is called an *equiaffine structure* on M.

Proposition **([103])** *An affine connection* ∇ *with zero torsion has symmetric Ricci tensor if and only if it is locally equiaffine.*

Thus we have the following result for statistical manifolds having equiaffine connection.

Lemma 2.97 **([7])** *Let* $\tilde{M}(c)$ *be an* $(n+k)$*-dimensional statistical manifold of constant curvature* $c\in\mathbb{R}$ *and* M *an* n*-dimensional submanifold of* $\tilde{M}(c)$. *Assume that the affine connection* ∇ *of* M *is equiaffine. Then one verifies*

$$\sum_{i=1}^{k}\left[A_{n_i},A_{n_i}^{*}\right]=0.$$

Corollary 2.98 **([7])** *Let* $\tilde{M}(c)$ *be an* $(n+k)$*-dimensional statistical manifold of constant curvature* $c\in\mathbb{R}$ *and* M *an* n*-dimensional equiaffine submanifold* M *of* $\tilde{M}(c)$. *Let* S *and* S^{*} *denote the dual Ricci tensors of* M. *Then we have*

$$\left(S-S^{*}\right)(X,Y)=\sum_{i=1}^{k}g\left(\left(A_{n_i}-A_{n_i}^{*}\right)X,Y\right)tr\left(A_{n_i}^{*}-A_{n_i}\right).$$

We established an estimate of the scalar curvature of a statistical submanifold in terms of its mean curvature vectors and the lengths of the imbedding curvature tensors.

Proposition 2.99 **([7])** *Let* $\tilde{M}(c)$ *be an* $(n+k)$*-dimensional statistical manifold of constant curvature* $c\in\mathbb{R}$ *and* M *an* n*-dimensional statistical submanifold of* $\tilde{M}(c)$. *We have*

$$2\tau\geq n(n-1)c+n^{2}\tilde{g}(H,H^{*})-\|h\|\left\|h^{*}\right\|, \tag{2.178}$$

where τ *is the scalar curvature of* M.

Proof From (2.147), we have the Gauss equation as follows:

$$R(X, Y, Z, W) = c\left[g(X, Z)\, g(Y, W) - g(X, W)\, g(Y, Z)\right]$$
$$+ \tilde{g}\left(h^*(X, Z), h(Y, W)\right) - \tilde{g}\left(h(X, W), h^*(Y, Z)\right),$$

where X, Y, Z, and $W \in \Gamma(TM)$. Putting $X = Z = e_i$ and $Y = W = e_j$, $i, j = 1, \ldots, n$, we write

$$R\left(e_i, e_j, e_i, e_j\right) = c\left[g(e_i, e_i)\, g\left(e_j, e_j\right) - g\left(e_i, e_j\right)^2\right] + \tilde{g}\left(h^*(e_i, e_i), h\left(e_j, e_j\right)\right)$$
$$- \tilde{g}\left(h\left(e_i, e_j\right), h^*\left(e_j, e_i\right)\right).$$

By summing over $1 \leq i, j \leq n$, it follows that

$$2\tau = \left(n^2 - n\right) c + n^2 g\left(H, H^*\right) - \sum_{i,j=1}^{n} \sum_{\alpha=1}^{k} h_{ij}^{\alpha} h_{ij}^{*\alpha} \tag{2.179}$$

$$\geq n(n-1)\, c + n^2 \tilde{g}\left(H, H^*\right) - \|h\|\, \left\|h^*\right\|,$$

which gives (2.178).

Remark On any statistical submanifold M of $\tilde{M}(c)$ the equality $\tau = \tau^*$ holds.

Let ∇^0 be the Levi-Civita connection of an n-dimensional submanifold M in an $(n+k)$-dimensional statistical manifold $\tilde{M}(c)$ of constant curvature c. Denote by H^0 the mean curvature vector field. Then a sharp relationship between the Ricci curvature and the squared mean curvature obtained by B.Y. Chen [29] is the following:

$$\mathrm{Ric}^0(X) \leq \frac{n^2}{4} \left\|H^0\right\|^2 + (n-1)\, c, \tag{2.180}$$

which is known as the *Chen-Ricci inequality*.

From (2.138) we get $2H^0 = H + H^*$ and thus

$$\left\|H^0\right\|^2 = \frac{1}{4}\left(\|H\|^2 + \left\|H^*\right\|^2 + 2g\left(H, H^*\right)\right). \tag{2.181}$$

Therefore, from the last two equations we derive

$$\mathrm{Ric}^0(X) \leq \frac{n^2}{16}\|H\|^2 + \frac{n^2}{16}\left\|H^*\right\|^2 + \frac{n^2}{8}\tilde{g}\left(H, H^*\right) + (n-1)\, c. \tag{2.182}$$

For statistical hypersurfaces we also obtained some inequalities.

By analogy with Proposition 2.99, we have an inequality for statistical hypersurfaces as follows:

Proposition 2.100 ([7]) *Let M be a statistical hypersurface of an $(n+1)$-dimensional statistical manifold $\tilde{M}(c)$ of constant curvature $c \in \mathbb{R}$. We have*

$$2\tau \geq n(n-1)c + n^2 \|H\| \left\|H^*\right\| - \|h\| \left\|h^*\right\|. \tag{2.183}$$

The proof uses the well-known Cauchy-Buniakowski-Schwarz inequality.

Proposition 2.101 ([7]) *Let M be a statistical hypersurface of an $(n+1)$-dimensional statistical manifold $\tilde{M}(c)$. For each $X \in T_p(M)$ we have*

$$Ric(X) = (n-1)c + n\tilde{g}\left(h^*(X,X), H\right) - \sum_{i=1}^{n} h_{i1}h^*_{i1},$$

$$Ric^*(X) = (n-1)c + n\tilde{g}\left(h(X,X), H^*\right) - \sum_{i=1}^{n} h_{i1}h^*_{i1}.$$

Example Recall Example 5.4 from [58]. Let $(\mathbb{H}, \tilde{g})$ be the upper half space of constant curvature -1

$$\mathbb{H} := \left\{y = \left(y^1, \ldots, y^{n+1}\right) \in \mathbb{R}^{n+1} | y^{n+1} > 0\right\}, \tilde{g} := \left(y^{n+1}\right)^{-2} \sum_{k=1}^{n+1} dy^k dy^k.$$

An affine connection $\tilde{\nabla}$ on $\mathbb{H}$ is given by

$$\tilde{\nabla}_{\frac{\partial}{\partial y^{n+1}}} \frac{\partial}{\partial y^{n+1}} = \left(y^{n+1}\right)^{-1} \frac{\partial}{\partial y^{n+1}},$$

$$\tilde{\nabla}_{\frac{\partial}{\partial y^i}} \frac{\partial}{\partial y^j} = 2\delta_{ij} \left(y^{n+1}\right)^{-1} \frac{\partial}{\partial y^{n+1}},$$

$$\tilde{\nabla}_{\frac{\partial}{\partial y^i}} \frac{\partial}{\partial y^{n+1}} = \tilde{\nabla}_{\frac{\partial}{\partial y^{n+1}}} \frac{\partial}{\partial y^j} = 0,$$

where $i, j = 1, \ldots, n$. The curvature tensor field $\tilde{R}$ of $\tilde{\nabla}$ is identically zero, i.e., $c = 0$. Thus $\left(\mathbb{H}, \tilde{\nabla}, \tilde{g}\right)$ is a Hessian manifold of constant Hessian curvature 4.

For a constant $y_0 > 0$, we get the following immersion by f_0:

$$f_0 : \mathbb{R}^n \longrightarrow \mathbb{H}, f_0\left(y^1, \ldots, y^n\right) = \left(y^1, \ldots, y^n, y_0\right).$$

Let (∇, g) be the statistical structure on $\mathbb{R}^n$ induced by f_0 from $\left(\tilde{\nabla}, \tilde{g}\right)$. We then get that (∇, g) is a Hessian structure and $K^{(\nabla,g)} = 0$. In other words, f_0 is a statistical immersion of the trivial Hessian manifold $(\mathbb{R}^n, \nabla, g)$ into the upper half Hessian space $\left(\mathbb{H}, \tilde{\nabla}, \tilde{g}\right)$. It is easy to calculate that

$$\xi = y_0 \frac{\partial}{\partial y^{n+1}}, \quad h = 2g, \quad h^* = 0, \quad \left\| H^* \right\| = 0, \tag{2.184}$$

which means that the equality case of (2.183) is satisfied for $(\mathbb{R}^n, \nabla, g)$ and $\left(\mathbb{H}, \tilde{\nabla}, \tilde{g}\right)$.

On the other hand this example can be generalized by using Lemma 5.3 of [58]. Let $\left(\mathbb{H}, \tilde{\nabla}, \tilde{g}\right)$ be a Hessian manifold of constant Hessian curvature $\tilde{c} \neq 0$, (M, ∇, g) a trivial Hessian manifold, and $f : M \longrightarrow \mathbb{H}$ a statistical immersion of codimension one. Then the following expressions hold:

$$A^* = 0, \quad h^* = 0, \quad \left\| H^* \right\| = 0;$$

thus the immersion f has codimension one and satisfies the equality case of (2.183).

Next, we give the complete proof of the Chen-Ricci inequalities for statistical submanifolds (of arbitrary codimension) in statistical manifold of constant curvature.

Let $\tilde{M}(c)$ be an $(n+k)$-dimensional statistical manifold of constant curvature $c \in \mathbb{R}$ and M an n-dimensional statistical submanifold of $\tilde{M}(c)$. Recall the Gauss equation:

$$\begin{aligned} \tilde{R}(X, Y, Z, W) = R(X, Y, Z, W) + \tilde{g}\left(h(X, Z), h^*(Y, W)\right) \\ - \tilde{g}\left(h^*(X, W), h(Y, Z)\right). \end{aligned}$$

By setting $X = Z = e_i$ and $Y = W = e_j$, $i, j = 1, \ldots, n$ and summing over $1 \leq i, j \leq n$, then we have

$$n(n-1)c = 2\tau - n^2 \tilde{g}\left(H, H^*\right) + \sum_{i,j=1}^{n} \tilde{g}\left(h^*\left(e_i, e_j\right), h\left(e_i, e_j\right)\right),$$

where H and H^* are the mean curvature vector fields defined by (2.174) and (2.175).

From this, we get

$$n(n-1)c = 2\tau - \frac{n^2}{2}\left[\tilde{g}\left(H + H^*, H + H^*\right) - \tilde{g}(H, H) - g\left(H^*, H^*\right)\right]$$

$$+\frac{1}{2}\sum_{i,j=1}^{n}\left[\tilde{g}\left(h^{*}\left(e_i,e_j\right)+h\left(e_i,e_j\right),h^{*}\left(e_i,e_j\right)+h\left(e_i,e_j\right)\right)\right.$$

$$\left.-\tilde{g}\left(h\left(e_i,e_j\right),h\left(e_i,e_j\right)\right)-\tilde{g}\left(h^{*}\left(e_i,e_j\right),h^{*}\left(e_i,e_j\right)\right)\right].$$

From $2H^0 = H + H^*$ it follows that

$$n(n-1)c = 2\tau - 2n^2\tilde{g}\left(H^0, H^0\right) + \frac{n^2}{2}\tilde{g}(H,H) + \frac{n^2}{2}\tilde{g}\left(H^*,H^*\right) \tag{2.185}$$

$$+2\sum_{i,j=1}^{n}\tilde{g}\left(h^0\left(e_i,e_j\right),h^0\left(e_i,e_j\right)\right)-\frac{1}{2}\left(\|h\|^2+\|h^*\|^2\right).$$

On the other hand we can write

$$\|h\|^2 = \sum_{\alpha=1}^{k}\left\{\left(h_{11}^{\alpha}\right)^2+\left(h_{22}^{\alpha}+\ldots+h_{nn}^{\alpha}\right)^2+2\sum_{1\le i<j\le n}\left(h_{ij}^{\alpha}\right)^2\right\}$$

$$-\sum_{\alpha=1}^{k}\sum_{2\le i\ne j\le n}h_{ii}^{\alpha}h_{jj}^{\alpha}$$

$$=\frac{1}{2}\sum_{\alpha=1}^{k}\left\{\left(h_{11}^{\alpha}+h_{22}^{\alpha}+\ldots+h_{nn}^{\alpha}\right)^2+\left(h_{11}^{\alpha}-h_{22}^{\alpha}-\ldots-h_{nn}^{\alpha}\right)^2\right\}$$

$$+2\sum_{\alpha=1}^{k}\sum_{1\le i<j\le n}\left(h_{ij}^{\alpha}\right)^2-\sum_{\alpha=1}^{k}\sum_{2\le i\ne j\le n}h_{ii}^{\alpha}h_{jj}^{\alpha}\ge\frac{1}{2}n^2\|H\|^2$$

$$-\sum_{\alpha=1}^{k}\sum_{2\le i\ne j\le n}\left[h_{ii}^{\alpha}h_{jj}^{\alpha}-(h_{ij}^{\alpha})^2\right].$$

We similarly derive

$$\left\|h^*\right\|^2\ge\frac{1}{2}n^2\left\|H^*\right\|^2-\sum_{\alpha=1}^{k}\sum_{2\le i\ne j\le n}\left[h_{ii}^{*\alpha}h_{jj}^{*\alpha}-(h_{ij}^{*\alpha})^2\right]. \tag{2.186}$$

Thus we have the following inequality:

$$\|h\|^2+\|h^*\|^2 \geq \frac{1}{2}n^2\|H\|^2+\frac{1}{2}n^2\|H^*\|^2-\sum_{\alpha=1}^{k}\sum_{2\leq i\neq j\leq n}\left(h_{ii}^{\alpha}+h_{ii}^{*\alpha}\right)\left(h_{jj}^{\alpha}+h_{jj}^{*\alpha}\right)$$
$$+2\sum_{\alpha=1}^{k}\sum_{2\leq i\neq j\leq n}h_{ii}^{\alpha}h_{jj}^{*\alpha}+\sum_{\alpha=1}^{k}\sum_{2\leq i\neq j\leq n}\left[(h_{ij}^{\alpha})^2+(h_{ij}^{*\alpha})^2\right]. \tag{2.187}$$

Substituting (2.187) into (2.185), we obtain

$$n(n-1)c \leq 2\tau-2n^2\tilde{g}(H^0,H^0)+\frac{n^2}{2}\tilde{g}(H,H)+\frac{n^2}{2}\tilde{g}(H^*,H^*)+2||h^0||^2$$
$$+2\sum_{\alpha=1}^{k}\sum_{2\leq i\neq j\leq n}h_{ii}^{0\alpha}h_{jj}^{0\alpha}-\frac{n^2}{4}\tilde{g}(H,H)-\frac{n^2}{4}\tilde{g}(H^*,H^*)-\sum_{\alpha=1}^{k}\sum_{2\leq i\neq j\leq n}h_{ii}^{\alpha}h_{jj}^{*\alpha}$$
$$-\frac{1}{2}\sum_{\alpha=1}^{k}\sum_{2\leq i\neq j\leq n}\left[(h_{ij}^{\alpha})^2+(h_{ij}^{*\alpha})^2\right].$$

Since

$$\sum_{2\leq i\neq j\leq n}R\left(e_i,e_j,e_i,e_j\right)=(n-1)(n-2)c+\sum_{\alpha=1}^{k}\sum_{2\leq i\neq j\leq n}\left(h_{ii}^{\alpha}h_{jj}^{*\alpha}-h_{ij}^{\alpha}h_{ij}^{*\alpha}\right),$$

the previous inequality becomes

$$n(n-1)c \leq 2\tau-2n^2\tilde{g}(H^0,H^0)+\frac{n^2}{4}\tilde{g}(H,H)+\frac{n^2}{4}\tilde{g}(H^*,H^*)+2|||h^0||^2$$

$$+2\sum_{\alpha=1}^{k}\sum_{2\leq i\neq j\leq n}h_{ii}^{0\alpha}h_{jj}^{0\alpha}-\sum_{2\leq i\neq j\leq n}R(e_i,e_j,e_i,e_j)+(n-1)(n-2)c$$
$$-\frac{1}{2}\sum_{\alpha=1}^{k}\sum_{2\leq i\neq j\leq n}(h_{ij}^{\alpha}+h_{ij}^{*\alpha})^2.$$

Then we get

$$\operatorname{Ric}(X) \geq n^2\tilde{g}(H^0,H^0)-\frac{n^2}{8}\tilde{g}(H,H)+\frac{n^2}{8}\tilde{g}(H^*,H^*)+(n-1)c \tag{2.188}$$

$$-\|h^0\|^2-\sum_{\alpha=1}^{k}\sum_{2\le i\ne j\le n}\left[h_{ii}^{0\alpha}h_{jj}^{0\alpha}-(h_{ij}^{0\alpha})^2\right].$$

By the Gauss equation with respect to the Levi-Civita connection, we have

$$\sum_{1\le i\ne j\le n}\tilde{R}^0(e_i,e_j,e_i,e_j)=2\tau^0-n^2\tilde{g}(H^0,H^0)+\|h^0\|^2,$$

and, respectively,

$$\begin{aligned}\sum_{2\le i\ne j\le n}\tilde{R}^0(e_i,e_j,e_i,e_j)&=\sum_{2\le i\ne j\le n}R^0(e_i,e_j,e_i,e_j)\\&-\sum_{\alpha=1}^{k}\sum_{2\le i\ne j\le n}\left[h_{ii}^{0\alpha}h_{jj}^{0\alpha}-(h_{ij}^{0\alpha})^2\right].\end{aligned}$$

Substituting in (2.188) it follows that

$$\begin{aligned}\mathrm{Ric}(X)&\ge 2\tau^0-\sum_{1\le i\ne j\le n}\tilde{R}^0(e_i,e_j,e_i,e_j)-\frac{n^2}{8}\tilde{g}(H,H)-\frac{n^2}{8}\tilde{g}(H^*,H^*)\\&+(n-1)c-\sum_{2\le i\ne j\le n}R^0(e_i,e_j,e_i,e_j)+\sum_{2\le i\ne j\le n}\tilde{R}^0(e_i,e_j,e_i,e_j).\end{aligned}$$

Finally we obtain

$$\mathrm{Ric}(X)\ge 2\mathrm{Ric}^0(X)-\frac{n^2}{8}\tilde{g}(H,H)-\frac{n^2}{8}\tilde{g}(H^*,H^*)+(n-1)c-2\sum_{i=2}^{n}\tilde{K}^0(X\wedge e_i).$$

We denote by $\max\tilde{K}^0(X\wedge\cdot)$ the maximum of the sectional curvature function of $\tilde{M}(c)$ with respect to $\tilde{\nabla}$ restricted to 2-plane sections of the tangent space T_pM which are tangent to X.

Summing up, we can state the following Chen-Ricci inequality:

Theorem 2.102 ([7]) *Let M be an n-dimensional statistical submanifold of an $(n+k)$-dimensional statistical manifold $\tilde{M}(c)$. For each $X\in T_p(M)$ unit, we have*

$$\begin{aligned}Ric(X)&\ge 2Ric^0(X)-\frac{n^2}{8}\tilde{g}(H,H)-\frac{n^2}{8}\tilde{g}(H^*,H^*)\\&+(n-1)c-2(n-1)\max\tilde{K}^0(X\wedge\cdot).\end{aligned}$$

Particular Case M is a minimal submanifold. Because $H^0 = 0$, we have $H + H^* = 0$. Then the previous inequality implies the following:

Corollary 2.103 ([7]) *Let M be a minimal n-dimensional statistical submanifold of an $(n+k)$-dimensional statistical manifold $\tilde{M}(c)$. For each $X \in T_p(M)$ unit, we have*

$$Ric(X) \geq 2Ric^0(X) + \frac{n^2}{4}\tilde{g}(H, H^*) + (n-1)c - 2(n-1)\max \tilde{K}^0(X \wedge \cdot).$$

Remark Similar inequalities can be stated for the Ricci curvature Ric*.

In 2017, in [8], we proved the generalized Wintgen inequality for statistical submanifolds in statistical manifolds of constants curvature. The Wintgen inequality is a sharp geometric inequality for surfaces in the four-dimensional Euclidean space involving the Gauss curvature (intrinsic invariant) and the normal curvature and squared mean curvature (extrinsic invariants), respectively.

Recall that De Smet, Dillen, Verstraelen, and Vrancken [54] conjectured a generalized Wintgen inequality for submanifolds of arbitrary dimension and codimension in Riemannian space forms. This conjecture was proved by Lu [72] and by Ge and Tang [60], independently.

For surfaces M^2 of the Euclidean space $\mathbb{E}^3$, the Euler inequality $G \leq \|H\|^2$ is fulfilled, where G is the (intrinsic) Gauss curvature of M^2 and $\|H\|^2$ is the (extrinsic) squared mean curvature of M^2.

Furthermore, $G = \|H\|^2$ everywhere on M^2 if and only if M^2 is totally umbilical, or still, by a theorem of Meusnier, if and only if M^2 is (a part of) a plane $\mathbb{E}^2$ or, it is (a part of) a round sphere S^2 in $\mathbb{E}^3$.

In 1979, P. Wintgen [120] proved that the Gauss curvature G, the squared mean curvature $\|H\|^2$, and the normal curvature $G^{\perp}$ of any surface M^2 in $\mathbb{E}^4$ always satisfy the inequality

$$G \leq \|H\|^2 - |G^{\perp}|;$$

the equality holds if and only if the ellipse of curvature of M^2 in $\mathbb{E}^4$ is a circle.

The Whitney 2-sphere satisfies the equality case of the Wintgen inequality identically.

A survey containing recent results on surfaces satisfying identically the equality case of Wintgen inequality can be read in [41].

Later, the Wintgen inequality was extended by B. Rouxel [110] and by I.V. Guadalupe and L. Rodriguez [63] independently, for surfaces M^2 of arbitrary codimension m in real space forms $\widetilde{M}^{2+m}(c)$, namely

$$G \leq \|H\|^2 - |G^{\perp}| + c.$$

The equality case was also investigated.

A corresponding inequality for totally real surfaces in n-dimensional complex space forms was obtained in [74]. The equality case was studied, and a nontrivial example of a totally real surface satisfying the equality case identically was given.

In 1999, P.J. De Smet, F. Dillen, L. Verstraelen, and L. Vrancken [54] formulated the conjecture on Wintgen inequality for submanifolds of real space forms, which is also known as the *DDVV conjecture*.

This conjecture was proven by the authors for submanifolds M^n of arbitrary dimension $n \geq 2$ and codimension 2 in real space forms $\tilde{M}^{n+2}(c)$ of constant sectional curvature c. The DDVV conjecture was finally settled for the general case by Z. Lu [72] and independently by J. Ge and Z. Tang [60].

Generalized Wintgen inequalities for Lagrangian submanifolds in complex space forms [96] and Legendrian submanifolds in Sasakian space forms [97] were obtained, respectively. Moreover, in [5] a version of the Euler inequality and the Wintgen inequality for statistical surfaces in statistical manifolds of constant curvature was stated.

By using the sectional curvature K on M^n defined in [5] and also in [105]:

$$K(X \wedge Y) = \frac{1}{2}\left[g(R(X, Y)X, Y) + g\left(R^*(X, Y)X, Y\right)\right],$$

for any orthonormal vectors $X, Y \in T_pM^n$, $p \in M^n$, we derive a generalized Wintgen inequality for statistical submanifolds in statistical manifolds of constant curvature:

Theorem 2.104 ([8]) *Let M^n be a submanifold in a statistical manifold $\left(\tilde{M}^{n+m}, c\right)$ of constant curvature c. Then*

$$\rho^{\perp} + 3\rho \leq \frac{15}{2}\|H\|^2 + \frac{15}{2}\left\|H^*\right\|^2 + 12g\left(H, H^*\right) - 3c + 30\left(\tilde{\rho}^0 - \rho^0\right).$$

3 Warped Product Submanifolds

We recall the results obtained in [75–77] on warped product submanifolds in complex space forms, generalized complex space forms, and quaternion space forms, respectively.

B.Y. Chen [37] established a sharp inequality for the warping function of a warped product submanifold in a Riemannian space form in terms of the squared mean curvature. In [36], he studied warped product submanifolds in complex hyperbolic spaces.

The notion of *warped product* plays some important role in Differential Geometry and physics [35]. For instance, the best relativistic model of the Schwarzschild space-time that describes the out space around a massive star or a black hole is given as a warped product.

One of the fundamental problems in the theory of submanifolds is the immersibility (or non-immersibility) of a Riemannian manifold in a Euclidean space (or, more generally, in a space form). According to a well-known theorem on Nash, every Riemannian manifold can be isometrically immersed in some Euclidean spaces with sufficiently high codimension.

Nash's theorem implies, in particular, that every warped product $M_1 \times_f M_2$ can be immersed as a Riemannian submanifold in some Euclidean space. Moreover, many important submanifolds in real and complex space forms are expressed as a warped product submanifold.

Every Riemannian manifold of constant curvature c can be locally expressed as a warped product whose warping function satisfies $\Delta f = cf$. For example, $S^n(1)$ is locally isometric to $(0, \pi) \times_{\cos t} S^{n-1}(1)$, $\mathbf{E}^n$ is locally isometric to $(0, \infty) \times_x S^{n-1}(1)$, and $H^n(-1)$ is locally isometric to $\mathbf{R} \times_{e^x} \mathbf{E}^{n-1}$ (see [35]).

Let (M_1, g_1) and (M_2, g_2) be two Riemannian manifolds and f a positive differentiable function on M_1. The *warped product* of M_1 and M_2 is the Riemannian manifold

$$M_1 \times_f M_2 = (M_1 \times M_2, g),$$

where $g = g_1 + f^2 g_2$ (see, for instance, [37]).

Let $x : M_1 \times_f M_2 \to \widetilde{M}(c)$ be an isometric immersion of a warped product $M_1 \times_f M_2$ into a complex space form $\widetilde{M}(c)$. We denote by h the second fundamental form of x and $H_i = \frac{1}{n_i} \text{trace}\, h_i$, where trace h_i is the trace of h restricted to M_i and $n_i = \dim M_i$ $(i = 1, 2)$.

Recall that for a warped product $M_1 \times_f M_2$, we denote by $\mathcal{D}_1$ and $\mathcal{D}_2$ the distributions given by the vectors tangent to leaves and fibers, respectively. Thus, $\mathcal{D}_1$ is obtained from the tangent vectors of M_1 via the horizontal lift and $\mathcal{D}_2$ by tangent vectors of M_2 via the vertical lift.

In [75] we established an inequality between the warping function f (intrinsic structure) and the squared mean curvature $\|H\|^2$ and the holomorphic sectional curvature c (extrinsic structures) for warped product submanifolds $M_1 \times_f M_2$ with $J\mathcal{D}_1 \perp \mathcal{D}_2$ (in particular, CR-warped product submanifolds and CR-Riemannian products) in any complex space form $\widetilde{M}(c)$. Examples of such submanifolds which satisfy the equality case are given.

Recall that a submanifold N in a Kaehler manifold $\widetilde{M}$ is called a *CR-submanifold* (see [83]) if there exists on N a holomorphic distribution $\mathcal{D}$ whose orthogonal complementary distribution $\mathcal{D}^\perp$ is a totally real distribution, i.e., $J\mathcal{D}_x^\perp \subset T_p^\perp N$. A CR-submanifold of a Kaehler manifold $\widetilde{M}$ is called a *CR- product* if it is a Riemannian product of a Kaehler submanifold and a totally real submanifold. There do not exist warped product CR-submanifolds of the form $M_\perp \times_f M_\top$, with $M_\perp$ a totally real submanifold and $M_\top$ a complex submanifold, other than CR-products. A *CR-warped product* is a warped product CR-submanifold of the form $M_\top \times_f M_\perp$, by reversing the two factors [33].

As applications we will give some non-immersions theorems.

Theorem 3.1 ([75]) *Let $x : M_1 \times_f M_2 \to \widetilde{M}(c)$ be an isometric immersion of an n-dimensional warped product with $J\mathcal{D}_1 \perp \mathcal{D}_2$ into a $2m$-dimensional complex space form $\widetilde{M}(c)$. Then,*

$$\frac{\Delta f}{f} \leq \frac{n^2}{4n_2}\|H\|^2 + n_1\frac{c}{4}, \tag{3.1}$$

where $n_i = \dim M_i, i = 1, 2$, and Δ is the Laplacian operator of M_1. Moreover, the equality case of holds identically if and only if x is a mixed totally geodesic immersion and $n_1 H_1 = n_2 H_2$, where $H_i, i = 1, 2$, are the partial mean curvature vectors.

Proof Let $M_1 \times_f M_2$ be a warped product submanifold into a complex space form $\widetilde{M}(c)$ of constant holomorphic sectional curvature c.

Since $M_1 \times_f M_2$ is a warped product, it is known that

$$\nabla_X Z = \nabla_Z X = \frac{1}{f}(Xf)Z, \tag{3.2}$$

for any vector fields X, Z tangent to M_1, M_2, respectively.

If X and Z are unit vector fields, it follows that the sectional curvature $K(X \wedge Z)$ of the plane section spanned by X and Z is given by

$$K(X \wedge Z) = g(\nabla_Z \nabla_X X - \nabla_X \nabla_Z X, Z) = \frac{1}{f}\{(\nabla_X X)f - X^2 f\}. \tag{3.3}$$

We choose a local orthonormal frame $\{e_1, \ldots, e_n, e_{n+1}, \ldots, e_{2m}\}$, such that $e_1, \ldots, e_{n_1}$ are tangent to M_1, $e_{n_1+1}, \ldots, e_n$ are tangent to M_2, and e_{n+1} is parallel to the mean curvature vector H.

Then, using (3.3), we get

$$\frac{\Delta f}{f} = \sum_{j=1}^{n_1} K(e_j \wedge e_s), \tag{3.4}$$

for each $s \in \{n_1 + 1, \ldots, n\}$.

From the equation of Gauss, we have

$$n^2\|H\|^2 = 2\tau + \|h\|^2 - n(n-1)\frac{c}{4} - 3\|P\|^2\frac{c}{4}. \tag{3.5}$$

We set

$$\delta = 2\tau - n(n-1)\frac{c}{4} - 3\|P\|^2\frac{c}{4} - \frac{n^2}{2}\|H\|^2. \tag{3.6}$$

Then, (3.5) can be written as

$$n^2\|H\|^2 = 2(\delta + \|h\|^2). \tag{3.7}$$

With respect to the above orthonormal frame, (3.7) takes the following form:

$$\left(\sum_{i=1}^{n} h_{ii}^{n+1}\right)^2 = 2\left\{\delta + \sum_{i=1}^{n}(h_{ii}^{n+1})^2 + \sum_{i\neq j}(h_{ij}^{n+1})^2 + \sum_{r=n+2}^{2m}\sum_{i,j=1}^{n}(h_{ij}^r)^2\right\}.$$

If we put $a_1 = h_{11}^{n+1}, a_2 = \sum_{i=2}^{n_1} h_{ii}^{n+1}$, and $a_3 = \sum_{t=n_1+1}^{n} h_{tt}^{n+1}$, the above equation becomes

$$\left(\sum_{i=1}^{3} a_i\right)^2 = 2\left\{\delta + \sum_{i=1}^{3} a_i^2 + \sum_{1\leq i\neq j\leq n}(h_{ij}^{n+1})^2 + \sum_{r=n+2}^{2m}\sum_{i,j=1}^{n}(h_{ij}^r)^2\right.$$

$$\left. - \sum_{2\leq j\neq k\leq n_1} h_{jj}^{n+1}h_{kk}^{n+1} - \sum_{n_1+1\leq s\neq t\leq n} h_{ss}^{n+1}h_{tt}^{n+1}\right\}.$$

Thus a_1, a_2, a_3 satisfy the lemma of Chen (for $n = 3$), i.e.,

$$\left(\sum_{i=1}^{3} a_i\right)^2 = 2\left(b + \sum_{i=1}^{3} a_i^2\right).$$

Then $2a_1a_2 \geq b$, with equality holding if and only if $a_1 + a_2 = a_3$.
In the case under consideration, this means

$$\sum_{1\leq j<k\leq n_1} h_{jj}^{n+1}h_{kk}^{n+1} + \sum_{n_1+1\leq s<t\leq n} h_{ss}^{n+1}h_{tt}^{n+1} \tag{3.8}$$

$$\geq \frac{\delta}{2} + \sum_{1\leq\alpha<\beta\leq n}(h_{\alpha\beta}^{n+1})^2 + \frac{1}{2}\sum_{r=n+2}^{2m}\sum_{\alpha,\beta=1}^{n}(h_{\alpha\beta}^r)^2.$$

The equality holds if and only if

$$\sum_{i=1}^{n_1} h_{ii}^{n+1} = \sum_{t=n_1+1}^{n} h_{tt}^{n+1}. \tag{3.9}$$

Using again the Gauss equation, we have

$$n_2\frac{\Delta f}{f} = \tau - \sum_{1\leq j<k\leq n_1} K(e_j \wedge e_k) - \sum_{n_1+1\leq s<t\leq n} K(e_s \wedge e_t) \tag{3.10}$$

$$= \tau - \frac{n_1(n_1-1)(c)}{8} - \sum_{r=n+1}^{2m} \sum_{1\le j<k\le n_1} (h^r_{jj}h^r_{kk} - (h^r_{jk})^2) - 3\frac{c}{4} \sum_{1\le j<k\le n_1} g^2(Je_j, e_k)$$

$$- \frac{n_2(n_2-1)(c)}{8} - \sum_{r=n+1}^{2m} \sum_{n_1+1\le s<t\le n} (h^r_{ss}h^r_{tt} - (h^r_{st})^2) - 3\frac{c}{4} \sum_{n_1+1\le s<t\le n} g^2(Je_s, e_t).$$

Combining (3.8) and (3.10) and taking account of (3.4), we obtain

$$n_2\frac{\Delta f}{f} \le \tau - \frac{n(n-1)(c)}{8} + n_1n_2\frac{c}{4} - \frac{\delta}{2} - 3\frac{c}{4} \sum_{1\le j<k\le n_1} g^2(Je_j, e_k) \tag{3.11}$$

$$- 3\frac{c}{4} \sum_{n_1+1\le s<t\le n} g^2(Je_s, e_t) - \sum_{1\le j\le n_1; n_1+1\le t\le n} (h^{n+1}_{jt})^2 - \frac{1}{2}\sum_{r=n+2}^{2m}\sum_{\alpha,\beta=1}^{n} (h^r_{\alpha\beta})^2$$

$$+ \sum_{r=n+2}^{2m} \sum_{1\le j<k\le n_1} ((h^r_{jk})^2 - h^r_{jj}h^r_{kk}) + \sum_{r=n+2}^{2m} \sum_{n_1+1\le s<t\le n} ((h^r_{st})^2 - h^r_{ss}h^r_{tt})$$

$$= \tau - \frac{n(n-1)(c)}{8} + n_1n_2\frac{c}{4} - \frac{\delta}{2} - \sum_{r=n+1}^{2m} \sum_{1\le j\le n_1; n_1+1\le t\le n} (h^r_{jt})^2$$

$$- 3\frac{c}{4} \sum_{1\le j<k\le n_1} g^2(Je_j, e_k)$$

$$- 3\frac{c}{4} \sum_{n_1+1\le s<t\le n} g^2(Je_s, e_t) - \frac{1}{2}\sum_{r=n+2}^{2m}\left(\sum_{j=1}^{n_1} h^r_{jj}\right)^2 - \frac{1}{2}\sum_{r=n+2}^{2m}\left(\sum_{t=n_1+1}^{n} h^r_{tt}\right)^2$$

$$\le \tau - \frac{n(n-1)(c)}{8} + n_1n_2\frac{c}{4} - \frac{\delta}{2} - 3\frac{c}{4} \sum_{1\le j<k\le n_1} g^2(Je_j, e_k)$$

$$- 3\frac{c}{4} \sum_{n_1+1\le s<t\le n} g^2(Je_s, e_t).$$

Since we assume that $J\mathcal{D}_1 \perp \mathcal{D}_2$, the last relation implies the inequality (3.1). We see that the equality sign of (3.11) holds if and only if

$$h^r_{jt} = 0, \quad 1 \le j \le n_1, n_1+1 \le t \le n, n+1 \le r \le 2m, \tag{3.12}$$

and

$$\sum_{i=1}^{n_1} h_{ii}^r = \sum_{t=n_1+1}^{n} h_{tt}^r = 0, \quad n+2 \leq r \leq 2m. \tag{3.13}$$

Obviously (3.12) is equivalent to the mixed totally geodesics of the warped product $M_1 \times_f M_2$ (i.e., $h(X, Z) = 0$, for any X in $\mathcal{D}_1$ and Z in $\mathcal{D}_2$), and (3.9) and (3.13) imply $n_1 H_1 = n_2 H_2$.

The converse statement is straightforward. □

Remark For $c \leq 0$ the inequality is true, without the condition $J\mathcal{D}_1 \perp D_2$ (see [36]).

As applications, we derive certain obstructions to the existence of minimal warped product submanifolds in complex hyperbolic spaces [36].

Let $x : M_1 \times_f M_2 \to \widetilde{M}(c)$ be an isometric minimal immersion. Then the above theorem implies

$$\frac{\Delta f}{f} \leq n_1 \frac{c}{4}.$$

Thus, if $c < 0$, f cannot be a harmonic function or an eigenfunction of Laplacian with positive eigenvalue.

One resumes this remark into the following.

Proposition 3.2 ([36]) *If f is a harmonic function, then $M_1 \times_f M_2$ does not admit any isometric minimal immersion into a complex hyperbolic space.*

Proposition 3.3 ([36]) *If f is an eigenfunction of Laplacian on M_1 with corresponding eigenvalue $\lambda > 0$, then $M_1 \times_f M_2$ does not admit any isometric minimal immersion into a complex hyperbolic space or a complex Euclidean space.*

Next, we will give some **examples** that satisfy the equality case of the inequality (3.1).

Recall that the *Hopf submersion* is the canonical projection of $\mathbf{C}^{n+1} - \{0\} \to P^n(\mathbf{C})$, restricted to S^{2n+1} (where S^{2n+1} is regarded as the set $\{z \in \mathbf{C}^{n+1}; \sum_{j=1}^{n+1} |z^j|^2 = 1\}$).

Example 3.4 Let us consider the following immersion: $\psi : M \to S^7$, where $M = (-\pi/2, \pi/2) \times_{\cos t} N^2$, with N^2 a minimal C-totally real submanifold in S^7, defined by

$$\psi(t, p) = (\cos t)p + (\sin t)v,$$

where v is a vector tangent to S^7, but normal to S^5.

Let $\pi : S^7 \to P^3(\mathbf{C})$ be the Hopf submersion. Then $\pi \circ \psi : M \to P^3(\mathbf{C})$ is a Lagrangian minimal immersion which satisfies the equality case.

Example 3.5 Let $\psi : S^n \to S^{2n+1}$ be an immersion defined by

$$\psi(x^1, \ldots, x^{n+1}) = (x^1, 0, x^2, 0, \ldots, x^{n+1}, 0)$$

and $\pi : S^{2n+1} \to P^n(\mathbf{C})$ the Hopf submersion.

Then $\pi \circ \psi : S^n \to P^n(\mathbf{C})$ satisfies the equality case.

Example 3.6 On $S^{n_1+n_2}$ let us consider the spherical coordinates $u_1, \ldots, u_{n_1+n_2}$ and on S^{n_1} the function

$$f(u_1, \ldots u_{n_1})) = \cos u_1 \ldots \cos u_{n_1}$$

(f is an eigenfunction of Δ).

Then $S^{n_1+n_2} = S^{n_1} \times_f S^{n_2}$.

Let $i : S^{n_1+n_2} \to S^{n_1+n_2+1}$ be the standard immersion and π the Hopf submersion.

Then $\pi \circ i : S^{n_1+n_2} \to P^{n_1+n_2}(\mathbf{C})$ satisfies the equality case.

Moreover, the examples given by B.Y. Chen in [37] for $c = 0$ in the real case are true in the complex case too, for $c = 0$.

In [76] we established an inequality between the warping function f (intrinsic structure) and the squared mean curvature $\|H\|^2$ and the holomorphic sectional curvature c (extrinsic structures) for warped product submanifolds $M_1 \times_f M_2$ in any generalized complex space form $\widetilde{M}(c, \alpha)$.

We shall consider a class of almost Hermitian manifolds, called *RK-manifolds*, which contains nearly Kaehler manifolds.

Definition **([117])** An *RK-manifold* $(\widetilde{M}, J, g)$ is an almost Hermitian manifold for which the curvature tensor $\widetilde{R}$ is invariant by J, i.e.,

$$\widetilde{R}(JX, JY, JZ, JW) = \widetilde{R}(X, Y, Z, W),$$

for any $X, Y, Z, W \in \Gamma T\widetilde{M}$.

An almost Hermitian manifold $\widetilde{M}$ is of *pointwise constant type* if for any $p \in \widetilde{M}$ and $X \in T_p\widetilde{M}$ we have $\lambda(X, Y) = \lambda(X, Z)$, where

$$\lambda(X, Y) = \widetilde{R}(X, Y, JX, JY) - \widetilde{R}(X, Y, X, Y),$$

and Y and Z are unit tangent vectors on $\widetilde{M}$ at p, orthogonal to X and JX, i.e., $g(Z, Z) = g(Y, Y) = 1$, $g(X, Y) = g(JX, Y) = g(X, Z) = g(JX, Z) = 0$.

The manifold $\widetilde{M}$ is said to be of *constant type* if for any unit $X, Y \in \Gamma T\widetilde{M}$ with $g(X, Y) = g(JX, Y) = 0$, $\lambda(X, Y)$ is a constant function.

Recall the following result.

Theorem **([117])** *Let $\widetilde{M}$ be an RK-manifold. Then $\widetilde{M}$ is of pointwise constant type if and only if there exists a function α on $\widetilde{M}$ such that*

$$\lambda(X, Y) = \alpha\big[g(X, X)g(Y, Y) - (g(X, Y))^2 - (g(X, JY))^2\big],$$

for any $X, Y \in \Gamma T\widetilde{M}$. *Moreover,* $\widetilde{M}$ *is of constant type if and only if the above equality holds good for a constant* α.

In this case, α is the *constant type* of $\widetilde{M}$.

Definition A *generalized complex space form* is an RK-manifold of constant holomorphic sectional curvature and of constant type.

We denote a generalized complex space form by $\widetilde{M}(c, \alpha)$, where c is the constant holomorphic sectional curvature and α the constant type, respectively.

Each complex space form is a generalized complex space form. The converse statement is not true. The sphere S^6 endowed with the standard nearly-Kaehler structure is an example of generalized complex space form which is not a complex space form.

Let $\widetilde{M}(c, \alpha)$ be a generalized complex space form of constant holomorphic sectional curvature c and of constant type α. Then the curvature tensor $\widetilde{R}$ of $\widetilde{M}(c, \alpha)$ has the following expression [117]:

$$\widetilde{R}(X, Y)Z = \frac{c+3\alpha}{4}[g(Y, Z)X - g(X, Z)Y] \tag{3.14}$$

$$+\frac{c-\alpha}{4}[g(X, JZ)JY - g(Y, JZ)JX + 2g(X, JY)JZ].$$

Let M be an n-dimensional submanifold of a $2m$-dimensional generalized complex space form $\widetilde{M}(c, \alpha)$ of constant holomorphic sectional curvature c and constant type α. One denotes by $K(\pi)$ the sectional curvature of M associated with a plane section $\pi \subset T_pM$, $p \in M$, and ∇ the Riemannian connection of M, respectively. Also, let h be the second fundamental form and R the Riemann curvature tensor of M.

Lemma 3.7 ([76]) *Let* $x : M_1 \times_f M_2 \to \widetilde{M}(c, \alpha)$ *be an isometric immersion of an* n*-dimensional warped product into a* $2m$*-dimensional generalized complex space form* $\widetilde{M}(c, \alpha)$. *Then*

$$\frac{\Delta f}{f} \le \frac{n^2}{4n_2}\|H\|^2 + n_1\frac{c+3\alpha}{4} + 3\frac{c-\alpha}{4n_2}\sum_{1\le i\le n_1}\sum_{n_1+1\le s\le n} g^2(Je_i, e_s), \tag{3.15}$$

where $n_i = \dim M_i$, $i = 1, 2$, *and* Δ *is the Laplacian operator of* M_1.

From the above lemma, it follows the theorem:

Theorem 3.8 ([76]) *Let* $x : M_1 \times_f M_2 \to \widetilde{M}(c, \alpha)$ *be an isometric immersion of an* n*-dimensional warped product into a* $2m$*-dimensional generalized complex space form* $\widetilde{M}(c, \alpha)$. *Then:*

(i) If $c < \alpha$*, then*

$$\frac{\Delta f}{f} \le \frac{n^2}{4n_2}\|H\|^2 + n_1\frac{c+3\alpha}{4}. \tag{3.16}$$

Moreover, the equality case of (3.16) *holds identically if and only if* x *is a mixed totally geodesic immersion,* $n_1 H_1 = n_2 H_2$*, where* $H_i, i = 1, 2$*, are the partial mean curvature vectors and* $J\mathcal{D}_1 \perp \mathcal{D}_2$*.*

(ii) If $c = \alpha$*, then*

$$\frac{\Delta f}{f} \le \frac{n^2}{4n_2}\|H\|^2 + n_1\frac{c+3\alpha}{4}. \tag{3.17}$$

Moreover, the equality case of (3.17) *holds identically if and only if* x *is a mixed totally geodesic immersion and* $n_1 H_1 = n_2 H_2$*, where* $H_i, i = 1, 2$*, are the partial mean curvature vectors.*

(iii) If $c > \alpha$*, then*

$$\frac{\Delta f}{f} \le \frac{n^2}{4n_2}\|H\|^2 + n_1\frac{c+3\alpha}{4} + 3\frac{c-\alpha}{8}\|P\|^2. \tag{3.18}$$

Moreover, the equality case of (3.18) *holds identically if and only if* x *is a mixed totally geodesic immersion,* $n_1 H_1 = n_2 H_2$*, where* $H_i, i = 1, 2$*, are the partial mean curvature vectors and both* M_1 *and* M_2 *are totally real submanifolds.*

Corollary 3.9 ([76]) *Let* M *be an* n*-dimensional CR-warped product submanifold of a* $2m$*-dimensional generalized complex space form* $\widetilde{M}(c, \alpha)$*. Then,*

$$\frac{\Delta f}{f} \le \frac{n^2}{4n_2}\|H\|^2 + n_1\frac{c+3\alpha}{4}. \tag{3.19}$$

Moreover, the equality case holds identically if and only if x *is a mixed totally geodesic immersion,* $n_1 H_1 = n_2 H_2$*, where* $H_i, i = 1, 2$*, are the partial mean curvature vectors.*

We derive the following nonexistence results.

Corollary 3.10 ([76]) *Let* $\widetilde{M}(c, \alpha)$ *be a generalized complex space form,* M_1 *an* n_1*-dimensional Riemannian manifold, and* f *a differentiable function on* M_1*. If there is a point* $p \in M_1$ *such that* $(\Delta f)(p) > n_1\frac{c+3\alpha}{4} f(p)$*, then there does not exist any minimal CR-warped product submanifold* $M_1 \times_f M_2$ *in* $\widetilde{M}(c, \alpha)$*.*

Corollary 3.11 ([76]) *Let* $\widetilde{M}(c, \alpha)$ *be a generalized complex space form, with* $c > \alpha$*,* M_1 *an* n_1*-dimensional totally real submanifold of* $\widetilde{M}(c, \alpha)$*, and* f *a differentiable function on* M_1*. If there is a point* $p \in M_1$ *such that* $(\Delta f)(p) > n_1\frac{c+3\alpha}{4} f(p)$*, then* *there does not exist any totally real submanifold M_2 in $\widetilde{M}(c,\alpha)$ such that $M_1 \times_f M_2$ is a minimal warped product submanifold in $\widetilde{M}(c,\alpha)$.*

In [77] we studied warped product submanifolds in quaternion space forms.

Definition Let $\overline{M}^m$ be a $4m$-dimensional Riemannian manifold with metric g. $\overline{M}^m$ is called a *quaternion Kaehlerian manifold* if there exists a three-dimensional vector space E of tensors of type (1, 1) with local basis of almost Hermitian structures ϕ_1, ϕ_2, and ϕ_3, such that:

(i) $\phi_1\phi_2 = -\phi_2\phi_1 = \phi_3$, $\phi_2\phi_3 = -\phi_3\phi_2 = \phi_1$, $\phi_3\phi_1 = -\phi_1\phi_3 = \phi_2$.
(ii) For any local cross-section ϕ of E and any vector X tangent to $\overline{M}$, $\overline{\nabla}_X\phi$ is also a cross-section in E (where $\overline{\nabla}$ denotes the Riemannian connection in $\overline{M}$), or, equivalently, there exist local 1-forms p, q, r such that

$$\begin{cases} \overline{\nabla}_X\phi_1 = r(X)\phi_2 - q(X)\phi_3, \\ \overline{\nabla}_X\phi_2 = -r(X)\phi_1 + p(X)\phi_3, \\ \overline{\nabla}_X\phi_3 = q(X)\phi_1 - p(X)\phi_2. \end{cases}$$

If X is a unit vector in $\overline{M}$, then X, ϕ_1X, ϕ_2X, and ϕ_3X form an orthonormal set on $\overline{M}$, and one denotes by $Q(X)$ the 4-plane spanned by them. For any orthonormal vectors X, Y on $\overline{M}$, if $Q(X)$ and $Q(Y)$ are orthogonal, the 2-plane $\pi(X,Y)$ spanned by X, Y is called a *totally real plane*. Any 2-plane in $Q(X)$ is called a *quaternionic plane*. The sectional curvature of a quaternionic plane π is called a *quaternionic sectional curvature*. A quaternion Kaehler manifold $\overline{M}$ is a *quaternion space form* if its quaternionic sectional curvatures are constant.

It is well known that a quaternion Kaehlerian manifold $\overline{M}$ is a quaternion space form $\overline{M}(c)$ if and only if its curvature tensor $\overline{R}$ has the following form (see [67]):

$$\overline{R}(X,Y)Z = \frac{c}{4}\{g(Y,Z)X - g(X,Z)Y \tag{3.20}$$

$$+ g(\phi_1Y,Z)\phi_1X - g(\phi_1X,Z)\phi_1Y + 2g(X,\phi_1Y)\phi_1Z$$

$$+ g(\phi_2Y,Z)\phi_2X - g(\phi_2X,Z)\phi_2Y + 2g(X,\phi_2Y)\phi_2Z$$

$$+ g(\phi_3Y,Z)\phi_3X - g(\phi_3X,Z)\phi_3Y + 2g(X,\phi_3Y)\phi_3Z\},$$

for vectors X, Y, Z tangent to $\overline{M}$.

A submanifold M of a quaternion Kaehler manifold $\overline{M}$ is called *quaternion* (respectively, *totally real*) submanifold if each tangent space of M is carried into itself (respectively, the normal space) by each section in E.

The curvature tensor R of M is related to the curvature tensor $\overline{R}$ of $\overline{M}$ by the Gauss equation

$$\overline{R}(X, Y, Z, W) = R(X, Y, Z, W)$$

$$- g(h(X, Z), h(Y, W)) + g(h(X, W), h(Y, Z)),$$

where h is the second fundamental form of M.

Definition ([10]) A submanifold M of a quaternion Kaehler manifold $\overline{M}$ is called a *quaternion CR-submanifold* if there exist two orthogonal complementary distributions D and $D^{\perp}$ such that D is invariant under quaternion structures, that is, $\phi_i(D_x) \subseteq D_x, i = 1, 2, 3, \forall x \in M$, and $D^{\perp}$ is totally real, that is, $\phi_i(D_x^{\perp}) \subseteq T_x^{\perp}M$, $i = 1, 2, 3, \forall x \in M$.

A submanifold M of a quaternion Kaehler manifold $\overline{M}$ is a quaternion submanifold (resp. totally real submanifold) if dim $D^{\perp} = 0$ (respectively, dim $D = 0$).

In this context, the following results were proved.

Lemma 3.12 ([77]) *Let $x : M_1 \times_f M_2 \to \overline{M}(c)$ be an isometric immersion of an n-dimensional warped product into a $4m$-dimensional quaternion space form $\overline{M}(c)$. Then*

$$n_2 \frac{\Delta f}{f} \leq \frac{n^2}{4} \|H\|^2 + n_1 n_2 \frac{c}{4} + 3\frac{c}{4} \sum_{\alpha=1}^{3} \sum_{i=1}^{n_1} \sum_{s=n_1+1}^{n} g^2(\phi_\alpha e_i, e_s).$$

Theorem 3.13 ([77]) *Let $x : M_1 \times_f M_2 \to \overline{M}(c)$ be an isometric immersion of an n-dimensional warped product into a $4m$-dimensional quaternion space form $\overline{M}(c)$ with $c < 0$. Then*

$$\frac{\Delta f}{f} \leq \frac{n^2}{4n_2} \|H\|^2 + n_1 \frac{c}{4}.$$

Moreover, the equality case holds identically if and only if x is a mixed totally geodesic immersion, $n_1 H_1 = n_2 H_2$ and $\phi_k \mathcal{D}_1 \perp \mathcal{D}_2$, for any $k = 1, 2, 3$.

As applications, one derives certain obstructions to the existence of minimal warped product submanifolds in quaternion hyperbolic space.

Corollary 3.14 ([77]) *If f is a harmonic function on M_1, then the warped product $M_1 \times_f M_2$ does not admit any isometric minimal immersion into a quaternion hyperbolic space.*

Corollary 3.15 ([77]) *There do not exist minimal warped product submanifolds in a quaternion hyperbolic space with M_1 compact.*

Theorem 3.16 ([77]) *Let $x : M_1 \times_f M_2 \to \overline{M}(c)$ be an isometric immersion of an n-dimensional warped product into a $4m$-dimensional flat quaternion space form. Then*

$$\frac{\Delta f}{f} \leq \frac{n^2}{4n_2}\|H\|^2.$$

Moreover, the equality case holds identically if and only if x is a mixed totally geodesic immersion and $n_1 H_1 = n_2 H_2$.

Corollary 3.17 ([77]) *If f is an eigenfunction of Laplacian on M_1 with corresponding eigenvalue $\lambda > 0$, then the warped product $M_1 \times_f M_2$ does not admit any isometric minimal immersion into a quaternion hyperbolic space or a quaternion Euclidean space.*

A warped product is said to be *proper* if the warping function is nonconstant.

Corollary 3.18 ([77]) *There does not exist minimal proper warped product submanifold in the quaternion Euclidean space $\mathbf{Q}^m$ with M_1 compact.*

Theorem 3.19 ([77]) *Let $x : M_1 \times_f M_2 \to \overline{M}(c)$ be an isometric immersion of an n-dimensional warped product into a $4m$-dimensional quaternion space form $\overline{M}(c)$ with $c > 0$. Then*

$$\frac{\Delta f}{f} \leq \frac{n^2}{4n_2}\|H\|^2 + n_1\frac{c}{4} + 3\frac{c}{4}\min\left\{\frac{n_1}{n_2}, 1\right\}.$$

Moreover, the equality case holds identically if and only if x is a mixed totally geodesic immersion, $n_1 H_1 = n_2 H_2$ and $\phi_k \mathcal{D}_1 \perp \mathcal{D}_2$, for any $k = 1, 2, 3$.

Also, Lemma 3.12 implies another inequality for certain submanifolds (in particular quaternion CR-submanifolds) in quaternion space forms with $c > 0$.

Theorem 3.20 ([77]) *Let $x : M_1 \times_f M_2 \to \overline{M}(c)$ be an isometric immersion of an n-dimensional warped product into a $4m$-dimensional quaternion space form $\overline{M}(c)$ with $c > 0$, such that $\phi_k \mathcal{D}_1 \perp \mathcal{D}_2$, for any $k = 1, 2, 3$. Then*

$$\frac{\Delta f}{f} \leq \frac{n^2}{4n_2}\|H\|^2 + n_1\frac{c}{4}.$$

Moreover, the equality case holds identically if and only if x is a mixed totally geodesic immersion and $n_1 H_1 = n_2 H_2$.

Next, we will give some examples which satisfy identically the equality case of the inequality given in Theorem 3.20.

Example 3.21 Let $\psi : S^n \to S^{4n+3}$ be an immersion defined by

$$\psi(x^1, \ldots, x^{n+1}) = (x^1, 0, 0, 0, x^2, 0, 0, 0, \ldots, x^{n+1}, 0, 0, 0),$$

and $\pi : S^{4n+3} \to P^n(\mathbf{Q})$ the Hopf submersion.

Then $\pi \circ \psi : S^n \to P^n(\mathbf{Q})$ satisfies the equality case.

Example 3.22 On $S^{n_1+n_2}$ let us consider the spherical coordinates $u_1, \ldots, u_{n_1+n_2}$ and on S^{n_1} the function

$$f(u_1, \ldots u_{n_1}) = \cos u_1 \ldots \cos u_{n_1}$$

(f is an eigenfunction of Δ).

Then $S^{n_1+n_2} = S^{n_1} \times_f S^{n_2}$.

Let $\psi : S^{n_1+n_2} \to S^{4(n_1+n_2)+3}$ be the above standard immersion and π the Hopf submersion $\pi : S^{4(n_1+n_2)+3} \to P^{n_1+n_2}(\mathbf{Q})$.

Then $\pi \circ \psi : S^{n_1+n_2} \to P^{n_1+n_2}(\mathbf{Q})$ satisfies the equality case.

For a comprehensive study on the differential geometry of warped product manifolds and submanifolds see the book by B.Y. Chen [42].

4 Curvature Symmetries Characterizing Einstein Spaces

We recall the well-known definition of an Einstein space:

A Riemannian manifold (M, g) of dimension $n \geq 3$ is called an *Einstein space* if $\mathcal{R}ic = \lambda \cdot id$, where trivially $\lambda = \kappa$; in this case one easily proves that $\lambda = \kappa = const$.

We recall the fact that any two-dimensional Riemannian n-manifold satisfies the relation $\mathcal{R}ic = \lambda \cdot id$, but for $n = 2$ the function $\lambda = \kappa$ is not necessarily a constant. It is well known that any three-dimensional Einstein space is of constant curvature. Thus the interest in Einstein spaces starts with dimension $n = 4$.

We give three concrete examples of Einstein spaces:

Example 4.1 Any Riemannian space form of arbitrary dimension $n \geq 2$ is an Einstein manifold. In particular, certain warped product manifolds are

$$\left(-\frac{\pi}{2}, \frac{\pi}{2}\right) \times_{\cos x} S^{n-1}, \ c = 1,$$

$$(0, \infty) \times_r S^{n-1}, \ c = 0,$$

$$\mathbf{R} \times_{e^x} \mathbf{E}^{n-1}, \ c = -1.$$

Example 4.2 The Schwarzschild space-time is an example of Einstein manifold ($\mathcal{R}ic = 0$), which has no constant sectional curvature.

Example 4.3 Let $N_1^{k(a)}$, $N_2^p(b)$ $(k, p \geq 2)$ be Riemannian space forms;

$$M = N_1^k(a) \times N_2^p(b)$$

and $\{e_1, \ldots, e_k, e_{k+1}, \ldots, e_{k+p}\}$ orthonormal frame on M, such that $e_1, \ldots, e_k$ tangent to $N_1^k(a)$, $e_{k+1}, \ldots, e_{k+p}$ tangent to $N_2^p(b)$. Then $\forall i = 1, \ldots, k$, $\mathcal{R}ic(e_i) = (k-1)a$; $\forall j = 1, \ldots, p$, $\mathcal{R}ic(e_{k+j}) = (p-1)b$. Then M is Einstein space if and only if $(k-1)a = (p-1)b$.

In particular:

- $a = 0 \Leftrightarrow b = 0$ trivial (Euclidean space, of null sectional curvature).
- $k = p \Leftrightarrow a = b$ (Einstein space of even dimension and nonconstant sectional curvature).
- $a \neq 0 \Rightarrow b = \frac{k-1}{p-1}a$; remark $ab > 0$ (Einstein space of arbitrary dimension and nonconstant sectional curvature).

In their famous paper [113] Singer and Thorpe considered four-dimensional Einstein spaces and started the study of two interesting topics, more precisely:

(i) The irreducible decomposition of the Riemannian (0, 4) curvature tensor; this study initialized generalizations first to algebraic curvature tensors, see [12], and later to other types of curvature tensors, see, e.g., [16] and [61].
(ii) Symmetry properties of certain curvature functions

Here we are interested in the second topic (ii). We recall the result of Singer and Thorpe. For $span(e_i, e_j)$ the orthogonal complement $span(e_i, e_j)^{\perp}$ is well defined; for simplicity we denote its sectional curvature by $\kappa_{ij}^{\perp}$.

Singer and Thorpe proved the following:

Theorem **([113])** *Let (M, g) be a Riemannian manifold of dimension $n = 4$. Then the following assertions are equivalent:*

(i) (M, g) is an Einstein space.
(ii) At any point $p \in M$ and for every 2-plane $span\{e_i, e_j\} \subset T_pM$, where $i, j \in \{1, 2, 3, 4\}$ and $i \neq j$, we have the following equality of sectional curvatures:

$$\kappa_{ij} = \kappa_{ij}^{\perp}.$$

The result of Singer and Thorpe was generalized to even-dimensional Riemannian manifolds in [44]. For a precise formulation we adopt some notational conventions from [44].

Let (M, g) be as before, and consider a k-dimensional subspace $L \subset T_pM$ for $k > 1$. Let $\{e_1, \ldots, e_k\}$ be an orthonormal basis of L. Define

$$\tau_k(L) = \sum_{1 \leq \alpha < \beta \leq k} \kappa(e_\alpha \wedge e_\beta);$$

in analogy to the above notation we write

$$\kappa_{\alpha\beta} = \kappa(e_\alpha \wedge e_\beta).$$

$2\tau_k(L)$ is called the *scalar k-curvature* of the subspace L. For $k = n$ we get the scalar curvature of (M, g) at p:

$$2\tau_n(L) = R,$$

while for $k = 2$ the expression $\tau_2(L)$ gives the sectional curvature of the plane $L = \text{span}(e_\alpha, e_\beta) \subset T_p(M)$.

The authors of [44] proved the following:

Theorem ([44]) *Let $\dim M = n = 2k$ for $k \geq 2$. Then the following statements are equivalent:*

- *(M, g) is an Einstein space.*
- *For any $p \in M$ and any subspace $L \subset T_pM$ with $dim(L) = k = dim(L^\perp)$ and $L \oplus L^\perp = T_pM$, we have*

$$\tau_k(L) = \tau_k(L^\perp).$$

It is well known that three-dimensional Einstein spaces are of constant sectional curvature (see [12]). Thus we consider arbitrary dimensions $n \geq 4$.

Our main Theorem 4.4 generalizes the results cited above, being valid in any dimension.

Notations

(i) Let (V, g) be a Euclidean vector space of dimension $n \geq 4$ with inner product g. Let $L \subset V$ be a subspace of dimension r, where $2 \leq r = \dim L \leq n - 2$, and consider the orthogonal decomposition

$$L \oplus L^\perp = V,$$

with $2 \leq s = \dim L^\perp \leq n - 2$, thus $r + s = n$.

(ii) Additionally we introduce the following notation:
Set $N = \{1, \ldots, n\}$, and let $\sigma : N \to N$ with $\{1, \ldots, n\} \mapsto \{i_1, \ldots, i_n\}$ be a permutation. We consider a disjoint decomposition

$$N = N_r^\sigma \cup N_s^\sigma$$

with $N_r^\sigma = \{i_1, \ldots, i_r\}$ and $N_s^\sigma := \{i_{r+1}, \ldots, i_{r+s}\}$.

(iii) On a Riemannian manifold (M, g) with $p \in M$ consider an r-dimensional subspace $L \subset T_pM$ with orthonormal basis $\{e_{i_1}, \ldots, e_{i_r}\}$; we extend it to an orthonormal basis $\{e_{i_1}, \ldots, e_{i_r}, e_{i_{r+1}}, \ldots, e_{i_n}\}$ of T_pM; thus $\{e_{i_{r+1}}, \ldots, e_{i_n}\}$ is an orthonormal basis of the subspace $L^\perp \subset T_pM$. From the foregoing sections, the curvature invariants

$$2\tau(L) = \sum_{p,q\in N_r^\sigma} \kappa_{pq} \quad \text{and} \quad 2\tau(L^\perp) = \sum_{p,q\in N_s^\sigma} \kappa_{pq}$$

are well defined.

Calculations To relate the scalar curvatures of subspaces we add up the Ricci curvatures:

$$\sum_{p\in N_r^\sigma} \rho_p = \sum_{p\in N_r^\sigma} \sum_{i\in N} \kappa_{pi},$$

$$\sum_{p\in N_s^\sigma} \rho_p = \sum_{p\in N_s^\sigma} \sum_{i\in N} \kappa_{pi}.$$

We have

$$\sum_{p\in N_r^\sigma} \rho_p - \sum_{p\in N_s^\sigma} \rho_p = \sum_{p\in N_r^\sigma} \sum_{i\in N_r^\sigma} \kappa_{pi} + \sum_{p\in N_r^\sigma} \sum_{i\in N_s^\sigma} \kappa_{pi} - \sum_{p\in N_s^\sigma} \sum_{i\in N_s^\sigma} \kappa_{pi}$$
$$- \sum_{p\in N_s^\sigma} \sum_{i\in N_r^\sigma} \kappa_{pi} = 2\big(\tau_r(L) - \tau_s(L^\perp)\big);$$

to prove the last equality we interchanged some subindices and used the symmetry $\kappa_{pi} = \kappa_{ip}$.

Theorem 4.4 ([92]) *Let (M, g) be a Riemannian manifold of dimension n. Using the notations of the foregoing subsections, we have equivalence of the following two conditions:*

(I) (M, g) is an Einstein space.
(II) Let $r, s \in \mathbb{N}$ satisfy the relations $r + s = n$ and $2 \leq r, s \leq n - 2$. For any $p \in M$ there exists a real $c = c_p$, independent of r, such that for any subspace $L \subset T_pM$ of dimension r we have

$$(r - s)c_p = 2\big(\tau_r(L) - \tau_s(L^\perp)\big).$$

Proof For an Einstein space all Ricci curvatures have the same value $\rho_i = \frac{1}{n}R$, where as before R denotes the scalar curvature of (M, g). Then $c_p = \frac{1}{n}R$, and the preceding calculation shows that (I) implies (II).

Vice versa, for fixed r consider two r-dimensional subspaces

$$L = \text{span}\{e_1, e_2, \ldots, e_r\} \text{ and } M = \text{span}\{e_{r+1}, e_2, \ldots, e_r\}$$

having the $(r-1)$-dimensional intersection span$\{e_2, \ldots, e_r\}$; together with their orthogonal complements both satisfy (II), respectively. The above calculation gives

$$(r-s)c_p = \left(\rho_1 + \sum_{2,\ldots,r} \rho_p\right) - \left(\rho_{r+1} + \sum_{r+2,\ldots,n} \rho_p\right),$$

and

$$(r-s)c_p = \left(\rho_{r+1} + \sum_{2,\ldots,r} \rho_p\right) - \left(\rho_1 + \sum_{r+2,\ldots,n} \rho_p\right).$$

A comparison of both equations gives $\rho_1 = \rho_{r+1}$. Analogously we have $\rho_1 = \rho_{r+2}$; thus also $\rho_{r+1} = \rho_1 = \rho_{r+2}$. In the same way we prove that all Ricci curvatures coincide at p, and as p is arbitrary, (M, g) is Einstein.

Remark 4.5 If $n = 2r = 2s$, we get the even-dimensional result from [44]; in the case $r = s = 2$ we get the result of Singer and Thorpe.

Remark 4.6

(i) Recall the fact that, for any Riemannian manifold, the sectional curvature function at $p \in M$ determines the Riemannian curvature tensor at p (see [68], p.198, Proposition 1.2).
(ii) From the preceding Theorem 4.4, for an Einstein space, at any $p \in M$, the scalar $(n-2)$-curvatures $\tau_{n-2}(T_pM)$ and the scalar curvature $R = \tau_n(T_pM)$ together determine the scalar 2-curvatures $\tau_2(T_pM)$, which means the sectional curvatures at p.

(i) and (ii) together imply the following:

Corollary ([92]) *Let (M, g) be an Einstein space of dimension $n \geq 5$. At any $p \in M$, the scalar $(n-2)$-curvatures $\tau_{n-2}(T_pM)$ together with the scalar curvature $R = \tau_n(T_pM)$ determine the Riemannian curvature tensor.*

Remark 4.7 Let (M, g) be an Einstein space of dimension $n = 4$. Following the result of Singer and Thorpe, at any $p \in M$ we have six sectional curvatures, which means three pairs as $\kappa_{ij} = \kappa_{ij}^{\perp}$; as three representatives of the three pairs we can fix an arbitrary index $i \in \{1, 2, 3, 4\}$ and consider the three sectional curvatures κ_{ij}, where $i \neq j \in \{1, 2, 3, 4\}$. Thus, for arbitrary $i \in \{1, 2, 3, 4\}$, these three representatives determine the Riemannian curvature tensor at $p \in M$.
For $n = 4$ this fact suggests the following procedure:
Choose $i = 1$ for a curve starting at p with e_1 as prescribed tangent vector at p. By a parallel displacement a frame $\{e_1, \ldots, e_4\}$ moves from p along the curve. The sectional curvatures κ_{1j} for $j = 2, 3, 4$ together determine the curvature tensor along the curve.

5 Short Review of the Chapter

Beside the classical Riemannian invariants, as sectional, scalar and Ricci curvatures, a crucial role in this topic is played by Chen invariants and, of course, by Chen-type inequalities involving them.

The definition of Chen invariants given by Professor B.Y. Chen and their study represented a huge contribution in Submanifold Theory, opened many interesting directions and new geometrical interpretations. The author is very indebted to Professor B.Y. Chen for the impact of his work in her research and for the opportunity to collaborate.

This chapter represents a collection of results from the author's papers on this topic; remark that the proofs are given in detail, so the reader can follow the techniques. Results from this chapter were included in the author's Habilitation Thesis, which has not been published anywhere.

In the first section the basics of submanifolds in complex space forms and Sasakian space forms are recalled. We then started to present Chen-type inequalities for different submanifolds in complex and Sasakian space forms.

In the second section, we first stated the most important Chen inequalities in real space forms. We gave a general construction method for purely real submanifolds and presented the geometric inequalities for purely submanifolds in complex space forms. We obtained an improved Chen-Ricci inequality for Kaehlerian slant submanifolds in complex space forms. Works on DDVV conjecture are also presented. Next subsection contains results on submanifolds in Sasakian manifolds. We proved first Chen inequality for contact slant submanifolds in Sasakian space forms. We defined Chen-type Sasakian invariants, obtained sharp inequalities for these invariants, and derived characterizations of the equality case in terms of the shape operator. We generalized a result of Chen and obtained a Chen-Ricci inequality for purely real submanifolds with T parallel with respect to the Levi-Civita connection. The third subsection presented the results obtained for submanifolds with semi-symmetric metric (respectively, nonmetric) connections. Subsection 2.4 dealt with statistical submanifolds, and their behavior in statistical manifolds of constant curvature is studied.

Section 3 presented results on warped product submanifolds in complex space form, generalized complex space forms, and quaternion space forms.

In Sect. 4 we gave a new characterization of Einstein spaces by using their curvatures symmetries.

Proofs were written explicitly. We would like to point out that, even the technique seems similar, each case has its particularity and geometrical meaning, and for this reason we gave significant proofs to almost each situation.

We intended to organize this contribution as a monograph, and, having a lot of complete proofs and examples, we hope it will be useful for the researchers in Submanifold Theory.

References

1. Agashe, N.S., Chafle, M.R.: A semi-symmetric non-metric connection on a Riemannian manifold. Ind. J. Pure Appl. Math. **23**(6), 399–409 (1992)
2. Agashe, N.S., Chafle, M.R.: On submanifolds of a Riemannian manifold with a semi-symmetric non-metric connection. Tensor (N.S.) **55**(2), 120–130 (1994)
3. Alegre, P., Carriazo, A., Kim, Y.H., Yoon, D.W.: B.Y. Chen's inequality for submanifolds of generalized space forms. Ind. J. Pure Appl. Math. **38**(3), 185–201 (2007)
4. Amari, S.: Differential-Geometrical Methods in Statistics. Springer, New York (1985)
5. Aydin, M.E., Mihai, I.: Wintgen inequality for statistical surfaces. Math. Ineq. Appl. **22**(1), 123–132 (2019)
6. Aydin, M.E., Lopez, R., Mihai, A.: Classification of translation surfaces in $\mathbf{R}^3$ with constant sectional curvature. https://arxiv.org/abs/2405.12825
7. Aydin, M.E., Mihai, A., Mihai, I.: Some inequalities on submanifolds in statistical manifolds of constant curvature. Filomat **29**(3), 465–476 (2015)
8. Aydin, M.E., Mihai, A., Mihai, I.: Generalized Wintgen inequality for statistical submanifolds in statistical manifolds of constant curvature. Bull. Math. Sci. **7**, 155–166 (2017)
9. Aydin, M.E., Lopez, R., Mihai, A.: Constant sectional curvature surfaces with a semi-symmetric non-metric connection. J. Math. Anal. Appl. **542**(2), 128795 (2025)
10. Barros, M., Chen, B.Y., Urbano, F.: Quaternion CR-submanifold of quaternion manifold. Kodai Math. J. **4**, 399–417 (1981)
11. Bejancu, A.: Geometry of CR-Submanifolds. D. Reidel Publishing Co., Dordrecht (1986)
12. Besse, A.: Einstein Manifolds. Springer, Berlin (1987)
13. Blair, D.E.: Contact Manifolds in Riemannian Geometry. Lecture Notes in Mathematics, vol. **509**. Springer, Berlin (1976)
14. Blair, D.E.: Riemannian Geometry of Contact and Symplectic Manifolds. Birkhäuser, Boston (2010)
15. Blair, D.E., Carriazo, A.: The contact Whitney sphere. Note Mat. **20**, 125–133 (2000/2001)
16. Blažić, N., Gilkey, P., Nikćević, St., Simon, U.: Algebraic theory of affine curvature tensors. Arch. Math. (Brno) **42**, 147–168 (2006)
17. Bolton, J., Dillen, F., Fastenakels, J., Vrancken, L.: A best possible inequality for curvature-like tensor fields. Math. Ineq. Appl. **12**, 663–681 (2009)
18. Bolton, J., Dillen, F., Vrancken, L.: Lagrangian submanifolds attaining equality in a basic inequality. In: Proceedings of Symposium on the Differential Geometry of Submanifolds, Valenciennes, pp. 37–42 (2007)
19. Cabrerizo, J.L., Carriazo, A., Fernandez, L.M., Fernandez, M.: Slant submanifolds in Sasakian space forms. Glasg. Math. J. **42**, 125–138 (2000)
20. Carriazo, A.: A contact version of B.-Y. Chen's inequality and its applications to slant immersions. Kyungpook Math. J. **39**, 465–476 (1999)
21. Chen, B.Y.: Geometry of Submanifolds. M. Dekker, New York (1973)
22. Chen, B.Y.: CR-submanifolds of a Kaehler manifold. J. Diff. Geom. **16**, 305–322 (1981)
23. Chen, B.Y.: Geometry of Slant Submanifolds. K. U. Leuven, Leuven (1990)
24. Chen, B.Y.: Some pinching and classification theorems for minimal submanifolds. Arch. Math. (Basel) **60**, 568–578 (1993)
25. Chen, B.Y.: Jacobi's elliptic functions and Lagrangian immersions. Proc. R. Soc. Edinb. **126**, 687–704 (1996)
26. Chen, B.Y.: Mean curvature and shape operator of isometric immersions in real-space-forms. Glasg. Math. J. **38**, 87–97 (1996)
27. Chen, B.Y.: Special slant surfaces and a basic inequality. Results Math. **33**, 65–78 (1998)
28. Chen, B.Y.: Intrinsic and extrinsic structures of Lagrangian surfaces in complex space forms. Tsukuba J. Math. **22**, 657–680 (1998)
29. Chen, B.Y.: Relations between Ricci curvature and shape operator for submanifolds with arbitrary codimensions. Glasg. Math. J. **41**, 33–44 (1999)

30. Chen, B.Y.: Some new obstructions to minimal and Lagrangian isometric immersions. Jpn. J. Math. **26**, 105–127 (2000)
31. Chen, B.Y.: On Ricci curvature of isotropic and Lagrangian submanifolds in complex space forms. Arch. Math. **74**, 154–160 (2000)
32. Chen, B.Y.: A general inequality for Kaehlerian slant submanifolds and related results. Geom. Dedicata **25**, 253–271 (2001)
33. Chen, B.Y.: Geometry of warped product CR-submanifolds in Kaehler manifolds. Monatsh. Math. **133**, 177–195 (2001); **134**, 103–119 (2001)
34. Chen, B.Y.: A series of Kählerian invariants and their applications to Kählerian geometry. Beitrage Alg. Geom. (Contributions to Algebra and Geometry) **42**, 165–178 (2001)
35. Chen, B.Y.: Geometry of warped products as Riemannian manifolds and related problems. Soochow J. Math. **28**, 125–156 (2002)
36. Chen, B.Y.: Non-immersions theorems for warped products in complex hyperbolic spaces. Proc. Jpn. Acad. Ser. A Math. Sci. **78**(6), 96–100 (2002)
37. Chen, B.Y.: On isometric minimal immersions from warped products into real space forms. Proc. Edinb. Math. Soc. **45**, 579–587 (2002)
38. Chen, B.Y.: δ-invariants, inequalities of submanifolds and their applications. In: Mihai, A., Mihai, I., Miron, R. (eds.) Topics in Differential Geometry. Ed. Academiei Române, Bucharest (2008)
39. Chen, B.Y.: On purely real surfaces in Kaehler surfaces and Lorentz surfaces in Lorentzian Kaehler surfaces. Bull. Transilvania Univ. Brasov, Ser. III, Math. Inform. Phys. **1**(50), 33–58 (2008)
40. Chen, B.Y.: On purely real surfaces in Kaehler surfaces. Turk. J. Math. **34**, 275–292 (2010)
41. Chen, B.Y.: On Wintgen ideal surfaces. In: Mihai, A., Mihai, I. (eds.) Riemannian Geometry and Applications – Proceedings RIGA 2011, pp. 59–74. Ed. Univ. Bucureşti, Bucharest (2011)
42. Chen, B.Y.: Differential Geometry of Warped Product Manifolds and Submanifolds. World Scientific Publishing, Hackensack (2017)
43. Chen, B.Y., Dillen, F., Van der Veken, J.: Complete classification of parallel Lorentzian surfaces in Lorentzian complex space forms. Int. J. Math. **21**, 665–686 (2010)
44. Chen, B.Y., Dillen, F., Verstraelen, L., Vrancken, L.: Characterizations of Riemannian space forms, Einstein spaces and conformally flat spaces. Proc. AMS **128**, 589–598 (2000)
45. Chen, B.Y., Ludden, G.D., Montiel, S.: Real submanifolds of a Kaehler manifold. Algebras, Groups Geom. **1**, 176–212 (1984)
46. Chen, B.Y., Mihai, A.: On purely real surfaces with harmonic Wirtinger function in complex space forms. Kyushu J. Math. **64**, 153–168 (2010)
47. Chen, B.Y., Mihai, A., Mihai, I.: Purely real surfaces in non-flat complex space forms satisfying a basic equality. Results Math. **56**, 453–472 (2009)
48. Chen, B.Y., Morvan, J.M.: Cohomologie des souts-variétés α-obliques. C. R. Acad. Sci. Paris **314**, 931–934 (1992)
49. Chen, B.Y., Ogiue, K.: On totally real submanifolds. Trans. Am. Math. Soc. **193**, 257–266 (1974)
50. Chen, B.Y., Tazawa, Y.: Slant submanifolds of complex projective and complex hyperbolic spaces. Glasg. Math. J. **42**, 439–454 (2000)
51. Chen, B.Y., Vrancken, L.: Lagrangian submanifolds satisfying a basic equality. Math. Proc. Camb. Philos. Soc. **120**, 291–307 (1996)
52. Choi, T., Lu, Z.: On the DDVV conjecture and the compass in calibrated geometry. Math. Z. **260**, 409–429 (2008)
53. Cioroboiu, D., Oiagă, A.: B. Y. Chen inequalities for slant submanifolds in Sasakian space forms. Rend. Circ. Mat. Palermo **52**, 367–381 (2003)
54. De Smet, P.J., Dillen, F., Verstraelen, L., Vrancken, L.: A pointwise inequality in submanifold theory. Arch. Math. **35**, 115–128 (1999)
55. Deng, S.: An improved Chen-Ricci inequality. Int. Electron. J. Geom. **2**, 39–45 (2009)

56. Dillen, F., Fastenakels, J., Van der Veken, J.: A pinching theorem for the normal scalar curvature of invariant submanifolds. J. Geom. Phys. **57**, 833–840 (2007)
57. Ejiri, E.: Totally real minimal immersions of n-dimensional totally real space forms into n-dimensional complex space forms. Proc. Am. Math. Soc. **84**, 243—246 (1982)
58. Furuhata, H.: Hypersurfaces in statistical manifolds. Diff. Geom. Appl. **27**, 420–429 (2009)
59. Furuhata, H.: Statistical hypersurfaces in the space of Hessian curvature zero. Diff. Geom. Appl. **29**, 586–590 (2011)
60. Ge, J., Tang, Z.: A proof of the DDVV conjecture and its equality case. Pac. J. Math. **237**, 87–95 (2008)
61. Gilkey, P., Nikćević, St., Simon, U.: Geometric theory of equiaffine curvature tensors. Results Math. **56**, 275–317 (2009)
62. Gmira, B., Verstraelen, L.: A curvature inequality for Riemannian submanifolds in a semi-Riemannian space forms. In: Geometry and Topology of Submanifolds, vol. IX, pp. 148–159. World Scientific, Singapore (1999)
63. Guadalupe, I.V., Rodriguez, L.: Normal curvature of surfaces in space forms. Pac. J. Math. **106**, 95–103 (1983)
64. Hayden, H.A.: Subspaces of a space with torsion. Proc. Lond. Math. Soc. **34**, 27–50 (1932)
65. Imai, T.: Hypersurfaces of a Riemannian manifold with semi-symmetric metric connection. Tensor (N.S.) **23**, 300–306 (1972)
66. Imai, T.: Notes on semi-symmetric metric connections. Tensor (N.S.) **24**, 293–296 (1972)
67. Ishihara, S.: Quaternion Kaehlerian manifolds. J. Diff. Geom. **9**, 483–500 (1974)
68. Kobayashi, S., Nomizu, N.: Foundations of Differential Geometry, vol. 1. Interscience, New York (1963)
69. Kobayashi, S., Nomizu, K.: Foundations of Differential Geometry, vol. 2. Interscience, New York (1969)
70. Levi-Civita, T., Ricci, G.: Methodes de calcul differential absolu et leurs applications. Math. Ann. B. **54**, 125–201 (1900)
71. Lotta, A.: Slant submanifolds in contact geometry. Bull. Math. Soc. Roumanie **39**, 183–198 (1996)
72. Lu, Z.: Normal scalar curvature conjecture and its applications. J. Funct. Anal. **261**, 1284–1308 (2011)
73. Matsumoto, K., Mihai, I., Oiagă, A.: Ricci curvature of submanifolds in complex space forms. Rev. Roumaine Math. Pures Appl. **46**, 775–782 (2001)
74. Mihai, A.: An inequality for totally real surfaces in complex space forms. Kragujevac J. Math. **26**, 83–88 (2004)
75. Mihai, A.: Warped product submanifolds in complex space forms. Acta Sci. Math. (Szeged) **70**, 419–427 (2004)
76. Mihai, A.: Warped product submanifolds in generalized complex space forms. Acta Math. Acad. Paedagog. Nyházi. (N.S.) **21**, 79–87 (2005)
77. Mihai, A.: Warped product submanifolds in quaternion space forms. Rev. Roumaine Math. Pures Appl. **50**, 283–291 (2005)
78. Mihai, A.: Modern Topics in Submanifold Theory. University of Bucharest, Bucharest (2006)
79. Mihai, A.: Scalar normal curvature of Lagrangian 3-dimensional submanifolds in complex space forms. In: Pure and Applied Differential Geometry – PADGE 2007. Berichte aus der Mathematik, pp. 171–177. Shaker Verlag, Aachen (2007)
80. Mihai, A.: Geometric inequalities for purely real submanifolds in complex space forms. Results Math. **55**, 457–468 (2009)
81. Mihai, A.: Inequalities on the Ricci curvature. J. Math. Inequal. **9**, 811–822 (2015)
82. Mihai, A.: A note on derived connections from semi-symmetric metric connections. Math. Slovaca **67**, 221–226 (2017)
83. Mihai, A., Mihai, I.: CR-submanifolds in complex and Sasakian space forms. In: Dragomir, S., et al. Geometry of Cauchy-Riemann Submanifolds. Springer, Singapore (2016)
84. Mihai, A., Mihai, I.: Some basic inequalities on slant submanifolds in space forms. In: Chen, B.-Y., et al. (eds.) Differential Geometry of Slant Submanifolds. Springer, Berlin (2022)

85. Mihai, A., Mihai, I.: A note on a well-defined sectional curvature of a semi-symmetric non-metric connection. Intern. Electron. J. Geom. **17**(1), 15–23 (2024)
86. Mihai, A., Özgür, C.: Chen inequalities for submanifolds of real space forms with a semi-symmetric metric connection. Taiwan. J. Math. **14**, 1465–1477 (2010)
87. Mihai, A., Özgür, C.: Corrigendum of Chen inequalities for submanifolds of real space forms with a semi-symmetric metric connection. Taiwan. J. Math. **14**, 1465–1477 (2010)
88. Mihai, A., Özgür, C.: Chen inequalities for submanifolds of complex space forms and Sasakian space forms endowed with semi-symmetric metric connections. Rocky Mt. J. Math. **41**(5), 1653–1673 (2011)
89. Mihai, A., Rădulescu, I.N.: An improved Chen-Ricci inequality for Kaehlerian slant submanifolds in complex space forms. Taiwan. J. Math. **16**, 761–770 (2012)
90. Mihai, A., Rădulescu, I.N.: Certain Riemannian invariants for Sasakian submanifolds. Publ. Math. Debrecen **82**, 685–695 (2013)
91. Mihai, A., Rădulescu, I.N.: Scalar and Ricci curvatures of special contact slant submanifolds in Sasakian space forms. Adv. Geom. **14**, 147–159 (2014)
92. Mihai, A., Simon, U.: Curvature symmetries characterizing Einstein spaces. Colloq. Math. **152**(1), 23–28 (2018)
93. Mihai, I.: Ricci curvature of submanifolds in Sasakian space forms. J. Aust. Math. Soc. **72**, 247–256 (2002)
94. Mihai, I.: Geometria Subvarietăţilor în Varietăţi Complexe. Ed. Univ. Bucureşti (2002)
95. Mihai, I.: Ideal Kaehlerian slant submanifolds in complex space forms. Rocky Mt. J. Math. **35**, 941–951 (2005)
96. Mihai, I.: On the generalized Wintgen inequality for Lagrangian submanifolds in complex space forms. Nonlinear Anal. **95**, 714–720 (2014)
97. Mihai, I.: On the generalized Wintgen inequality for Legendrian submanifolds in Sasakian space forms. Tohoku J. Math. **69**, 43–53 (2017)
98. Mihai, I., Ghişoiu, V.: Minimality of certain contact slant submanifolds of complex in Sasakian space forms. Int. J. Pure Appl. Math. Sci. **12**, 95–99 (2004)
99. Mihai, I., Rădulescu, I.N.: An improved Chen-Ricci inequality for Legendrian submanifolds in Sasakian space forms. J. Adv. Math. Stud. **4**, 51–58 (2011)
100. Mihai, I., Rouxel, B.: Tensor product surfaces of Euclidean plane curves. Results Math. **27**, 308–315 (1995)
101. Morvan, J.M.: Classe de Maslov d'une immersion Lagranienne et minimalité. C. R. Acad. Sc. Paris **292**, 633–636 (1981)
102. Nakao, Z.: Submanifolds of a Riemannian manifold with semisymmetric metric connections. Proc. Am. Math. Soc. **54**, 261–266 (1976)
103. Nomizu, K., Sasaki, T.: Affine Differential Geometry. Cambridge University Press, Cambridge (2000)
104. Oiagă, A., Mihai, I.: B.Y. Chen inequalities for slant submanifolds in complex space forms. Demonstratio Math. **32**, 835–846 (1999)
105. Opozda, B.: Bochner's technique for statistical structures. Ann. Global Anal. Geom. **48**, 357–395 (2015)
106. Opozda, B.: A sectional curvature for statistical structures. Linear Algebra Appl. **497**, 134–161 (2016)
107. Özgür, C., Mihai, A.: Chen inequalities for submanifolds of real space forms with a semi-symmetric non-metric connection. Can. Math. Bull. **55**, 611–622 (2012)
108. Özgür, C., Mihai, A.: Corrigendum to "Chen inequalities for submanifolds of real space forms with a semi-symmetric non-metric connection". Can. Math. Bull. **58**, 158–159 (2015)
109. Rouxel, B.: A-submanifolds in Euclidean space. Kodai Math. J. **4**, 181–188 (1981)
110. Rouxel, B.: Sur une famille des A-surfaces d'un espace Euclidien E^4, p. 185. Österreischer Mathematiker Kongress, Insbruck (1981)
111. Schouten, J.A.: Ricci-Calculus: An Introduction to Tensor Analysis and Its Geometrical Applications. Springer, Berlin (1954)

112. Senturk, Z., Verstraelen, L.: Chen ideal Kaehler hypersurfaces. Taiwan. J. Math. **12**, 1597–1608 (2008)
113. Singer, I.M., Thorpe, J.A.: The curvature of 4-dimensional Einstein spaces. In: Global Analysis (Papers in Honor of K. Kodaira), pp. 355–365. University of Tokyo Press, Tokyo (1969)
114. Shima, H.: The Geometry of Hessian Structures. World Scientific, Singapore (2007)
115. Suceavă, B.: Some remarks on B.Y. Chens's inequality involving classical invariants. An. St. Univ. Al. I. Cuza Iaşi Mat. **45**, 405–412 (1999)
116. Uohashi, K., Ohara, A., Fujii, T.: 1-conformally flat statistical submanifolds. Osaka J. Math. **37**, 501–507 (2000)
117. Urbano, F.: CR-submanifolds of nearly-Kaehler manifolds. Doctoral Thesis, Granada (1980)
118. Vaisman, I.: Symplectic Geometry and Secondary Characteristic Classes. Progress in Mathematics, vol. 72. Birkhäuser, Boston (1987)
119. Vos, P.W.: Fundamental equations for statistical submanifolds with applications to the Bartlett correction. Ann. Inst. Stat. Math. **41**(3), 429–450 (1989)
120. Wintgen, P.: Sur l'inégalité de Chen-Willmore. C.R. Acad. Sci. Paris **288**, 993–995 (1979)
121. Yano, K.: On semi-symmetric metric connection. Rev. Roumaine Math. Pures Appl. **15**, 1579–1586 (1970)
122. Yano, K., Kon, M.: Anti-invariant Submanifolds. M. Dekker, New York (1976)
123. Yano, K., Kon, M.: CR-submanifolds of Kaehlerian and Sasakian Manifolds. Birkhäuser, Boston (1983)
124. Yano, K., Kon, M.: Structures on Manifolds. World Scientific, Singapore (1984)

Chen Inequalities for Submanifolds of Kenmotsu Space Forms

İnan Ünal, Ajit Barman, and D. G. Prakasha

Abstract This chapter explores Chen-Ricci inequalities for submanifolds of Kenmotsu space forms endowed with a $\phi - \eta$-connection, a special type of quarter-symmetric metric connection. The geometric background of Kenmotsu space forms and the properties of the $\phi - \eta$-connection are briefly introduced, followed by the derivation of relevant curvature relations. The study proceeds by examining submanifolds within this framework, focusing on the behavior of curvature tensors and associated Riemannian invariants. Using the properties of the ambient space and the chosen connection, several inequalities involving Chen invariants and Ricci curvatures are obtained.

Keywords Chen inequality · Ricci inequality · Kenmotsu manifold · ϕ–η-connection

1 Introduction

Structures on manifolds have become a central topic of interest among differential geometers in recent years. These structures are typically defined on Riemannian or, more generally, semi-Riemannian manifolds. When associated with a metric, a structure on a manifold imparts significant geometric properties to the space. The most fundamental of these is the complex structure. A complex structure is an endomorphism of the tangent bundle, represented by a $(1, 1)$-tensor field J.

İ. Ünal (✉)
Department of Computer Engineering, Munzur University, Tunceli, Turkey
e-mail: inanunal@munzur.edu.tr

A. Barman
Department of Mathematics, Ramthakur College, Agartala, West Tripura, India

D. G. Prakasha
Department of Mathematics, Davangere University, Davangere, India

B.-Y. Chen, M. A. Choudhary (eds.), *Geometric Inequalities and Applications*, Infosys Science Foundation Series in Mathematical Sciences,
https://doi.org/10.1007/978-981-95-5148-4_4

A manifold endowed with such a structure is called an almost complex manifold. In contrast, a manifold that is locally homeomorphic to complex Euclidean space is called a complex manifold. An almost complex manifold becomes a complex manifold when the complex structure J is integrable, a condition characterized by a specific tensorial relation. This framework for defining complex structures has been widely adopted in the development of various geometric structures on manifolds.

A contact structure on a manifold is a significant geometric framework defined by a contact form. In the 1960s, the tensorial perspective on contact manifolds greatly expanded the field and sparked extensive research activity. One of the most influential contributions came from Sasaki, who introduced the notion of normality by establishing the integrability conditions for a contact structure on a Riemannian manifold. Contact manifolds satisfying these conditions are known as Sasakian manifolds, which are often considered the one-dimensional analogs of Kähler manifolds. Following these developments, substantial efforts have been made to classify contact manifolds. In particular, Kenmotsu [20] introduced a new class of contact manifolds that are normal but not Sasakian—now known as Kenmotsu manifolds. This class exhibits a rich and distinctive geometry. A notable contribution to the study of these manifolds is the comprehensive monograph by Pitiş [26], which explores the geometry of Kenmotsu manifolds in depth.

Even though a manifold may be locally similar to Euclidean space, determining whether it possesses a specific global structure remains a fundamental question in differential geometry. One of the most powerful tools for investigating the global geometry of a manifold is its curvature and associated curvature invariants. Certain manifolds exhibit specific types of curvature that facilitate their classification. Such manifolds are known as space forms. A Riemannian manifold is considered a space form if it has constant sectional curvature. If a manifold is equipped with a complex structure and has constant holomorphic sectional curvature with respect to that structure, it is referred to as a complex space form. In contrast, a Kenmotsu space form is a Kenmotsu manifold with constant ϕ-sectional curvature, which is intrinsically linked to its underlying contact structure.

Riemannian curvature invariants, described by Chen as the "DNA" of Riemannian manifolds [13], are intrinsic quantities that fundamentally influence the geometry and behavior of manifolds. These invariants establish deep connections between intrinsic properties—such as scalar curvature—and extrinsic measures, such as mean curvature. These relationships lead to sharp inequalities and significant insights in areas like minimal immersions, rigidity theorems, and eigenvalue estimates. Their applications extend across both mathematics and physics, serving as powerful tools for analyzing the geometric and physical structures of various spaces.

In submanifold theory, a fundamental question is *"what are the simple relationships between the main extrinsic and intrinsic invariants of a submanifold?"* To address this problem, Chen [10] introduced a novel Riemannian invariant, defined as $\delta(2) = \tau - \inf K$, where τ denotes the scalar curvature and K represents the sectional curvature. Using this invariant, Chen established the following inequality for submanifolds in a real space form $R^m(c)$ with constant sectional curvature c:

$$\delta(2) \leq \frac{n-2}{2}\left\{\frac{n^2}{n-1}\|H\|^2 + (n+1)c\right\}, \quad n = \dim M \geq 3, \tag{1.1}$$

where $\|H\|^2$ is the squared mean curvature of the submanifold M. This inequality, known as Chen's first inequality, has inspired extensive research across various geometric structures. A comprehensive review on Chen inequalities has been conducted by Chen and Vilcu [15]. If equality holds in (1.1), the submanifold is called a $\delta(2)$-ideal submanifold [15]. An important classification theorem for such submanifolds is given below:

Theorem 1.1 ([10]) *Let M be an n-dimensional minimal submanifold of $\mathbb{R}^m$. Then M is $\delta(2)$-ideal if and only if, locally, M is one of the following:*

1. *A totally geodesic submanifold of $\mathbb{E}^m$*
2. *A spherical cylinder $\mathbb{R} \times S^{n-1}(c)$*
3. *A direct product of a Euclidean k-space $\mathbb{E}^\ell$ and a Riemannian $(n-\ell)$-manifold $P^{n-\ell}$ satisfying $\delta(2) = 0$, such that $M = \mathbb{E}^k \times P^{n-\ell} \subset \mathbb{E}^m$ is minimally immersed as a direct product submanifold.*

Furthermore, in [11, 12], Chen proposed another approach for n-dimensional submanifolds within a real space form $R^m(c)$, leading to the following inequality:

$$\mathrm{Ric}(X) \leq \frac{n^2}{4}\|H\|^2 + (n-1)c, \quad n \geq 2, \tag{1.2}$$

where $\mathrm{Ric}(X)$ denotes the Ricci curvature along a unit vector $X \in TM$. This result, now known as the Chen-Ricci inequality, has been widely studied and applied in differential geometry.

In Riemannian geometry, the Levi-Civita connection is a derivative operator associated with the metric that enables calculus on manifolds. It is a torsion-free and metric-compatible connection. Beyond the Levi-Civita connection, there exist various connections that, although lacking its symmetric and metric properties, yield significant geometric results. On a Riemannian manifold M, such a general connection $\mathcal{D}$ is defined by

$$\mathcal{D}_{V_1}V_2 = \nabla_{V_1}V_2 + \mathcal{T}(V_1, V_2),$$

for all vector fields $V_1, V_2 \in \Gamma(TM)$, where ∇ is the Levi-Civita connection and $\mathcal{T}$ is a $(1, 2)$-type tensor field. Depending on the specific properties of $\mathcal{D}$, it is classified into various types, such as semi-symmetric metric, semi-symmetric nonmetric, semi-symmetric quarter-metric, etc. A generalization of these connections was provided in [31]. Several authors have explored the theory and applications of different connections [1, 3, 8, 17–19, 21, 29, 34]. Studies on Kenmotsu manifolds with different types of connections can be found in [2, 6, 9, 28, 30, 32]. Barman and Ünal [7] introduced the $\phi - \eta$-connection, a special type of connection, and applied it to Kenmotsu manifolds.

In this study, Chen-Ricci inequalities for submanifolds of a Kenmotsu space form are examined using the $\phi - \eta$-connection. In the second section of the study, the definition and properties of Kenmotsu space forms are provided, and the curvature properties associated with the $\phi - \eta$-connection are examined, resulting in various equations. The third section contains general information about submanifolds and discusses the curvature properties and Riemannian invariants of submanifolds in Kenmotsu space forms equipped with the $\phi - \eta$-connection. Finally, the last section presents results concerning Chen-Ricci inequalities for submanifolds of Kenmotsu space forms admitting the $\phi - \eta$-connection.

2 Kenmotsu Space Forms

In this section, the concept of Kenmotsu space forms are introduced, and the ϕ-η connection on these structures is examined.

An almost contact structure on a $(2m + 1)$-dimensional Riemann manifold $\bar{M}$ is a triple (ϕ, η, ξ) such that

$$\phi^2 = -I + \eta \otimes \xi,$$

where ϕ is a $(1, 1)$-type tensor field, ξ is a vector field, and η is a 1-form on $\overline{M}$. A Riemannian metric related with structure (ϕ, η, ξ) is given by

$$g(\phi V_1, \phi V_2) = g(V_1, V_2) - \eta(V_1)\eta(V_2), \quad \eta(V_1) = g(V_1, \xi),$$

for any vector fields $V_1, V_2 \in \Gamma(T\overline{M})$, where $\Gamma(T\overline{M})$ is the vector field space of $\overline{M}$. On an almost contact metric manifold $(\overline{M}, \phi, \eta, \xi, g)$ we have the following relations:

$$\eta(\xi) = 1, \quad \phi\xi = 0, \quad \eta \circ \phi = 0.$$

An almost contact metric manifold recalls as *normal* if ϕ is *integrable*, i.e., $N_\varphi(V_1, V_2) = [\varphi V_1, \varphi V_2] - \varphi[\varphi V_1, V_2] - \varphi[V_1, \varphi V_2] + \varphi^2[V_1, V_2] = 0$, for all $V_1, V_2 \in \Gamma(T\bar{M})$. A Kenmotsu manifold is a class of almost contact metric manifolds which is normal and

$$d\omega = 2\omega \wedge \eta, \quad d\eta = 0,$$

where $\omega(V_1, V_2) = g(\phi V_1, V_2)$ recall as the fundamental 2-form of the contact structure and $d\eta$ is differential of η. The following theorem states when an almost contact metric manifold is Kenmotsu.

Theorem 2.1 *An almost contact metric manifold* $(\overline{M}, \phi, \eta, \xi, g)$ *is a Kenmotsu manifold if and only if*

$$(\overline{\nabla}_{V_1}\phi)V_2 = -g(V_1, \phi V_2)\xi - \eta(V_2)\phi V_1 \tag{2.1}$$

is satisfied for any vector fields V_1, V_2 on $\overline{M}$, where $\overline{\nabla}$ is Levi-Civita connection on $\overline{M}$ [20].

From (2.1) we have

$$\overline{\nabla}_{V_1}\xi = V_1 - \eta(V_1)$$

for any vector fields V_1 on $\overline{M}$.

Sectional curvature is a significant parameter in the context of Riemannian geometry, as it provides direct information about the curvature of a Riemann manifold. Manifolds with constant sectional curvature are referred to as space forms and are classified based on this constant curvature. In addition to the sectional curvature derived from the curvature of the manifold, there are different kinds of sectional curvatures that arise due to specific structures on the manifold. The sectional curvature associated with an almost complex structure is termed holomorphic sectional curvature. If the holomorphic sectional curvature of a Hermitian manifold is constant, the manifold is called a complex space form. Similarly, the curvature associated with a contact structure is referred to as the ϕ-holomorphic sectional curvature. A Kenmotsu manifold $\overline{M}$ with constant ϕ-holomorphic sectional curvature c is denoted by $\overline{M}(c)$ and is called a *Kenmotsu space form*. The Riemannian curvature tensor of a Kenmotsu space form is expressed as follows [26]:

$$\begin{aligned}\overline{R}(V_1, V_2, V_3, V_4) = {} & \frac{c-3}{4}\big(g(V_2, V_3)g(V_1, V_4) - g(V_1, V_3)g(V_2, V_4)\big) \qquad (2.2)\\ & + \frac{c+1}{4}\Big[\eta(V_1)\eta(V_3)g(V_2, V_4) - \eta(V_2)\eta(V_3)g(V_1, V_4)\\ & + \eta(V_2)\eta(V_4)g(V_1, V_3) - \eta(V_1)\eta(V_4)g(V_2, V_3)\\ & + g(\phi V_2, V_3)g(\phi V_1, V_4) - g(\phi V_1, V_3)g(\phi V_2, V_4)\\ & - 2g(\phi V_1, V_2)g(\phi V_3, V_4)\Big].\end{aligned}$$

Here, $\overline{R}$ denotes the Riemannian curvature tensor of the Kenmotsu space form.

In general, a connection on a Riemannian manifold $\overline{M}$ is described by a mapping $\widehat{\nabla} : \Gamma(T\overline{M}) \times \Gamma(T\overline{M}) \to \Gamma(T\overline{M})$, defined as

$$\widehat{\overline{\nabla}}_{V_1} V_2 = \overline{\nabla}_{V_1} V_2 + \mathcal{T}(V_1, V_2),$$

for any $V_1, V_2 \in \Gamma(T\overline{M})$, where $\overline{\nabla}$ represents the Levi-Civita connection, and $\mathcal{T}$ is a tensor field of type $(1, 2)$. The connection is classified and named by the definition of $\mathcal{T}$. A specialized quarter-symmetric nonmetric connection was introduced in [7] specifically for Kenmotsu manifolds as follows:

$$\widehat{\overline{\nabla}}_{V_1} V_2 = \overline{\nabla}_{V_1} V_2 - \eta(V_1)\phi V_2 + g(V_1, V_2)\xi - \eta(V_2)V_1 - \eta(V_1)V_2 + \eta(V_1)\eta(V_2)\xi. \qquad (2.3)$$

$\widehat{\overline{\nabla}}$ is called the *quarter-symmetric nonmetric $\phi - \eta$-connection* [7]. In this study we recall $\widehat{\overline{\nabla}}$ as a $\phi - \eta$-connection.

Let $\widehat{\overline{R}}$ denote the curvature tensor of $\overline{M}^{n+p}$ with respect to $\widehat{\overline{\nabla}}$, where $n + p = 2m + 1$. In [7], the relation between the curvature tensors of $\widehat{\overline{\nabla}}$ and $\overline{\nabla}$ is given by

$$\widehat{\overline{R}}(V_1, V_2, V_3, V_4) = \overline{R}(V_1, V_2)g(V_3, V_4) + g(V_2, V_3)g(V_1, V_4) - g(V_1, V_3)g(V_2, V_4). \tag{2.4}$$

Using (2.2) in (2.4), we have

$$\begin{aligned} \widehat{\overline{R}}(V_1, V_2, V_3, V_4) = \frac{c+1}{4}\Big[& g(V_2, V_3)g(V_1, V_4) - g(V_1, V_3)g(V_2, V_4) \\ &+\eta(V_1)\eta(V_3)g(V_2, V_4) - \eta(V_2)\eta(V_3)g(V_1, V_4) \\ &+\eta(V_2)\eta(V_4)g(V_1, V_3) - \eta(V_1)\eta(V_4)g(V_2, V_3) \\ &+g(\phi V_2, V_3)g(\phi V_1, V_4) - g(\phi V_1, V_3)g(\phi V_2, V_4) \\ &-2g(\phi V_1, V_2)g(\phi V_3, V_4)\Big], \end{aligned} \tag{2.5}$$

for all $V_1, V_2, V_3, V_4 \in \Gamma(T\overline{M})$.

3 Submanifolds of Kenmotsu Space Forms

Submanifolds can exhibit geometric properties that differ from those of the ambient manifold in which they are embedded. While some properties can be derived from the ambient manifold, others may not be reducible. Moreover, the intrinsic geometry of submanifolds can possess characteristics distinct from the ambient manifold. Some equations and results originally introduced by Gauss for surfaces have been extended to submanifolds, providing relationships between the tangential and normal components of a vector field on a manifold. On the other hand, the properties of submanifolds can also vary depending on the structure (complex, contact, etc.) of the ambient manifold. In fact, there exist different classes (invariant, anti-invariant, etc.) of submanifolds associated with structured manifolds. In this section, we will present the general equations for submanifolds of a Kenmotsu space form and relate them with the $\phi - \eta$-connection.

Let $\overline{M}$ be a Kenmotsu space form and M be an n-dimensional submanifold tangent to ξ. Let us denote:

- ∇ the induced connection on M from the Levi-Civita connection $\overline{\nabla}$
- $\widehat{\nabla}$ the induced $\phi - \eta$-connection from $\widehat{\overline{\nabla}}$

Then, the Gauss formulas are given as follows:

$$\overline{\nabla}_{V_1} V_2 = \nabla_{V_1} V_2 + h(V_1, V_2), \tag{3.1}$$

$$\widehat{\overline{\nabla}}_{V_1} V_2 = \widehat{\nabla}_{V_1} V_2 + \widehat{h}(V_1, V_2), \tag{3.2}$$

for $V_1, V_2 \in \Gamma(T\overline{M})$, where h is a (0, 2)-tensor second fundamental form of M in $\overline{M}^{n+p}$, and $\widehat{h}$ is the second fundamental form on M with $\phi - \eta$-connection. By using (2.3) and (3.1) and considering tangential and normal parts, we get $\widehat{h} = h$. A submanifold of a Riemannian manifold is called as *totally geodesic* if $h = 0$. We observe that the property of being totally geodesic is the same with respect to both ∇ and $\widehat{\nabla}$. This situation could be changed by the definition of $\mathcal{T}$ in the connection. Crucially, the equality $\check{h} = h$ is determined by whether the structure vector field ξ is tangent to the submanifold M. The norm square of h is given by

$$\|h\|^2 = \sum_{i,j=1}^{n} g\left(h\left(e_i, e_j\right), h\left(e_i, e_j\right)\right). \tag{3.3}$$

By the definition of the direct sum of vector spaces, the space of vector fields on the ambient manifold can be expressed as

$$\Gamma(T\overline{M}) = \Gamma(TM) \oplus \Gamma(TM)^{\perp}.$$

In this context, the bases of the manifold M could be given by

$$T_pM = \{e_1, e_2, \ldots, e_n, \xi\}, \quad T_pM^{\perp} = \{e_{n+1}, e_{n+2}, \ldots, e_{n+p}\}.$$

Thus, any vector field V_1 can be written as

$$V_1 = \sum_{i=1}^{n-1} V_1^i e_i + \eta(V_1)\xi + \sum_{r=n+1}^{n+p} V_1^r e_r,$$

where this representation makes the tangential and normal components of the vector field more explicit. By consider this decomposition and from (3.3), we obtain

$$\|h\|^2 = \sum_{r=n+2}^{2m+1} \sum_{i,j=1}^{n+1} \left(h_{ij}^r\right)^2, \tag{3.4}$$

where $h_{ij}^r = g\left(h\left(e_i, e_j\right), e_r\right), i, j = 1, \ldots, n, r \in \{n+1, \ldots, n+p\}$, the components of the second fundamental form.

The trace of the second fundamental form is defined as the mean curvature which is denoted by H and given by

$$H(p) = \frac{1}{n} \sum_{i=1}^{n} h\left(e_i, e_i\right), \tag{3.5}$$

where $\{e_1, \ldots, e_n\}$ is an orthonormal basis of the tangent space T_pM, $p \in M$. The square norm of $||H||$ is given by

$$\|H\|^2 = g(H, H) = g\left(\frac{1}{n}\sum_{i=1}^{n} h\left(e_i, e_i\right), \frac{1}{n}\sum_{j=1}^{n} h\left(e_j, e_j\right)\right),$$

and so, we have

$$\|H\|^2 = \frac{1}{n^2}\sum_{i=1}^{n}\sum_{j=1}^{n} g\left(h\left(e_i, e_i\right), h\left(e_j, e_j\right)\right).$$

Let us take

$$B\left(e_i, e_i\right) = \sum_{r=n+1}^{n+p} B_{ii}^r e_r,$$

where $h_{ii}^r = g\left(h\left(e_i, e_i\right), e_r\right)$ and $\left\{e_{n+1}, \ldots, e_{n+p}\right\}$ is an orthonormal basis of the normal space $T_pM^{\perp}$. Substituting $h\left(e_i, e_i\right)$ into the expression for $\|H\|^2$,

$$\|H\|^2 = \frac{1}{n^2}\sum_{i=1}^{n}\sum_{j=1}^{n}\sum_{r=n+1}^{n+p}\sum_{s=n+1}^{n+p} h_{ii}^r h_{jj}^s \left\langle e_r, e_s\right\rangle.$$

Since $\{e_r\}$ is an orthonormal basis, $\langle e_r, e_s\rangle = \delta_{rs}$, where δ_{rs} is the Kronecker delta. This simplifies the expression to

$$\|H\|^2 = \frac{1}{n^2}\sum_{r=n+1}^{n+p}\left(\sum_{i=1}^{n} h_{ii}^r\right)^2. \tag{3.6}$$

The squared norm of the mean curvature vector, $\|H\|^2$, frequently appears in the study of geometric inequalities. It quantifies the total contribution of the mean curvature H to the curvature in all normal directions. This formula establishes a fundamental connection between the intrinsic geometry (characterized by the dimension n) and the extrinsic geometry (represented by the second fundamental form h) of the submanifold. Let us give a useful expression of (3.6). The expanded form is given by

$$\|H\|^2 = \frac{1}{n^2}\sum_{r=n+1}^{n+p}\left(h_{11}^r + h_{22}^r + \cdots + h_{nn}^r\right)^2.$$

In a fully expanded form it is like

$$\|H\|^2 = \frac{1}{n^2} \sum_{r=n+1}^{n+p} \left(\sum_{i=1}^{n} h_{ii}^r \cdot \sum_{j=1}^{n} h_{jj}^r \right),$$

which can also be written as

$$\|H\|^2 = \frac{1}{n^2} \sum_{r=n+1}^{n+p} \left(\sum_{i=1}^{n} (h_{ii}^r)^2 + \sum_{1 \le i < j \le n} 2h_{ii}^r h_{jj}^r \right). \tag{3.7}$$

Also, from (3.3) and (3.7) we have the following relation between the squared second fundamental form and the squared mean curvature:

$$\|h\|^2 = \frac{1}{2} n^2 \|H\|^2 + \frac{1}{2} \sum_{r=n+1}^{m} (h_{11}^r - h_{22}^r - \cdots - h_{nn}^r)^2 \tag{3.8}$$

$$+2 \sum_{r=n+1}^{m} \sum_{j=2}^{n} \left(h_{1j}^r\right)^2 - 2 \sum_{r=n+1}^{m} \sum_{2 \le i < j \le n} \left(h_{ii}^r h_{jj}^r - \left(h_{ij}^r\right)^2 \right).$$

The contact structure on a manifold may behave differently on a submanifold. This behavior is determined by the endomorphism ϕ that defines the contact structure. If V_1 is a vector field on the submanifold, it is not immediately clear where ϕV_1 will project. For any vector field V_1, let us write

$$\phi V_1 = P V_1 + F V_1, \tag{3.9}$$

where $P V_1$ (respectively, $F V_1$) denotes the tangential (respectively, normal) component of ϕV_1. Here, P acts as an endomorphism on the tangent bundle TM, while F represents a normal bundle-valued 1-form on TM. The norms of the projections P and F are defined as

$$\|P\|^2 = \sum_{i,j=1}^{n} g^2 \left(e_i, P e_j\right), \quad \|F\|^2 = \sum_{i=1}^{n} \|F e_i\|^2,$$

where $\|P\|^2$ and $\|F\|^2$ are independent of the choice of $\{e_1, \ldots, e_n\}$ orthonormal basis.

One of the most significant results in submanifold theory is the Gauss equation, which establishes a relationship between the curvature of the ambient manifold and that of the submanifold. We present the Gauss equation, which describes these curvatures, in different forms corresponding to the various connections discussed above. Subsequently, we will relate these forms to one another and ultimately derive the curvature associated with the connection induced from the $\phi - \eta$-connection.

The Gauss Equations with Respect to $\overline{\nabla}$ and ∇

$$\overline{R}(V_1, V_2, V_3, V_4) = R(V_1, V_2, V_3, V_4) + g(h(V_1, V_4), h(V_2, V_3)) - g(h(V_1, V_3), h(V_2, V_4)), \quad (3.10)$$

where $\overline{R}$ and R are the curvature tensors of $\overline{M}$ and M, respectively.

The Gauss Equation with Respect to $\widehat{\overline{\nabla}}$ and $\widehat{\nabla}$

$$\widehat{\overline{R}}(V_1, V_2, V_3, V_4) = \widehat{R}(V_1, V_2, V_3, V_4) + g(h(V_1, V_4), h(V_2, V_3)) - g(h(V_1, V_3), h(V_2, V_4)), \quad (3.11)$$

for any vector fields V_1, V_2, V_3, V_4 tangent to M. Here, $\widehat{\overline{R}}$ is the curvature tensor of $\overline{M}$ with respect to the $\phi - \eta$-connection, and $\widehat{R}$ is the induced curvature tensor of the $\phi - \eta$-connection on M.

Combining (2.4), (3.10), and (3.11), we derive

$$\widehat{R}(V_1, V_2, V_3, V_4) = R(V_1, V_2, V_3, V_4) + \big(g(V_2, V_3)g(V_1, V_4) - g(V_1, V_3)g(V_2, V_4)\big). \quad (3.12)$$

We now present the following lemma:

Lemma 3.1 *A submanifold M cannot be simultaneously* ∇*-flat and* $\widehat{\nabla}$*-flat.*

Using (3.13) in (3.11) we get the Gauss equation with respect to $\phi - \eta$-connection as follows:

$$\begin{aligned} \widehat{\overline{R}}(V_1, V_2, V_3, V_4) = {} & R(V_1, V_2, V_3, V_4) \\ & + \big(g(V_2, V_3)g(V_1, V_4) - g(V_1, V_3)g(V_2, V_4)\big) \\ & + g(h(V_1, V_4), h(V_2, V_3)) - g(h(V_1, V_3), h(V_2, V_4)). \end{aligned} \quad (3.13)$$

We mentioned that the decomposition of the vector field ϕV_1 into its tangential and normal components on a submanifold is not always clear. This situation can vary depending on whether the vector field ξ is tangent to the submanifold. In this study, we assume that ξ is tangent to the submanifold. The tangent space of the submanifold can be decomposed into different subspaces, referred to as distributions, which allow for a classification. Below, we present some of these classifications. A submanifold M of a Kenmotsu space form $\overline{M}$ is defined as follows:

- **Invariant submanifolds:** $\phi\left(T_pM\right) \subset T_pM$, i.e., $FV_1 = 0$ for any vector field V_1 on M
- **Anti-invariant submanifolds:** $\phi\left(T_pM\right) \subset T_p^{\perp}M$, i.e., $PV_1 = 0$ for any vector field V_1 on M
- **CR-submanifold:** There exists a pair of orthogonal differentiable distributions D and $D^{\perp}$ on M, such that $TM = D \oplus D^{\perp} \oplus \{\xi\}$, where $\{\xi\}$ is the one-dimensional distribution spanned by ξ:
  - D is invariant by ϕ, i.e., $\phi\left(D_p\right) \subset D_p$, for all $p \in M$. That is, $FV_1 = 0$ if V_1 on D.
  - $D^{\perp}$ is anti-invariant by ϕ, i.e., $\phi\left(D_p^{\perp}\right) \subset T_p^{\perp}M$, for all $p \in M$. That is, $PV_1 = 0$ if V_1 on $D^{\perp}$.

  In particular, if $D^{\perp} = \{0\}$ (respectively, $D = \{0\}$), M is an invariant (respectively, anti-invariant) submanifold.

The relationships between the intrinsic invariants of submanifolds and the invariants of the ambient manifold are among the key topics of study in submanifold theory. These invariants are often referred to as Riemannian invariants or curvature invariants. One of the most significant of these is sectional curvature. The tangent space of a submanifold is decomposed into two-dimensional sections, and their sectional curvatures are studied. The sum of these curvatures gives the scalar curvature. Below, further details regarding these concepts are provided; for more details see [14].

Let $\pi \subset T_xM^n, x \in M^n$, be a 2-plane section. Denote by $K(\pi)$ the sectional curvature of M^n $\phi - \eta$-connection $\hat{\hat{\nabla}}$. The scalar curvature of M is given by

$$\tau(x) = \sum_{1\leq i<j\leq n} K\left(e_i \wedge e_j\right)$$

for any orthonormal basis $\{e_1, \ldots, e_m\}$ of the tangent space T_xM^n.

Let us consider a k-dimensional subspace L of the tangent space T_xM^n at a point $x \in M^n$ and X be a unit vector lying within L. Let L be a k-dimensional subspace of the tangent space T_xM^n at a point $x \in M^n$, and let X be a unit vector lying within L. The k-Ricci curvature of L along X is then defined by

$$\mathrm{Ric}_L(X) = K_{12} + K_{13} + \cdots + K_{1k}, \tag{3.14}$$

where we choose an orthonormal basis as $\{e_1 = X, e_2, \ldots, e_k\}$, where K_{ij} denotes the sectional curvature associated with the two-dimensional plane spanned by e_i and e_j.

The scalar curvature $\tau(L)$ of L is defined as

$$\tau(L) = \sum_{i<j} K\left(e_i \wedge e_j\right), \quad 1 \leq i, j \leq k, \tag{3.15}$$

where $K\left(e_i \wedge e_j\right)$ denotes the sectional curvature of the 2-plane spanned by e_i and e_j. We denote the scalar curvature of the k-plane section spanned by $\{e_1, \ldots, e_r\}$ as $\tau_{1\cdots k}$. The scalar curvature $\tau(p)$ of M at a point $p \in M$ is simply the scalar curvature of the tangent space T_pM. When L is a two-dimensional plane section, $\tau(L)$ reduces to the sectional curvature $K(L)$ of L. Geometrically, $\tau(L)$ corresponds to the scalar curvature at p of the image of L under the exponential map $\exp_p$ [14].

For a fixed integer k such that $2 \leq k \leq n$, a Riemannian invariant Θ_k on M^n is defined as

$$\Theta_k(x) = \frac{1}{k-1} \inf_{L,X} \mathrm{Ric}_L(X), \quad x \in M^n, \tag{3.16}$$

where the infimum is taken over all k-dimensional subspaces $L \subset T_xM^n$ and all unit vectors $X \in L$.

The invariant $\Theta_k(x)$ generalizes the concept of Ricci curvature by considering k-dimensional subspaces of the tangent space instead of the entire tangent space. It measures the "worst-case" k-dimensional Ricci curvature, providing valuable insights into the interaction between the intrinsic and extrinsic curvatures of submanifolds. For $k = 2$, $\Theta_2(x)$ corresponds to the minimum sectional curvature at x, and for $k = n$, where n is the dimension of M^n, $\Theta_n(x)$ corresponds to the traditional Ricci curvature. The total scalar curvature $\tau(p)$ at a point p can be expressed in terms of $\Theta_k(p)$ by summing or averaging over all k-dimensional plane sections. $\Theta_k(x)$ plays a crucial role in various geometric inequalities and is particularly significant in the study of submanifolds.

The concept of the *relative null space* is a crucial tool in submanifold theory, providing insights into the relationship between the intrinsic geometry of a submanifold and its embedding in the ambient Riemannian manifold. For a submanifold M of a Riemannian manifold $(\overline{M}, \overline{g})$, the relative null space of M at a point $p \in M$ is defined as

$$\mathcal{N}_p = \left\{X \in T_pM \mid h(X, Y) = 0, \ \forall Y \in T_pM\right\}.$$

The relative null space $\mathcal{N}_p$ consists of all tangent vectors at p that are "flat" in terms of their interaction with the second fundamental form. That is, vectors in $\mathcal{N}_p$ do not contribute to the bending of the submanifold in any direction. For a totally geodesic submanifold, $h = 0$, and hence $\mathcal{N}_p = T_pM$ for all $p \in M$. The dimension of $\mathcal{N}_p$ depends on the point p and the geometric properties of M. It reflects how the submanifold bends within the ambient manifold. If M is minimal, the mean curvature vector H vanishes, which implies certain symmetry properties in h and affects $\mathcal{N}_p$. $\mathcal{N}_p$ provides a way to characterize submanifolds with specific curvature properties, such as null 2-type submanifolds or submanifolds with harmonic curvature. The size and structure of $\mathcal{N}_p$ influence the relationship between the intrinsic geometry of M (e.g., Ricci curvature) and its extrinsic properties (e.g., mean curvature).

4 Chen-Ricci Inequalities on Kenmotsu Space Forms Admitting $\phi - \eta$-Connection

In this section, results related to Chen-Ricci inequalities for submanifolds of Kenmotsu space forms with a $\phi - \eta$-connection are presented. Various inequalities for submanifolds of Kenmotsu space forms have been studied by several authors [4, 5, 16, 23–25, 27, 33].

The following lemma, proven by Chen [10], provides a useful relationship among real numbers, which play a crucial role in deriving inequalities:

Lemma 4.1 ([10]) *Let $a_1, a_2, \ldots, a_n$ and b be $(n+1)$ real numbers with $n \geq 2$, satisfying the equation:*

$$\left(\sum_{i=1}^{n} a_i\right)^2 = (n-1)\left(\sum_{i=1}^{n} a_i^2 + b\right).$$

Then the inequality $2a_1a_2 \geq b$ holds, with equality if and only if

$$a_1 + a_2 = a_3 = \ldots = a_n.$$

For submanifolds of Kenmotsu space forms endowed with a $\phi - \eta$-connection, we establish the Chen first inequality.

Theorem 4.2 *Let $M^n, n \geq 3$, be an n-dimensional submanifold of an $(n+p)$-dimensional Kenmotsu space form $\overline{M}^{n+p}(c)$ of constant sectional curvature c, endowed with $\phi - \eta$-connection $\widehat{\overline{\nabla}}$. Then we have*

$$\delta_M(x) \leq \left(n^2 - 3n + 3||P||^2\right)\frac{c+1}{8} - 3\frac{c-1}{4} + \frac{(n-2)n^2}{2(n-1)}||H||^2. \tag{4.1}$$

Here, π denotes a two-dimensional plane section of T_xM^n, where $x \in M^n$. At a point $x \in M^n$, the equality holds if and only if there exist:

- *An orthonormal basis $\{e_1, e_2, \ldots, e_n\}$ for the tangent space T_xM^n and*
- *An orthonormal basis $\{e_{n+1}, \ldots, e_{n+p}\}$ for the normal space $T_x^{\perp}M^n$*

such that the shape operators of M^n in $\overline{M}^{n+p}(c)$ at x satisfy the following conditions:

$$A_{e_{n+1}} = \begin{pmatrix} a & 0 & 0 & \cdots & 0 \\ 0 & b & 0 & \cdots & 0 \\ 0 & 0 & \gamma & \cdots & 0 \\ \vdots & \vdots & \vdots & \ddots & \vdots \\ 0 & 0 & 0 & \cdots & \gamma \end{pmatrix}, \quad a + b = \gamma,$$

$$A_{e_{n+i}} = \begin{pmatrix} h_{11}^{n+i} & h_{12}^{n+i} & 0 \cdots 0 \\ h_{12}^{n+i} & -h_{11}^{n+i} & 0 \cdots 0 \\ 0 & 0 & 0 \cdots 0 \\ \vdots & \vdots & \vdots \ddots \vdots \\ 0 & 0 & 0 \cdots 0 \end{pmatrix}, \quad 2 \leq i \leq p,$$

where we define $h_{ij}^r = g\left(h\left(e_i, e_j\right), e_r\right)$ *for* $1 \leq i, j \leq n$ *and* $n + 1 \leq r \leq n + p$.

Proof Let $x \in M^n$, $\{e_1, e_2, \ldots, e_n\}$ and $\left\{e_{n+1}, \ldots, e_{n+p}\right\}$ be orthonormal basis of $T_x M^n$ and $T_x^{\perp} M^n$, respectively. For $V_1 = V_4 = e_i$, $V_2 = V_3 = e_j$, $i \neq j$, from Eq. (2.4), and by summation on $1 \leq i, j \leq n$ we get

$$\widehat{\overline{R}}(e_i, e_j, e_j, e_i) = \left(n^2 - 3n + 2 - 3||P||^2\right) \frac{c+1}{4}.$$

Similarly from (3.13) and by summation on $1 \leq i, j \leq n$, it follows from (3.4) and (3.6) that

$$\left(n^2 - 3n + 2 - 3||P||^2\right) \frac{c+1}{4} = 2\tau + n^2 - n + ||h||^2 - n^2||H||^2. \tag{4.2}$$

First, multiplying Eq. (4.2) by $(n-1)$ and then adding and subtracting $n^2\|H\|^2$ to both sides, we get

$$n^2||H||^2 = (n-1)\left(||h||^2 + 2\tau - (n^2 - 3n + 2 - 3||P||^2)\frac{c+1}{4} + n^2 \right.$$
$$\left. -n - \frac{(n-2)n^2}{n-1}||H||^2\right).$$

Let us take

$$\vartheta(\tau, c) = 2\tau - \left(n^2 - 3n + 2 - 3||P||^2\right) \frac{c+1}{4} + n^2 - n - \frac{(n-2)n^2}{n-1}||H||^2,$$

and thus we can write

$$n^2||H||^2 = (n-1)\left(||h||^2 + \vartheta(\tau, c)\right). \tag{4.3}$$

Let $x \in M^n$, $\pi \subset T_x M^n$, $\dim \pi = 2$, $\pi = \text{sp}\{e_1, e_2\}$, and define $e_{n+1} = \frac{H}{\|H\|}$. By using (3.4) and (3.5) and from the relation (4.3), we obtain

$$\left(\sum_{i=1}^n h_{ii}^{n+1}\right)^2 = (n-1)\left(\sum_{i,j=1}^n \sum_{r=n+1}^{n+p} \left(h_{ij}^r\right)^2 + \vartheta(\tau, c)\right)$$

or equivalently

$$\left(\sum_{i=1}^{n} h_{ii}^{n+1}\right)^2 = (n-1)\left\{\sum_{i=1}^{n}\left(h_{ii}^{n+1}\right)^2 + \sum_{i\neq j}\left(h_{ij}^{n+1}\right)^2 + \sum_{i,j=1}^{n}\sum_{r=n+2}^{n+p}\left(h_{ij}^{r}\right)^2 + \vartheta(\tau, c)\right\}. \tag{4.4}$$

By using Lemma 4.1 and from (4.4), we have

$$2h_{11}^{n+1}h_{22}^{n+1} \geq \sum_{i\neq j}\left(h_{ij}^{n+1}\right)^2 + \sum_{i,j=1}^{n}\sum_{r=n+2}^{n+p}\left(h_{ij}^{r}\right)^2 + \vartheta(\tau, c). \tag{4.5}$$

On the other hand, from the Gauss equation we get

$$\begin{aligned} R(V_1, V_2, V_3, V_4) &= \widehat{\widehat{R}}(V_1, V_2, V_3, V_4) \\ &\quad - (g(V_2, V_3)g(V_1, V_4) + g(V_1, V_3)g(V_2, V_4)) \\ &\quad - g(h(V_1, V_4), h(V_2, V_3)) + g(h(V_1, V_3), h(V_2, V_4)). \end{aligned} \tag{4.6}$$

Let us take $V_1 = V_4 = e_1$ and $V_2 = V_3 = e_2$. Since $e_1, e_2 \in \Gamma(TM)$, we have $Pe_1 = e_1$, $Pe_2 = e_2$ and so we obtain

$$K(\pi) = \frac{c-1}{2} + \sum_{r=n+1}^{2m+1}\sum_{2\leq i<j\leq n}\left[h_{ii}^{r}h_{jj}^{r} - \left(h_{ij}^{r}\right)^2\right].$$

Thus considering (4.5) we have

$$\begin{aligned} K(\pi) &\geq \frac{c-1}{2} + \frac{1}{2}\left[\sum_{i\neq j}\left(h_{ij}^{n+1}\right)^2 + \sum_{i,j=1}^{n}\sum_{r=n+2}^{n+p}\left(h_{ij}^{r}\right)^2 + \vartheta(\tau, c)\right] \\ &\quad + \sum_{r=n+2}^{n+p} h_{11}^{r}h_{22}^{r} - \sum_{r=n+1}^{n+p}\left(h_{12}^{r}\right)^2 \\ &= \frac{c-1}{2} + \frac{1}{2}\sum_{i\neq j}\left(h_{ij}^{n+1}\right)^2 + \frac{1}{2}\sum_{i,j=1}^{n}\sum_{r=n+2}^{n+p}\left(h_{ij}^{r}\right)^2 + \frac{1}{2}\vartheta(\tau, c) \\ &\quad + \sum_{r=n+2}^{n+p} h_{11}^{r}h_{22}^{r} - \sum_{r=n+1}^{n+p}\left(h_{12}^{r}\right)^2, \end{aligned}$$

and thus, we get

$$
\begin{aligned}
K(\pi) \geq{}& \frac{c-1}{2} + \frac{1}{2}\sum_{i\neq j}\left(h_{ij}^{n+1}\right)^2 + \frac{1}{2}\sum_{r=n+2}^{n+p}\sum_{i,j>2}\left(h_{ij}^{r}\right)^2 \\
&+\frac{1}{2}\sum_{r=n+2}^{n+p}(h_{11}^r + h_{22}^r)^2 + \sum_{j>2}\left[\left(h_{1j}^{n+1}\right)^2 + \left(h_{2j}^{n+1}\right)^2\right] + \frac{1}{2}\vartheta(\tau, c) \\
={}& \frac{c-1}{2} + \frac{1}{2}\vartheta(\tau, c).
\end{aligned}
$$

Let us set the expression of $\vartheta(\tau, c)$ in the last inequality; then we get

$$
\begin{aligned}
\widehat{K}(\pi) \geq \tau - \frac{1}{8}\left(n^2 - 3n - 6 + 3||P||^2\right)c - \frac{1}{8}\left(n^2 - 3n + 6 + 3||P||^2\right) \\
-\frac{(n-2)n^2}{2(n-1)}||H||^2,
\end{aligned}
$$

and so, the inequality has the following form:

$$
\tau - \widehat{K}(\pi) \leq \left(n^2 - 3n + 3||P||^2\right)\frac{c+1}{8} - 3\frac{c-1}{4} + \frac{(n-2)n^2}{2(n-1)}||H||^2.
$$

In the inequality obtained above, a necessary and sufficient condition for the equality to hold at an arbitrary point $x \in M^n$ is that equality must occur in all the previous inequalities, as well as in the equality given in Lemma 4.1.

$$
\begin{aligned}
h_{ij}^{n+1} &= 0, \forall i \neq j, i, j > 2, \\
h_{ij}^{r} &= 0, \forall i \neq j, i, j > 2, r = n+1, \ldots, n+p, \\
h_{11}^r + h_{22}^r &= 0, \forall r = n+2, \ldots, n+p, \\
h_{1j}^{n+1} = h_{2j}^{n+1} &= 0, \forall j > 2, \\
h_{11}^{n+1} + h_{22}^{n+1} = h_{33}^{n+1} &= \cdots = h_{nn}^{n+1}.
\end{aligned}
$$

Choosing $\{e_1, e_2\}$ to be $h_{12}^{n+1} = 0$ and denoting $a = h_{11}^r, b = h_{22}^r, \mu = h_{33}^{n+1} = \cdots = h_{nn}^{n+1}$, the shape operators are obtained as a desired form. □

Every invariant submanifold of a Kenmotsu manifold is minimal [22]. We show that $h = \widehat{h}$; thus from the definition of mean curvature, it is evident that if the submanifold of a Kenmotsu manifold is minimal, then the submanifold of a Kenmotsu manifold admitting a $\phi - \eta$-connection is also minimal. On the other hand, for an invariant submanifold and any unit tangent vector $X \in T_pM$ orthogonal to ξ, since the manifold is invariant, we have $\|PX\| = \|\phi X\| = \|X\| = 1$. If M is anti-invariant, then $PX = 0$. Thus, we can state the following result:

Corollary 4.3 *Let M be a submanifold of Kenmotsu space form $\overline{M}(c)$ admitting a $\phi - \eta$-connection.*

(i) If M is an invariant submanifold, then we have

$$\delta_M \leq \left(n^2 - 3n + 3\right) \frac{c+1}{8} - 3\frac{c-1}{4}.$$

(ii) If M is an anti-invariant submanifold, then we have

$$\delta_M \leq \left(n^2 - 3n\right) \frac{c+1}{8} - 3\frac{c-1}{4} + \frac{(n-2)n^2}{2(n-1)}||H||^2.$$

(iii) If M is CR-submanifold, then we have:

(a) For each unit vector $X \in D_p$,

$$\delta_M \leq \left(n^2 - 3n\right) \frac{c+1}{8} - 3\frac{c-1}{4} + \frac{(n-2)n^2}{2(n-1)}||H||^2.$$

(b) For each unit vector $X \in D_p^{\perp}$, we have

$$\delta_M \leq \left(n^2 - 3n + 3\right) \frac{c+1}{8} - 3\frac{c-1}{4} + \frac{(n-2)n^2}{2(n-1)}||H||^2.$$

Let $X \in T_x M$ be a unit vector in the tangent space at x. We select an orthonormal basis $\{e_1, e_2, \ldots, e_n, e_{n+1}, \ldots, e_{n+p}\}$, where $\{e_1, e_2, \ldots, e_n\}$ spans the tangent space $T_x M$ with $e_1 = X$. Consider Eq. (4.2). From this equation we get

$$n^2||H||^2 = 2\tau + n^2 - n + ||h||^2 - \left(n^2 - 3n + 2 - 3||P||^2\right) \frac{c+1}{4}. \quad (4.7)$$

From (4.6), $V_1 = e_i$, $V_2 = V_3 = e_j$, we get

$$K_{ij} = \frac{c-1}{2} + \sum_{r=n+1}^{n+p} \left[h_{ii}^r h_{jj}^r - \left(h_{ij}^r\right)^2\right].$$

By summation, it follows that

$$\sum_{2\leq i<j\leq n} K_{ij} = \sum_{r=n+1}^{n+p} \sum_{2\leq i<j\leq n} \left[h_{ii}^r h_{jj}^r - \left(h_{ij}^r\right)^2\right] + \sum_{2\leq i<j\leq n} \left[\frac{c-1}{2}\right]$$

$$= \sum_{r=n+1}^{n+p} \sum_{2 \le i < j \le n} \left[h_{ii}^r h_{jj}^r - \left(h_{ij}^r\right)^2 \right] + \frac{(n-2)(n-1)(c-1)}{2}.$$

Thus, using (4.7) and (3.8) we get

$$\begin{aligned} n^2 \|H\|^2 = 2\tau + \frac{1}{2} n^2 \|H\|^2 + \frac{1}{2} \sum_{r=n+1}^{m} \left(h_{11}^r - h_{22}^r - \cdots - h_{nn}^r\right)^2 \\ +2 \sum_{r=n+1}^{m} \sum_{j=2}^{n} \left(h_{1j}^r\right)^2 - 2 \sum_{r=n+1}^{m} \sum_{2 \le i < j \le n} \left(h_{ii}^r h_{jj}^r - \left(h_{ij}^r\right)^2\right) \\ +n^2 - n - \left(n^2 - 3n + 2 - 3||P||^2\right) \frac{c+1}{4}. \end{aligned}$$

Hence we obtain

$$\begin{aligned} \frac{1}{2} n^2 \|H\|^2 &= 2\,\mathrm{Ric}(e_1) + 2 \sum_{2 \le i < j \le n} K_{ij} + \frac{1}{2} \sum_{r=n+1}^{m} \left(h_{11}^r - h_{22}^r - \cdots - h_{nn}^r\right)^2 \\ &\quad +2 \sum_{r=n+1}^{m} \sum_{j=2}^{n} \left(h_{1j}^r\right)^2 - 2 \sum_{r=n+1}^{m} \sum_{2 \le i < j \le n} \left(h_{ii}^r h_{jj}^r - \left(h_{ij}^r\right)^2\right) \\ &\quad +n^2 - n - (n^2 - 3n + 2 - 3||Pe_1||^2) \frac{c+1}{4} \\ &= \mathrm{Ric}(e_1) + \frac{(n-2)(n-1)(c-1)}{2} \\ &\quad + \frac{1}{2} \sum_{r=n+1}^{m} \left(h_{11}^r - h_{22}^r - \cdots - h_{nn}^r\right)^2 \\ &\quad +n^2 - n - \left(n^2 - 3n + 2 - 3||P||^2\right) \frac{c+1}{8} \\ &\ge \mathrm{Ric}(e_1) + \frac{(n-2)(n-1)(c-1)}{2} + n^2 - n \\ &\quad - \left(n^2 - 3n + 2 - 3||Pe_1||^2\right) \frac{c+1}{8}. \end{aligned}$$

Finally we get

$$\begin{aligned} \mathrm{Ric}(X) \le \frac{1}{4} \Big[n^2 \|H\|^2 - 2(n-1)((n-2)c+2) \\ - \left(n^2 - 3n + 2 - 3||X||^2\right) \frac{c+1}{2} \Big]. \end{aligned} \tag{4.8}$$

Assume that $H(p) = 0$. Equality in (4.8) holds if and only if

$$h^r_{12} = \cdots = h^r_{1n} = 0, \quad h^r_{11} = h^r_{22} + \cdots + h^r_{nn}, r \in \{n+1, \ldots, n+p\}.$$

This implies that $h^r_{1j} = 0$ for all $j \in \{1, \ldots, n\}$ and $r \in \{n+1, \ldots, 2m\}$, which means $X \in \mathcal{N}_p$. Equality in (4.8) for all unit tangent vectors at p holds if and only if

$$\begin{aligned} h^r_{ij} &= 0, \quad i \neq j, r \in \{n+1, \ldots, n+p\}, \\ h^r_{11} + \cdots + h^r_{nn} - 2h^r_{ii} &= 0, \quad i \in \{1, \ldots, n\}, r \in \{n+1, \ldots, n+p\}. \end{aligned}$$

Consequently, the point p is totally geodesic. The converse is straightforward and follows directly. Finally we state the following result.

Theorem 4.4 *Let $\bar{M}(c)$ be an $(n+p)$-dimensional Kenmotsu space form admitting $\phi - \eta$-connection and M an n-dimensional submanifold tangent to ξ. Then:*

(i) For each unit vector $X \in T_pM$ orthogonal to ξ, we have (4.8).
(ii) If $H(p) = 0$, then a unit tangent vector $X \in T_pM$ orthogonal to ξ satisfies the equality case of (4.8) *if and only if $X \in \mathcal{N}_p$.*
(iii) The equality case of (4.8) *holds identically for all unit tangent vectors orthogonal to ξ at p if and only if p is a totally geodesic point.*

For special submanifolds we have the following result.

Corollary 4.5 *Let M be a submanifold of Kenmotsu space form $\overline{M}(c)$ admitting a $\phi - \eta$-connection.*

(i) If M is an invariant submanifold, then we have

$$\mathrm{Ric}(X) \leq \frac{1}{8}\Big[(n^2 - 3n + 5)c - n^2 + 11n - 7\Big].$$

(ii) If M is an anti-invariant submanifold, then we have

$$\mathrm{Ric}(X) \leq \frac{1}{8}\Big[\Big(n^2 - 3n + 2\Big)c - n^2 + 12n - 8 + n^2\|H\|^2\Big].$$

(iii) If M is CR-submanifold, then we have:

(a) For each unit vector $X \in D_p$,

$$\mathrm{Ric}(X) \leq \frac{1}{4}\Big[\Big(n^2 - 3n + 5\Big)c - n^2 + 11n - 7 + n^2\|H\|^2\Big].$$

(b) For each unit vector $X \in {D_p}^{\perp}$, we have

$$\mathrm{Ric}(X) \leq \frac{1}{4}\Big[\Big(n^2 - 3n + 2\Big)c - n^2 + 12n - 8 + n^2 + n^2\|H\|^2\Big].$$

5 k-Ricci Curvature

k-Ricci curvature of M^n is an intrinsic geometric invariant, and the squared mean curvature $\|H\|^2$ is an extrinsic invariant. The k-Ricci curvature of M^n is an intrinsic geometric invariant, while the squared mean curvature H^2 is an extrinsic invariant. A fundamental question in submanifold theory is to establish a sharp relation between these invariants. In this subsection, we provide a new result concerning the interplay between intrinsic and extrinsic properties.

Theorem 5.1 *Let $M^n, n \geq 3$, be an n-dimensional submanifold M tangent to ξ of an $(n+p)$-dimensional Kenmotsu space form $\overline{M}^{n+p}(c)$ of constant sectional curvature c endowed with $\phi - \eta$-connection $\hat{\overline{\nabla}}$. Then we have*

$$\|H\|^2 \geq \frac{2\tau}{n(n-1)} + \frac{\left(n^2 - 3n + 2 - 3||PX||^2\right)(c+1) + 4(n^2 - n)}{4(n^2 - n)}. \tag{5.1}$$

Proof We begin by selecting an orthonormal basis $\{e_1, \ldots, e_n, e_{n+1}, \ldots, e_{2m+1} = \xi\}$ at the point $p \in M$, such that:

- e_{n+1} is aligned with the mean curvature vector $H(p)$.
- $e_1, \ldots, e_n$ diagonalize the shape operator A_{n+1}.

Under these assumptions, the shape operators are expressed as

$$A_{n+1} = \begin{pmatrix} a_1 & 0 & 0 \cdots & 0 \\ 0 & a_2 & 0 \cdots & 0 \\ \vdots & \vdots & \vdots \cdots & \vdots \\ 0 & 0 & 0 \cdots & a_n \end{pmatrix},$$

$$A_r = \left(h_{ij}^r\right), \quad i, j = 1, \ldots, n, \quad r = n+2, \ldots, 2m, \quad \operatorname{trace} A_r = \sum_{i=1}^{n} h_{ii}^r = 0.$$

Using the Gauss equation (4.2), we derive the following relationship:

$$n^2\|H\|^2 = 2\tau + \sum_{i=1}^{n} a_i^2 + \sum_{r=n+2}^{2m} \sum_{i,j=1}^{n} \left(h_{ij}^r\right)^2 + (n^2 - 3n + 2 - 3||PX||^2)\frac{c+1}{4} + n^2 - n. \tag{5.2}$$

To simplify further, we note the inequality

$$0 \leq \sum_{i<j} (a_i - a_j)^2 = (n-1)\sum_{i=1}^{n} a_i^2 - 2\sum_{i<j} a_i a_j.$$

From this, it follows that

$$n^2\|H\|^2 = \left(\sum_{i=1}^{n} a_i\right)^2 = \sum_{i=1}^{n} a_i^2 + 2\sum_{i<j} a_i a_j \leq n\sum_{i=1}^{n} a_i^2.$$

Thus, we deduce

$$\sum_{i=1}^{n} a_i^2 \geq n\|H\|^2.$$

Substituting this inequality into (5.2), we obtain

$$n^2\|H\|^2 \geq 2\tau + n\|H\|^2 + (n^2 - 3n + 2 - 3||PX||^2)\frac{c+1}{4} + n^2 - n.$$

This completes the proof. □

Using this result, we derive the following theorem.

Theorem 5.2 *Let $M^n, n \geq 3$, be an n-dimensional submanifold M tangent to ξ of an $(n+p)$-dimensional Kenmotsu space form $\overline{M}^{n+p}(c)$ of constant sectional curvature c endowed with $\phi - \eta$-connection $\overline{\overline{\nabla}}$. Then we have*

$$\|H\|^2(p) \geq \Theta_k(p) + \frac{(n^2 - 3n + 2 - 3||PX||^2)(c+1) + 4(n^2 - n)}{4(n^2 - n)}. \tag{5.3}$$

Proof Consider an orthonormal basis $\{e_1, \ldots, e_n\}$ of T_pM. Let $L_{i_1 \cdots i_k}$ represent the k-dimensional plane section spanned by the vectors $e_{i_1}, \ldots, e_{i_k}$. From (3.14) and (3.15), the following relations hold:

$$\tau\left(L_{i_1 \cdots i_k}\right) = \frac{1}{2}\sum_{i \in \{i_1, \ldots, i_k\}} \mathrm{Ric}_{L_{i_1 \cdots i_k}}(e_i), \tag{5.4}$$

$$\tau(p) = \frac{1}{C_{n-2}^{k-2}}\sum_{1 \leq i_1 < \cdots < i_k \leq n} \tau\left(L_{i_1 \cdots i_k}\right).$$

By combining (3.16) with (5.4), we derive the inequality:

$$\tau(p) \geq \frac{n(n-1)}{2}\Theta_k(p). \tag{5.5}$$

From (5.1) and (5.5), we obtain (5.3). □

For special submanifolds of Kenmotsu space form we get the following corollary.

Corollary 5.3 *Let M be a submanifold of Kenmotsu space form $\overline{M}(c)$ admitting a $\phi-\eta$-connection. Then, for any integer k, $2 \leq k \leq n$, and any point $p \in M$:*

(i) If M is an invariant submanifold, we have

$$\Theta_k(p) \leq \frac{(1+3n-n^2(c+1))+4(n^2-n)}{4(n^2-n)}.$$

(ii) If M is an anti-invariant submanifold, we have

$$\|H\|^2(p) \geq \Theta_k(p) + \frac{(n^2-3n+2)(c+1)+4(n^2-n)}{4(n^2-n)}.$$

(iii) If M is CR-submanifold, we have

$$\|H\|^2(p) \geq \Theta_k(p) + \frac{(n^2-3n+2-6h)(c+1)+4(n^2-n)}{4(n^2-n)},$$

where $2h = \dim D$.

6 Conclusion

In this study, we addressed the fundamental problem in submanifold theory, namely finding simple relationships between the main extrinsic and intrinsic invariants of a submanifold, within the framework of Kenmotsu space forms. We investigated Chen invariants and Ricci inequalities for submanifolds of Kenmotsu space forms endowed with a special quarter-symmetric connection. The inequalities involving Chen invariants on Kenmotsu space forms equipped with this special connection yielded significant geometric results.

These results were applied to invariant, anti-invariant, and CR-submanifolds, which are special classes of submanifolds. The findings of this study provide valuable contributions to ongoing research in this field and offer a foundation for further exploration. Additionally, the proposed connection allows for potential investigations into other special submanifolds of Kenmotsu space forms. Moreover, extending this connection to other geometric structures could lead to novel insights and diverse applications in differential geometry.

This work underscores the importance of the proposed approach and invites future studies to build upon these results, extending the scope of submanifold theory in the context of Kenmotsu geometry and beyond.

References

1. Agashe, N.S., Chafle, M.R.: A semi-symmetric non-metric connection on a Riemannian manifold. Indian J. Pure Appl. Math. **23**(1), 399–409 (1992)
2. Alghamdi, M.A., Bahadır, O.: Some characterizations in Kenmotsu manifolds with a new connection. BSG Proc. **27**(1), 1–16 (2020)
3. Andonie, O.C.: On semi-symmetric non-metric connections on a Riemannian manifold. Ann. Fac. Sci. Kinshasa **2**, 1–14 (1976)
4. Aquib, M., Shahid, M.H.: Generalized Wintgen inequality for submanifolds in Kenmotsu space forms. Tamkang J. Math. **50**(2), 155–164 (2019)
5. Arslan, K., Ezentas, R., Mihai, I., Murathan, C., Özgür, C.: Ricci curvature of submanifolds in Kenmotsu space forms. Int. J. Math. Math. Sci. **29**(12), 719–726 (2002)
6. Barman, A.: On a type of quarter-symmetric non-metric ϕ-connection on a Kenmotsu manifold. Bull. Math. Anal. Appl. **4**(1), 1–11 (2012)
7. Barman, A., Unal, I.: Geometry of Kenmotsu manifolds admitting Z-tensor. Bull. Transilv. Univ. Brasov. Ser. III: Math. Comput. Sci. **15**(1), 23–40 (2022)
8. Barua, B., Mukhopadhyay, S.: A sequence of semi-symmetric connections on a Riemannian manifold. In: Proceedings of the Seventh National Seminar on Finsler, Lagrange and Hamiltonian Spaces, Brasov (1992)
9. Chaubey, S., Yadav, S.K.: Study of Kenmotsu manifolds with semi-symmetric metric connection. Universal J. Math. Appl. **1**(2), 89–97 (2018)
10. Chen, B.Y.: Some pinching and classification theorems for minimal submanifolds. Arch. Math. **60**(6), 568–578 (1993)
11. Chen, B.Y.: Mean curvature and shape operator of isometric immersions in real-space forms. Glasg. Math. J. **38**(1), 87–97 (1996)
12. Chen, B.Y.: Relations between Ricci curvature and shape operator for submanifolds with arbitrary codimensions. Glasg. Math. J. **41**(1), 33–41 (1999)
13. Chen, B.Y.: Riemannian DNA, inequalities, and their applications. J. Appl. Sci. Eng. **3**(3), 123–130 (2000)
14. Chen, B.Y.: Pseudo-Riemannian Geometry, [delta]-invariants and Applications. World Scientific, Singapore (2011)
15. Chen, B.Y., Vilcu, G.E.: Recent developments on the first Chen inequality in differential geometry. Mathematics **11**(19), 4186 (2023)
16. Costache, S., Zamfir, I.: An improved Chen-Ricci inequality for special slant submanifolds in Kenmotsu space forms. Ann. Pol. Math. **110**(1), 81–89 (2014)
17. Friedman, A.A., Schouten, J.A.: Über die Geometrie der halbsymmetrischen Übertragung. Math. Z. **21**(1), 211–223 (1924)
18. Golab, S.: On semi-symmetric and quarter-symmetric linear connections. Tensor, N. S. **29**(1), 249–254 (1975)
19. Hayden, H.A.: Subspaces of space with torsion. Proc. Lond. Math. Soc. **34**(1), 27–50 (1932)
20. Kenmotsu, K.: A class of almost contact Riemannian manifolds. Tohoku Math. J. **24**(1), 93–103 (1972)
21. Liang, Y.: On semi-symmetric recurrent-metric connection. Tensor, N. S. **55**(1), 107–112 (1994)
22. Matsumoto, K., Mihai, I., Roşca, R.: A certain locally conformal almost cosymplectic manifold and its submanifolds. Tensor, N. S. **51**(1), 91–102 (1992)
23. Mohammed, M.: Geometric inequalities for bi-slant submanifolds in Kenmotsu space forms. Rom. J. Math. Comput. Sci. **12**(2), 52–63 (2022)
24. Mustafa, A., Uddin, S., Al-Solamy, F.R.: Chen–Ricci inequality for warped products in Kenmotsu space forms and its applications. Rev. R. Acad. Cienc. Exactas Fis. Nat. Ser. A Mat. **113**(4), 3585–3602 (2019)
25. Pandey, P.K., Gupta, R.S., Sharfuddin, A.: B. Y. Chen's inequalities for bi-slant submanifolds in Kenmotsu space forms. Demonstr. Math. **43**(4), 887–898 (2010)

26. Pitiş, G.: Geometry of Kenmotsu Manifolds. Publishing House of Transilvania University of Braşov, Braşov (2007)
27. Polat, M.: B. Y. Chen-Ricci inequalities for anti-invariant Riemannian submersions in Kenmotsu space forms. Arab. J. Math. **13**(1), 181–196 (2024)
28. Prakash, A., Pandey, V.K.: On a quarter symmetric non-metric connection in a Kenmotsu manifold. Int. J. Pure Appl. Math. **83**(2), 271–278 (2013)
29. Prvanović, M.: On pseudo-metric semi-symmetric connections. Publ. Inst. Math. **18**(1), 157–164 (1975)
30. Sular, S., Özgür, C., De, U.C.: Quarter-symmetric metric connection in a Kenmotsu manifold. Tensor, N. S. **36**(2), 189–199 (2008)
31. Tripathi, M.M.: A new connection in a Riemannian manifold. Int. Electron. J. Geom. **1**(1), 15–24 (2008)
32. Vanlı, A.T., Sari, R.: On semi-invariant submanifolds of a generalized Kenmotsu manifold admitting a semi-symmetric metric connection. Pure Appl. Math. J. **4**(1), 14–18 (2015)
33. Vilcu, G.E.: Curvature inequalities for slant submanifolds in pointwise Kenmotsu space forms. In: Contact Geometry of Slant Submanifolds, pp. 13–37. Springer, Berlin (2022)
34. Yano, K.: On semi-symmetric connection. Rev. Roum. Math. Pures Appl. **15**(1), 1570–1586 (1970)

Improved Chen-Ricci Inequalities for Semi-slant $\xi^{\perp}$-Riemannian Submersions from Sasakian Space Forms

Mehmet Akif Akyol and Nergiz (Önen) Poyraz

Abstract This study is devoted to find Chen and Chen-Ricci -type inequalities for semi-slant $\xi^{\perp}$-Riemannian submersions from Sasakian space forms onto Riemannian manifolds. These inequalities reveal relationships between intrinsic invariants, such as Ricci and scalar curvature, and extrinsic invariants like the second fundamental form. Finally, we analyze the conditions under which equality is attained, and we present illustrative examples.

Keywords Riemannian submersion · Semi-slant $\xi^{\perp}$-Riemannian submersion · Chen inequality · Sasakian manifold · Vertical distribution

1 Introduction

In a series of pioneering works [13, 14], B. Y. Chen explored meaningful connections between the intrinsic curvature quantities (such as Ricci, scalar, and k-Ricci curvature) and the extrinsic curvature quantities (such as the squared mean curvature and shape operator) in submanifold geometry. These led to the development of what are now known as the Chen inequalities, which provide sharp bounds involving curvature invariants for Riemannian submanifolds in space forms (see: [13–16]). Since then, extensive studies have expanded upon these results, leading to a wide array of inequalities applicable to various classes of submanifolds and Riemannian submersions in diverse ambient geometries (e.g., [1, 3–5, 8–11, 18, 21–25, 27, 28, 30, 32–34, 37–39, 44, 48–53], among others).

M. A. Akyol (✉)
Faculty of Engineering and Natural Sciences, Department of Mathematics, Uşak University, Uşak, Turkey
e-mail: mehmet.akyol@usak.edu.tr

N. (Önen) Poyraz
Faculty of Arts and Sciences, Department of Mathematics, Çukurova University, Adana, Turkey
e-mail: nonen@cu.edu.tr

B.-Y. Chen, M. A. Choudhary (eds.), *Geometric Inequalities and Applications*, Infosys Science Foundation Series in Mathematical Sciences,
https://doi.org/10.1007/978-981-95-5148-4_5

The theory of Riemannian submersions, introduced by O'Neill [29] and further developed by Gray [20], has proven to be a useful tool in the study of the geometric structure of manifolds. Later, Şahin [40] extended this framework by introducing anti-invariant Riemannian submersions from almost Hermitian manifolds. Motivated by these developments, researchers have proposed several generalizations such as slant, semi-invariant, and semi-slant submersions, each offering novel insights into submersion geometry [2, 6, 7, 17, 26, 31, 36, 41–43, 45–47].

Specifically, Lee [26] introduced the concept of anti-invariant $\xi^{\perp}$-Riemannian submersions from almost contact metric manifolds. This notion was later generalized by Akyol et al. [7] to semi-invariant $\xi^{\perp}$-Riemannian submersions and subsequently extended to slant submersions from Sasakian manifolds by Erken and Murathan [17]. Building on these contributions, Akyol and Sarı [6] proposed the notion of semi-slant $\xi^{\perp}$-Riemannian submersions, which generalizes the previous frameworks.

In this chapter, we examine Chen-Ricci inequalities for semi-slant $\xi^{\perp}$-Riemannian submersions from Sasakian space forms. We derive new inequalities, study their equality cases, and offer explicit examples to illustrate the results.

This chapter is organized as follows. In Sect. 2, we review some fundamental geometric properties of Riemannian submersions, Sasakian manifolds, and Sasakian space forms. In Sect. 3, we derive the Chen-Ricci inequality and the Chen inequality for semi-slant $\xi^{\perp}$-Riemannian submersions from Sasakian space forms. We also discuss the equality cases. Finally, we provide some illustrative examples.

2 Preliminaries

Definition 2.1 Let (M, g) and (M_1, g_1) be Riemannian manifolds, where $dim(M) = m$, $dim(M_1) = m_1$, and $m > m_1$. A Riemannian submersion $\psi : M \to M_1$ is a map of M onto M_1 satisfying the following axioms:

(i) ψ has maximal rank.
(ii) The differential ψ_* preserves the lengths of horizontal vectors, that is, ψ_* is a linear isometry [29].

The geometry of Riemannian submersions is characterized by O'Neill's tensors $\mathcal{T}$ and $\mathcal{A}$, defined as follows [19]:

$$\mathcal{T}_{U_1}U_2 = \mathcal{V}\nabla_{\mathcal{V}U_1}\mathcal{H}U_2 + \mathcal{H}\nabla_{\mathcal{V}U_1}\mathcal{V}U_2, \tag{2.1}$$

$$\mathcal{A}_{U_1}U_2 = \mathcal{V}\nabla_{\mathcal{H}U_1}\mathcal{H}U_2 + \mathcal{H}\nabla_{\mathcal{H}U_1}\mathcal{V}U_2 \tag{2.2}$$

for any vector fields U_1 and U_2 on M, where ∇ is the Levi-Civita connection of g. It is easy to see that $\mathcal{T}_{U_1}$ and $\mathcal{A}_{U_1}$ are skew-symmetric operators on the tangent bundle of M reversing the vertical and the horizontal distributions. We now summarize the properties of the tensor fields $\mathcal{T}$ and $\mathcal{A}$. Let [19] V_1, V_2 be vertical and X_1, X_2 be horizontal vector fields on M, and then we have

$$\mathcal{T}_{V_1} V_2 = \mathcal{T}_{V_2} V_1, \tag{2.3}$$

$$\mathcal{A}_{X_1} X_2 = -\mathcal{A}_{X_2} X_1 = \frac{1}{2}\mathcal{V}[X_1, X_2]. \tag{2.4}$$

By virtue of (2.1) and (2.2), we get

$$\nabla_{V_1} V_2 = \mathcal{T}_{V_1} V_2 + \hat{\nabla}_{V_1} V_2, \tag{2.5}$$

$$\nabla_{V_1} X_1 = \mathcal{T}_{V_1} X_1 + \mathcal{H}\nabla_{V_1} X_1, \tag{2.6}$$

$$\nabla_{X_1} V_1 = \mathcal{A}_{X_1} V_1 + \mathcal{V}\nabla_{X_1} V_1, \tag{2.7}$$

$$\nabla_{X_1} X_2 = \mathcal{H}\nabla_{X_1} X_2 + \mathcal{A}_{X_1} X_2, \tag{2.8}$$

for all $V_1, V_2 \in \Gamma(ker\psi_*)$ and $X_1, X_2 \in \Gamma((ker\psi_*)^{\perp})$, where $\hat{\nabla}_{V_1} V_2 = \mathcal{V}\nabla_{V_1} V_2$. If X_1 is basic, $\mathcal{H}\nabla_{V_1} X_1 = \mathcal{A}_{X_1} V_1$ [19]. We also note that:

(i) For $U_1, U_2 \in \mathcal{V}(M)$, $\mathcal{T}_{U_1} U_2 = \mathcal{T}_{U_2} U_1$.
(ii) For $X_1, X_2 \in \mathcal{H}(M)$, $\mathcal{A}_{X_1} X_2 = -\mathcal{A}_{X_2} X_1$.

Denote by R, R', $\hat{R}$, and R_1 the Riemannian curvature tensor of Riemannian manifolds M, M_1, the vertical distribution $\mathcal{V}$, and the horizontal distribution $\mathcal{H}$, respectively. Then the Gauss-Codazzi-type equations are given by

$$R(U_1, U_2, U_3, U_4) = \hat{R}(U_1, U_2, U_3, U_4) + g(\mathcal{T}_{U_1} U_4, \mathcal{T}_{U_2} U_3) - g(\mathcal{T}_{U_2} U_4, \mathcal{T}_{U_1} U_3) \tag{2.9}$$

$$\begin{aligned} R(X_1, X_2, X_3, X_4) = {} & R^*(X_1, X_2, X_3, X_4) - 2g(\mathcal{A}_{X_1} X_2, \mathcal{A}_{X_3} X_4) \\ & + g(\mathcal{A}_{X_2} X_3, \mathcal{A}_{X_1} X_4) - g(\mathcal{A}_{X_1} X_3, \mathcal{A}_{X_2} X_4) \end{aligned} \tag{2.10}$$

$$\begin{aligned} R(X_1, U_2, X_2, U_4) = {} & g((\nabla_X \mathcal{T})(U_2, U_4), X_2) + g((\nabla_V \mathcal{A})(X_1, X_2), U_4) \\ & - g(\mathcal{T}_{U_2} X_1, \mathcal{T}_{U_4} X_2) + g(\mathcal{A}_{X_2} U_4, \mathcal{A}_{X_1} U_2), \end{aligned} \tag{2.11}$$

where

$$\psi_*(R^*(X_1, X_2)X_3)) = R'(\psi_* X_1, \psi_* X_2)\psi_* X_3 \tag{2.12}$$

for all $U_1, U_2, U_3, U_4 \in \mathcal{V}(M)$ and $X_1, X_2, X_3, X_4 \in \mathcal{H}(M)$ [19, 29].

The mean curvature vector field H of any fiber of Riemannian submersion ψ is given by

$$N = rH, N = \sum_{j=1}^{r} \mathcal{T}_{U_j} U_j, \tag{2.13}$$

where $\{U_1, \ldots, U_r\}$ is an orthonormal basis of the vertical distribution $\mathcal{V}$. Moreover, ψ has totally geodesic fibers if $\mathcal{T} = 0$ on $\mathcal{H}(M)$ and $\mathcal{V}(M)$ [19].

Now, we have the following lemma which shows that $\mathcal{A}$ and $\mathcal{T}$ are antisymmetric with respect to g.

Lemma 2.2 ([29]) *Let* (M, g) *and* (M_1, g_1) *be Riemannian manifolds and* $\psi : M \to M_1$ *Riemannian submersion. For* $E_1, E_2, E_3 \in \chi(M)$*, we have*

$$g(\mathcal{T}_{E_1} E_2, E_3) = -g(E_2, \mathcal{T}_{E_1} E_3), \tag{2.14}$$

$$g(\mathcal{A}_{E_1} E_2, E_3) = -g(E_2, \mathcal{A}_{E_1} E_3). \tag{2.15}$$

A $(2m + 1)$-dimensional Riemannian manifold (M, g) is said to be a Sasakian manifold if it admits an endomorphism ϕ of its tangent bundle TM, a vector field U, and a 1-form η satisfying

$$\phi^2 = -Id + \eta \otimes \xi, \eta(U) = 1, \phi\xi = 0, \eta \circ \xi = 0,$$
$$g(\phi X_1, \phi X_2) = g(X_1, X_2) - \eta(X_1)\eta(X_2), \eta(X_1) = g(X_1, \xi),$$
$$(\nabla^M_{X_1} \phi) X_2 = g(X_1, X_2)\xi - \eta(X_2) X_1, \nabla^M_{X_1} \xi = -\phi X_1,$$

for any vector fields X_1, X_2 on TM, where ∇^M denotes the Riemannian connection with respect to g.

Definition 2.3 Let (M, ϕ, ξ, η, g) be a Sasakian manifold and (M_1, g_1) be a Riemannian manifold. A Riemannian submersion $\psi : (M, \phi, \xi, \eta, g) \to (M_1, g_1)$ is said to be semi-slant $\xi^\perp$-Riemannian submersion if there is a distribution $\tilde{\mathcal{D}} \subset \ker \psi_*$ such that

$$\ker \psi_* = \tilde{\mathcal{D}} \oplus \bar{\mathcal{D}} \oplus \langle \xi \rangle, J(\tilde{\mathcal{D}}) = \tilde{\mathcal{D}},$$

and the angle $\theta = \theta(X)$ between ϕX and the space $\bar{\mathcal{D}}_q$ is constant for nonzero $X \in \Gamma(\bar{\mathcal{D}})_q$ and $q \in M$, where $\bar{\mathcal{D}}$ is the orthogonal complement of $\tilde{\mathcal{D}}$ in $\ker \psi_*$ [6]. Here we call the angle θ a semi-slant angle.

Remark 2.4 In this chapter, we suppose that the Reeb vector field ξ is vertical.

From now on, we will assume that ψ be a semi-slant $\xi^\perp$-Riemannian submersion from a Sasakian manifold (M, ϕ, ξ, η, g) onto a Riemannian manifold (M_1, g_1).

Now [6], let ψ be a semi-slant $\xi^{\perp}$-Riemannian submersion. Then, for $U_1 \in \Gamma(ker\psi_*)$, we get

$$\phi U_1 = \varphi U_1 + \omega U_1,$$

where φU_1 and ωU_1 are vertical and horizontal components of ϕU_1, respectively. Similarly, for any $X_1 \in \Gamma((ker\psi_*)^{\perp})$, we have

$$\phi X_1 = \mathcal{B}X_1 + \mathcal{C}X_1,$$

where $\mathcal{B}X_1$ (resp. $\mathcal{C}X_1$) is the vertical part (resp. horizontal part) of ϕX_1. Then, the horizontal distribution $(ker\psi_*)^{\perp}$ is decomposed as

$$(\ker\psi_*)^{\perp} = \omega\bar{\mathcal{D}} \oplus \mu.$$

Theorem 2.5 *Let ψ be a semi-slant $\xi^{\perp}$-Riemannian submersion. For any $U_2 \in \Gamma\left(\bar{\mathcal{D}}\right)$, we have*

$$\phi^2 U_2 = -\cos^2\theta U_2,$$

where θ denotes the semi-slant angle of $\bar{\mathcal{D}}$ [6].

Lemma 2.6 ([6]) *Let ψ be a semi-slant $\xi^{\perp}$-Riemannian submersion. For any $U_1, U_2 \in \Gamma\left(\bar{\mathcal{D}}\right)$, we have*

$$g\left(\varphi U_1, \varphi U_2\right) = \cos^2\theta g\left(U_1, U_2\right),$$
$$g\left(\omega U_1, \omega U_2\right) = \sin^2\theta g\left(U_1, U_2\right).$$

A plane section π in T_pM is called a ϕ-section if it is spanned by X_1 and ϕX_1, where X_1 is a unit tangent vector orthogonal to U_1. The sectional curvature of a ϕ-section is called a ϕ- sectional curvature. A Sasakian manifold with constant ϕ-sectional curvature c is said to be a Sasakian space form and is denoted by $M(c)$ [12, 35]. The curvature tensor of $M(c)$ of a Sasakian space form $M(c)$ is given by

$$\begin{aligned}\tilde{R}(X_1, X_2)X_3 =& \frac{(c+3)}{4}\{g(X_2, X_3)X_1 - g(X_1, X_3)X_2\} \\ &+ \frac{(c-1)}{4}\{\eta(X_1)\eta(X_3)X_2 \\ &- \eta(X_2)\eta(X_3)X_1 + g(X_1, X_3)\eta(X_2)\xi - g(X_2, X_3)\eta(X_1)\xi \\ &+ g(\phi X_2, X_3)\phi X_1 - g(\phi X_1, X_3)\phi X_2 - 2g(\phi X_1, X_2)\phi X_3\}\end{aligned} \tag{2.16}$$

for any tangent vector fields X_1, X_2, X_3 on $M(c)$.

Let $(M(c), g)$, (M_1, g_1) be a Sasakian space form and a Riemannian manifold, respectively, and $\psi : M(c) \to M_2$ a semi-slant $\xi^{\perp}$-Riemannian submersion. Furthermore, let $\{E_1, \ldots, E_r, F_1, \ldots, F_n\}$ be an orthonormal basis of $T_pM(c)$ such that $\mathcal{V} = span\{E_1, \ldots, E_r = \xi\}$, $\mathcal{H} = span\{F_1, \ldots, F_n\}$, and $r = 2d_1 + 2d_2 + 1$. Then we consider adapted semi-slant orthonormal frames

$$\begin{aligned} &E_1, E_2 = \varphi E_1, \ldots, E_{2d_1-1}, E_{2d_1-1} = \varphi E_{2d_1}, E_{2d_1+1}, \\ &E_{2d_1+2} = \frac{1}{cos\theta}\varphi E_{2d_1+1}, \ldots, E_{2d_1+2d_2-1}, E_{2d_1+2d_2} \\ &= \frac{1}{cos\theta}\varphi E_{2d_1+2d_2-1}, E_{2d_1+2d_2+1} = \xi. \end{aligned}$$

Here, we have

$$g^2(\phi E_i, E_{i+1}) = \begin{cases} 1, & \text{for } i \in \{1, \ldots, 2d_1 - 1\}, \\ cos^2\theta, & \text{for } i \in \{2d_1 + 1, \ldots, 2d_1 + 2d_2 - 1\}, \end{cases}$$

and then

$$\sum_{i,j=1}^{r} g^2(\phi E_i, E_j) = 2(d_1 + d_2 cos^2\theta).$$

3 Chen-Ricci Inequality and Chen Inequalities

$\psi : M(c) \to M_1$ a semi-slant $\xi^{\perp}$-Riemannian submersion. Furthermore, let $\{E_1, \ldots, E_r, F_1, \ldots, F_n\}$ be an orthonormal basis of $T_pM(c)$ such that $\mathcal{V} = span\{E_1, \ldots, E_r = \xi\}$, $\mathcal{H} = span\{F_1, \ldots, F_n\}$. By virtue of (2.9), (2.10), and (2.16), we have

$$\begin{aligned} \hat{R}(U_1, U_2, U_3, U_4) &= \frac{(c+3)}{4}\{g(U_2, U_3)g(U_1, U_4) - g(U_1, U_3)g(U_2, U_4)\} \\ &+ \frac{(c-1)}{4}\{\eta(U_1)\eta(U_3)g(U_2, U_4) - \eta(U_2)\eta(U_3)g(U_1, U_4) \\ &+ \eta(U_2)\eta(U_4)g(U_1, U_3) - \eta(U_1)\eta(U_4)g(U_2, U_3) \\ &+ g(\phi U_2, U_3)g(\phi U_1, U_4) \\ &- g(\phi U_1, U_3)g(\phi U_2, U_4) - 2g(\phi U_1, U_2)g(\phi U_3, U_4)\} \\ &- g(\mathcal{T}_{U_1}U_4, \mathcal{T}_{U_2}U_3) + g(\mathcal{T}_{U_2}U_4, \mathcal{T}_{U_1}U_3), \end{aligned} \tag{3.1}$$

$$
\begin{aligned}
R^*(F_1, F_2, F_3, F_4) = \frac{(c+3)}{4}&\{g(F_2, F_3)g(F_1, F_4) - g(F_1, F_3)g(F_2, F_4)\}\\
&+ \frac{(c-1)}{4}\{\eta(F_1)\eta(F_3)g(F_2, F_4) - \eta(F_2)\eta(F_3)g(F_1, F_4)\\
&+ \eta(F_2)\eta(F_4)g(F_1, F_3) - \eta(F_1)\eta(F_4)g(F_2, F_3)\\
&+ g(\phi F_2, F_3)g(\phi F_1, F_4) - g(\phi F_2, F_4)g(\phi F_1, F_3)\\
&- 2g(\phi F_1, F_2)g(F_4, \phi F_3)\}\\
&+ 2g(\mathcal{A}_{F_1}F_2, \mathcal{A}_{F_3}F_4) - g(\mathcal{A}_{F_2}F_3, \mathcal{A}_{F_1}F_4)\\
&+ g(\mathcal{A}_{F_1}F_3, \mathcal{A}_{F_2}F_4). \qquad (3.2)
\end{aligned}
$$

Theorem 3.1 *Let $\psi : M(c) \to N$ be a semi-slant $\xi^{\perp}$-Riemannian submersion such that ξ is vertical. Then, the following statements are true:*

(i) For any unit vector field $E \in \Gamma(\tilde{\mathcal{D}})$, it follows that

$$
\widehat{Ric}(E) \geq \frac{(c+3)}{4}(r-1) + \frac{(c-1)}{2} - rg(\mathcal{T}_E E, H). \qquad (3.3)
$$

The equality case of (3.3) holds for a unit vertical vector $E \in \Gamma(\tilde{\mathcal{D}})$ if and only if each fiber is totally geodesic.

(ii) For any unit vector field $E \in \Gamma(\bar{\mathcal{D}})$, it follows that

$$
\widehat{Ric}(E) \geq \frac{(c+3)}{4}(r-1) + \frac{(c-1)}{4}(-1 + 3\cos^2\theta) - rg(\mathcal{T}_E E, H). \qquad (3.4)
$$

The equality case of (3.4) holds for a unit vertical vector $E \in \Gamma(\bar{\mathcal{D}})$ if and only if each fiber is totally geodesic.

Proof From (3.1), we have

$$
\begin{aligned}
\widehat{Ric}(E) = \frac{(c+3)}{4}&(r-1)g(E, E) + \frac{(c-1)}{2}\{(2-r)(\eta(E))^2\\
&- g(E, E) + 3\sum_{i=1}^{r} g^2(\phi E, E_i)\} - rg(\mathcal{T}_E E, H) + \|\mathcal{T}_E E_i\|^2, \qquad (3.5)
\end{aligned}
$$

where

$$
\widehat{Ric}(E) = \sum_{i=1}^{r} \hat{R}(E, E_i, E_i, E).
$$

If we get $E \in \Gamma(\tilde{\mathcal{D}})$, we have

$$\sum_{i=1}^{r} g^2(\phi E, E_i) = 1. \tag{3.6}$$

Similarly, if we get $E \in \Gamma(\bar{\mathcal{D}})$, we obtain

$$\sum_{i=1}^{r} g^2(\phi E, E_i) = \cos^2\theta. \tag{3.7}$$

By virtue of the last two equations in (3.5), we get (3.3) and (3.4). □

Theorem 3.2 *Let $\psi : M(c) \to M_1$ be a semi-slant $\xi^{\perp}$-Riemannian submersion such that ξ is vertical. Then*

$$2\hat{\tau} \geq \frac{(c+3)}{4} r(r-1) + \frac{(c-1)}{4}\{2(1-r) + 6(d_1 + d_2\cos^2\theta)\} - r^2\,\|H\|^2. \tag{3.8}$$

The equality case of (3.8) holds if and only if each fiber is totally geodesic.

Proof By using the symmetry of $\mathcal{T}$ in (3.1), we have

$$2\hat{\tau} = \frac{(c+3)}{4} r(r-1) + \frac{(c-1)}{4}\left\{2(1-r) + 3\sum_{i=1}^{r} g^2(\phi E, E_i)\right\} - r^2\,\|H\|^2$$
$$+ \sum_{i,j=1}^{r} g(\mathcal{T}_{E_i} E_j, \mathcal{T}_{E_i} E_j), \tag{3.9}$$

where

$$\hat{\tau} = \sum_{1 \leq i < j \leq r} \hat{R}(E_i, E_j, E_j, E_i).$$

Since

$$\sum_{i=1}^{r} g^2(\phi E, E_i) = 2d_1 + 2d_2\cos^2\theta, \tag{3.10}$$

then by using last two equations in (3.5), we obtain (3.8). □

For the horizontal distribution, in view of (3.2), since ψ is semi-slant $\xi^{\perp}$-Riemannian submersion and ξ is vertical, using the antisymmetry of A, we find

$$2\tau^* = \frac{(c+3)}{4}n(n-1) + 3\sum_{i,j=1}^{n}\left\{\frac{(c-1)}{4}g(\mathcal{C}F_i, F_j)g(\mathcal{C}F_i, F_j)\right.$$
$$\left. -g(\mathcal{A}_{F_i}F_j, \mathcal{A}_{F_i}F_j)\right\}, \tag{3.11}$$

where

$$\tau^* = \sum_{1\le i<j\le r} R^*(F_i, F_j, F_j, F_i). \tag{3.12}$$

Now we define

$$\|\mathcal{C}\|^2 = \sum_{i=1}^{n} g^2(\mathcal{C}F_i, F_j), \tag{3.13}$$

and then from (3.11) and (3.13), we obtain

$$2\tau^* = \frac{(c+3)}{4}n(n-1) + \frac{3(c-1)}{4}\|\mathcal{C}\|^2 - 3\sum_{i,j=1}^{n} g(\mathcal{A}_{F_i}F_j, \mathcal{A}_{F_i}F_j). \tag{3.14}$$

From (3.14) we have:

Theorem 3.3 *Let $\psi : M(c) \to M_1$ be a semi-slant $\xi^{\perp}$-Riemannian submersion such that ξ is vertical. Then*

$$2\tau^* \le \frac{(c+3)}{4}n(n-1) + \frac{3(c-1)}{4}\|\mathcal{C}\|^2. \tag{3.15}$$

The equality case of (3.15) holds if and only if $\mathcal{H}(M)$ is integrable.

Let $\psi : M(c) \to M_1$ be a semi-slant $\xi^{\perp}$-Riemannian submersion such that ξ is vertical and $\{E_1, \ldots, E_r, F_1, \ldots, F_n\}$ is an orthonormal basis of $TpM(c)$ such that $\mathcal{V}p(M) = span\{E_1, \ldots, E_r\}$, $\mathcal{H}p(M) = span\{F_1, \ldots, F_n\}$. Now we denote $\mathcal{T}_{ij}^{s}$ by

$$\mathcal{T}_{ij}^{s} = g(\mathcal{T}_{E_i}E_j, F_s), \tag{3.16}$$

where $1 \le i, j \le r$ and $1 \le s \le n$ (see [22]).

Similarly, we denote $\mathcal{A}_{ij}^{\alpha}$ by

$$\mathcal{A}_{ij}^{\alpha} = g(\mathcal{A}_{E_i}E_j, E_\alpha), \tag{3.17}$$

where $1 \le i, j \le n$ and $1 \le \alpha \le r$. From [22], we use

$$\delta(N) = \sum_{i=1}^{n}\sum_{k=1}^{r} g((\nabla_{F_i}\mathcal{T})_{E_k}E_k, F_i). \tag{3.18}$$

From the binomial theorem, there is such as the following equation between the tensor fields $\mathcal{T}$:

$$\sum_{s=1}^{n}\sum_{i,j=1}^{r}(\mathcal{T}_{ij}^{s})^2 = \frac{1}{2}r^2\,\|H\|^2 + \frac{1}{2}(\mathcal{T}_{11}^{s} - \mathcal{T}_{22}^{s} - \ldots - \mathcal{T}_{rr}^{s})^2$$

$$+2\sum_{s=1}^{n}\sum_{j=2}^{r}(\mathcal{T}_{1j}^{s})^2 - 2\sum_{s=1}^{n}\sum_{2\le i<j\le r}(\mathcal{T}_{ii}^{s}\mathcal{T}_{jj}^{s} - (\mathcal{T}_{ij}^{s})^2). \tag{3.19}$$

Theorem 3.4 *Let $\psi : M(c) \to M_1$ be a semi-slant $\xi^{\perp}$-Riemannian submersion such that ξ is vertical. Then, the following statements are true:*

(i) For any unit vector field $E_1 \in \Gamma(\tilde{\mathcal{D}})$, it follows that

$$\widehat{Ric}(E_1) \ge \frac{(c+3)}{4}(r-1) + \frac{(c-1)}{2} - \frac{1}{4}r^2\,\|H\|^2. \tag{3.20}$$

(ii) For any unit vector field $E_1 \in \Gamma(\bar{\mathcal{D}})$, it follows that

$$\widehat{Ric}(E_1) \ge \frac{(c+3)}{4}(r-1) + \frac{(c-1)}{4}(-1+3\cos^2\theta) - \frac{1}{4}r^2\,\|H\|^2. \tag{3.21}$$

The equality case of (3.20) and (3.21) holds if and only if

$$\mathcal{T}_{11}^{s} = \mathcal{T}_{22}^{s} + \ldots + \mathcal{T}_{rr}^{s},$$
$$\mathcal{T}_{1j}^{s} = 0, \quad j = 2, \ldots, r.$$

Proof Let $\{E_1, \ldots, E_{2d_1}, E_{2d_1+1}, E_{2d_1+2}, \ldots, E_{2d_1+2d_2-1}, E_{2d_1+2d_2}, E_{2d_1+2d_2+1}\}$ be an adapted semi-slant basis of $\mathcal{V}p(M)$.

(i) Due to the fact that one can choose the above adapted semi-slant basis such that $E_1 = E$, it suffices to prove (3.20) for $E = E_1$. Considering (3.9) and (3.10), we have

$$2\hat{\tau} = \frac{(c+3)}{4}r(r-1) + \frac{(c-1)}{4}\{2(1-r) + 6(d_1 + d_2\cos^2\theta)\} - r^2\,\|H\|^2$$
$$+ \sum_{i,j=1}^{r} g(\mathcal{T}_{E_i}E_j, \mathcal{T}_{E_i}E_j). \tag{3.22}$$

By using (3.16) in (3.22) and the symmetry of $\mathcal{T}$, we can write

$$2\hat{\tau} = \frac{(c+3)}{4}r(r-1) + \frac{(c-1)}{4}\{2(1-r) + 6(d_1 + d_2\cos^2\theta)\} - r^2\,\|H\|^2$$

$$+\sum_{s=1}^{n}\sum_{i,j=1}^{r}(\mathcal{T}_{ij}^{s})^{2}. \tag{3.23}$$

Hence using (3.19) in (3.23), we obtain

$$2\hat{\tau}=\frac{(c+3)}{4}r(r-1)+\frac{(c-1)}{4}\{2(1-r)+6(d_1+d_2\cos^2\theta)\}-\frac{1}{2}r^2\left\|H\right\|^2$$
$$+\frac{1}{2}(\mathcal{T}_{11}^{s}-\mathcal{T}_{22}^{s}-\ldots-\mathcal{T}_{rr}^{s})^2+2\sum_{s=1}^{n}\sum_{j=2}^{r}(\mathcal{T}_{1j}^{s})^2$$
$$-2\sum_{s=1}^{n}\sum_{2\leq i<j\leq r}(\mathcal{T}_{ii}^{s}\mathcal{T}_{jj}^{s}-(\mathcal{T}_{ij}^{s})^2). \tag{3.24}$$

Then from (3.24), we have

$$2\hat{\tau}\geq\frac{(c+3)}{4}r(r-1)+\frac{(c-1)}{4}\{2(1-r)+6(d_1+d_2\cos^2\theta)\}-\frac{1}{2}r^2\left\|H\right\|^2$$
$$-2\sum_{s=1}^{n}\sum_{2\leq i<j\leq r}(\mathcal{T}_{ii}^{s}\mathcal{T}_{jj}^{s}-(\mathcal{T}_{ij}^{s})^2). \tag{3.25}$$

Besides, taking $U_1=U_4=E_i$, $U_2=U_3=E_j$ in (3.1) and using (3.16), we obtain

$$2\sum_{2\leq i<j\leq r}R(E_i,E_j,E_j,E_i)=2\sum_{2\leq i<j\leq r}\hat{R}(E_i,E_j,E_j,E_i)$$
$$+2\sum_{s=1}^{n}\sum_{2\leq i<j\leq r}(\mathcal{T}_{ii}^{s}\mathcal{T}_{jj}^{s}-(\mathcal{T}_{ij}^{s})^2). \tag{3.26}$$

Considering (3.26) in (3.25), we get

$$2\hat{\tau}\geq\frac{(c+3)}{4}r(r-1)+\frac{(c-1)}{4}\{2(1-r)+6(d_1+d_2\cos^2\theta)\}-\frac{1}{2}r^2\left\|H\right\|^2$$
$$+2\sum_{2\leq i<j\leq r}\hat{R}(E_i,E_j,E_j,E_i)-2\sum_{2\leq i<j\leq r}R(E_i,E_j,E_j,E_i). \tag{3.27}$$

Moreover, we have

$$2\hat{\tau}=2\sum_{2\leq i<j\leq r}\hat{R}(E_i,E_j,E_j,E_i)+2\sum_{j=1}^{r}\hat{R}(E_1,E_j,E_j,E_1). \tag{3.28}$$

By using (3.28) in (3.27), we get

$$2\widehat{Ric}(E_1) \geq \frac{(c+3)}{4}r(r-1) + \frac{(c-1)}{4}\{2(1-r) + 6(d_1 + d_2\cos^2\theta)\}$$
$$-\frac{1}{2}r^2\,\|H\|^2 - 2\sum_{2\leq i<j\leq r} R(E_i, E_j, E_j, E_i). \tag{3.29}$$

Since M is a Sasakian space form, its curvature tensor R satisfies the equality (2.16), we have

$$\sum_{2\leq i<j\leq r} R(E_i, E_j, E_j, E_i) = \frac{(c+3)}{8}(r-2)(r-1)$$
$$+\frac{(c-1)}{4}\left\{(2-r)\sum_{j=2}^{r}(\eta(U_j))^2\right.$$
$$\left.+3\sum_{2\leq i<j\leq r} g^2(\phi E_i, E_j)\right\}. \tag{3.30}$$

If we get $E_1 \in \Gamma(\tilde{\mathcal{D}})$ in (3.30), we obtain

$$\sum_{2\leq i<j\leq r} R(E_i, E_j, E_j, E_i) = \frac{(c+3)}{8}(r-2)(r-1) + \frac{(c-1)}{4}\{(2-r)$$
$$+3((d_1-1) + d_2\cos^2\theta)\}. \tag{3.31}$$

Taking into account of the last equation in (3.29), we get (3.20).

(ii) Due to the fact that in this case one can choose the adapted semi-slant basis $\{E_1, \ldots, E_{2d_1}, E_{2d_1+1}, E_{2d_1+2}, \ldots, E_{2d_1+2d_2-1}, E_{2d_1+2d_2}\}$ such that $E_{2d_1+1} = U$, it suffices to prove (3.21) for $E = E_{2d_1+1}$. If we get $E_1 \in \Gamma(\bar{\mathcal{D}})$ in (3.30), we obtain:

With similar arguments as in case (i), we obtain

$$2\widehat{Ric}(E_{2d_1+1}) \geq \frac{(c+3)}{4}r(r-1) + \frac{(c-1)}{4}\{2(1-r) + 6(d_1 + d_2\cos^2\theta)\}$$
$$-\frac{1}{2}r^2\,\|\mathcal{H}\|^2$$
$$-2\sum_{1\leq k<s\leq r;\ k,s\neq 2d_1+1} R(E_k, E_s, E_s, E_k). \tag{3.32}$$

If we get $E_{2d_1+1} \in \Gamma(\tilde{\mathcal{D}})$ in (3.30), we obtain

$$\sum_{1\leq k<s\leq r;\ k,s\neq 2d_1+1} R(E_k, E_s, E_s, E_k) = \frac{(c+3)}{8}(r-2)(r-1) + \frac{(c-1)}{4}\{(2-r) + 3(d_1 + (d_2-1)\cos^2\theta)\}. \tag{3.33}$$

Taking into account of the last equation in (3.32), we get (3.21).

□

Theorem 3.5 *Let $\psi : M(c) \to M_1$ be a semi-slant $\xi^{\perp}$-Riemannian submersion such that ξ is vertical. Then we have*

$$Ric^*(F_1) \leq \frac{(c+3)}{4}(n-1) + \frac{3(c-1)}{4}\|\mathcal{C}F_1\|^2. \tag{3.34}$$

The equality case of (3.34) holds if and only if

$$\mathcal{A}^{\alpha}_{1j} = 0, \quad j = 2, \ldots, n.$$

Proof From (3.14) and (3.17), we have

$$2\tau^* = \frac{(c+3)}{4}n(n-1) + \frac{3(c-1)}{4}\|\mathcal{C}\|^2 - 3\sum_{\alpha=1}^{r}\sum_{i,j=1}^{n}(\mathcal{A}^{\alpha}_{ij})^2. \tag{3.35}$$

Since $\mathcal{A}$ is anti-symmetric on $\mathcal{H}(M(c))$, (3.35) can be written as

$$2\tau^* = \frac{(c+3)}{4}n(n-1) + \frac{3(c-1)}{4}\|\mathcal{C}\|^2 - 6\sum_{\alpha=1}^{r}\sum_{j=2}^{n}(\mathcal{A}^{\alpha}_{1j})^2 - 6\sum_{\alpha=1}^{r}\sum_{2\leq i<j\leq n}(\mathcal{A}^{\alpha}_{ij})^2. \tag{3.36}$$

Moreover, taking $X_1 = X_4 = F_i$, $X_2 = X_3 = F_j$ in (3.2) and using (3.17), we get

$$2\sum_{2\leq i<j\leq n} R(F_i, F_j, F_j, F_i) = 2\sum_{2\leq i<j\leq n} R^*(F_i, F_j, F_j, F_i) + 6\sum_{\alpha=1}^{r}\sum_{2\leq i<j\leq n}(\mathcal{A}^{\alpha}_{ij})^2. \tag{3.37}$$

By using (3.37) in (3.36), we get

$$2\tau^* = \frac{(c+3)}{4} n(n-1) + \frac{3(c-1)}{4} \|\mathcal{C}\|^2 - 6 \sum_{\alpha=1}^{r} \sum_{j=2}^{n} (\mathcal{A}_{1j}^{\alpha})^2$$

$$+ 2 \sum_{2 \leq i < j \leq n} R^*(F_i, F_j, F_j, F_i) - 2 \sum_{2 \leq i < j \leq n} R(F_i, F_j, F_j, F_i). \tag{3.38}$$

Besides, from (3.2), we obtain

$$\sum_{2 \leq i < j \leq n} R(F_i, F_j, F_j, F_i) = \frac{(c+3)}{8} (n-2)(n-1)$$

$$+ \frac{3(c-1)}{4} \sum_{2 \leq i < j \leq n} g^2(\mathcal{C} F_i, F_j). \tag{3.39}$$

Then from (3.38) and (3.39), we derive

$$2Ric^*(F_1) = \frac{(c+3)}{2} (n-1) + \frac{3(c-1)}{2} \|\mathcal{C} F_1\|^2 - 6 \sum_{\alpha=1}^{r} \sum_{j=2}^{n} (\mathcal{A}_{1j}^{\alpha})^2. \tag{3.40}$$

This completes the proof. □

Now, by taking into account of ξ is vertical, we compute the Chen-Ricci inequality between the vertical and horizontal distributions. For the scalar curvature τ of $M(c)$, we have

$$2\tau = \sum_{s=1}^{n} Ric(F_s, F_s) + \sum_{k=1}^{r} Ric(E_k, E_k), \tag{3.41}$$

$$2\tau = \sum_{j,k=1}^{r} R(E_j, E_k, E_k, E_j) + \sum_{i=1}^{n} \sum_{k=1}^{r} R(F_i, E_k, E_k, F_i)$$

$$+ \sum_{i,s=1}^{n} R(F_i, F_s, F_s, F_i) + \sum_{s=1}^{n} \sum_{j=1}^{r} R(E_j, F_s, F_s, E_j). \tag{3.42}$$

Let us denote [22]

$$\left\| \mathcal{T}^{\mathcal{V}} \right\|^2 = \sum_{i=1}^{n} \sum_{k=1}^{r} g(\mathcal{T}_{E_k} F_i, \mathcal{T}_{E_k} F_i), \tag{3.43}$$

$$\left\|\mathcal{T}^{\mathcal{H}}\right\|^2 = \sum_{k,j=1}^{r} g(\mathcal{T}_{E_k} E_j, \mathcal{T}_{E_k} E_j), \tag{3.44}$$

$$\left\|\mathcal{A}^{\mathcal{V}}\right\|^2 = \sum_{i,j=1}^{n} g(\mathcal{A}_{F_i} F_j, \mathcal{A}_{F_i} F_j), \tag{3.45}$$

$$\left\|\mathcal{A}^{\mathcal{H}}\right\|^2 = \sum_{i=1}^{n}\sum_{k=1}^{r} g(\mathcal{A}_{F_i} E_k, \mathcal{A}_{F_i} E_k). \tag{3.46}$$

Theorem 3.6 *Let $\psi : M(c) \to M_1$ be a semi-slant $\xi^{\perp}$-Riemannian submersion such that ξ is vertical. Then we have:*

(i) For any unit vector field $E_1 \in \Gamma(\tilde{\mathcal{D}})$, it follows that

$$\frac{(c+3)}{4}(nr+n+r-2) + \frac{(c-1)}{4}[2-n+3(\|\mathcal{B}\|^2 + \|\mathcal{C}F_1\|^2)] \leq \widehat{Ric}(E_1)$$
$$+ Ric^*(F_1)$$
$$+ \frac{1}{4}r^2\|H\|^2 + 3\sum_{\alpha=1}^{r}\sum_{s=2}^{n}(\mathcal{A}^{\alpha}_{1s})^2 - \delta(N) + \left\|\mathcal{T}^{\mathcal{V}}\right\|^2 - \left\|\mathcal{A}^{\mathcal{H}}\right\|^2. \tag{3.47}$$

(ii) For any unit vector field $E_1 \in \Gamma(\bar{\mathcal{D}})$, it follows that

$$\frac{(c+3)}{4}(nr+n+r-2) + \frac{(c-1)}{4}[-1-n+3\cos^2\theta + 3(\|\mathcal{B}\|^2 + \|\mathcal{C}F_1\|^2)]$$
$$\leq \widehat{Ric}(E_1) + Ric^*(F_1)$$
$$+ \frac{1}{4}r^2\|H\|^2 + 3\sum_{\alpha=1}^{r}\sum_{s=2}^{n}(\mathcal{A}^{\alpha}_{1s})^2 - \delta(N) + \left\|\mathcal{T}^{\mathcal{V}}\right\|^2 - \left\|\mathcal{A}^{\mathcal{H}}\right\|^2. \tag{3.48}$$

The equality case of (3.47) and (3.48) holds if and only if

$$\mathcal{T}^s_{11} = \mathcal{T}^s_{22} + \ldots + \mathcal{T}^s_{rr},$$
$$\mathcal{T}^s_{1j} = 0, \quad j = 2, \ldots, r.$$

Proof Since $M(c)$ is a Sasakian space form, from (3.42) we obtain

$$2\tau = \frac{(c+3)}{4}(n+r)(n+r-1) + \frac{(c-1)}{4}[2(1-r-n)$$
$$+ 3\{2(d_1 + d_2\cos^2\theta) + \|\mathcal{C}\|^2 + 2\sum_{i=1}^{n}\sum_{k=1}^{r} g^2(\mathcal{B}F_i, E_k)\}]. \tag{3.49}$$

Now, we define

$$\|\mathcal{B}\|^2 = \sum_{i=1}^{n}\sum_{k=1}^{r} g^2(\mathcal{B}F_i, E_k). \tag{3.50}$$

On the other hand, using the Gauss-Codazzi-type equations (2.9), (2.10), and (2.11), we get

$$\begin{aligned}
2\tau &= 2\hat{\tau} + 2\tau^* + r^2\,\|H\|^2 - \sum_{k,j=1}^{r} g(\mathcal{T}_{E_k}E_j, \mathcal{T}_{E_k}E_j) + 3\sum_{i,s=1}^{n} g(\mathcal{A}_{F_i}F_s, \mathcal{A}_{F_i}F_s)\\
&- \sum_{i=1}^{n}\sum_{k=1}^{r} g((\nabla_{F_i}\mathcal{T})_{E_k}E_k, F_i) + \sum_{i=1}^{n}\sum_{k=1}^{r}(g(\mathcal{T}_{E_k}F_i, \mathcal{T}_{E_k}F_i)\\
&- g(\mathcal{A}_{F_i}E_k, \mathcal{A}_{F_i}E_k))\\
&- \sum_{s=1}^{n}\sum_{j=1}^{r} g((\nabla_{F_s}\mathcal{T})_{E_j}E_j, F_s) + \sum_{s=1}^{n}\sum_{j=1}^{r}(g(\mathcal{T}_{E_j}F_s, \mathcal{T}_{E_j}F_s)\\
&- g(\mathcal{A}_{F_s}E_j, \mathcal{A}_{F_s}E_j)).
\end{aligned} \tag{3.51}$$

Thus from (3.19) and (3.51), we derive

$$\begin{aligned}
2\tau &= 2\hat{\tau} + 2\tau^* + \frac{1}{2}r^2\,\|H\|^2 - \frac{1}{2}(\mathcal{T}_{11}^s - \mathcal{T}_{22}^s - \ldots - \mathcal{T}_{rr}^s)^2 - 2\sum_{s=1}^{n}\sum_{j=2}^{r}(\mathcal{T}_{1j}^s)^2\\
&+ 2\sum_{s=1}^{n}\sum_{2\le i<j\le r}(\mathcal{T}_{ii}^s\mathcal{T}_{jj}^s - (\mathcal{T}_{ij}^s)^2) + 6\sum_{\alpha=1}^{r}\sum_{s=2}^{n}(\mathcal{A}_{1s}^{\alpha})^2 + 6\sum_{\alpha=1}^{r}\sum_{2\le i<s\le n}(\mathcal{A}_{is}^{\alpha})^2\\
&+ \sum_{i=1}^{n}\sum_{k=1}^{r}(g(\mathcal{T}_{E_k}F_i, \mathcal{T}_{E_k}F_i) - g(\mathcal{A}_{F_i}E_k, \mathcal{A}_{F_i}E_k)) - 2\delta(N)\\
&+ \sum_{s=1}^{n}\sum_{j=1}^{r}(g(\mathcal{T}_{E_j}F_s, \mathcal{T}_{E_j}F_s) - g(\mathcal{A}_{F_s}E_j, \mathcal{A}_{F_s}E_j)).
\end{aligned} \tag{3.52}$$

Using (3.26), (3.37), (3.49), and (3.50) in (3.52), we obtain

$$\frac{(c+3)}{4}(n+r)(n+r-1) + \frac{(c-1)}{4}[2(1-r-n) + 3\{2(d_1 + d_2\cos^2\theta)$$

$$
\begin{aligned}
&+ 2\|\mathcal{B}\|^2 + \|\mathcal{C}\|^2\}] \\
&= 2\widehat{Ric}(E_1) + 2Ric^*(F_1) + \frac{1}{2}r^2\,\|H\|^2 - \frac{1}{2}(\mathcal{T}^s_{11} - \mathcal{T}^s_{22} - \ldots - \mathcal{T}^s_{rr})^2 \\
&- 2\sum_{s=1}^{n}\sum_{j=2}^{r}(\mathcal{T}^s_{1j})^2 \\
&+ 6\sum_{\alpha=1}^{r}\sum_{s=2}^{n}(\mathcal{A}^\alpha_{1s})^2 + \sum_{i=1}^{n}\sum_{k=1}^{r}\{g(\mathcal{T}_{E_k}F_i, \mathcal{T}_{E_k}F_i) - g(\mathcal{A}_{F_i}E_k, \mathcal{A}_{F_i}E_k)\} \\
&- 2\delta(N) + \sum_{s=1}^{n}\sum_{j=1}^{r}\{g(\mathcal{T}_{E_j}F_s, \mathcal{T}_{E_j}F_s) - g(\mathcal{A}_{F_s}E_j, \mathcal{A}_{F_s}E_j)\} \\
&+ \sum_{2\le i<j\le r} 2R(E_i, E_j, E_j, E_i) + \sum_{2\le i<j\le n} 2R(F_i, F_j, F_j, F_i).
\end{aligned}
\tag{3.53}
$$

If we take $U_1 \in \Gamma(\tilde{\mathcal{D}})$, considering (3.31), (3.39), (3.43), and (3.46) in (3.53), we obtain (3.47). In a similar way, if we take $U_1 \in \Gamma(\bar{\mathcal{D}})$, considering (3.33), (3.39), (3.43), and (3.46) in (3.53), we obtain (3.48). This completes the proof. □

From (3.49) and (3.51), we obtain

$$
\begin{aligned}
&\frac{(c+3)}{4}(n+r)(n+r-1) + \frac{(c-1)}{4}[2(1-r-n) + 3\{2(d_1 + d_2\cos^2\theta) \\
&+ 2\|\mathcal{B}\|^2 + \|\mathcal{C}\|^2\}] \\
&= 2\hat{\tau} + 2\tau^* + r^2\,\|H\|^2 - \left\|\mathcal{T}^{\mathcal{H}}\right\|^2 + 3\left\|\mathcal{A}^{\mathcal{V}}\right\|^2 - 2\delta(N) + 2\left\|\mathcal{T}^{\mathcal{V}}\right\|^2 - 2\left\|\mathcal{A}^{\mathcal{H}}\right\|^2.
\end{aligned}
\tag{3.54}
$$

From (3.54), we get:

Theorem 3.7 *Let $\psi : M(c) \to M_1$ be a semi-slant $\xi^\perp$-Riemannian submersion such that ξ is vertical. Then we have*

$$
\begin{aligned}
2\hat{\tau} + 2\tau^* \le\ & \frac{(c+3)}{4}(n+r)(n+r-1) + \frac{(c-1)}{4}[2(1-r-n) + 3\{2(d_1 + d_2\cos^2\theta) \\
&+ 2\|\mathcal{B}\|^2 + \|\mathcal{C}\|^2] - r^2\,\|H\|^2 + \left\|\mathcal{T}^{\mathcal{H}}\right\|^2 + 2\delta(N) - 2\left\|\mathcal{T}^{\mathcal{V}}\right\|^2 \\
&+ 2\left\|\mathcal{A}^{\mathcal{H}}\right\|^2,
\end{aligned}
\tag{3.55}
$$

$$2\hat{\tau}+2\tau^{*} \geq \frac{(c+3)}{4}(n+r)(n+r-1)+\frac{(c-1)}{4}[2(1-r-n)+3\{2(d_1+d_2\cos^2\theta) + 2\|\mathcal{B}\|^2+\|\mathcal{C}\|^2]-r^2\|H\|^2+\left\|\mathcal{T}^{\mathcal{H}}\right\|^2-3\left\|\mathcal{A}^{\mathcal{V}}\right\|^2+2\delta(N) - 2\left\|\mathcal{T}^{\mathcal{V}}\right\|^2. \tag{3.56}$$

Equality cases of (3.55) and (3.56) hold for all $p \in M$ if and only if horizontal distribution $\mathcal{H}$ is integrable.

From Theorem 3.7, we have:

Corollary 3.8 *Let $\psi : M(c) \to M_1$ be a semi-slant $\xi^{\perp}$-Riemannian submersion such that ξ is vertical and each fiber be totally geodesic. Then we have*

$$2\hat{\tau}+2\tau^{*} \leq \frac{(c+3)}{4}(n+r)(n+r-1)+\frac{(c-1)}{4}[2(1-r-n)+3\{2(d_1+d_2\cos^2\theta) + 2\|\mathcal{B}\|^2+\|\mathcal{C}\|^2]+2\left\|\mathcal{A}^{\mathcal{H}}\right\|^2, \tag{3.57}$$

$$2\hat{\tau}+2\tau^{*} \geq \frac{(c+3)}{4}(n+r)(n+r-1)+\frac{(c-1)}{4}[2(1-r-n)+3\{2(d_1+d_2\cos^2\theta) + 2\|\mathcal{B}\|^2+\|\mathcal{C}\|^2\}]-3\left\|\mathcal{A}^{\mathcal{V}}\right\|^2. \tag{3.58}$$

Equality cases of (3.57) and (3.58) hold for all $p \in M$ if and only if horizontal distribution $\mathcal{H}$ is integrable.

Theorem 3.9 *Let $\psi : M(c) \to M_1$ be a semi-slant $\xi^{\perp}$-Riemannian submersion such that ξ is vertical. Then we have*

$$2\hat{\tau}+2\tau^{*} \geq \frac{(c+3)}{4}(n+r)(n+r-1)+\frac{(c-1)}{4}[2(1-r-n)+3\{2(d_1+d_2\cos^2\theta) + 2\|\mathcal{B}\|^2+\|\mathcal{C}\|^2\}]-r^2\|H\|^2+2\delta(N)-2\left\|\mathcal{T}^{\mathcal{V}}\right\|^2+2\left\|\mathcal{A}^{\mathcal{H}}\right\|^2 - 3\left\|\mathcal{A}^{\mathcal{V}}\right\|^2, \tag{3.59}$$

$$2\hat{\tau}+2\tau^{*} \leq \frac{(c+3)}{4}(n+r)(n+r-1)+\frac{(c-1)}{4}[2(1-r-n)+3\{2(d_1+d_2\cos^2\theta) + 2\|\mathcal{B}\|^2+\|\mathcal{C}\|^2\}]-r^2\|H\|^2+\left\|\mathcal{T}^{\mathcal{H}}\right\|^2+2\delta(N)+2\left\|\mathcal{A}^{\mathcal{H}}\right\|^2-3\left\|\mathcal{A}^{\mathcal{V}}\right\|^2. \tag{3.60}$$

Equality cases of (3.59) and (3.60) hold for all $p \in M$ if and only if the fiber through p of ψ is a totally geodesic submanifold of M.

From Theorem 3.9, we have the following corollary.

Corollary 3.10 *Let $\psi : M(c) \to M_1$ be a semi-slant $\xi^{\perp}$-Riemannian submersion such that ξ is vertical and $\mathcal{H}$ is integrable. Then we have*

$$2\hat{\tau} + 2\tau^* \geq \frac{(c+3)}{4}(n+r)(n+r-1) + \frac{(c-1)}{4}[2(1-r-n) + 3\{2(d_1 + d_2\cos^2\theta) + 2\|\mathcal{B}\|^2 + \|\mathcal{C}\|^2\}] - r^2\|H\|^2 + 2\delta(N) - 2\left\|\mathcal{T}^{\mathcal{V}}\right\|^2, \quad (3.61)$$

$$2\hat{\tau} + 2\tau^* \leq \frac{(c+3)}{4}(n+r)(n+r-1) + \frac{(c-1)}{4}[2(1-r-n) + 3\{2(d_1 + d_2\cos^2\theta) + 2\|\mathcal{B}\|^2 + \|\mathcal{C}\|^2\}] - r^2\|H\|^2 + 2\delta(N) + \left\|\mathcal{T}^{\mathcal{H}}\right\|^2. \quad (3.62)$$

Equality cases of (3.61) and (3.62) hold for all $p \in M$ if and only if the fiber through p of ψ is a totally geodesic submanifold of M.

Lemma 3.11 *Let p and q be nonnegative real number, and then*

$$\frac{p+q}{2} \geq \sqrt{pq}$$

with equality iff $p = q$.

By virtue of Lemma 3.11 in (3.54), we get:

Theorem 3.12 *Let $\psi : M(c) \to M_1$ be a semi-slant $\xi^{\perp}$-Riemannian submersion such that ξ is vertical. Then we have*

$$\frac{(c+3)}{4}(n+r)(n+r-1) + \frac{(c-1)}{4}[2(1-r-n) + 3\{2(d_1 + d_2\cos^2\theta) + 2\|\mathcal{B}\|^2 + \|\mathcal{C}\|^2\}] \leq 2\hat{\tau} + 2\tau^* + r^2\|H\|^2 + 2\left\|\mathcal{T}^{\mathcal{V}}\right\|^2 + 3\left\|\mathcal{A}^{\mathcal{V}}\right\|^2 - 2\delta(N) - 2\sqrt{2}\left\|\mathcal{A}^{\mathcal{H}}\right\|\left\|\mathcal{T}^{\mathcal{H}}\right\|. \quad (3.63)$$

Equality cases of (3.63) hold for all $p \in M$ if and only if $\|\mathcal{A}^{\mathcal{H}}\| = \|\mathcal{T}^{\mathcal{H}}\|$.

Theorem 3.13 *Let $\psi : M(c) \to M_1$ be a semi-slant $\xi^{\perp}$-Riemannian submersion such that ξ is vertical. Then we have*

$$\frac{(c+3)}{4}(n+r)(n+r-1)+\frac{(c-1)}{4}[2(1-r-n)+3\{2(d_1+d_2\cos^2\theta)$$
$$+2\|\mathcal{B}\|^2+\|\mathcal{C}\|^2\}]\geq 2\hat{\tau}+2\tau^*+r^2\|H\|^2-\left\|\mathcal{T}^{\mathcal{H}}\right\|^2-2\delta(N)-2\left\|\mathcal{A}^{\mathcal{H}}\right\|^2$$
$$+2\sqrt{6}\left\|\mathcal{A}^{\mathcal{V}}\right\|\left\|\mathcal{T}^{\mathcal{V}}\right\|. \tag{3.64}$$

Equality cases of (3.64) hold for all $p\in M$ if and only if $\|\mathcal{A}^{\mathcal{V}}\|=\|\mathcal{T}^{\mathcal{V}}\|$.

Lemma 3.14 ([49]) *Let $p_1, p_2, \ldots, p_n$, be n-real number ($n>1$), and then*

$$\frac{1}{n}\left(\sum_{i=1}^{n}p_i\right)^2\leq\sum_{i=1}^{n}p_i^2$$

with equality iff $p_1=p_2=\ldots=p_n$.

Theorem 3.15 *Let $\psi: M(c)\to M_1$ be a semi-slant $\xi^{\perp}$-Riemannian submersion such that ξ is vertical. Then we have*

$$\frac{(c+3)}{4}(n+r)(n+r-1)+\frac{(c-1)}{4}[2(1-r-n)+3\{2(d_1+d_2\cos^2\theta)$$
$$+2\|\mathcal{B}\|^2+\|\mathcal{C}\|^2\}]\leq 2\hat{\tau}+2\tau^*+r(r-1)\|H\|^2+3\left\|\mathcal{A}^{\mathcal{V}}\right\|^2-2\delta(N)+2\left\|\mathcal{T}^{\mathcal{V}}\right\|^2$$
$$-2\left\|\mathcal{A}^{\mathcal{H}}\right\|^2. \tag{3.65}$$

Equality case of (3.65) holds for all $p\in M$ if and only if we have the following statements:

(i) ψ is a Riemannian submersion that has totally umbilical fibers.
(ii) $\mathcal{T}_{ij}=0$, for $i\neq j\in\{1,2,\ldots,r\}$.

Proof From (3.54), we have

$$\frac{(c+3)}{4}(n+r)(n+r-1)+\frac{(c-1)}{4}[2(1-r-n)+3\{2(d_1+d_2\cos^2\theta)$$
$$+2\|\mathcal{B}\|^2+\|\mathcal{C}\|^2\}]$$
$$=2\hat{\tau}+2\tau^*+r^2\|H\|^2-\sum_{i=1}^{n}\sum_{j=1}^{r}(\mathcal{T}_{jj}^{s})^2-\sum_{i=1}^{n}\sum_{j\neq k}^{r}(\mathcal{T}_{jk}^{s})^2+3\left\|\mathcal{A}^{\mathcal{V}}\right\|^2-2\delta(N)$$
$$+2\left\|\mathcal{T}^{\mathcal{V}}\right\|^2-2\left\|\mathcal{A}^{\mathcal{H}}\right\|^2. \tag{3.66}$$

Considering Lemma 3.11 in (3.66), we get

$$\frac{(c+3)}{4}(n+r)(n+r-1)+\frac{(c-1)}{4}[2(1-r-n)+3\{2(d_1+d_2\cos^2\theta)$$
$$+2\|\mathcal{B}\|^2+\|\mathcal{C}\|^2\}] \leq 2\hat{\tau}+2\tau^*+r^2\|H\|^2-\frac{1}{r}\sum_{s=1}^{n}\left(\sum_{j=1}^{r}\mathcal{T}_{jj}^{s}\right)^2-\sum_{s=1}^{n}\sum_{j\neq k}^{r}(\mathcal{T}_{jk}^{s})^2$$
$$+3\left\|\mathcal{A}^{\mathcal{V}}\right\|^2-2\delta(N)+2\left\|\mathcal{T}^{\mathcal{V}}\right\|^2-2\left\|\mathcal{A}^{\mathcal{H}}\right\|^2, \tag{3.67}$$

which is equivalent to (3.65). Equality case of (3.65) holds for all $p \in M$ if and only if

$$\mathcal{T}_{11}=\mathcal{T}_{22}=\ldots=\mathcal{T}_{rr} \text{ and } \sum_{s=1}^{n}\sum_{j\neq k}^{r}(T_{jk}^{s})^2,$$

which completes proof of the theorem. □

The same proof way of Theorem 3.15, we have:

Theorem 3.16 *Let $\pi : M(c) \to M_1$ be a semi-slant $\xi^{\perp}$-Riemannian submersion such that ξ is vertical. Then we have*

$$\frac{(c+3)}{4}(n+r)(n+r-1)+\frac{(c-1)}{4}[2(1-r-n)+3\{2(d_1+d_2\cos^2\theta)$$
$$+2\|\mathcal{B}\|^2+\|\mathcal{C}\|^2\}]$$
$$\geq 2\hat{\tau}+2\tau^*+r^2\|H\|^2-\left\|\mathcal{T}^{\mathcal{H}}\right\|^2+\frac{3}{n}tr(\mathcal{A}^{\mathcal{V}})^2-2\delta(N)+2\left\|\mathcal{T}^{\mathcal{V}}\right\|^2-2\left\|\mathcal{A}^{\mathcal{H}}\right\|^2. \tag{3.68}$$

Equality case of (3.68) holds for all $p \in M$ if and only if $\mathcal{A}_{11}=\mathcal{A}_{22}=\ldots=\mathcal{A}_{nn}$ and $\mathcal{A}_{ij}=0$, for $i \neq j \in \{1,2,\ldots,n\}$.

From Theorem 3.16, we get:

Corollary 3.17 *Let $\pi : M(c) \to M_1$ be a semi-slant $\xi^{\perp}$-Riemannian submersion such that ξ is vertical and each fiber is totally geodesic. Then we have*

$$\frac{(c+3)}{4}(n+r)(n+r-1)+\frac{(c-1)}{4}[2(1-r-n)+3\{2(d_1+d_2\cos^2\theta)$$
$$+2\|\mathcal{B}\|^2+\|\mathcal{C}\|^2\}] \geq 2\hat{\tau}+2\tau^*+r^2\|H\|^2+\frac{3}{n}tr(\mathcal{A}^{\mathcal{V}})^2-2\delta(N)-2\left\|\mathcal{A}^{\mathcal{H}}\right\|^2. \tag{3.69}$$

Equality case of (3.69) holds for all $p \in M$ if and only if $\mathcal{A}_{11}=\mathcal{A}_{22}=\ldots=\mathcal{A}_{nn}$ and $\mathcal{A}_{ij}=0$, for $i \neq j \in \{1,2,\ldots,n\}$.

Finally, in this section, we are going to provide some illustrative examples for semi-slant $\xi^{\perp}$-Riemannian submersion from a Sasakian manifold. We first have the following trivial examples:

- Every invariant submersion from a Sasakian manifold to a Riemannian manifold is a semi-slant $\xi^{\perp}$-Riemannian submersion with $\bar{\mathcal{D}} = 0$ and $\theta = 0$.
- Every anti-invariant submersion from a Sasakian manifold to a Riemannian manifold is a semi-slant $\xi^{\perp}$-Riemannian submersion with $\bar{\mathcal{D}} = 0$ and $\theta = \frac{\pi}{2}$ [26].
- Every slant Riemannian submersion from a Sasakian manifold to a Riemannian manifold is a semi-slant $\xi^{\perp}$-Riemannian submersion with $\tilde{\mathcal{D}} = 0$ [17].

The following example is a nontrivial example for semi-slant $\xi^{\perp}$-Riemannian submersion from a Sasakian manifold.

Example 3.18 Let be $(\mathbb{R}^{11}, g_1, \phi, \xi, \eta)$ almost contact metric with Sasakian metric structure manifold and $(\mathbb{R}^6, g_2)$ be Riemannian manifold. Here

$$\eta = \frac{1}{4}\{2dz + u^2du^6 - u^5du^1 - u^7du^5\}, \xi = 2\frac{\partial}{\partial z},$$

$$g_{\mathbb{R}^{11}} = \eta \otimes \eta + \frac{1}{4}(du^1 \otimes du^1 + du^2 \otimes du^2 + \ldots + du^{10} \otimes du^{10})$$

and

$$g_{\mathbb{R}^6} = \frac{1}{4}\sum_{i=1}^{n}(du^i \otimes du^i).$$

$$\phi(u_1, \ldots, u_{10}, z) = (-u_4, -u_{10}, u_7, u_1, u_6, -u_5, -u_3, -u_9, u_8, u_2, \frac{u^2u_5 + u^7u_6 - u^5u_4}{2}\Big).$$

Let ψ be a submersion defined by

$$\psi : (\mathbb{R}^{11}, g_{\mathbb{R}^{11}}, \phi, \xi, \eta) \to (\mathbb{R}^6, g_{\mathbb{R}^6}),$$

$$\psi(u_1, \ldots, u_{10}, z) = \left(\frac{-1}{\sqrt{2}}u_2 + \frac{1}{\sqrt{2}}u_7, u_1, u_6, \frac{1}{\sqrt{2}}u_3 - \frac{1}{\sqrt{2}}u_{10}, u_5, \frac{-1}{\sqrt{2}}u_4 - \frac{1}{\sqrt{2}}u_8\right).$$

Then it follows that

$$ker\psi_* = \left\{ E_1 = \frac{1}{\sqrt{2}}\frac{\partial}{\partial u_2} + \frac{1}{\sqrt{2}}\frac{\partial}{\partial u_7}, E_2 = \frac{-1}{\sqrt{2}}\frac{\partial}{\partial u_3} - \frac{1}{\sqrt{2}}\frac{\partial}{\partial u_{10}}, \right.$$
$$E_3 = \frac{1}{\sqrt{2}}\frac{\partial}{\partial u_4} - \frac{1}{\sqrt{2}}\frac{\partial}{\partial u_8},$$
$$\left. E_4 = \frac{\partial}{\partial u_9}, \xi = 2\frac{\partial}{\partial z} \right\}$$

and

$$(ker\psi_*)^{\perp} = \left\{ F_1 = \frac{-1}{\sqrt{2}}\frac{\partial}{\partial u_2} + \frac{1}{\sqrt{2}}\frac{\partial}{\partial u_7}, F_2 = \frac{\partial}{\partial u_1} + \frac{u^5}{2}\frac{\partial}{\partial z}, F_3 = \frac{\partial}{\partial u_6} - \frac{u^2}{2}\frac{\partial}{\partial z}, \right.$$
$$\left. F_4 = \frac{1}{\sqrt{2}}\frac{\partial}{\partial u_3} - \frac{1}{\sqrt{2}}\frac{\partial}{\partial u_{10}}, F_5 = \frac{\partial}{\partial u_5} + \frac{u^7}{2}\frac{\partial}{\partial z}, F_6 = \frac{-1}{\sqrt{2}}\frac{\partial}{\partial u_4} - \frac{1}{\sqrt{2}}\frac{\partial}{\partial u_8} \right\};$$

hence we have $\phi(E_1) = -E_2$. Thus it follows that $\tilde{\mathcal{D}} = span\{E_1, E_2\}$ and $\bar{\mathcal{D}} = span\{E_3, E_4\}$ is a slant distribution with semi-slant angle $\theta = \frac{\pi}{4}$. In this case ψ is a semi-slant $\xi^{\perp}$ submersion. Also by direct computations, we obtain $g_{\mathbb{R}^6}(F_i, F_i) = g_{\mathbb{R}^{11}}(\phi F_i, \phi F_i)$; $i = 1, \ldots, 6$, which show that ψ is a semi-slant $\xi^{\perp}$-Riemannian submersion, where $(u_1, \ldots, u_{10}, z)$ are the Cartesian coordinates.

References

1. Agarwal, R., Fatima, T., Yadav, S.K., Ali, S.: Geometric analysis of Riemannian doubly warped product submersions. Phys. Scr. **99**, 075271 (2024)
2. Akyol, M.A.: Conformal semi-slant submersions. Int. J. Geom. Methods Mod. Phys. **14**(7), 1750114 (2017)
3. Akyol, M.A., Demir, R., Poyraz, N.Ö.: Ricci and scalar curvature inequalities for semi-slant Riemannian submersions in complex space forms. Int. J. Geom. Methods Mod. Phys. **20**(08), 2350133 (2023)
4. Akyol, M.A., Demir, R., Poyraz, N. Ö., Vîlcu, G.E.: Optimal inequalities for hemi-slant Riemannian submersions. Mathematics **10**(21), 3993 (2022)
5. Akyol, M.A., Poyraz, N. Ö.: Sharp inequalities involving the Chen-Ricci inequality for slant Riemannian submersions. Bull. Korean Math. Soc. **60**(5), 1155–1179 (2023)
6. Akyol, M.A., Sarı, R.: On semi-slant $\xi^{\perp}$-Riemannian submersions. Mediterr. J. Math. **14**, 234 (2017). https://doi.org/10.1007/s00009-017-1035-2
7. Akyol, M.A., Sarı, R., Aksoy, E.: Semi-invariant $\xi^{\perp}$-Riemannian submersions from almost contact metric manifolds. Int. J. Geom. Methods Mod. Phys. **14**(5), 1750074 (2017)
8. Alegre, P., Chen, B.Y., Munteanu, M.I.: Riemannian submersions, δ-invariants, and optimal inequality. Ann. Glob. Anal. Geom. **42**(3), 317–331 (2012)
9. Aydın, M., Mihai, A., Mihai, I.: Some inequalities on submanifolds in statistical manifolds of constant curvature. Filomat **29**(3), 465–477 (2015). https://doi.org/10.2298/fil1503465a
10. Aytimur, H., Özgür, C.: Inequalities for submanifolds in statistical manifolds of quasiconstant curvature. Ann. Pol. Math. **121**(3), 197–215 (2018). https://doi.org/10.4064/ap171106-27-6
11. Aytimur, H., Özgür, C.: Sharp inequalities for anti-invariant Riemannian submersions from Sasakian space forms. J. Geom. Phys. **166**, 104251 (2021)

12. Blair, D.E.: Contact manifold in riemannian geometry. In: Lecture Notes in Mathematics, vol. 509, Springer, Berlin (1976)
13. Chen, B.Y.: A general inequality for submanifolds in complex-space-forms and its applications. Arch. Math. **67**, 519–528 (1996)
14. Chen, B.Y.: Relations between Ricci curvature and shape operator for submanifolds with arbitrary codimensions. Glasg. Math. J. **41**(1), 33–41 (1999)
15. Chen, B.Y.: Examples and classification of Riemannian submersions satisfying a basic equality. Bull. Aust. Math. Soc. **72**(3), 391–402 (2005)
16. Chen, B.Y.: A general optimal inequality for arbitrary Riemannian submanifolds. J. Inequal. Pure Appl. Math. **6**(3), 77 (2005)
17. Erken, I.K., Murathan, C.: Slant Riemannian submersions from Sasakian manifolds. Arab J. Math. Sci. **22**(2), 250–2264 (2016)
18. Faghfouri, M., Ghaffarzadeh, N.: Chen's inequality for invariant submanifolds in generalized (κ, μ)- space forms. Global J. Adv. Res. Class. Mod. Geom. **101**(4), 86 (2015)
19. Falcitelli, M., Ianus, S., Pastore, A.M.: Riemannian Submersions and Related Topics. World Scientific, River Edge (2004)
20. Gray, A.: Pseudo-Riemannian almost product manifolds and submersions. J. Math. Mech. **16**, 715–737 (1967)
21. Gupta, R.S.: B.Y. Chen's inequalities for bi-slant submanifolds in cosymplectic space forms. Saraj. J. Math. **9**(1), 117–128 (2013). https://doi.org/10.5644/sjm.09.1.11
22. Gülbahar, M., Meriç, E.Ş., Kılıç, E.: Sharp inequalities involving the Ricci curvature for Riemannian submersions. Kragujevac J. Math. **41**(2), 279–293 (2017)
23. Gülbahar, M., Meriç, Ş.E., Kılıç, E.: Chen invariants for Riemannian submersions and their applications. Commun. Fac. Sci. Univ. Ankara Ser. A1 Math. Stat. **71**(4), 1007–1022 (2022)
24. Gündüzalp, Y.: Semi-slant submersions from almost product Riemannian manifolds. Demonstratio Math. **49**(3), 345–356 (2016)
25. Kılıç, E., Tripathi, M.M., Gülbahar, M.: Chen–Ricci inequalities for submanifolds of Riemannian and Kaehlerian product manifolds. Ann. Pol. Math. **116**, 37–56 (2016)
26. Lee, J.W.: Anti-invariant $\xi^{\perp}$-Riemannian submersions from almost contact manifolds. Hacettepe J. Math. Stat. **42**(3), 231–241 (2013)
27. Liang, Z., Pan, Z.: Notes on Chen's inequalities for submanifolds of real space forms with a semi-symmetric non-metric connection. J. East China Normal Univ. (Nat. Sci.) **2015**(1), 6 (2015)
28. Mihai, A., Özgür, C.: Chen inequalities for submanifolds of real space forms with a semi-symmetric metric connection. Taiwan. J. Math. **14**(4), 1465–1477 (2010)
29. O'Neill, B.: The fundamental equations of a submersion. Mich. Math. J. **13**, 458–469 (1966)
30. Özgür, C., Murathan, C.: Chen inequalities for submanifolds of a locally conformal almost cosymplectic manifold with a semi-symmetric metric connection. An. St. Univ. Ovidius Constunta **18**(1), 239–254 (2010)
31. Park, K.S., Prasad, R.: Semi-slant submersions. Bull. Korean Math. Soc. **50**(3), 951–962 (2013)
32. Poyraz, N.Ö.: Chen inequalities on spacelike hypersurfaces of a GRW spacetime. Differ. Geom. Appl. **81**, 101863 (2022)
33. Poyraz, N.Ö., Akyol, M.A., Demir, R.: Ricci and scalar curvature inequalities for semi-slant Riemannian submersions in complex space forms. Int. J. Geom. Methods Mod. Phys. **20**(08), 2350133 (2023)
34. Poyraz, N.Ö., Gündüzalp, Y., Akyol, M.A.: Some inequalities for bi-slant Riemannian submersions in complex space forms
35. Sasaki, S., Hatakeyama, Y.: On differentiable manifolds with contact metric structure. J. Math. Soc. Japan **14**, 249–271 (1961)
36. Siddiqi, M.D., Akyol, M.A.: Anti-invariant $\xi^{\perp}$-Riemannian Submersions From Almost Hyperbolic Contact Manifolds. Int. Electron. J. Geom. **12**(1), 32–42 (2019)
37. Siddiqui, A.N., Chen, B.Y., Siddiqi, M.D.: Chen inequalities for statistical submersions between statistical manifolds. Int. J. Geom. Methods Mod. Phys. **18**(04), 2150049 (2021)

38. Sikander, F., Fatima, T., Alharbi, S.A.: A study on curvature relations of conformal generic submersions. J. King Saud Univ. Sci. **35**(39), 102526 (2023)
39. Sikander, F., Fatima, T., Deshmukh, S., Elsharkawy, A.: Curvature analysis of concircular trajectories in doubly warped product manifolds. AIMS Math. **9**(8), 21940–21951 (2024)
40. Şahin, B.: Anti-invariant Riemannian submersions from almost Hermitian manifolds. Cent. Eur. J. Math. **8**(3), 437–447 (2010)
41. Şahin, B.: Slant submersions from almost Hermitian manifolds. Bull. Math. Soc. Sci. Math. Roumanie **54**(1), 93–105 (2011)
42. Şahin, B.: Semi-invariant Riemannian submersions from almost Hermitian manifolds. Can. Math. Bull. **56**, 173–183 (2011)
43. Şahin, B.: Riemannian submersions from almost Hermitian manifolds. Taiwan. J. Math. **17**(2), 629–659 (2013)
44. Şahin, B.: Chen's first inequality for Riemannian maps. Ann. Pol. Math. **117**(3), 249–258 (2016)
45. Şahin, B.: Riemannian Submersions, Riemannian Maps in Hermitian Geometry, and their Applications. Elsevier, Academic, Amsterdam (2017)
46. Taştan, H.M.: Lagrangian submersions. Hacettepe J. Math. Stat. **43**(6), 993–1000 (2014)
47. Taştan, H.M., Şahin, B., Yanan, Ş.: Hemi-slant submersions. Mediterr. J. Math. **13**(4), 2171–2184 (2016)
48. Tripathi, M.M.: Certain basic inequalities for submanifolds in (κ, μ) space. Recent Adv. Riemannian and Lorentzian Geom. **337**, 187 (2003)
49. Tripathi, M.M.: Improved Chen–Ricci inequality for curvature-like tensors and its applications. Differ. Geom. Appl. **29**(5), 685–698 (2011)
50. Vîlcu, G.E.: On Chen invariants and inequalities in quaternionic geometry. J. Inequal. Appl. **2013**(1), 1–14 (2013)
51. Vîlcu, A.D., Vîlcu, G.E.: Statistical manifolds with almost quaternionic structures and quaternionic Kahler-like statistical submersions. Entropy **17**(9), 6213–6228 (2015)
52. Zhang, P.: Remarks on Chen's inequalities for submanifolds of a Riemannian manifold of nearly quasi-constant curvature. Vietnam J. Math. **43**, 557–569 (2015)
53. Zhang, P., Zhang, L., Song, W.: Chen's inequalities for submanifolds of a Riemannian manifold of quasi-constant curvature with a semi-symmetric metric connection (2014)

Characterizations of Perfect Fluid and Generalized Robertson-Walker Space-Times Admitting k-Almost Ricci-Yamabe Solitons

Krishnendu De and Uday Chand De

Abstract This chapter concerns with the investigation of k-almost Ricci-Yamabe solitons and gradient k-almost Ricci-Yamabe solitons in perfect fluid space-times and generalized Robertson-Walker space-times. First, we deduce the criterion for which the k-almost Ricci-Yamabe solitons in a perfect fluid space-time is steady, expanding or shrinking. Then we establish that if perfect fluid space-times admit a gradient k-almost Ricci-Yamabe soliton with Killing velocity vector, then either it represents phantom era, or the gradient k-almost Ricci-Yamabe soliton is expanding or shrinking under some condition. Moreover, we illustrate that a generalized Robertson-Walker space-time represents a perfect fluid space-time if it admits a k-almost Ricci-Yamabe soliton. Next, we establish that if a generalized Robertson-Walker space-time allows a k-almost Ricci-Yamabe soliton of gradient type with constant scalar curvature, then it also represents a perfect fluid space-time.

Keywords PF space-times · GRW space-times · k-almost Ricci-Yamabe solitons

1 Introduction

Einstein's general relativity (GR) theory is usually called the gravitation theory of geometry. GR, the finest well-known physics theories of this century, has established the fundamental relationship between the geometry of space-time and physics. It has been the areas of greatest interest in both mathematics and physics

K. De (✉)
Department of Mathematics, Kabi Sukanta Mahavidyalaya, The University of Burdwan, Bhadreswar, West Bengal, India
e-mail: krishnendu.de@outlook.in

U. C. De
Department of Pure Mathematics, University of Calcutta, Kolkata, West Bengal, India

B.-Y. Chen, M. A. Choudhary (eds.), *Geometric Inequalities and Applications*, Infosys Science Foundation Series in Mathematical Sciences,
https://doi.org/10.1007/978-981-95-5148-4_6

during the last century. Today, one of the most significant issue is trying to solve Einstein's field equation (EFE) in many different approaches.

The most straightforward answer of the above issue is the Minkowski space-time (four-dimensional Euclidean space R^4 with a Lorentzian metric). Further nontrivial solutions include the Kerr, de-Sitter, and Schwartzchild solutions. In GR theory, Lorentzian warped product manifolds were modified to acquire a general solution to EFEs. Standard static space-time [1] and generalized Robertson-Walker space-time (GRW) [2, 3] are two prominent instances.

From GR we know that a space-time is a Lorentzian manifold M^4 that allows for a globally time-oriented vector and has the metric g (Lorentzian) of signature $(+, +, +, -)$. The notion of GRW space-times was first invented by Alias et al. [4]. If a Lorentzian n-manifold M with $n \geq 3$ can be formed as a warped product $M = -I \times \rho^2 M^*$, where $\rho > 0$ indicates a scale factor and M^* denotes an $(n-1)$-dimensional Riemannian manifold, then it is referred to as a GRW space-time. The GRW space-time reduces to a Robertson-Walker (RW) space-time if the Riemannian manifold M^* is a 3-dimensional manifold of constant curvature.

In [5], the authors define pseudo-Einstein space as a Riemannian space whose Ricci tensor S fulfills the condition

$$S(X_1, Y_1) = a_1 g(X_1, Y_1) + b_1 A(X_1) A(Y_1), \tag{1.1}$$

where $a_1, b_1 \in \mathbb{R}$ and A is a nonzero 1-form such that $A(X_1) = g(X_1, \rho)$, ρ is a unit vector field.

Later, in [6], Duggal and Sharma defined pseudo-Einstein space as a semi-Riemannian space whose Ricci tensor obeys the relation (1.1), where a_1 and b_1 are scalar functions.

Subsequently, in 1991, Deszcz and Verstraelen [7] studied hypersurfaces of pseudo-Riemannian conformally flat spaces and named a semi-Riemannian space that satisfies the Eq. (1.1), a quasi-Einstein space.

In [8], Chaki and Maity introduced the notion of a quasi-Einstein manifold whose Ricci tensor is not identically zero and satisfies the condition (1.1) in which a_1 and b_1 are scalar functions and ρ is a unit vector field.

In 2004, the authors [9] studied a 2-quasi umbilical hypersurface of a Euclidean space, and they obtained the following expression for Ricci tensor S as

$$S(X_1, Y_1) = a_1 g(X_1, Y_1) + b_1 A(X_1) A(Y_1) + c_1 B(X_1) B(Y_1) \tag{1.2}$$

in which a_1, b_1, c_1 are certain nonzero scalars and A_1, B_1 are two nonzero 1-forms. The unit vector fields ρ and μ corresponding to the 1-forms A and B, respectively, defined by

$$A(X_1) = g(X_1, \rho) \quad \text{and} \quad B(X_1) = g(X_1, \mu), \tag{1.3}$$

are orthogonal, that is, $A(\mu) = B(\rho) = g(\rho, \mu) = 0$. The vector fields ρ, μ are called the generators of the manifold and a_1, b_1, c_1 are the associated scalars. The authors [9] named such a manifold generalized quasi-Einstein manifold.

The nonvanishing Ricci tensor S in a perfect fluid (PF) space-time is presented by

$$S = a_1 g + b_1 \eta \otimes \eta, \tag{1.4}$$

ϱ is demonstrated by $g(X_1, \varrho) = \eta(X_1)$ for any X_1, and a_1, b_1 are scalars. Also, ϱ is a time-like unit vector field of the PF space-time, that is, $g(\varrho, \varrho) = -1$. Each and every RW space-time is a PF space-time [10]. In dimension 4, the GRW space-time represents a PF space-time iff it is a RW space-time. The characteristics of GRW space-times and PF space-times have been found in [2–14].

For a gravitational constant k, the EFEs with vanishing cosmological constant are described by

$$S - \frac{r}{2} g = kT, \tag{1.5}$$

in which r represents the scalar curvature and the energy-momentum tensor (EMT) is denoted by T.

For a PF space-time, T is described by

$$T = pg + (p + \sigma)\eta \otimes \eta, \tag{1.6}$$

p denotes isotropic pressure, and σ indicates the energy density. The Eq. (1.4) can be obtained from the Eqs. (1.5) and (1.6) [11].

Jointly using the Eqs. (1.4)–(1.6), we provide

$$a_1 = -\frac{k(p - \sigma)}{n - 2}, \quad b_1 = \kappa(p + \sigma). \tag{1.7}$$

Besides, $p = p(\sigma)$ interconnects p and σ known as equation of state (EOS), and in this case, the PF space-time is called isentropic. For $p = \sigma$, this space-time is named as stiff matter. As per [15], the radiation era, the dust matter fluid, and the dark energy epoch are represented by the PF space-time if $p = \frac{\sigma}{3}$, $p = 0$, and if $p + \sigma = 0$, respectively. It also covers the phantom era in which $\frac{p}{\sigma} < -1$.

On the contrary, the conformal curvature tensor, also named as the Weyl tensor, is important in relativity theory and geometry. Weyl tensors have been used by many researchers to characterize space-times. The conformal curvature tensor C is described by

$$\begin{aligned} C(X_1, Y_1)Z_1 = R(X_1, Y_1)Z_1 &- \frac{1}{n-2}[g(QY_1, Z_1)X_1 \\ &- g(QX_1, Z_1)Y_1 + g(Y_1, Z_1)QX_1 - g(X_1, Z_1)QY_1] \\ &+ \frac{r}{(n-1)(n-2)}[g(Y_1, Z_1)X_1 - g(X_1, Z_1)Y_1], \end{aligned}$$

R stands for the Riemann curvature tensor, and the Ricci operator Q is presented by $g(QX_1, Y_1) = S(X_1, Y_1)$.

In addition, we are aware that

$$(div\, C)(X_1, Y_1)Z_1 = \frac{n-3}{n-2}[\{(\nabla_{X_1} S)(Y_1, Z_1) - (\nabla_{Y_1} S)(X_1, Z_1)\} - \frac{1}{2(n-1)}\{(X_1 r)g(Y_1, Z_1) - (Y_1 r)g(X_1, Z_1)\}], \tag{1.8}$$

"div" denotes the divergence, and if divergence vanishes, then it is called harmonic.

The theory of geometric flows has given rise to some of the foremost interesting mathematical methods used in the last few decades to illustrate the geometric structures in differential geometry. A part of solutions in which the metric transforms through diffeomorphisms has a substantial impact on the understanding of flow singularities, as they appear to be realistic models of singularities. They are referred to as soliton solutions in general.

In [16], Hamilton simultaneously introduced the Ricci and Yamabe flow. The Ricci solitons (RSs) and Yamabe solitons (YSs) are the specific solutions of the Ricci and Yamabe flow, respectively. Lately, geometric flows, such the Ricci and Yamabe flows, have attracted the theoretical attention of many geometers. In 2019, the Ricci-Yamabe (RY) map was presented in [17] as a novel geometric flow. This map is just a scalar combination of Ricci and Yamabe flow. A flow like this improves the metrics on the semi-Riemannian manifold M, which are given by Guler and Crasmareanu [17]

$$\frac{\partial}{\partial t} g(t) = \beta_2 r(t) g(t) - 2\beta Ric(t), \quad g_0 = g(0). \tag{1.9}$$

Based on the signs of the related scalars, β and β_2, anyone can interpret the RS flow as a Riemannian, semi-Riemannian, or singular Riemannian flow. Certain mathematical or physical models require this range of choices. The fact that RSs and YSs are essentially distinct in higher dimensions, even if they are equal in two dimensions, is another compelling reason to start investigating Ricci-Yamabe solitons (RYSs).

The investigation of space-time symmetries is crucial for figuring out EFEs. Geometry is characterized by symmetry, which reveals the physics. Space-time geometry exhibits numerous symmetries. The equations of metric are advantageous since they simplify several solutions. They are mostly used in GR to classify solutions to EFEs. The soliton is a type of symmetry that includes the geometrical flow of space-time geometry. Consequently, the flows of RS and YS are helpful, since they make the theories of energy and entropy easier to understand.

A k-almost RYS on (M, g) is a data $(g, Z_1, k, \lambda_1, \alpha_1, \beta_1)$ fulfilling

$$k\$_{Z_1} g = -2\alpha_1 S - (2\lambda_1 - \beta_1 r)g, \tag{1.10}$$

in which $k, \lambda_1, \alpha_1, \beta_1$ are smooth functions on M, S represents the Ricci tensor, r indicates the scalar curvature, and the Lie derivative is represented by $\$$.

The previously mentioned concept is known as gradient k-almost RYS if f denotes a smooth function and Z_1 is the gradient of f on M, and then Eq. (1.10) transforms into

$$k\nabla^2 f = -\alpha_1 S - \left(\lambda_1 - \frac{1}{2}\beta_1 r\right) g, \tag{1.11}$$

in which $\nabla^2 f$ indicates the Hessian of f.

The k-almost RYS (or gradient k-almost RYS) is called expanding for $\lambda_1 > 0$, steady for $\lambda_1 = 0$ and shrinking when $\lambda_1 < 0$. If $\beta_1 = 0$, $\alpha_1 = 1$, then k-almost RYS (or gradient k-almost RYS) reduces to k-almost RS (or gradient k-almost RS). Similarly, it turns into k-almost YS (or gradient k-almost YS) if $\beta_1 = 1$, $\alpha_1 = 0$. Also, if $\beta_1 = -1$, $\alpha_1 = 1$, it reduces to a k-almost Einstein soliton (or gradient k-almost Einstein soliton).

The k-almost RYS (or gradient k-almost RYS) is named proper if $\alpha_1 \neq 0, 1$.

Venkatesh et al. [18] have investigated $*$-Ricci solitons and gradient almost $*$-Ricci solitons on Kenmotsu manifolds. In 2022, Blaga and Özgür [19] have studied almost η-Ricci and η-Yamabe solitons with the help of torse-forming vector field where as in [20] the Ricci-Yamabe solitons have examined. Also in [21], the authors have investigated $*$-η-Ricci-Yamabe Solitons on Sasakian manifolds.

Very recently, there is a notable increase of fascination in researching solitons in numerous geometrical contexts because of their connection to GR. Many geometers have recently studied many sorts of solitons in PF space-times including RSs and gradient type RSs [22, 23], η-RSs [24], YSs [22, 25], k-almost YSs [26], η-Einstein solitons of gradient type [23], gradient ϱ-Einstein solitons [27], gradient Schouten solitons [23], m-quasi Einstein solitons of gradient type [22], RYSs [28], respectively.

The research mentioned above motivates us to explore k-almost RYSs in PF space-times and GRW space-times. Specifically, we arrive at the following conclusions:

Theorem 1.1 *Let the PF space-time admit a k-almost RYS, and then its potential vector field Z_1 is Killing if $divZ_1 = 0$ and $\alpha_1 = 0$. Also, the soliton is expanding for $r < \frac{6\alpha_1 a_1}{\beta_1}$ or $\beta_1 > -\frac{\alpha_1}{2}$, steady if $r = \frac{6\alpha_1 a_1}{\beta_1}$ or $\beta_1 = -\frac{\alpha_1}{2}$, and shrinking for $r > \frac{6\alpha_1 a_1}{\beta_1}$ or $\beta_1 < -\frac{\alpha_1}{2}$, provided $div\varrho = 0$.*

Theorem 1.2 *If the PF space-times of dimensions 4 allow a gradient k-almost RYS with Killing velocity vector field ϱ, then either the space-time represents a phantom era, or the gradient k-almost RYS is expanding, steady or shrinking if $r > \frac{2\alpha_1(a_1-b_1)}{\beta_1}$, $r = \frac{2\alpha_1(a_1-b_1)}{\beta_1}$ or $r < \frac{2\alpha_1(a_1-b_1)}{\beta_1}$, respectively.*

Theorem 1.3 *Let a GRW space-time admit a k-almost RYS, and then the space-time becomes a PF space-time. Also, the soliton is expanding for $r > \frac{2\alpha_1(1-n)\mu_1}{\beta_1}$, steady if $r = \frac{2\alpha_1(1-n)\mu_1}{\beta_1}$, and shrinking for $r < \frac{2\alpha_1(1-n)\mu_1}{\beta_1}$.*

Corollary 1.4 *In dimension 4, a GRW space-time admitting a k-almost RYS is of Petrov type I, D, or O and the space-time reduces to a RW space-time.*

Theorem 1.5 *If the GRW space-time admits a k-almost RYS of gradient type with $r =$ constant, then it becomes a PF space-time.*

As a result of the aforementioned theorem, we establish:

Corollary 1.6 *In dimension 4, if a GRW space-time allows a gradient k-almost RYS with $r =$ constant, then the space-time belongs to Petrov classification I, D, or O and the space-time reduces to a RW space-time.*

2 PF Space-Times and GRW Space-Times

From the PF Eq. (1.4), we provide

$$QX_1 = a_1X_1 + b_1\eta(X_1)\varrho, \tag{2.1}$$

Q indicates the Ricci operator demonstrated by $g(QX_1, Y_1) = S(X_1, Y_1)$, and contracting the above equation gives

$$r = \sum_j \epsilon_i Qe_j = na_1 - b_1, \tag{2.2}$$

in which at every point of the space-time $\{e_j\}$ represents the orthonormal basis of the tangent space and $\epsilon_j = g(e_j, e_j) = \pm 1$. The covariant differentiation of Eq. (2.1) yields

$$(\nabla_{X_1}Q)(Y_1) = (X_1a_1)Y_1 + (X_1b_1)\eta(Y_1)\varrho + b_1(\nabla_{X_1}\eta)(Y_1)\varrho + b_1\eta(Y_1)\nabla_{X_1}\varrho. \tag{2.3}$$

Theorem 2.1 ([11]) *An n ($n \geq 3$)-dimensional Lorentzian manifold represents a GRW space-time iff it allows a time-like and unit torse-forming vector field: $\nabla_{X_1}v = \Psi[X_1 + A(X_1)v]$, A denotes a one-form demonstrated as $g(X_1, v) = A(X_1)$ for any X_1, which is also an eigenvector of the Ricci tensor.*

Let us assume that the velocity vector field ϱ is a torse-forming vector field. Therefore, by using Theorem I, we obtain

$$\nabla_{X_1}\varrho = \Psi[X_1 + \eta(X_1)\varrho] \tag{2.4}$$

and

$$S(X_1, \varrho) = \phi\eta(X_1), \tag{2.5}$$

ϕ is a nonzero eigenvalue, and Ψ denotes a scalar.

Lemma 2.2 *For any GRW space-time, we provide* [29]

$$R(X_1, Y_1)\varrho = \mu_1[\eta(Y_1)X_1 - \eta(X_1)Y_1] \tag{2.6}$$

and

$$S(X_1, \varrho) = (n-1)\mu_1\eta(X_1), \tag{2.7}$$

where we set $\mu_1 = (\varrho\Psi + \Psi^2)$.

Lemma 2.3 *Any GRW space-time satisfies the following* [29]

$$(X_1\mu_1) + (\varrho\mu_1)\eta(X_1) = 0. \tag{2.8}$$

3 Proof of the Main Results

Proof of Theorem 1.1 Suppose the PF space-time admits a k-almost RY soliton $(g, X, k, \lambda_1, \alpha_1, \beta_1)$. Then from Eq. (1.10), we acquire

$$k(\$_{Z_1}g)(X_1, Y_1) + 2\alpha_1 S(X_1, Y_1) + (2\lambda_1 - \beta_1 r)g(X_1, Y_1) = 0. \tag{3.1}$$

Using the Lie differentiation's explicit form, the foregoing equation yields

$$\alpha_1 S(X_1, Y_1) = -\frac{k}{2}[g(\nabla_{X_1} Z_1, Y_1) + g(X_1, \nabla_{Y_1} Z_1)] - \left(\lambda_1 - \frac{\beta_1 r}{2}\right) g(X_1, Y_1). \tag{3.2}$$

Contracting Eq. (3.2) gives

$$\alpha_1 r = -k div Z_1 - 4\left(\lambda_1 - \frac{\beta_1 r}{2}\right),$$

which implies

$$\lambda_1 - \frac{\beta_1 r}{2} = \frac{k div Z_1}{4} + \frac{\alpha_1 r}{4}. \tag{3.3}$$

Using Eq. (3.3) in Eq. (3.1), we obtain

$$\frac{k}{2}(\pounds_{Z_1}g)(X_1, Y_1) + \alpha_1 S(X_1, Y_1) + \left(\frac{kdivZ_1}{4} + \frac{\alpha_1 r}{4}\right) = 0. \tag{3.4}$$

If we take $divZ_1 = 0$ and $\alpha_1 = 0$, then from the last equation Theorem follows. Again contracting the PF equation (1.4) provides

$$r = -b_1 + 4a_1. \tag{3.5}$$

From the Eqs. (3.3) and (3.5), we acquire

$$(\alpha_1 + 2\beta_1)(-b_1 + 4a_1) = 4\lambda_1 - kdivZ_1. \tag{3.6}$$

Again comparing Eqs. (1.4) and (3.2), we infer

$$\begin{aligned} a_1 g(X_1, Y_1) + b_1\eta(X_1)\eta(Y_1) &= -\frac{k}{2\alpha_1}[g(\nabla_{X_1}\varrho, Y_1) + g(X_1, \nabla_{Y_1}\varrho)] \\ &\quad - \frac{1}{\alpha_1}\left(\lambda_1 - \frac{\beta_1 r}{2}\right) g(X_1, Y_1). \end{aligned} \tag{3.7}$$

Putting $X_1 = Y_1 = \varrho$ in the previous equation gives

$$a_1 - b_1 = \frac{1}{\alpha_1}\left(\lambda_1 - \frac{\beta_1 r}{2}\right). \tag{3.8}$$

Comparing the Eqs. (3.6) and (3.8) and taking $div\varrho = 0$, we achieve

$$\lambda_1 = \frac{(\alpha_1 + 2\beta_1)}{(6\alpha_1 - 4\beta_1)}[6\alpha_1 a_1 - \beta_1 r]. \tag{3.9}$$

Therefore, the soliton is expanding for $r < \frac{6\alpha_1 a_1}{\beta_1}$ or $\beta_1 > -\frac{\alpha_1}{2}$, steady if $r = \frac{6\alpha_1 a_1}{\beta_1}$ or $\beta_1 = -\frac{\alpha_1}{2}$, and shrinking for $r > \frac{6\alpha_1 a_1}{\beta_1}$ or $\beta_1 < -\frac{\alpha_1}{2}$, provided $div\varrho = 0$. □

Proof of the Theorem 1.2 Choose a PF space-time which admits a k-almost RY soliton of gradient type, and therefore from Eq. (1.11), we get

$$k\nabla_{X_1} Df + \alpha_1 Q X_1 = -\left(\lambda_1 - \frac{\beta_1 r}{2}\right) X_1. \tag{3.10}$$

Differentiating the Eq. (3.10), we provide

$$\begin{aligned} k\nabla_{Y_1}\nabla_{X_1} Df + (Y_1 k)\nabla_{X_1} Df &= -\alpha_1 \nabla_{Y_1} Q X_1 \\ &\quad - \left(\lambda_1 - \frac{\beta_1 r}{2}\right)\nabla_{Y_1} X_1 + \frac{\beta_1}{2}(Y_1 r) X_1. \end{aligned} \tag{3.11}$$

From the Eq. (3.11), interchanging X_1 and Y_1, we infer

$$k\nabla_{X_1}\nabla_{Y_1}Df + (X_1k)\nabla_{Y_1}Df = -\alpha_1\nabla_{X_1}QY_1 - \left(\lambda_1 - \frac{\beta_1 r}{2}\right)\nabla_{X_1}Y_1 + \frac{\beta_1}{2}(X_1r)Y_1. \quad (3.12)$$

Again, from Eq. (3.10), we acquire

$$k\nabla_{[X_1,Y_1]}Df = -\alpha_1 Q([X_1, Y_1]) - \left(\lambda_1 - \frac{\beta_1 r}{2}\right)[X_1, Y_1]. \quad (3.13)$$

From Eqs. (3.11), (3.12), and (3.13), we reveal

$$\begin{aligned} kR(X_1, Y_1)Df &= -\alpha_1[(\nabla_{X_1}Q)Y_1 + (\nabla_{Y_1}Q)X_1] \quad (3.14)\\ &+\frac{\alpha_1}{k}[(X_1k)QY_1 - (Y_1k)QX_1]\\ &+\frac{1}{k}\left(\lambda_1 - \frac{\beta_1 r}{2}\right)[(X_1k)Y_1 - (Y_1k)X_1]\\ &+\frac{\beta_1}{2}[(X_1r)Y_1 - (Y_1r)X_1]. \end{aligned}$$

The covariant differentiation of Eq. (2.1) yields

$$(\nabla_{X_1}Q)(Y_1) = (X_1a_1)Y_1 + (X_1b_1)\eta(Y_1)\varrho + b_1(\nabla_{X_1}\eta)(Y_1)\varrho + b_1\eta(Y_1)\nabla_{X_1}\varrho. \quad (3.15)$$

The Eqs. (3.14) and (3.15) yield

$$\begin{aligned} kR(X_1, Y_1)Df = \alpha_1[&(X_1a_1)Y_1 - (Y_1a_1)X_1 + \{(X_1b_1)\eta(Y_1) - (Y_1b_1)\eta(X_1)\\ &+b_1(\nabla_{X_1}\eta)(Y_1) - b_1(\nabla_{Y_1}\eta)(X_1)\}\varrho + b_1\{\eta(Y_1)\nabla_{X_1}\varrho\\ &-\eta(X_1)\nabla_{Y_1}\varrho\}]\\ &+\frac{\alpha_1}{k}[(X_1k)QY_1 - (Y_1k)QX_1] \quad (3.16)\\ &+\frac{1}{k}\left(\lambda_1 - \frac{\beta_1 r}{2}\right)[(X_1k)Y_1 - (Y_1k)X_1]\\ &+\frac{\beta_1}{2}[(X_1r)Y_1 - (Y_1r)X_1]. \end{aligned}$$

Now contracting the Eq. (3.16), we provide

$$\begin{aligned} kS(Y_1, Df) = \alpha_1[&(1-n)(Y_1a_1) + (Y_1b_1) + (\varrho b_1)\eta(Y_1)\\ &+b_1[(\nabla_\varrho\eta)(Y_1) - (\nabla_{Y_1}\eta)(\varrho) + \eta(Y_1)\, div\varrho] \end{aligned}$$

$$+\frac{\alpha_1}{k}[\{a_1(1-n)+b_1\}(Y_1k)+b_1(\varrho k)\eta(Y_1)]$$
$$+\frac{1}{k}\left(\lambda_1-\frac{\beta_1 r}{2}\right)(1-n)(Y_1k)+\frac{\beta_1}{2}(1-n)(Y_1r). \quad (3.17)$$

Also the PF equation (1.4) gives

$$S(Y_1, Df)=a_1(Y_1f)+b_1\eta(Y_1)(\varrho f). \quad (3.18)$$

Putting $Y_1=\varrho$ in Eqs. (3.17) and (3.18) and then comparing, we acquire

$$k(a_1-b_1)(\varrho f)=\alpha_1[(1-n)(\varrho a_1)-b_1\, div\varrho]$$
$$+\frac{\alpha_1}{k}[a_1(1-n)+b_1-b_1](\varrho k)$$
$$+\frac{1}{k}\left(\lambda_1-\frac{\beta_1 r}{2}\right)(1-n)(\varrho k)+\frac{\beta_1}{2}(1-n)(\varrho r). \quad (3.19)$$

Let $(\varrho k)=0$ and ϱ be Killing; therefore, we acquire (see, [30], p. 89) $\$_\varrho p=0$ and $\$_\varrho\sigma=0$. It is known that $a_1=\frac{k(p-\sigma)}{n-2}$ and $b_1=k(p+\sigma)$. Thus, we infer

$$(\varrho a_1)=(\varrho b_1)=0.$$

Again, from (2.2), we obtain

$$r=na_1-b_1.$$

Hence, we get $(\varrho r)=0$. Because the hypothesis ϱ is Killing, then div$\varrho=0$. Thus, using the foregoing result, Eq. (3.19) yields

$$k(a_1-b_1)(\varrho f)=0. \quad (3.20)$$

This reflects that either $a_1=b_1$ or $(\varrho f)=0$, since $k\neq 0$ on a PF space-time with the gradient k-almost RY soliton. Here, we consider the following two cases:

Case (i): Let $a_1=b_1$ and $(\varrho f)\neq 0$, and hence the Eq. (1.7) gives

$$p=-\frac{n-3}{n-1}\sigma,$$

which provides the EOS in a PF space-time equipped with a gradient k-almost RY soliton. For $n=4$, the EOS is $3p+\sigma=0$, which entails that the PF space-time represents phantom era.

Case (ii): Let $(\varrho f)=0$ and $a_1\neq b_1$. The covariant differentiation of $g(\varrho, Df)=0$ produces

$$g(\nabla_{X_1}\varrho, Df) = \left[\frac{\alpha_1}{k}(a_1 - b_1) + \frac{1}{k}\left(\lambda_1 - \frac{\beta_1 r}{2}\right)\right]\eta(X_1), \tag{3.21}$$

in which we have used Eqs. (2.1) and (3.10). Since here ϱ is Killing, we infer $g(\nabla_{X_1}\varrho, Y_1) + g(X_1, \nabla_{Y_1}\varrho) = 0$. Now putting $Y_1 = \varrho$ in the last relation, we acquire that $g(X_1, \nabla_\varrho \varrho) = 0$, since $g(\nabla_{X_1}\varrho, \varrho) = 0$. Therefore, we state that $\nabla_\varrho \varrho = 0$. Using the previous relation, putting $X_1 = \varrho$ in Eq. (3.21), we find that

$$\lambda_1 = \frac{\beta_1 r}{2} + \alpha_1(b_1 - a_1), \tag{3.22}$$

which shows that the k-almost RY soliton of gradient type in a PF space-time is expanding, steady, or shrinking if $r > \frac{2\alpha_1(a_1-b_1)}{\beta_1}$, $r = \frac{2\alpha_1(a_1-b_1)}{\beta_1}$ or $r < \frac{2\alpha_1(a_1-b_1)}{\beta_1}$, respectively. □

Remark 3.1 divϱ = 0 implies the space-time is expansion free. It may be mentioned that div$\varrho = 0$ plays a significant role in Raychaudhuri's equation.

Proof of the Theorem 1.3 Suppose the GRW space-time admits a k-almost RY soliton $(g, \varrho, k, \alpha_1, \lambda_1, \beta_1)$, and hence the Eq. (1.10) provides

$$k(\$_V g)(X_1, Y_1) + 2\alpha_1 S(X_1, Y_1) + (2\lambda_1 - \beta_1 r)g(X_1, Y_1) = 0, \tag{3.23}$$

which entails

$$\begin{aligned} &k\{g(\nabla_{X_1}\varrho, Y_1) + g(X_1, \nabla_{Y_1}\varrho)\} \\ &+2\alpha_1 S(X_1, Y_1) + (2\lambda_1 - \beta_1 r)g(X_1, Y_1) = 0. \end{aligned} \tag{3.24}$$

Using the Eq. (2.4) in Eq. (3.24), we provide

$$S(X_1, Y_1) = -\frac{1}{\alpha_1}\left\{(\lambda_1 - \frac{\beta_1 r}{2} + k\Psi\right\} g(X_1, Y_1) - \frac{k\Psi}{\alpha_1}\eta(X_1)\eta(Y_1), \tag{3.25}$$

which represents PF space-time.

Putting $X_1 = Y_1 = \varrho$ in the previous equation yields

$$\lambda_1 = -\alpha_1(n-1)\mu_1 + \frac{\beta_1 r}{2}.$$

Therefore, the soliton is shrinking for $r < \frac{2\alpha_1(1-n)\mu_1}{\beta_1}$, steady if $r = \frac{2\alpha_1(1-n)\mu_1}{\beta_1}$, and expanding for $r > \frac{2\alpha_1(1-n)\mu_1}{\beta_1}$. □

Proof of the Corollary 1.4. In [31], Mantica et al. proved that a GRW space-time becomes a PF space-time iff $(div\, C)(X_1, Y_1)Z_1 = 0$. Also, we know that in a GRW space-time, $C(X_1, Y_1)\varrho = 0$ iff $(div\, C)(X_1, Y_1)Z_1 = 0$. Also,

$C(X_1, Y_1)\varrho = 0$ tells us that the Weyl conformal curvature tensor is purely electric[32]. In four dimensions, the space-times are of Petrov types I, D, or O if C is purely electric ([33], p. 73).

For dimension 4, $C(Y_1, X_1)\varrho = 0$ is identical to ([34], p. 128)

$$\eta(U_1)C(Y_1, X_1, W_1, Z_1) + \eta(Y_1)C(X_1, U_1, W_1, Z_1) \\ +\eta(X_1)C(U_1, Y_1, W_1, Z_1) = 0, \tag{3.26}$$

in which $\eta(Y_1) = g(Y_1, \varrho)$ and $C(Y_1, X_1, W_1, Z_1) = g(C(Y_1, X_1)W_1, Z_1)$ for any Y_1, X_1, W_1, Z_1, U_1.

Now, replacing U_1 by ϱ yields

$$C(Y_1, X_1, W_1, Z_1) = 0, \tag{3.27}$$

from which we say that the space-time is conformally flat.

A GRW space-time has been found to be conformally flat iff it is a RW space-time [35].

Hence, the proof. □

In one specific instance, we get the following:

Corollary 3.2 *The GRW space-time allowing a k-almost Ricci soliton represents a PF space-time. Also, the soliton is steady if $\mu_1 = 0$, expanding for $\mu_1 < 0$, and shrinking for $\mu_1 > 0$.*

Proof In particular, if we take $\beta_1 = 0$ and $\alpha_1 = 1$, then the Eq. (3.25) entails

$$S(X_1, Y_1) = -\{\lambda_1 + k\Psi\}g(X_1, Y_1) - k\Psi\eta(X_1)\eta(Y_1), \tag{3.28}$$

which means that it is a PF space-time.

Putting $X_1 = Y_1 = \varrho$ in Eq. (3.28) yields

$$\lambda_1 = -(n-1)\mu_1.$$

Thus, the soliton is steady if $\mu_1 = 0$, expanding for $\mu_1 < 0$, and shrinking for $\mu_1 > 0$.

Hence the result follows.

□

Proof of the Theorem 1.5 Assume that GRW space-time allows a k-almost RY soliton of gradient type. Then, the Eq. (1.11) yields

$$k\nabla_{X_1} Df + \alpha_1 QX_1 = -\left(\lambda_1 - \frac{\beta_1 r}{2}\right) X_1. \tag{3.29}$$

Then the Eq. (3.14) tells that

$$\begin{aligned} kg(R(X_1, Y_1)Df, \varrho) = & -\alpha_1[g((\nabla_{X_1} Q)Y_1, \varrho) + g((\nabla_{Y_1} Q)X_1, \varrho)] \\ & +\frac{\alpha_1}{k}[(X_1 k)\eta(QY_1) - (Y_1 k)\eta(QX_1)] \\ & +\frac{1}{k}\left(\lambda_1 - \frac{\beta_1 r}{2}\right)[(X_1 k)\eta(Y_1) - (Y_1 k)\eta(X_1)] \\ & +\frac{\beta_1}{2}[(X_1 r)\eta(Y_1) - (Y_1 r)\eta(X_1)]. \end{aligned} \tag{3.30}$$

From Eq. (2.7), we get

$$Q\varrho = (n-1)\mu_1 \varrho. \tag{3.31}$$

Differentiating (3.31), we reach

$$\begin{aligned} (\nabla_{X_1} Q)\varrho = & (n-1)(X_1 \mu_1)\varrho \\ & +(n-1)\Psi\mu_1[X_1 + \eta(X_1)\varrho] \\ & -\Psi Q X_1 - (n-1)\Psi\mu_1\eta(X_1)\varrho. \end{aligned} \tag{3.32}$$

Using Eqs. (2.6) and (3.32) in Eq. (3.30), we obtain

$$\begin{aligned} k\mu_1[\eta(X_1)Y_1 f - \eta(Y_1)X_1 f] = & \alpha_1(n-1)[(Y_1\mu_1)\eta(X_1) - (X_1\mu_1)\eta(Y_1)] \\ & +\frac{\alpha_1}{k}[(X_1 k)\eta(QY_1) - (Y_1 k)\eta(QX_1)] \\ & +\frac{1}{k}\left(\lambda_1 - \frac{\beta_1 r}{2}\right)[(X_1 k)\eta(Y_1) - (Y_1 k)\eta(X_1)] \\ & +\frac{\beta_1}{2}[(X_1 r)\eta(Y_1) - (Y_1 r)\eta(X_1)]. \end{aligned} \tag{3.33}$$

Putting $Y_1 = \varrho$ in Eq. (3.33), we have

$$\begin{aligned} k\mu_1[X_1 f + (\varrho f)\eta(X_1)] = & \alpha_1(n-1)[(X_1\mu_1) + (\varrho\mu_1)\eta(X_1)] \\ & +\frac{\alpha_1}{k}[(X_1 k)\eta(Q\varrho) - (\varrho k)\eta(QX_1)] \\ & -\frac{1}{k}\left(\lambda_1 - \frac{\beta_1 r}{2}\right)[(X_1 k) + (\varrho k)\eta(X_1)] \\ & -\frac{\beta_1}{2}[(X_1 r) + (\varrho r)\eta(X_1)]. \end{aligned} \tag{3.34}$$

With the use of Lemma 2.3 and $(X_1 k) = 0$, the Eq. (3.34) entails that

$$k\mu_1[X_1 f + (\varrho f)\eta(X_1)] = -\frac{\beta_1}{2}[(X_1 r) + (\varrho r)\eta(X_1)]. \tag{3.35}$$

Let r be constant, and then from the Eq. (3.35), we obtain

$$k\mu_1[X_1 f + (\varrho f)\eta(X_1)] = 0, \tag{3.36}$$

which implies

$$X_1 f = -(\varrho f)\eta(X_1), \quad \text{since } \mu_1 \text{ and } k \neq 0. \tag{3.37}$$

The above equation reduces to

$$Df = -(\varrho f)\varrho, \tag{3.38}$$

which reflects

$$\nabla_{X_1} Df = -\{X_1(\varrho f)\}\varrho - \Psi(\varrho f)\{X_1 + \eta(X_1)\varrho\}. \tag{3.39}$$

The Eqs. (3.29) and (3.39) jointly entail

$$\begin{aligned}&\{X_1(\varrho f)\}\eta(Y_1) + \Psi(\varrho f)[g(X_1, Y_1) + \eta(X_1)\eta(Y_1)]\\ &= \frac{\alpha_1}{k} S(X_1, Y_1) + \frac{1}{k}\left(\lambda_1 - \frac{\beta_1 r}{2}\right) g(X_1, Y_1).\end{aligned} \tag{3.40}$$

Setting $Y_1 = \varrho$ in Eq. (3.40) yields

$$\{X_1(\varrho f)\} = -\left\{\frac{\alpha_1}{k}(n-1)\mu_1 + \frac{1}{k}\left(\lambda_1 - \frac{\beta_1 r}{2}\right)\right\}\eta(X_1). \tag{3.41}$$

With the help of the above two equations, we provide

$$\begin{aligned}S(X_1, Y_1) = &\frac{1}{\alpha_1}\left\{k\Psi(\varrho f) - \left(\lambda_1 - \frac{\beta_1 r}{2}\right)\right\} g(X_1, Y_1)\\ &+\frac{1}{\alpha_1}\left\{k\Psi(\varrho f) - \alpha_1(n-1)\mu_1 - \left(\lambda_1 - \frac{\beta_1 r}{2}\right)\right\}\eta(X_1)\eta(Y_1),\end{aligned} \tag{3.42}$$

which implies that it is a PF space-time.

This finishes our proof. □

Proof of the Corollary 1.6. The proof is the same as that of previous Corollary.

Corollary 3.3 *Let a GRW space-time admit a k-almost gradient-type Ricci soliton. Then the GRW space-time reduces to a PF space-time.*

Proof In particular, if $\beta_1 = 0$ and $\alpha_1 = 1$, then Eq. (3.35) implies

$$k\mu_1[X_1 f + (\varrho f)\eta(X_1)] = 0. \tag{3.43}$$

Using the aforementioned theorem's analogous calculations, we obtain

$$S(X_1, Y_1) = \{k\Psi(\varrho f) - \lambda_1\}g(X_1, Y_1) + \{k\Psi(\varrho f) - (n-1)\mu_1 - \lambda_1\}\eta(X_1)\eta(Y_1), \quad (3.44)$$

which represents PF space-time.
Therefore, the corollary follows. □

4 Discussions

The current stage of the physical world's predictive models is space-time. The proper EMT may be used in GR theory to estimate the Cosmos's matter content, which is acknowledged to act like a PF space-time in cosmological models. The simplest kind of fluid, which is incapable of transferring heat, is called a PF. A perfect fluid cannot resist a tangential force since it lacks viscosity. In GR, perfect fluids are used to simulate distributions of matter (idealized), like an isotropic universe or the inside of a star.

A wave packet known as a soliton or solitary wave keeps its form while moving at a steady speed. Gradient is commonly used in physics and mathematics to denote the direction and magnitude of a force acting on a particle or field. Gradients are often used in chemistry and engineering, among other fields, to characterize how a substance's or system's property varies in response to its location or other factors.

In this chapter, we determine the condition under which the k- almost RYSs and gradient k-almost RYSs are expanding, stable, or shrinking in a PF space-time. Also, we derive that if a GRW space-time admits a k-almost RYS, then the space-time represents a PF space-time. Also, if it allows a k-almost RYS of gradient type with $r =$ constant, then it represents a PF space-time.

In future, we or perhaps other researchers will look at the characteristics of various solitons in cosmological models and GR theory.

5 Declarations

5.1 Funding

NA.

5.2 Code Availability

NA.

5.3 Availability of Data

NA.

5.4 Conflicts of Interest

The authors confirm that they do not have any competing interests.

Acknowledgments The authors are thankful to the referees for their valuable suggestions toward the improvement of this book chapter.

References

1. Sánchez, M.: On the geometry of static space-times. Nonlinear Anal. Theory Methods Appl. **63**, 455–463 (2005)
2. Chen, B.Y.: Pseudo-Riemannian Geometry, δ-invariants and Applications. World Scientific, Singapore (2011)
3. Chen, B.Y.: A simple characterization of generalized Robertson-Walker space-times. Gen. Relativ. Gravit. **46**(1833), 5 (2014)
4. Alías, L., Romero, A., Sánchez, M.: Uniqueness of complete spacelike hypersurfaces of constant mean curvature in generalized Robertson-Walker space-times. Gen. Relativ. Gravit. **27**, 71–84 (1995)
5. Yano, K., Kon, M.: CR submanifolds of Kaehlerian and Sasakian manifolds. In: Progress in Mathematics, vol. 30. Birkhäuser, Boston (1983)
6. Duggal, K.L., Sharma, R.: Hypersurfaces in a conformally flat space with curvature collineation. Int. J. Math. Math. Sci. **14**, 595–604 (1991)
7. Deszcz, R., Verstraelen, L.: Hypersurfaces of semi-Riemannian conformally flat manifolds. In: Geometry and Topology of Submanifolds, III, pp. 131–147. World Science Publishing, River Edge (1991)
8. Chaki, M.C., Maity, R.K.: On quasi Einstein manifolds. Publ. Math. Debr. **57**, 297–306 (2000)
9. De, U.C., Ghosh, G.C.: On generalized quasi Einstein manifolds. Kyungpook Math. J. **44**, 607–615 (2004)
10. O'Neill, B.: Semi-Riemannian Geometry with Applications to Relativity. Academic Press, New York (1983)
11. Mantica, C.A., Molinari, L.G.: Generalized Robertson-Walker space-times-A survey. Int. J. Geom. Methods Mod. Phys. **14**, 1730001 (2017)
12. De, K., De, U.C., Velimirovic, L.: Some curvature properties of perfect fluid space-times. Quaest. Math. (2023). https://doi.org/10.2989/16073606.2023.2255933
13. De, K., De, U.C.: Perfect fluid space-times obeying certain restrictions on the energy-momentum tensor. Filomat **37**(11), 3483–3492 (2023)
14. De, U.C., Gezer, A.: A short note on generalized Robertson Walker spacetimes. Turk. J. Math. **48**(5), Article 11 (2024). https://doi.org/10.55730/1300-0098.3552
15. Chavanis, P.H.: Cosmology with a stiff matter era. Phys. Rev. D **92**, 103004 (2015)
16. Hamilton, R.S.: The Ricci flow on surfaces, mathematics and general relativity. Contemp. Math. **71**, 237–262 (1998)

17. Guler, S., Crasmareanu, M.: Ricci-Yamabe maps for Riemannian flow and their volume variation and volume entropy. Turk. J. Math. **43**, 2631–2641 (2019)
18. Venkatesh, V., Naik, D.M., Kumara, H.A.: $*$-Ricci solitons and gradient almost $*$-Ricci solitons on Kenmotsu manifolds. Math. Slovaca **69**, 1447–1458 (2019)
19. Blaga, A.M., Özgür, C.: Almost η-Ricci and almost η-Yamabe solitons with torse-forming potential vector field. Quaest. Math. **45**(1), 143–163 (2022)
20. De, U.C., Sardar, A., De, K.: Ricci-Yamabe solitons and 3-dimensional Riemannian manifolds. Turk. J. Math. **46**, 1078–1088 (2022)
21. Haseeb, A., Prasad, R., Mofarreh, F.: Sasakian Manifolds Admitting $*$-η-Ricci-Yamabe Solitons. Adv. Math. Phys. **2022**(5718736), 7 (2022). https://doi.org/10.1155/2022/5718736
22. De, K., De, U.C., Syied, A.A., Turki, N.B., Alsaeed, S.: Perfect fluid spacetimes and gradient solitons. J. Nonlinear Math. Phys. **29**, 843–858 (2022)
23. De, U.C., Mantica, C.A., Suh, Y.J.: Perfect fluid spacetimes and gradient solitons. Filomat **36**, 829–842 (2022)
24. Blaga, A.M.: Solitons and geometrical structures in a perfect fluid space-time. Rocky Mt. J. Math. **50**, 41-53 (2020)
25. De, U.C., Chaubey, S.K., Shenawy, S.: Perfect fluid space-times and Yamabe solitons. J. Math. Phys. **62**, 032501 (2021)
26. De, K., De, U.C., Gezer, A.: Perfect fluid space-times and k-almost Yamabe solitons. Turk. J. Math. **47**, 1236–1246 (2023)
27. De, K., De, U.C.: Investigation on gradient solitons in perfect fluid space-times. Rep. Math. Phys. **91**, 277–289 (2023)
28. Siddiqi, M.D., De, U.C.: Relativistic perfect fluid space-times and Ricci-Yamabe solitons. Lett. Math. Phys. **112**, 1 (2022). https://doi.org/10.1007/s11005-021-01493-z
29. De, K., Khan, M.N., De, U.C.: Characterizations of GRW space-times concerning gradient solitons. Heliyon (2024). https://doi.org/10.1016/j.heliyon.2024.e25702
30. Duggal, K.L., Sharma, R.: Symmetries of space-times and Riemannian manifolds. Kluwer Academic Press, Boston, London (1999)
31. Mantica, C.A., De, U.C., Suh, Y.J., Molinari, L.G.: Perfect fluid spacetimes with harmonic generalized curvature tensor. Osaka J.Math. **56**, 173–182 (2019)
32. Hervik, S., Ortaggio, M., Wylleman, L.: Minimal tensors and purely electric or magnetic space-times of arbitrary dimension. Class. Quantum Grav. **30**, 165014 (2013)
33. Stephani, H., Kramer, D., Mac-Callum, M., Hoenselaers, C., Herlt, E.: Exact Solutions of Einstein's Field Equations. Cambridge University Press, Cambridge (2009)
34. Lovelock, D., Rund, H.: Tensors, Differential Forms, and Variational Principles. Courier Corporation, North Chelmsford (1989)
35. Brozos-Vazquez, M., Garcia-Rio, E., Vazquez-Lorenzo, R.: Some remarks on locally conformally flat static space–times. J. Math. Phys. **46**, 022501 (2005)

Riemannian Concircular Structure Manifolds and Solitons

Sudhakar Kumar Chaubey and Abdul Haseeb

Abstract Ricci flows are used as a powerful tool to address several problems of information technology, engineering, medical science, and other allied areas. A self-similar solution of the Ricci flow on a Riemannian manifold is named as the Ricci soliton. The Ricci soliton becomes an almost Ricci soliton if we think of the soliton constant as a smooth function in the Ricci soliton equation. This chapter explores the properties of almost Ricci solitons within the framework of Riemannian concircular structure manifolds (briefly, $(RCS)_n$-manifolds). We establish the conditions for which the $(RCS)_n$-manifolds to be quasi-Einstein manifolds, and the solitons are expanding, shrinking, and steady. We also provide the restrictions for the soliton function of almost Ricci solitons to be harmonic, strictly super-harmonic, and strictly subharmonic. The existence of projectively semisymmetric $(RCS)_n$-manifolds is ensured. Some geometrical properties of $(RCS)_n$-manifolds satisfying $Q \cdot \mathcal{P} = 0$ are investigated, and the definition of extended Ricci recurrent manifolds is encoded.

Keywords Riemannian manifolds · $(RCS)_n$-manifolds · Curvature tensors · Symmetric spaces · Torse-forming vector field · Concircular vector field · Generalized soliton

S. K. Chaubey (✉)
Unit of Sciences and Mathematics, DSR, University of Technology and Applied Sciences, Shinas, OMAN
e-mail: sudhakar.chaubey@utas.edu.om

A. Haseeb
Department of Mathematics, College of Science, Jazan University, Jazan, Kingdom of Saudi Arabia
e-mail: haseeb@jazanu.edu.sa; malikhaseeb80@gmail.com

B.-Y. Chen, M. A. Choudhary (eds.), *Geometric Inequalities and Applications*,
Infosys Science Foundation Series in Mathematical Sciences,
https://doi.org/10.1007/978-981-95-5148-4_7

1 Introduction

Let (M, g) be a Riemannian manifold of dimension n. Vector fields play a significant role in understanding the geometry and topology of Riemannian manifolds, and they are indeed central to various aspects of differential geometry and theoretical physics. For example:

- Torse-forming vector fields: These vector fields generate torsion in the manifold, and they are important in the study of connections and the curvature of the manifold.
- Torqued vector fields: Related to the concept of torque, these vector fields influence how the manifold might twist or rotate, and they have applications in studying dynamics, fluid flows, and electromagnetism on manifolds.
- Concircular vector fields: These fields preserve the shape of a curve under parallel transport, meaning they maintain the geometry of paths on the manifold and have applications in understanding symmetries and geodesics.
- Recurrent vector fields: These fields are of interest in dynamical systems, where they correspond to vector fields whose flow returns to its initial configuration after a certain time, and they are essential in studying periodic or quasiperiodic phenomena.
- Parallel vector fields: These are vector fields that are "constant" in some sense, meaning that they remain unchanged under parallel transport along any curve on the manifold. They are related to the concept of curvature and can help characterize the manifold's geometry.

These vector fields are used to explore and understand the curvature, topology, and other intrinsic properties of Riemannian manifolds. They also show up in the study of Einstein's theory of general relativity, fluid dynamics, and even string theory, where the geometry of space-time is often modeled as a Riemannian manifold. The behavior and classification of these vector fields allow mathematicians and physicists to draw significant conclusions about the manifold's global properties.

Riemannian manifolds admitting concircular, special concircular, non-isometric concircular, non-isometric conformal, and non-affine projective vector fields were classified by Tashiro [68] in 1963. The features of Riemannian manifolds with exterior concurrent vector fields and torse-forming were investigated by Mihai and Mihai [55]. Chen [25] established several results of Ricci solitons and investigated the properties of concircular vector fields. In [26] and [27], he examined the characteristics of torqued and parallel vector fields in collaboration with his coauthors. Various authors have conducted studies on Riemannian manifolds featuring distinct vector fields. For example, we cite [5, 19, 33, 37, 45, 48, 49, 53, 56, 68, 70, 71] along with their respective references. In [17], Chaubey and Suh have started their study by considering n-dimensional Riemannian manifolds admitting a torse-forming vector field and introduced the notion of Riemannian concircular structure (briefly, $(RCS)_n$-manifolds). They have validated the existence of such structures by proving nontrivial examples. They also proved that the $(RCS)_n$-manifolds are integrable and established some curvature identities. Several authors have explored the properties of Riemannian concircular structure manifolds by considering the different values of potential function α (see Eq. 2.4). For instance, the authors of [4, 10, 11, 14, 15, 29, 44, 47] have assumed $\alpha = -1$ and investigated their findings. This chapter explores the properties of $(RCS)_n$-manifolds if the Riemannian metric is an almost Ricci soliton.

The exploration of symmetric spaces represents a compelling and significant area within the realm of differential geometry. Semisymmetric space $(R(U, V) \cdot R = 0)$ [67] is the generalization of locally symmetric space $(\nabla R = 0)$, and it has been studied by several geometers. Here R denotes the non-vanishing curvature tensor of the Riemannian manifold M, ∇ is the Levi-Civita connection of the Riemannian metric g, and $R(U, V)$ acts as a derivation on R for all vector fields U and V on M. A Riemannian manifold M is referred to as a Ricci semisymmetric if its non-vanishing Ricci tensor S satisfies the curvature condition $R(U, V) \cdot S = 0$. Note that while the class of Ricci symmetric manifold $(\nabla S = 0)$ includes the class of Ricci semisymmetric manifold $(R(U, V) \cdot S = 0)$, the converse is generally not true. It is well known that every semisymmetric manifold is Ricci semisymmetric, but its converse part is not true (in general). Numerous authors have examined the characteristics of symmetric spaces in Riemannian and semi-Riemannian settings, including locally symmetric, semisymmetric, Ricci semisymmetric, and others. For further information, see [2, 3, 21, 51, 56, 66], and their references.

As a generalization of the Kähler-Einstein metric, Koiso [50] introduced the concept of quasi-Einstein metric on Fano manifolds in 1987. Numerous scholars have examined the characteristics of quasi-Einstein metrics. Chaki and Maity [6] investigate the characteristics of the Ricci tensor S of an n-dimensional Riemannian manifold M that satisfies the following relation after being inspired by the work by Chave and Valent [23] on quasi-Einstein metrics.

$$S = b\eta \otimes \eta + ag. \tag{1.1}$$

Here a and b are nonzero smooth functions on M, and η is a nonzero 1-form associated with the vector field ξ, that is, $g(\cdot, \xi) = \eta(\cdot)$. M is referred to as a quasi-Einstein manifold (briefly, $(QE)_n$-manifold) [6] if the non-vanishing Ricci tensor S of M satisfies Eq. (1.1). In the general theory of relativity, the space-times satisfy Eq. (1.1) and is termed as perfect fluid space-times provided that vector field ξ is a unit timelike vector field, that is, $g(\xi, \xi) = -1$ (see [7, 9, 18, 30, 64, 65]). Specifically, M with $b = 0$ recovers the Einstein manifold, while $a = b = 0$ recovers the Ricci-flat manifold. Chaki and Maity examined the following presumptions in [6]:

- a and b are constants, and the generator of $(QE)_n$-manifold is recurrent
- $a + b = 0$, $W = \frac{1}{2a} grad\, a$, and $\nabla_U W = -U + A(U)W$, where A is a dual 1-form of W and U is an arbitrary vector field of M

for $(QE)_n$-manifolds, and they proved that in both the cases $(QE)_n$-manifolds are conformally conservative. In [17], the authors have generalized results of Chaki and Maity's work [6]. They gave a clue to evaluate the smooth functions a and b on M.

In 1822, Joseph Fourier introduced the concept of a heat flow equation, which is a nonlinear partial differential equation. In 1964, Eells and Sampson [36] introduced a similar nonlinear variant of the heat flow equation, known as harmonic map heat flow. This inspired Hamilton to introduce the concept of Ricci flow in 1982 [39, 40]. A Ricci flow is defined by the equation: $\frac{\partial g}{\partial t} = -2Ric,\ g(0) = g_0$. In this nonlinear partial differential equation, the variables g, t, and Ric represent the Riemannian metric, time, and Ricci tensor of the Riemannian n-manifold, respectively. In 2002 and 2003, respectively, Perelman used the Ricci flow to solve the Poincaré Conjecture (one of the millennium problems) and the Geometrization Conjecture [58–60]. Using the Ricci flow, M. T. Anderson [54] provided the geometrization of 3-manifolds. Note that a number of long-standing, unresolved issues in mathematics, physics, medicine, engineering, and technology have been addressed with the help of the Ricci flow. A Ricci soliton is a self-similar solution of the Ricci flow. In order to address a number of problems in the mathematical sciences and related fields, the Ricci soliton has been employed. The equation

$$\frac{1}{2}\pounds_V g + S + \lambda g = 0 \tag{1.2}$$

represents the Ricci soliton equation on an n-dimensional Riemannian manifold M, where $\pounds$ is the Lie derivative operator of the Riemannian metric g. S is the Ricci tensor, V is a soliton vector, and λ is a soliton constant. The Ricci soliton is represented by the symbol (g, V, λ). The Ricci soliton equation (1.2) becomes an almost Ricci soliton if we select λ as a smooth function on M. If λ is positive, negative, or zero, then an almost Ricci soliton (g, V, λ) is expanding, shrinking, or steady. For more detailed information on solitons, we refer to [1, 13, 22, 34, 35, 41–43, 46, 57, 62].

The following is how we set up our work. We define the Riemannian concircular structure manifold and list some of its fundamental characteristics in Sect. 2. Section 3 deals with the study of $(RCS)_n$-manifolds admitting almost Ricci solitons. Section 4 ensures the existence of projectively semisymmetric $(RCS)_n$-manifolds admitting almost Ricci solitons. Next, Sect. 5 deals with the study of $(RCS)_n$-manifolds admitting Ricci solitons and satisfying the condition $\mathcal{Q} \cdot \mathcal{P} = 0$. Here $\mathcal{Q}$ and $\mathcal{P}$ denote the Ricci operator and projective curvature tensor of M.

2 Riemannian Manifolds and Torse-Forming Vector Field

In this section, we consider the Riemannian manifolds endowed with a torse-forming vector field and encode the basic results of Riemannian concircular structure manifolds ($(RCS)_n$-manifolds).

Yano [71] introduced the idea of a torse-forming vector field on Riemannian spaces, and numerous scholars have examined its characteristics in Riemannian and semi-Riemannian settings (see [5, 53, 55, 56]). In [71], he states that a smooth vector field ξ defined on M is a torse-forming vector field if

$$(\nabla_U \eta)(V) = \alpha g(U, V) + \pi(U)\eta(V), \ \forall \ U, V \in \mathfrak{X}(M), \tag{2.1}$$

where $\eta(\cdot) = g(\cdot, \xi)$ is a 1-form associated with ξ and π is a 1-form. Here $\mathfrak{X}(M)$ is the collection of all smooth vector fields of M. If 1-form π is closed on M, then ξ is said to be a concircular vector field [37, 70]. The torse-forming vector field ξ on M, in particular, reduces to the:

* Torqued vector field [26] if $\pi(\xi) = 0$
* Concircular vector field (in Fialkow's sense) [25, 37] if $\pi = 0$
* Concircular vector field (in Yano's sense) [70] if the 1-form π is closed
* Recurrent vector field [61] if $\alpha = 0$
* Concurrent vector field [27] if $\pi = 0$, and $\alpha = 1$
* Parallel vector field [27, 37] if $\pi = 0$, and $\alpha = 0$

Researchers are drawn to the study of geometric structures using these vectors because they can address a number of scientific and technological issues, particularly because they play a unique role in geometry and physics.

These vectors are capable to address several issues of science and technology, especially they play a peculiar role in geometry and physics, and therefore the study of geometric structures with these vectors attracts researchers. We categorize Riemannian manifolds with concircular vector fields (in Yano's sense) in this chapter. The geometrical and physical properties of Lorentzian manifolds endowed with concircular vector fields (in Yano's sense) have been explored by Mantica and Molinari [52] and then proved that the Lorentzian manifolds are the generalized Robertson-Walker space-times. For instance, we refer to [5, 8, 9, 12, 16, 18, 21, 28, 30–32, 53, 63–65, 69].

Assume that M admits a unit torse-forming vector field ξ, that is, $g(\xi, \xi) = 1 \implies g(\nabla_U \xi, \xi) = 0$. Using Eq. (2.1) with $V = \xi$, we discover

$$\alpha\eta(U) + \pi(U) = 0, \tag{2.2}$$

since $g(\xi, \xi) = \eta(\xi) = 1$ and $g(\nabla_U \eta)(\xi) = 0$. In Eq. (2.1), we apply Eq. (2.2) to get

$$(\nabla_U \eta)(V) = \alpha\{g(U, V) - \eta(U)\eta(V)\}, \tag{2.3}$$

which implies that

$$\nabla_U \xi = \alpha\{U - \eta(U)\xi\}. \tag{2.4}$$

Here α is a nonzero scalar, and for some smooth function μ on M, $\nabla_U \alpha = g(U, D\alpha) = U(\alpha) = \mu\eta(U)$. The gradient operator of g in this case is D. It is clear from Eq. (2.1) that the 1-form η is closed. By taking the covariant derivative of Eq. (5) along V and applying Eq. (2.1) and the fact that $U(\alpha) = \mu\eta(U)$, we can also conclude that π is closed. According to Yano, the unit torse-forming vector field ξ defined in (2.1) is a unit concircular vector field on M. The smooth function α on M is the potential function of the concircular vector field. Equation $U(\alpha) = \mu\eta(U)$ gives that $\xi(\alpha) = \mu \implies \xi(\xi(\alpha)) = \xi(\mu)$. Again $U(\alpha) = \mu\eta(U)$ infers that $D\alpha = \mu\xi$. Along U, the covariant derivative of $D\alpha = \mu\xi$ yields

$$\nabla_U D\alpha = U(\mu)\xi + \mu\alpha(U - \eta(U)\xi).$$

Examining an orthonormal frame field on M, we can then contract the aforementioned equation over U to arrive at

$$\Delta\alpha = \xi(\xi(\alpha)) + \alpha(n-1)\xi(\alpha),$$

where Δ represents the Laplace operator of g. A smooth function Ψ on M is regarded as harmonic if and only if $\Delta\Psi = 0$. Assuming that $\xi = \frac{\partial}{\partial t}$ on M, the equation above takes the following form:

$$\Delta\alpha = \frac{\partial}{\partial t}\left(\frac{\partial\alpha}{\partial t} + \frac{n-1}{2}\alpha^2\right). \tag{2.5}$$

Thus, we state the following result:

Lemma 2.1 ([17]) *The partial differential equation (2.5) is satisfied by the potential function α of ξ if a unit concircular vector field ξ is admitted by an n-dimensional Riemannian manifold.*

Equation (2.5) also allows us to say:

Lemma 2.2 ([17]) *On an n-dimensional Riemannian manifold endowed with a unit concircular vector field ξ, the potential function α of ξ is harmonic if and only if $\frac{\partial\alpha}{\partial t} + \frac{n-1}{2}\alpha^2 = constant$.*

Let a $(1, 1)$ tensor field ϕ be admitted to the Riemannian manifold M such that

$$\alpha\phi U = \nabla_U \xi, \quad \alpha \neq 0,$$

which provides

$$\phi U = U - \eta(U)\xi, \tag{2.6}$$

where (2.4) is applied. Following (2.6) and $\eta(\xi) = 1$ and operating ϕ on either side of Eq. (2.6), we get

$$\phi^2 = I - \eta \otimes \xi,$$

where I stands for identity transformation and $\otimes$ denotes the tensor product on M. In view of Eq. (2.6), we have

$$g(\phi U, \phi V) = g(U, V) - \eta(U)\eta(V), \quad \eta(V) = g(V, \xi).$$

Remark that $g(\phi U, \phi V) = g(\phi U, V) = g(U, \phi V)$, $\forall\, U, V \in \mathfrak{X}(M)$. Therefore, we deduce that if M admits a 1-form η, a (1, 1) tensor field ϕ, and a unit concircular vector field ξ, then we have

$$\phi^2 = I - \eta \otimes \xi, \quad \eta(\xi) = 1, \quad g(\phi\, \cdot, \phi\, \cdot) = g(\cdot\,, \cdot) - \eta \otimes \eta. \tag{2.7}$$

The authors of [17] provided the following definition after taking into account all of the aforementioned facts.

Definition 2.3 Assume that the data (ϕ, ξ, η, g) on an n-dimensional Riemannian manifold M satisfies (2.7). A Riemannian concircular structure manifold, or $(RCS)_n$-manifold, is then defined as M equipped with (ϕ, ξ, η, g). On M, the structure (ϕ, ξ, η, g) is defined as a Riemannian concircular structure.

A few fundamental $(RCS)_n$-manifold results are encoded as follows:

Proposition 2.4 ([17]) An n-dimensional $(RCS)_n$-manifold satisfies:

(i) $\phi\xi = 0$,
(ii) $\eta(\phi U) = 0$,
(iii) $\operatorname{rank}(\phi) = n - 1$,
(iv) $(\nabla_U \phi)(V) = \alpha[2\eta(U)\eta(V)\xi - g(U, V)\xi - \eta(V)U]$, $\forall\, U, V \in \mathfrak{X}(M)$.

Proposition 2.5 ([17]) In an $(RCS)_n$-manifold, we have:

(i) $R(U, V)\xi = (\alpha^2 + \mu)\{\eta(U)V - \eta(V)U\}$,
(ii) $R(\xi, U)V = (\alpha^2 + \mu)\{\eta(V)U - g(U, V)\xi\}$,
(iii) $\eta(R(U, V)W) = (\alpha^2 + \mu)\{\eta(V)g(U, W) - \eta(U)g(V, W)\}$,
(iv) $S(U, \xi) = -(n-1)(\alpha^2 + \mu)\eta(U) \iff Q\xi = -(n-1)(\alpha^2 + \mu)\xi$,
for all $U, V, W \in \mathfrak{X}(M)$, and $(\alpha^2 + \mu) \neq 0$.

Proposition 2.6 ([17]) In an $(RCS)_n$-manifold, we have:

(i) $'R(U, V, \phi W, Z) - {}'R(U, V, W, \phi Z) = (\alpha^2 + \mu)\{\eta(W)[\eta(V)g(U, Z) - \eta(U)g(V, Z)] + \eta(Z)[\eta(V)g(U, W) - \eta(U)g(V, W)]\}$,
(ii) $'R(U, V, \phi W, \phi Z) = {}'R(\phi U, \phi V, W, Z)$,
(iii) $'R(\phi U, \phi V, \phi W, \phi Z) = {}'R(U, V, W, Z) - (\alpha^2 + \mu)\{\eta(Z)[\eta(U)g(V, W) - \eta(V)g(U, W)] + \eta(W)[\eta(V)g(U, Z) - \eta(U)g(V, Z)]\}$,
(iv) $'R(\phi U, V, W, \phi Z) - {}'R(U, \phi V, \phi W, Z) = (\alpha^2 + \mu)\{\eta(U)\eta(W)g(V, Z) - \eta(V)\eta(W)g(U, Z)\}$,
(v) $S(\phi U, \phi V) = S(U, V) + (n-1)(\alpha^2 + \mu)\eta(U)\eta(V)$, for all $U, V, W, Z \in \mathfrak{X}(M)$, here $'R(U, V, W, Z) = g(R(U, V)W, Z)$,

3 $(RCS)_n$-Manifolds Admitting Almost Ricci Solitons

Let us consider that $(RCS)_n$-manifolds M admit an almost Ricci soliton (g, ξ, λ). Then by the equation of almost Ricci soliton (1.2), we have

$$(\pounds_\xi g)(U, V) + 2S(U, V) + 2\lambda g(U, V) = 0 \tag{3.1}$$

for arbitrary vector fields U and V on M. The definition of Lie derivative together with Eq. (2.4) assumes the following form:

$$(\pounds_\xi g)(U, V) = g(\nabla_U \xi, V) + g(U, \nabla_V \xi) = 2\alpha(g(U, V) - \eta(U)\eta(V)). \tag{3.2}$$

By making use of (3.2) in (3.1), we find

$$S(U, V) = -(\lambda + \alpha)g(U, V) + \alpha\eta(U)\eta(V), \tag{3.3}$$

which infers with the help of (1.1) that M endowed with (g, ξ, λ) is a quasi-Einstein manifold. The geometrical and physical properties of quasi-Einstein manifold have been studied by many researchers. In the general theory of relativity, the space-time possesses a non-vanishing Ricci tensor S that satisfies Eq. (3.3) and is termed as a perfect fluid space-time provided $g(\xi, \xi) = -1$.

Let $\{e_1, e_2, \cdots, e_n = \xi\}$ be a set of orthonormal vector fields $e_1, e_2, \cdots, e_n = \xi$ on M. Then the contraction of Eq. (3.3) over the vector fields U and V gives

$$r = -\lambda n + (1 - n)\alpha,$$

where $r = \sum_{i=1}^{n} S(e_i, e_i)$ is a scalar curvature of M. Next, taking $V = \xi$ in (3.3), we find

$$S(U, \xi) = -\lambda\eta(U) \Longleftrightarrow Q\xi = -\lambda\xi, \tag{3.4}$$

where Q is the Ricci operator of g corresponding to the Ricci tensor S such that $S(\cdot, \cdot) = g(Q\cdot, \cdot)$. Equation (3.4) shows that $-\lambda$ is the eigenvalue of Q corresponding to the eigenvector ξ.

From (3.4) and Proposition (2.5) (iv), we obtain

$$\lambda = (\alpha^2 + \mu)(n - 1), \tag{3.5}$$

which reduces to

$$\lambda = (n - 1)\left(\alpha^2 + \frac{\partial\alpha}{\partial t}\right),$$

where $\xi = \frac{\partial\alpha}{\partial t}$ and $\mu = \xi(\alpha)$ are used. Hence, we state the following:

Theorem 3.1 *Let M be an $(RCS)_n$-manifold admitting an almost Ricci soliton (g, ξ, λ). Then M is a quasi-Einstein manifold, and the soliton function λ is given by $\lambda = (n-1)(\alpha^2 + \frac{\partial\alpha}{\partial t})$.*

It is well known that an almost Ricci soliton (g, ξ, λ) on M is shrinking, expanding, and steady provided that λ is negative, positive, and zero, respectively. These facts together with Theorem 3.1 observe the following:

Corollary 3.2 *An almost Ricci soliton (g, ξ, λ) on $(RCS)_n$-manifolds is shrinking or expanding if $\frac{\partial\alpha}{\partial t} < -\alpha^2$ or $\frac{\partial\alpha}{\partial t} > -\alpha^2$, respectively.*

In consequence of Proposition 2.2 and Theorem 3.1, we have the following corollary:

Corollary 3.3 *An almost Ricci soliton (g, ξ, λ) on $(RCS)_n$-manifolds reduces to an expanding Ricci soliton (g, ξ, λ) if the potential function α of the concircular vector field ξ is a nonzero constant.*

Let us suppose that α is a smooth function on M. Then from Eq. (3.5), we have

$$D\lambda = (n-1)(2\alpha\mu + \sigma)\xi,$$

since $\mu = \xi(\alpha)$. The covariant derivative of the above equation along X gives

$$\nabla_X D\lambda = (n-1)\{X(2\alpha\mu + \sigma)\xi + (2\alpha\mu + \sigma)\nabla_X \xi\},$$

which, after contraction over X, gives

$$\Delta\lambda = (n-1)\{\xi(2\alpha\mu + \sigma) + (n-1)\alpha(2\alpha\mu + \sigma)\}.$$

A smooth function $\mathfrak{F}$ on a Riemannian manifold M of $dimM = n \geq 3$ is, respectively, named as harmonic, strictly super-harmonic, and strictly subharmonic if $\Delta\mathfrak{F} = 0$, $\Delta\mathfrak{F} < 0$, and $\Delta\mathfrak{F} > 0$.

Let $\xi = \frac{\partial}{\partial t}$. Therefore, the above definitions together with Eq. (3.6) reveal the following:

Corollary 3.4 *Let an $(RCS)_n$-manifold admit the almost Ricci soliton (g, ξ, λ). Then the soliton function λ of (g, ξ, λ) is:*

(i) Harmonic if $\frac{\partial}{\partial t}(2\alpha\mu + \sigma) + (n-1)\alpha(2\alpha\mu + \sigma) = 0$,
(ii) Strictly super-harmonic if $\frac{\partial}{\partial t}(2\alpha\mu + \sigma) < (1-n)\alpha(2\alpha\mu + \sigma)$,
(iii) Strictly subharmonic if $\frac{\partial}{\partial t}(2\alpha\mu + \sigma) > (1-n)\alpha(2\alpha\mu + \sigma)$.

Corollary 3.5 *Let an $(RCS)_n$-manifold admit the almost Ricci soliton (g, ξ, λ). If the potential function α of the concircular vector field ξ is a nonzero constant, then the soliton function λ is harmonic.*

Let an $(RCS)_n$-manifold M $(n > 3)$ admit an almost Ricci soliton (g, ξ, λ). Then M satisfies Eq. (3.3). Taking the covariant derivative of (3.3) along the vector field

X, we find

$$(\nabla_X S)(U, V) = -X(\lambda + \alpha)g(U, V) + X(\alpha)\eta(U)\eta(V) \\ +\alpha^2\{g(X, U)\eta(V) - 2\eta(U)\eta(V)\eta(X) + \eta(U)g(X, V)\},$$

where Eq. (2.1) is used.

$$(\nabla_U S)(X, V) = -U(\lambda + \alpha)g(X, V) + U(\alpha)\eta(X)\eta(V) \\ +\alpha^2\{g(X, U)\eta(V) - 2\eta(X)\eta(V)\eta(U) + \eta(X)g(U, V)\}.$$

The last two equations give

$$(\nabla_X S)(U, V) - (\nabla_U S)(X, V) = -U(\lambda + \alpha)g(X, V) - X(\lambda + \alpha)g(U, V) \\ +X(\alpha)\eta(U)\eta(V) - U(\alpha)\eta(X)\eta(V) + \alpha^2\{g(X, V)\eta(U) - \eta(X)g(U, V)\} \quad (3.6)$$

Contracting the above equation over U and V, we find

$$X(r) = -2(n-2)X(\alpha) - 2(n-1)X(\lambda) - \xi(\lambda)\eta(X) - (n-1)\alpha^2\eta(X). \quad (3.7)$$

Again, contraction of Eq. (3.3) over U and V gives

$$r = -\lambda n - (n-1)\alpha,$$

which becomes

$$X(r) = -nX(\lambda) - (n-1)X(\alpha). \quad (3.8)$$

In consequence of Eqs. (3.7) and (3.8) we have

$$(n-3)X(\alpha) + (n-2)X(\lambda) = -[\xi(\lambda) + (n-1)\alpha^2]\eta(X). \quad (3.9)$$

From Eq. (3.5) we obtain

$$X(\lambda) = (n-1)[2\alpha\mu + \sigma]\eta(X), \quad (3.10)$$

since $X(\alpha) = \mu\eta(X)$ and $X(\mu) = \sigma\eta(X)$.

In view of Eqs. (3.9) and (3.10), we lead to

$$(n-3)\mu + (n-1)[(n-1)(2\mu\alpha + \sigma) + \alpha^2] = 0.$$

Thus, we can state the following:

Theorem 3.6 *Let an $(RCS)_n$-manifold M $(n > 3)$ admit an almost Ricci soliton (g, ξ, λ). Then the functions α, μ, and σ on M satisfy the relation*

$$(n-3)\mu + (n-1)[(n-1)(2\mu\alpha + \sigma) + \alpha^2] = 0.$$

4 Projectively Semisymmetric $(RCS)_n$-Manifolds

Let M be an n-dimensional Riemannian concircular structure manifold admitting an almost Ricci soliton (g, ξ, λ). Suppose $\mathcal{P}$ denotes the projective curvature tensor on M; then it can be expressed as

$$\mathcal{P}(U, V)W = R(U, V)W - \frac{1}{n-1}(S(V, W)U - S(U, W)V), \tag{4.1}$$

where $dimM = n$.

A Riemannian manifold M is said to be projectively semisymmetric if and only if

$$R(X, Y) \cdot \mathcal{P} = 0, \tag{4.2}$$

for arbitrary vector fields X, Y on M. Equation (4.2) is equivalent to

$$\begin{aligned} &R(X, Y)\mathcal{P}(U, V)W - \mathcal{P}(R(X, Y)U, V)W \\ &- \mathcal{P}(U, R(X, Y)V)W - \mathcal{P}(U, V)R(X, Y)W = 0, \end{aligned}$$

which becomes

$$\begin{aligned} &R(\xi, X)\mathcal{P}(U, V)W - \mathcal{P}(R(\xi, X)U, V)W \\ &- \mathcal{P}(U, R(\xi, X)V)W - \mathcal{P}(U, V)R(\xi, X)W = 0, \end{aligned} \tag{4.3}$$

since $X = \xi$ is used.

Let M be an $(RCS)_n$-manifold. Then in view of Proposition (2.5)(ii), Eq. (4.3) takes the form

$$\begin{aligned} &(\alpha^2 + \mu)[\eta(\mathcal{P}(U, V)W)X - g(X, \mathcal{P}(U, V)W)\xi - \eta(U)\mathcal{P}(X, V)W \\ &+g(X, U)\mathcal{P}(\xi, V)W - \eta(V)\mathcal{P}(U, X)W + g(X, V)\mathcal{P}(U, \xi)W \\ &-\eta(W)\mathcal{P}(U, V)X + g(X, W)\mathcal{P}(U, V)\xi] = 0, \end{aligned} \tag{4.4}$$

where $\alpha^2 + \mu \neq 0$. Taking the inner product of (4.4) with ξ, we have

$$\begin{aligned} &(\alpha^2 + \mu)[\eta(\mathcal{P}(U, V)W)\eta(X) - g(X, \mathcal{P}(U, V)W) - \eta(U)\eta(\mathcal{P}(X, V)W) \\ &+g(X, U)\eta(\mathcal{P}(\xi, V)W) - \eta(V)\eta(\mathcal{P}(U, X)W) + g(X, V)\eta(\mathcal{P}(U, \xi)W) \\ &-\eta(W)\eta(\mathcal{P}(U, V)X) + g(X, W)\eta(\mathcal{P}(U, V)\xi)] = 0. \end{aligned} \tag{4.5}$$

From (4.1), we find

$$\eta(\mathcal{P}(U,V)W) = \left(\alpha^2 + \mu - \frac{\lambda+\alpha}{n-1}\right)(g(U,W)\eta(V) - g(V,W)\eta(U)), \quad (4.6)$$

$$\eta(\mathcal{P}(\xi,V)W) = -\left(\alpha^2 + \mu - \frac{\lambda+\alpha}{n-1}\right)(g(V,W) - \eta(V)\eta(W)), \quad (4.7)$$

$$\eta(\mathcal{P}(U,V)\xi) = 0. \quad (4.8)$$

In view of (4.6)–(4.8), Eq. (4.5) reduces to

$$g(X,\mathcal{P}(U,V)W) = \left(\alpha^2 + \mu - \frac{\lambda+\alpha}{n-1}\right)(g(U,W)g(X,V) - g(X,U)g(V,W)), \quad (4.9)$$

where $\alpha^2 + \mu \neq 0$.

In view of (3.3) and (4.1), Eq. (4.9) takes the form

$$g(R(U,V)W,X) = (\alpha^2+\mu)[g(U,W)g(X,V) - g(X,U)g(V,W)] - \frac{\alpha}{n-1}(g(X,V)\eta(U)\eta(W) - g(X,U)\eta(V)\eta(W)), \quad (4.10)$$

which is equivalent to

$$R(U,V)W = f_1[g(U,W)V - g(V,W)U] + f_2(\eta(U)V - \eta(V)U)\eta(W), \quad (4.11)$$

where $f_1 = \alpha^2 + \mu$ and $f_2 = -\frac{\alpha}{n-1}$.

A Riemannian manifold M of dimension n is said to be a manifold of quasi-constant curvature [24] if the curvature tensor R of M satisfies

$$R(X,Y)Z = a\{g(Y,Z)X - g(X,Z)Y\} + b\{g(Y,Z)\eta(X)\xi - g(X,Z)\eta(Y)\xi + \eta(Z)\eta(Y)X - \eta(Z)\eta(X)Y\},$$

for some smooth functions a and b on M. Motivated from the above definition, we say that an n-dimensional Riemannian manifold M is said to be a semi-quasi-constant curvature if its non-vanishing curvature tensor R satisfies Eq. (4.11). From the above definition and Eq. (4.11), we notice that if M is projectively semisymmetric, then it is a manifold of semi-quasi-constant curvature.

Let $\{e_i\}$, $i = 1, 2, 3 \ldots n$, be an orthonormal basis of the tangent space at any point of the manifold. If we put $V = W = e_i$ in (4.10) and taking summation with respect to $i (1 \leq i \leq n)$, then we get

$$S(U,X) = -\left[(\alpha^2+\mu)(n-1) - \frac{\alpha}{n-1}\right]g(U,X) - \frac{\alpha}{n-1}\eta(U)\eta(X).$$

From Eqs. (3.3) and (4.11), we have

$$-\left[(\alpha^2+\mu)(n-1)-\frac{\alpha}{n-1}\right]g(X,Y)-\frac{\alpha}{n-1}\eta(Y)\eta(X)=-(\lambda+\alpha)g(X,Y)$$
$$+\alpha\eta(X)\eta(Y).$$

Contracting the above equation over the vector fields X and Y, we get $\alpha = 0$, which is inadmissible. Hence, our hypothesis that M together with an almost Ricci soliton is projectively semisymmetric is not possible. Thus we can state the following:

Theorem 4.1 *Let M be an $(RCS)_n$-manifold admitting an almost Ricci soliton (g, ξ, λ). Then M cannot be a projectively semisymmetric manifold.*

5 $(RCS)_n$-Manifolds Admitting Almost Ricci Solitons Satisfying $\mathcal{Q}\cdot\mathcal{P}=0$

Let M be an $(RCS)_n$-manifold admitting almost Ricci solitons. If M satisfies the relation $\mathcal{Q}\cdot\mathcal{P}=0$, then we have

$$\mathcal{Q}(\mathcal{P}(U,V)W)-\mathcal{P}(QU,V)W-\mathcal{P}(U,QV)W-\mathcal{P}(U,V)QW=0 \tag{5.1}$$

for all $U, V, W \in \mathfrak{X}(M)$. In view of (4.1), (5.1) turns to

$$\mathcal{Q}(R(U,V)W)-R(QU,V)W-R(U,QV)W$$
$$-R(U,V)QW+\frac{2}{n-1}[S(QV,W)U-S(QU,W)V]=0,$$

which by taking the inner product with ξ takes the form

$$\eta(\mathcal{Q}(R(U,V)W))-\eta(R(QU,V)W)-\eta(R(U,QV)W) \tag{5.2}$$
$$-\eta(R(U,V)QW)+\frac{2}{n-1}[S(QV,W)\eta(U)-S(QU,W)\eta(V)]=0.$$

Putting $V=\xi$ in (5.2), we have

$$\eta(\mathcal{Q}(R(U,\xi)W))-\eta(R(QU,\xi)W)-\eta(R(U,Q\xi)W) \tag{5.3}$$
$$-\eta(R(U,\xi)QW)+\frac{2}{n-1}[S(Q\xi,W)\eta(U)-S(QU,W)]=0.$$

From Proposition (2.5) (ii) and (iv), we find

$$\begin{cases} \eta(Q(R(U,\xi)W)) = \eta(R(U,Q\xi)W) \\ \qquad\qquad = (\alpha^2+\mu)^2(n-1)(\eta(U)\eta(W)-g(U,W)), \\ \eta(R(QU,\xi)W) = \eta(R(U,\xi)QW) \\ \qquad\qquad = (\alpha^2+\mu)(S(U,W)+(\alpha^2+\mu)(n-1)\eta(U)\eta(W)), \\ S(Q\xi,W) = (\alpha^2+\mu)^2(n-1)^2\eta(W). \end{cases} \tag{5.4}$$

By the use of (5.4), Eq. (5.3) takes the form

$$S(QU,W) = -(\alpha^2+\mu)(n-1)S(U,W). \tag{5.5}$$

This implies

$$S^2(U,W) = -(\alpha^2+\mu)(n-1)S(U,W),$$

where $S^2(U,W) = S(QU,W)$. In view of (3.3), Eq. (5.5) leads to

$$S(U,W) = -\frac{\alpha(n-1)(\alpha^2+\mu)}{\lambda+\alpha-(n-1)(\alpha^2+\mu)}\eta(U)\eta(W), \tag{5.6}$$

provided $\lambda+\alpha-(n-1)(\alpha^2+\mu) \neq 0$. Equation (5.6) infers that M under consideration is a special type of quasi-Einstein manifold. By putting $U = W = \xi$ in (5.6), it follows that

$$(n-1)(\alpha^2+\mu)\{\lambda-(n-1)(\alpha^2+\mu)\} = 0. \tag{5.7}$$

This shows that either $\alpha^2+\mu = 0$ or $\lambda-(n-1)(\alpha^2+\mu) = 0$. If possible, we suppose that $\alpha^2+\mu = 0$, and hence Eq. (5.6) reflects that the $(RCS)_n$-manifolds under consideration are Ricci-flat.

In [38], Fischer and Wolf have studied the properties of compact Ricci-flat Riemannian manifolds and established several interesting results. They proved that a compact connected Ricci-flat n-manifold M^n has the expression $M^n = \Psi\backslash T^k \times M^{n-k}$, where k is the first Betti number $b_1(M^n)$, T^k is a flat Riemannian k-torus, M^{n-k} is a compact connected Ricci-flat $(n-k)$-manifold, and Ψ is a finite group of fixed-point-free isometries of $T^k \times M^{n-k}$ of a certain sort (see Theorem 4.1, [38] and Theorem 1.2, [20]).
Since $\alpha^2+\mu = 0$, the $(RCS)_n$-manifold admitting an almost Ricci soliton (g,ξ,λ) and satisfying the expression $Q\cdot\mathcal{P} = 0$ can be expressed as

$$M = \Psi\backslash T^k \times M^{n-k}.$$

Let $\alpha^2+\mu \neq 0$. Then from (5.7) we have

$$\lambda = (n-1)(\alpha^2 + \mu). \tag{5.8}$$

Equations (3.5), (5.6), and (5.8) lead to

$$S(U, W) = (n-1)(\alpha^2 + \mu)\eta(U)\eta(W), \tag{5.9}$$

since $\alpha^2 + \mu \neq 0$. Thus, we can state the following theorem:

Theorem 5.1 *Let a compact $(RCS)_n$-manifold M admit an almost Ricci soliton (g, ξ, λ). If M satisfies the relation $Q \cdot \mathcal{P} = 0$, then either $M = \Psi\backslash T^k \times M^{n-k}$ or M is a special type of quasi-Einstein manifold, and its Ricci tensor satisfies (5.9).*

Let ρ be the eigenvalue of the endomorphism Q corresponding to the eigenvector U, i.e., $QU = \rho U$. Then from (5.5), we have

$$\rho^2 g(U, W) = -\rho(\alpha^2 + \mu)(n-1)g(U, W). \tag{5.10}$$

By putting $U = W = \xi$ in (5.10), we have $\rho^2 + \rho(\alpha^2 + \mu)(n-1) = 0$. This gives that either $\rho = 0$ or $\rho = -(n-1)(\alpha^2 + \mu)$. Thus, we have the following corollary:

Corollary 5.2 *If an $(RCS)_n$-manifold admitting an almost Ricci soliton (g, ξ, λ) satisfies $Q \cdot \mathcal{P} = 0$, then the eigenvector of Q is either* 0 *or* $-(n-1)(\alpha^2 + \mu)$.

Let $\alpha^2 + \mu \neq 0$. The covariant derivative of (5.9) along the vector field V gives

$$(\nabla_V S)(U, W) = 2\alpha\eta(V)S(U, W) - \alpha(\alpha^2 + \mu)\{g(V, U)\eta(W) + g(V, W)\eta(U)\},$$

since Eq. (2.4) is used. This equation can be rewritten as

$$(\nabla_V S)(U, W) = A(V)S(U, W) + B(W)g(V, U) + B(U)g(V, W), \tag{5.11}$$

where $A(V) = 2\alpha\eta(V)$ and $B(U) = -\alpha(\alpha^2 + \mu)\eta(U)$ are 1-forms.
Inspired from Eq. (5.11) and the definition of Ricci recurrent and generalized Ricci recurrent manifolds, we define the following definition:

Definition 5.3 A complete Riemannian manifold M of dimension $n \geq 3$ is said to be an extended Ricci recurrent manifold if its non-vanishing Ricci tensor satisfies

$$(\nabla_X S)(Y, Z) = A(X)S(Y, Z) + B(Y)g(X, Z) + B(Z)g(X, Y)$$

for arbitrary vector fields X, Y, and Z on M, where A and B are 1-forms corresponding to the generators ρ_1 and ρ_2, that is, $A(\cdot) = g(\cdot, \rho_1)$ and $B(\cdot) = g(\cdot, \rho_2)$.

In particular, if we take $A = 0 = B$ and $B = 0$ in the above equation, then an extended Ricci recurrent manifold reduces to the Ricci symmetric manifold $((\nabla_X S)(Y, Z) = 0)$ and Ricci recurrent manifold $((\nabla_X S)(Y, Z) = A(X)S(Y, Z))$, respectively.

Equation (5.9) and Definition 5.3 state the following:

Corollary 5.4 *Let a complete $(RCS)_n$-manifold M admit a Ricci soliton (g, ξ, λ). Then M satisfies $\mathcal{Q}\cdot\mathcal{P} = 0$, and $\alpha^2+\mu \neq 0$ is an extended Ricci recurrent manifold.*

Data Availability

Data sharing is not applicable to this article as no datasets were generated or analyzed during the current study.

References

1. Ayar, G., Chaubey, S.K.: M-projective curvature tensor over cosymplectic manifolds. Differ. Geom. Dyn. Syst. **21**, 23–33 (2019)
2. Berndt, J., Tamaru, H.: Homogeneous codimension one foliations on noncompact symmetric spaces. J. Differ. Geom. **63**(1), 1–40 (2003)
3. Berndt, J., Tamaru, H.: Cohomogeneity one actions on noncompact symmetric spaces of rank one. Trans. Am. Math. Soc. **359**(7), 3425–3438 (2007)
4. Biswas, G.G., De, U.C.: Study of gradient solitons in three dimensional Riemannian manifolds. Commun. Korean Math. Soc. **37**(3), 825–837 (2022)
5. Capozziello, S., Mantica, C.A., Molinari, L.G.: Cosmological perfect-fluids in $f(R)$ gravity. Int. J. Geom. Methods Mod. Phys. **16**(1), 1950008, 14 (2019)
6. Chaki, M.C., Maity, R.K.: On quasi Einstein manifolds. Publ. Math. Debrecen **57**(3–4), 297–306 (2000)
7. Chaubey, S.K.: Certain results on $N(k)$-quasi Einstein manifolds. Afr. Mat. **30**(1–2), 113–127 (2019)
8. Chaubey, S.K.: Generalized Robertson-Walker space-times with W_1-curvature tensor. J. Phys. Math. **10**(2), 1000303 (2019)
9. Chaubey, S.K.: Characterization of perfect fluid spacetimes admitting gradient η-Ricci and gradient Einstein solitons. J. Geom. Phys. **162**, 104069, 10 (2021)
10. Chaubey, S.K., De, U.C.: Characterization of three-dimensional Riemannian manifolds with a type of semi-symmetric metric connection admitting Yamabe soliton. J. Geom. Phys. **157**, 103846, 8 (2020)
11. Chaubey, S.K., De, U.C.: Three-dimensional Riemannian manifolds and Ricci solitons. Quaest. Math. **45**(5), 765–778 (2022)
12. Chaubey, S.K., De, U.C., Siddiqi, M.D.: Characterization of Lorentzian manifolds with a semi-symmetric linear connection. J. Geom. Phys. **166**, 104269, 11 (2021)
13. Chaubey, S.K., De, U.C., Suh, Y.J.: Conformal vector field and gradient Einstein solitons on η-Einstein cosymplectic manifolds. Int. J. Geom. Methods Mod. Phys. **20**(8), 2350135, 16 (2023)
14. Chaubey, S.K., Haseeb, A.: Conformal η-Ricci-Yamabe solitons in the framework of Riemannian manifolds. In: Geometry of Submanifolds and Applications, pp. 209–224. Infosys Science Foundation Series in Mathematical Sciences. Springer, Singapore (2024)
15. Chaubey, S.K., Lee, J.W., Yadav, S.K.: Riemannian manifolds with a semi-symmetric metric P-connection. J. Korean Math. Soc. **56**(4), 1113–1129 (2019)
16. Chaubey, S.K., Shaikh, A.A.: On 3-dimensional Lorentzian concircular structure manifolds. Commun. Korean Math. Soc. **34**(1), 303–319 (2019)

17. Chaubey, S.K., Suh, Y.J.: Riemannian concircular structure manifolds. Filomat **36**, 6699–6711 (2022)
18. Chaubey, S.K., Suh, Y.J.: Generalized Ricci recurrent spacetimes and GRW spacetimes. Int. J. Geom. Methods Mod. Phys. **18**(13), 2150209, 16 (2021)
19. Chaubey, S.K., Suh, Y.J.: Characterizations of Lorentzian manifolds. J. Math. Phys. **63**(6), 062501, 10 (2022)
20. Chaubey, S.K., Suh, Y.J.: Ricci-Bourguignon solitons and Fischer-Marsden conjecture on generalized Sasakian-space-forms with β-Kenmotsu structure. J. Korean Math. Soc. **60**(2), 341–358 (2023)
21. Chaubey, S.K., Suh, Y.J., De, U.C.: Characterizations of the Lorentzian manifolds admitting a type of semi-symmetric metric connection. Anal. Math. Phys. **10**(4), 61, 15 (2020)
22. Chaubey, S.K., Vîlcu, G.E.: Gradient Ricci solitons and Fischer-Marsden equation on cosymplectic manifolds. Rev. R. Acad. Cienc. Exactas Fís. Nat. Ser. A Mat. RACSAM **116**(4), 186, 14 (2022)
23. Chave, T., Valent, G.: Quasi-Einstein metrics and their renormalizability properties. Journées relativistes 96, Part I (Ascona, 1996). Helv. Phys. Acta **69**(3), 344–347 (1996)
24. Chen, B.Y., Yano, K.: Hypersurfaces of a conformally flat space. Tensor **26**, 318–322 (1972)
25. Chen, B.Y.: Some results on concircular vector fields and their applications to Ricci solitons. Bull. Korean Math. Soc. **52**(5), 1535–1547 (2015)
26. Chen, B.Y.: Classification of torqued vector fields and its applications to Ricci solitons. Kragujevac J. Math. **41**(2), 239–250 (2017)
27. Chen, B.Y., Yano, K.: On submanifolds of submanifolds of a Riemannian manifold. J. Math. Soc. Japan **23**(3), 548–554 (1971)
28. Chen, B.Y.: A simple characterization of generalized Robertson-Walker spacetimes. Gen. Relativ. Gravit. **46**(12), 1833, 5 (2014)
29. De, U.C., Sardar, A., De, K.: Ricci-Yamabe solitons and 3-dimensional Riemannian manifolds. Turk. J. Math. **46**(3), 1078–1088 (2022)
30. De, U.C., Chaubey, S.K., Shenawy, S.: Perfect fluid spacetimes and Yamabe solitons. J. Math. Phys. **62**(3), 032501, 10 (2021)
31. De, U.C., Suh, Y.J., Chaubey, S.K.: Semi-symmetric curvature properties of Robertson-Walker spacetimes. J. Math. Phys. Anal. Geom. **18**(3), 368–381 (2022)
32. De, U.C., Suh, Y.J., Chaubey, S.K., Shenawy, S.: On pseudo $\mathcal{H}$-symmetric Lorentzian manifolds with applications to relativity. Filomat **34**(10), 3287–3297 (2020)
33. De, U.C., Suh, Y.J., Chaubey, S.K.: Conformal vector fields on almost co-Kähler manifolds. Math. Slovaca **71**(6), 1545–1552 (2021)
34. De, U.C., Chaubey, S.K., Suh, Y.J.: Gradient Yamabe and gradient m-quasi Einstein metrics on three-dimensional cosymplectic manifolds. Mediterr. J. Math. **18**(3), 80, 14 (2021)
35. De, U.C., Chaubey, S.K., Suh, Y.J.: A note on almost co-Kähler manifolds. Int. J. Geom. Methods Mod. Phys. **17**(10), 2050153, 14 (2020)
36. Eells, J., Sampson, J.H.: Harmonic mappings of Riemannian manifolds. Am. J. Math. **86**, 109–160 (1964)
37. Fialkow, A.: Conformal geodesics. Trans. Am. Math. Soc. **45**(3), 443–473 (1939)
38. Fischer, A.E., Wolf, J.A.: The structure of compact Ricci-flat Riemannian manifolds. J. Differ. Geom. **10**, 277–288 (1975)
39. Hamilton, R.S.: Three-manifolds with positive Ricci curvature. J. Differ. Geom. **17**, 255–306 (1982)
40. Hamilton, R.S.: The Ricci flow on surfaces. Mathematics and general relativity (Santa Cruz, CA, 1986). Contemp. Math. **71**, 237–262 (1988)
41. Haseeb, A., Bilal, M., Chaubey, S.K., Ahmadini, A.A.H.: ξ-Conformally Flat LP-Kenmotsu Manifolds and Ricci–Yamabe Solitons. Mathematics **11**(1), 212 (2023). https://doi.org/10.3390/math11010212
42. Haseeb, A., Bilal, M., Chaubey, S.K., Khan, M.N.I.: Geometry of Indefinite Kenmotsu Manifolds as $*\eta$-Ricci-Yamabe Solitons. Axioms **11**(9), 461 (2022). https://doi.org/10.3390/axioms11090461

43. Haseeb, A., Chaubey, S.K.: Lorentzian para-Sasakian manifolds and $*$-Ricci solitons. Kragujevac J. Math. **48**(2), 167–179 (2024)
44. Haseeb, A., Chaubey, S.K., Khan, M.A.: Riemannian 3-manifolds and Ricci-Yamabe solitons. Int. J. Geom. Methods Mod. Phys. **20**(1), 2350015, 13 (2023)
45. Haseeb, A., Chaubey, S.K., Mofarreh, F., Ahmadini, A.A.H.: A solitonic study of Riemannian manifolds equipped with a semi-symmetric metric ξ-connection. Axioms **12**(9), 809 (2023). https://doi.org/10.3390/axioms12090809
46. Haseeb, A., Mofarreh, F., Chaubey, S.K., Prasad, R.: A study of $*$-Ricci–Yamabe solitons on LP-Kenmotsu manifolds. AIMS Math. **9**(8), 22532–22546 (2024)
47. Khan, M.N.I.: Complete lifts of a semi-symmetric metric P-connection on a Riemannian manifold to its tangent bundle. Int. J. Math. Comput. Sci. **17**(2), 909–916 (2022)
48. Khan, M.A., Al-Dayel, I., Chaubey, S.K.: Semi-symmetric metric connections and homology of CR-warped product submanifolds in a complex space form admitting a concurrent vector field. Symmetry **16**, 719 (2024). https://doi.org/10.3390/sym16060719
49. Khan, M.A., Ozel, C., Al-Dayel, I., Chaubey, S.K.: Semi-symmetric metric connections and homology of warped product pointwise semi-slant submanifolds in a Kaehler manifold admitting a concurrent vector field. Filomat **38**(17), 6135–6146 (2024)
50. Koiso, N.: On rotationally symmetric Hamilton's equations for Kähler-Einstein metrics. Max-Planck-Institute preprint series **16**, 87 (1987)
51. Kwon, J.H., Pyo, Y.S., Suh, Y.J.: On semi-Riemannian manifolds satisfying the second Bianchi identity. J. Korean Math. Soc. **40**(1), 129–167 (2003)
52. Mantica, C.A., Molinari, L.G.: Generalized Robertson-Walker spacetimes—a survey. Int. J. Geom. Methods Mod. Phys. **14**(3), 1730001, 27 (2017)
53. Mantica, C.A., Molinari, L.G., Suh, Y.J., Shenawy, S.: Perfect-fluid, generalized Robertson-Walker space-times, and Gray's decomposition. J. Math. Phys. **60**(5), 052506, 9 (2019)
54. Michael, T.A.: Geometrization of 3-manifolds via the Ricci flow. Not. Am. Math. Soc. **51**, 184–193 (2004)
55. Mihai, A., Mihai, I.: Torse forming vector fields and exterior concurrent vector fields on Riemannian manifolds and applications. J. Geom. Phys. **73**, 200–208 (2013)
56. Mikeš, J., Rachunek, L.: Torse-forming vector fields in T-semisymmetric Riemannian spaces. In: Steps in Differential Geometry (Debrecen, 2000), pp. 219–229. Institute of Mathematics and Informatics, Debrecen (2001)
57. Chaubey, S.K., Prasad, R.: Three-dimensional Lorentzian para-Kenmotsu manifolds and Yamabe solitons. Honam Math. J. **43**(4), 613–626 (2021)
58. Perelman, G.: The entropy formula for the Ricci flow and its geometric applications. arXiv:math/0211159 (2002)
59. Perelman, G.: Finite extinction time for the solutions to the Ricci flow on certain three-manifolds. arXiv:math.DG/03072457 (2003)
60. Perelman, G.: Ricci flow with surgery on three-manifolds. arXiv:math/0303109 (2003)
61. Schouten, J.A.: Ricci-Calculus: An Introduction to Tensor Analysis and its Geometrical Applications, 2nd edn. Die Grundlehren der mathematischen Wissenschaften. Springer, Berlin/Göttingen/Heidelberg (1954)
62. Siddiqi, M.D., Chaubey, S.K., Khan, M.N.I.: $f(R, T)$-Gravity Model with Perfect Fluid Admitting Einstein Solitons. Mathematics **10**(1), 82 (2022). https://doi.org/10.3390/math10010082
63. Siddiqi, M.D., Mofarreh, F., Chaubey, S.K.: Solitonic aspect of relativistic magneto-fluid spacetime with some specific vector fields. Mathematics **11**, 1596 (2023)
64. Suh, Y.J., Chaubey, S.K.: Ricci solitons on general relativistic spacetimes. Phys. Scr. **98**, 065207 (2023)
65. Suh, Y.J., Chaubey, S.K., Khan, M.N.I.: Lorentzian manifolds: a characterization with a type of semi-symmetric non-metric connection. Rev. Math. Phys. **36**(3), 2450001, 16 (2024)
66. Suh, Y.J., Kwon, J.H., Yang, H.Y.: Conformally symmetric semi-Riemannian manifolds. J. Geom. Phys. **56**(5), 875–901 (2006)

67. Szabó, Z.I.: Structure theorems on Riemannian spaces satisfying $R(X, Y) \cdot R = 0$. I. The local version. J. Differ. Geom. **17**(4), 531–582 (1982)
68. Tashiro, Y.: Complete Riemannian manifolds and some vector fields. Trans. Am. Math. Soc. **117**, 251–275 (1965)
69. Yadav, S.K., Suthar, D.L., Chaubey, S.K.: Almost conformal Ricci solitons on generalized Robertson-Walker space-times. Res. Math. **10**(1), 2244748, 10 (2023)
70. Yano, K.: Concircular geometry I. Concircular transformations. Proc. Imp. Acad. Tokyo **16**, 195–200 (1940)
71. Yano, K.: On torse forming direction in a Riemannian space. Proc. Imp. Acad. Tokyo **20**, 340–345 (1944)

Statistical Maps and Chen's First Inequality for These Maps

Sema Kazan and Aliya Naaz Siddiqui

Abstract This chapter proposes the concept of a statistical map between two statistical manifolds and presents illustrative examples. Subsequently, we generalize Chen's first inequality for Riemannian maps to the framework of statistical maps by deriving the corresponding Gauss equation.

Keywords Statistical map · Statistical manifold · Chen's first inequality · Doubly totally geodesic map

1 Introduction

The theory of statistical manifolds, which originated with the seminal work of C.R. Rao in 1945 [18], forms the foundation of what is now known as information geometry. This field primarily investigates the differential-geometric structures associated with statistical models, particularly those defined on manifolds of probability distributions.

In recent years, information geometry has found diverse applications across several domains, including information theory, stochastic processes, dynamical systems and time series, statistical physics, quantum mechanics, and the mathematical modeling of neural networks [4]. Numerous studies have further explored the role of statistical manifolds in these contexts. For example, in [2], the authors analytically

---

The original version of the chapter has been revised. A correction to this chapter can be found at https://doi.org/10.1007/978-981-95-5148-4_16

S. Kazan
Faculty of Arts and Sciences, Department of Mathematics, Inonu University, Malatya, Turkey
e-mail: sema.bulut@inonu.edu.tr

A. N. Siddiqui (✉)
Division of Mathematics, School of Basic Sciences, Galgotias University, Greater Noida, Uttar Pradesh, India
e-mail: aliya.siddiqui@galgotiasuniversity.edu.in

B.-Y. Chen, M. A. Choudhary (eds.), *Geometric Inequalities and Applications*, Infosys Science Foundation Series in Mathematical Sciences,
https://doi.org/10.1007/978-981-95-5148-4_8

compute the asymptotic temporal behavior of the information-geometric complexity in finite-dimensional Gaussian statistical manifolds under the influence of microcorrelations. Similarly, the [9] presents an extension of the ergodic hierarchy (encompassing ergodic, mixing, and Bernoulli levels) for statistical models on curved manifolds using tools from information geometry.

A significant structural component in information geometry is the concept of dual connections (or conjugate connections) in affine differential geometry, introduced into statistics by S. Amari in 1985 [3]. A statistical manifold is defined as a differentiable manifold equipped with a Riemannian metric and a pair of dual torsion-free affine connections. For comprehensive treatments of statistical manifolds, one may refer to [7, 10, 16, 17, 22, 24], and [11], among others.

Building on these structures, the notion of statistical submersions was introduced by N. Abe and K. Hasegawa in 2001 [1], extending foundational results of B. O'Neill [13, 15] concerning Riemannian submersions and geodesics to the statistical setting. This topic has been further explored in subsequent works (see, e.g., [12, 21, 23, 25–30]), etc.

On a parallel track, the theory of Riemannian maps (a generalization of both isometric immersions and Riemannian submersions) has garnered considerable attention in Riemannian geometry. These maps provide a flexible framework for comparing geometric structures between manifolds. Let (M, g_M) and (N, g_N) be Riemannian manifolds. A smooth map $\pi : (M, g_M) \to (N, g_N)$ is called an isometric immersion if the differential π_* is injective and preserves the metric:

$$g_N(\pi_* X, \pi_* Y) = g_M(X, Y), \tag{1.1}$$

for all vector fields X, Y tangent to M. This theory traces back to Gauss's investigations on surfaces in Euclidean spaces.

A Riemannian submersion, on the other hand, is a smooth map $\pi : (M, g_M) \to (N, g_N)$ for which π_* is surjective and satisfies the above metric condition (1.1) on the horizontal distribution $(\ker \pi_*)^\perp$.

In 1992, Fischer introduced the broader notion of Riemannian maps [6]. For a smooth map $\pi : (M, g_M) \to (N, g_N)$ with $0 < \operatorname{rank} a\pi_* < \min\{m, n\}$, where $\dim M = m$ and $\dim N = n$, the tangent bundle TM decomposes as

$$TM = \ker \pi_* \oplus \mathcal{H},$$

where $\mathcal{H} = (\ker \pi_*)^\perp$.

Similarly, the tangent bundle TN decomposes as

$$TN = range\pi_* \oplus (range\pi_*)^\perp.$$

A map π is said to be a Riemannian map at $p_1 \in M$ if the horizontal restriction $\pi^h_{*p_1} : (\ker \pi_{*p_1})^\perp \longrightarrow range\pi_{*p_1}$ is a linear isometry with respect to the induced metrics. Thus, both isometric immersions and Riemannian submersions appear as special cases of Riemannian maps, corresponding to $\ker \pi_* = \{0\}$ and $(range\pi_*)^\perp = \{0\}$ [19, 20].

Furthermore, for a smooth map $\pi : (M, g_M) \to (N, g_N)$, the second fundamental form of π is given by

$$(\nabla\pi_*)(X, Y) = \nabla^{\pi}_X(\pi_*(Y)) - \pi_*(\nabla^M_X Y), \tag{1.2}$$

where ∇^{π} is the pullback connection. It is noted that the connection ∇ on the bundle $Hom(TM, \pi^{-1}TN)$ is induced by the Levi-Civita connection ∇^M and the pullback connection. This form is symmetric in its arguments and plays a crucial role in analyzing curvature relations via the Gauss and Codazzi equations. Here $\pi^{-1}TN$ is the pullback bundle which has fibers $(\pi^{-1}TN)_p = T_{\pi(p)}N,\ p \in M$.

In this chapter, we explore statistical maps by integrating the ideas outlined above. We begin by revisiting the foundational notions and properties of statistical manifolds and statistical submersions. We then introduce the definition of a statistical map, along with illustrative examples. Finally, we derive the Gauss equation for statistical maps and extend Chen's first inequality (originally formulated for Riemannian maps) into the statistical geometric setting.

2 Statistical Submersions

Let M be an n-dimensional smooth semi-Riemannian manifold equipped with a metric tensor g_M, where g_M is a symmetric nondegenerate $(0, 2)$-tensor field of constant index. The common value ν of index of the index of g_M on M is called the index of M with $0 \leq \nu \leq n$. We denote such a manifold by M^n_ν. When $\nu = 0$, M becomes a Riemannian manifold.

At any point $p \in M$, a tangent vector E to M is called:

1. Spacelike if $g_M(E, E) > 0$ or $E = 0$
2. Null (or lightlike) if $g_M(E, E) = 0$ or $E \neq 0$
3. Timelike if $g_M(E, E) < 0$

Let $\mathbb{R}^n_\nu$ be an n-dimensional real vector space endowed with an inner product of signature $(\nu, n - \nu)$, defined by

$$\langle x, x\rangle = -\sum_{i=1}^{\nu} x_i^2 + \sum_{i=\nu+1}^{n} x_i^2,$$

where $x = (x_1, \ldots, x_n)$ are the standard coordinates. This space is called the semi-Euclidean space of dimension n and index ν. In particular, $\mathbb{R}^n_0$ corresponds to the standard Euclidean space and $\mathbb{R}^n_1$ to the Lorentzian space [28].

Following [13, 14], a smooth map $\pi : (M, g_M) \to (N, g_N)$ is called a semi-Riemannian submersion if:

1. $d\pi_p$ is surjective for all $p \in M$.
2. Each fiber $\pi^{-1}(b)$, $b \in N$, is a semi-Riemannian submanifold of M.

3. The metric is preserved on the horizontal distribution, that is,

$$g_M(X, Y) = g_N(d\pi X, d\pi Y),$$

for all vectors X, Y normal to the fibers.

It is worth noting that (semi-)Riemannian submersions are of considerable importance not only in differential geometry but also in various scientific and technological domains. Numerous researchers have contributed to this area of study, for example, see [31].

Assume that $\pi : M^m \to N^n$ is such a semi-Riemannian submersion. For each $b \in N$, the fiber $\overline{M}_b = \pi^{-1}(b)$ inherits an induced metric $\overline{g}$ and forms an $r = (m-n)$-dimensional semi-Riemannian submanifold of M. A vector field on M is called vertical if it is always tangent to the fibers and horizontal if it is orthogonal to the fibers.

Let $\mathcal{V}_p(M)$ and $\mathcal{H}_p(M)$ denote the vertical and horizontal subspaces of T_pM, $p \in M$, respectively. Then the tangent bundle decomposes as

$$T_pM = \mathcal{H}_p(M) \oplus \mathcal{V}_p(M).$$

We denote the corresponding projection operators by $\mathcal{V} : TM \to \mathcal{V}(M)$ and $\mathcal{H} : TM \to \mathcal{H}(M)$.

A vector field X on M is called projectable if there exists a vector field X_* on N such that $d\pi(X_p) = X_{*\pi(p)}$ for every $p \in M$; in this case, X and X_* are said to be π-related. If in addition X is horizontal, it is called basic.

Lemma 2.1 ([13, 14]) *Let X and Y be basic vector fields on M, π-related to X_* and Y_* on N. Then:*

1. $g_M(X, Y) = g_N(X_*, Y_*) \circ \pi$.
2. *$\mathcal{H}[X, Y]$ is basic and π-related to $[X_*, Y_*]$.*

Let M be a semi-Riemannian manifold equipped with a torsion-free affine connection ∇^M. The triple (M, ∇^M, g_M) is called a statistical manifold if $\nabla^M g_M$ is symmetric. For such a manifold, the conjugate (or dual) connection ∇^{*^M} is defined by

$$Eg_M(F, G) = g_M(\nabla^M_E F, G) + g_M(F, \nabla^{*^M}_E G) \tag{2.1}$$

for vector fields E, F, and G on M. The connection ∇^{*^M} is torsion-free, and $\nabla^{*^M} g_M$ is symmetric. Moreover, the duality condition satisfies $(\nabla^{*^M})^* = \nabla^M$, implying that (M, ∇^{*^M}, g_M) is also a statistical manifold. Let R and R^* denote the curvature tensors corresponding to ∇^M and ∇^{*^M}, respectively. Then the following identity holds:

$$g_M(R(E, F)G, H) = -g_M(G, R^*(E, F)H),$$

where $R(E,F)G = [\nabla^M_E, \nabla^M_F]G - \nabla^M_{[E,F]}G$. Hence $R \equiv 0$ if and only if so is $R^* \equiv 0$. In this case, the manifold is said to be flat.

Define the difference tensor

$$K = \frac{1}{2}(\nabla^M - \nabla^{*M}) = \nabla^M - \hat{\nabla}^M, \tag{2.2}$$

where $\hat{\nabla}^M$ is the Levi-Civita connection. Then K is symmetric in the sense that $K_E F = K_F E$ and $g_M(K_E F, G) = g_M(F, K_E G)$ hold [28].

Let (M, ∇^M, g_M) be a statistical manifold and $\pi : M \to N$ be a semi-Riemannian submersion. Denote by $\overline{\nabla}$ and $\overline{\nabla}^*$ the affine connections induced on each fiber $\overline{M}$. For vertical vector fields U, V, these are given by

$$\overline{\nabla}_U V = \mathcal{V}\nabla^M_U V, \quad \overline{\nabla}^*_U V = \mathcal{V}\nabla^{*M}_U V,$$

and both are torsion-free and mutually dual with respect to the induced metric $\overline{g}$ on the fiber.

Let $\pi : (M, g_M) \to (N, g_N)$ be a smooth map between statistical manifolds with $0 < \operatorname{rank}\pi_* < \min\{m, n\}$. At each point $p \in M$, let $\mathcal{V}_p = \ker \pi_{*p}$ and $\mathcal{H}_p = (\ker \pi_{*p})^\perp$. Then we have the orthogonal decomposition:

$$T_p M = \ker \pi_{*p} \oplus (\ker \pi_{*p})^\perp = \mathcal{V}_p \oplus \mathcal{H}_p.$$

Similarly, at $\pi(p) \in N$, we define the range $range\pi_{*p} \subset T_{\pi(p)}N$ and its orthogonal complement $(range\pi_{*p})^\perp$. Since $\operatorname{rank}\pi_* \leq \min\{m, n\}$, this decomposition is nontrivial:

$$T_{\pi(p)}N = (range\pi_{*p}) \oplus (range\pi_{*p})^\perp.$$

3 Statistical Maps

In this section, we first define a statistical map.

A smooth map $\pi : (M, g_M) \to (N, g_N)$ is a statistical map at $p_1 \in M$ if the horizontal restriction

$$\pi^h_{*p_1} : (\ker \pi_{*p_1})^\perp \to (range\pi_{*p_1})$$

is a linear isometry between the inner product spaces $((\ker \pi_{*p_1})^\perp, g_M(p_1)|_{(\ker \pi_{*p_1})^\perp})$ and

$(range\pi_{*p_1}, g_N(p_2)|_{(range\pi_{*p_1})})$, $p_2 = \pi(p_1)$. Thus π_* satisfies the equation

$$g_N(\pi_* X, \pi_* Y) = g_M(X, Y), \tag{3.1}$$

which holds for all horizontal vector fields X, Y.

Thus, isometric immersions and statistical submersions appear as special cases of statistical maps, corresponding to $\ker \pi_* = \{0\}$ and $(range\pi_*)^\perp = \{0\}$, respectively. Moreover, a statistical map must be a submersion, implying that the rank of $\pi_* : T_pM \to T_{\pi(p)}N$ is constant on each connected component of M.

Definition 3.1 Let (M, ∇^M, g_M) and (N, ∇^N, g_N) be statistical manifolds and $\pi : M \to N$ a smooth map between them. If $0 \le \operatorname{rank} \pi_{*p_1} \le \min\{m, n\}$, π_{*p_1} maps the horizontal space $\mathcal{H}_{p_1} = (\ker(\pi_{*p_1}))^\perp$ isometrically onto $range(\pi_{*p_1})$, i.e.,

$$g_N(\pi_{*p_1}X, \pi_{*p_1}Y) = g_M(X, Y),$$

and the affine connections satisfy the relation $\overset{N}{\nabla}{}^{\pi}_X \pi_{*p_1}(Y) = \pi_{*p_1}(\nabla^M_X Y) + C(X, Y)_{\pi_{p_1}}$ for $X, Y \in \mathcal{H}_{p_1}$, then π is called a statistical map, where $C(X, Y)_{\pi_{p_1}} \in \Gamma((range\pi_*)^\perp)$.

If $C(X, Y)_{\pi_{p_1}} = 0$, then a statistical map becomes a statistical submersion.

Example 3.2 Let $\pi : (M, g_M) \to (N, g_N)$ be an isometric immersion between statistical manifolds. Then π is a statistical map with $\ker \pi_* = \{0\}$.

Example 3.3 Let $\pi : (M, g_M) \to (N, g_N)$ be a statistical submersion between statistical manifolds. Then π is a statistical map with $(range\pi_*)^\perp = \{0\}$.

Example 3.4 Let us take two statistical manifolds $(M = \mathbb{R}^2, \nabla^M, g_M)$ with $g_M = dx^2 + dy^2$ and affine connection ∇^M defined by

$$\nabla^M_{\partial_x} \partial_y = 0 = \nabla^M_{\partial_y} \partial_x,$$

$$\nabla^M_{\partial_x} \partial_x = \partial_y, \ \nabla^M_{\partial_y} \partial_y = 0$$

and $(N = \mathbb{R}^2, \nabla^N, g_N)$ with $g_N = dz^2 + dw^2$ and affine connection ∇^N defined as

$$\nabla^N_{\partial_z} \partial_w = 0 = \nabla^N_{\partial_w} \partial_z,$$

$$\nabla^N_{\partial_z} \partial_z = \partial_w, \quad \nabla^N_{\partial_w} \partial_w = 0.$$

Define the differentiable map $\pi : (M^m, \nabla^M, g_M) \to (N^n, \nabla^N, g_N)$ with Cartesian coordinates (x, y) by

$$\pi(x, y) = (x, 0).$$

Hence we get

$$\pi_* = \begin{pmatrix} 1 & 0 \\ 0 & 0 \end{pmatrix},$$

where $0 \leq \operatorname{rank} \pi_* = 1 \leq \min\{2, 2\}$. Then we have

$$\ker \pi_* = Span\{e_2 = \partial_y\}$$

and

$$(\ker \pi_*)^{\perp} = Span\{e_1 = \partial_x\},$$

where $\partial_x = \partial/\partial x$.

Thus it is easy to see that

$$g_N(\pi_*(e_i), \pi_*(e_i)) = g_M(e_i, e_i) = 1,$$

and

$$g_N(\pi_*(e_i), \pi_*(e_j)) = g_M(e_i, e_j) = 0,$$

$i \neq j$, for $i = 1, 2$.

On the other hand, if the expressions $C(X, Y)$ are calculated with respect to the bases, the following cases are obtained:

1. For $X = \partial_x$ and $Y = \partial_y$, using $\pi_*\partial_x = \partial_z$, $\pi_*\partial_y = 0$, we get $C(\partial_x, \partial_y) = \nabla^N_{\pi_*\partial_x}\pi_*\partial_y - \pi_*(\nabla^M_{\partial_x}\partial_y) = 0$.
2. For $X = \partial_x$ and $Y = \partial_x$, using $\pi_*\partial_x = \partial_z$, we have $C(\partial_x, \partial_x) = \nabla^N_{\pi_*\partial_x}\pi_*\partial_x - \pi_*(\nabla^M_{\partial_x}\partial_x) = \partial_w$.
3. For $X = \partial_y$ and $Y = \partial_x$, using $\pi_*\partial_y = 0$, $\pi_*\partial_x = \partial_z$, we get $C(\partial_y, \partial_x) = \nabla^N_{\pi_*\partial_y}\pi_*\partial_x - \pi_*(\nabla^M_{\partial_y}\partial_x) = 0$.
4. For $X = \partial_y$ and $Y = \partial_y$, using $\pi_*\partial_y = 0$, we put $C(\partial_y, \partial_y) = \nabla^N_{\pi_*\partial_y}\pi_*\partial_y - \pi_*(\nabla^M_{\partial_y}\partial_y) = 0$.

Then π is a statistical map with $range\pi_* = Span\{e_1^{'} = \partial_z\}$, $(range\pi_*)^{\perp} = Span\{e_2^{'} = \partial_w\}$.

Let π be a statistical map from a statistical manifold (M, ∇^M, g_M) to a statistical manifold (N, ∇^N, g_N). Then we define T and A as

$$T_E F = \mathcal{H}\nabla^M_{\mathcal{V}E}\mathcal{V}F + \mathcal{V}\nabla^M_{\mathcal{V}E}\mathcal{H}F, \qquad A_E F = \mathcal{H}\nabla^M_{\mathcal{H}E}\mathcal{V}F + \mathcal{V}\nabla^M_{\mathcal{H}E}\mathcal{H}F,$$

where ∇^M is the linear connection of g_M. We can easily obtain the tensors T^* and A^* corresponding to the conjugate connection by simply replacing ∇^M by ∇^{*^M} in the above equations. We note that $(T^*)^* = T$ and $(A^*)^* = A$. For vertical vector fields, T and T^* are symmetric. Also for $X, Y \in \mathcal{H}(M)$ and $U, V \in \mathcal{V}(M)$, we have

$$g_M(T_U V, X) = -g_M(V, T_U^* X) \quad \text{and} \quad g_M(A_X Y, U) = -g_M(Y, A_X^* U). \tag{3.2}$$

Thus, $T_U V \equiv 0$ (resp. $T_U X \equiv 0$) if and only if $T_U^* X \equiv 0$ (resp. $T_U^* V \equiv 0$).

On the other hand, from (3.2) we have

$$\begin{aligned}
\nabla_U^M V &= T_U V + \overline{\nabla}_U V & \nabla_U^{*M} V &= T_U^* V + \overline{\nabla}_U^* V,\\
\nabla_U^M X &= \mathcal{H}\nabla_U^M X + T_U X & \nabla_U^{*M} X &= \mathcal{H}\nabla_U^{*M} X + T_U^* X,\\
\nabla_X^M U &= A_X U + \mathcal{V}\nabla_X^M U & \nabla_X^{*M} U &= A_X^* U + \mathcal{V}\nabla_X^{*M} U,\\
\nabla_X^M Y &= \mathcal{H}\nabla_X^M Y + A_X Y & \nabla_X^{*M} Y &= \mathcal{H}\nabla_X^{*M} Y + A_X^* Y,
\end{aligned}$$

for $X, Y \in \mathcal{H}(M)$ and $U, V \in \mathcal{V}(M)$.

Furthermore, if X is basic, then $\mathcal{H}\nabla_U^M X = A_X U$, $\mathcal{H}\nabla_U^{*M} X = A_X^* U$, and $A_X Y = -A_Y^* X$ for horizontal vector fields X and Y. The tensor $A \equiv 0$ if and only if $A^* \equiv 0$. Since A characterizes the integrability of the horizontal distribution $\mathcal{H}(M)$, it is identically zero if and only if $\mathcal{H}(M)$ is integrable with respect to ∇^M.

Proposition 3.5 *Let $\pi : (M^m, \nabla^M, g_M) \to (N^n, \nabla^N, g_N)$ be a statistical map. Then we have*

$$g_N((\nabla\pi_*)(X, Y), \pi_*(Z)) = -g_N((\nabla^*\pi_*)(X, Y), \pi_*(Z)),$$

for $X, Y, Z \in \Gamma((\ker \pi_*)^{\perp})$.

Proof The proof is clear from Eqs. (1.1) and (2.2). □

Then we can give the following corollary:

Corollary 3.6 *Let $\pi : (M^m, \nabla^M, g_M) \to (N^n, \nabla^N, g_N)$ be a statistical map. Then, for all $X, Y, Z \in \Gamma((\ker \pi_*)^{\perp})$, $(\nabla\pi_*)(X, Y) \in \Gamma((range\pi_*)^{\perp})$ if and only if $(\nabla^*\pi_*)(X, Y) \in \Gamma((range\pi_*)^{\perp})$.*

Proposition 3.7 *Let $\pi : (M^m, \nabla^M, g_M) \to (N^n, \nabla^N, g_N)$ be a statistical map. Then we have*

$$g_N((\nabla^*\pi_*)(X, Y), \pi_*(Z)) = g_N(\pi_*(Y), (\nabla^*\pi_*)(X, Z)),$$

for $X, Y, Z \in \Gamma((\ker \pi_*)^{\perp})$.

Considering these results, we can give the following lemma.

Lemma 3.8 *Let $\pi : (M^m, \nabla^M, g_M) \to (N^n, \nabla^N, g_N)$ be a statistical map. If $K(X, \pi_* Y) = \pi_*(K(X, Y))$, we have*

$$g_N((\nabla\pi_*)(X, Y), \pi_*(Z)) = 0, \tag{3.3}$$

for $X, Y, Z \in \Gamma((\ker \pi_*)^{\perp})$.

Proof Since π is a statistical map, from (1.1), we get

$$g_N((\hat{\nabla}\pi_*)(X, Y), \pi_*(Z)) = g_N(\hat{\nabla}^{\pi}_X \pi_* Y, \pi_*(Z)) - g_M(\hat{\nabla}^M_X Y, Z), \tag{3.4}$$

where $\hat{\nabla}$ is the Levi-Civita connection. Using (2.2), we have

$$\begin{aligned} g_N((\hat{\nabla}\pi_*)(X, Y), \pi_*(Z)) &= g_N(\nabla^{\pi}_X \pi_* Y, \pi_*(Z)) - g_N(K(X, \pi_*(Y)), \pi_*(Z)) \\ &- g_N(\pi_*(\nabla^M_X Y), \pi_*(Z)) + g_N(\pi_*(K(X, Y)), \pi_*(Z)). \end{aligned}$$

So, considering the hypothesis, we obtain

$$g_N((\hat{\nabla}\pi_*)(X, Y), \pi_*(Z)) = g_N(C(X, Y), \pi_*(Z)).$$

We know that $g_N((\hat{\nabla}\pi_*)(X, Y), \pi_*(Z)) = 0$. In this case, Eq. (3.3) is obtained, where $C(X, Y) = (\nabla\pi_*)(X, Y)$. □

Similarly, we can also say that $g_N((\nabla^*\pi_*)(X, Y), \pi_*(Z)) = 0$ under the same condition.

We recall that the second fundamental form of π is the map $\nabla\pi_* : \Gamma(TM) \times \Gamma(TM) \to \Gamma(TN)$ defined by

$$(\nabla\pi_*)(X, Y) = \overset{N}{\nabla^{\pi}_X} \pi_* Y - \pi_*(\nabla^M_X Y), \tag{3.5}$$

where ∇ is a linear connection on M and $\overset{N}{\nabla^{\pi}_X} \pi_* Y \circ \pi = \overset{N}{\nabla^{\pi}_{\pi_* X}} \pi_* Y$. Considering Lemma 3.8, we have

$$(\nabla\pi_*)(X, Y) \in \Gamma((range\pi_*)^{\perp}),$$

for all $X, Y \in \Gamma((\ker \pi_*)^{\perp})$. Hence, we say that at $p \in M$

$$\overset{N}{\nabla^{\pi}_X} \pi_*(Y)(p) = \pi_*(\nabla^M_X Y)(p) + (\nabla\pi_*)(X, Y)(p), \tag{3.6}$$

where $\overset{N}{\nabla^{\pi}}{}_X \pi_*(Y) \in T_{\pi(p)}N$, $\pi_*(\nabla^M_X Y)(p) \in \pi_{*p}(T_pM)$ and $(\nabla\pi_*)(X, Y)(p) \in (\pi_{*p}(T_pM))^{\perp}$.

Also, we get the second fundamental form of π according to the dual connection ∇^* defined by

$$(\nabla^*\pi_*)(X, Y) = \overset{N}{\nabla^{*\pi}_X} \pi_* Y - \pi_*(\overset{*M}{\nabla_X} Y); \tag{3.7}$$

then we have $(\nabla^*\pi_*)(X, Y) \in \Gamma((range\pi_*)^{\perp})$.

Example 3.9 From Example 3.4, let $M = \mathbb{R}^2$ with coordinates (x, y) and the Euclidean metric $g_M = dx^2 + dy^2$ equipped with the Levi-Civita connection $\hat{\nabla}^M$ and a statistical connection $\nabla^M = \hat{\nabla}^M + K^M$, where K^M is a symmetric difference tensor defined by

$$K^M(\partial_x, \partial_x) = \lambda\partial_x, \quad K^M(\partial_x, \partial_y) = 0 = K^M(\partial_y, \partial_x)\,, \quad K^M(\partial_y, \partial_y) = 0.$$

Let $N = \mathbb{R}^2$ with coordinates (z, w) and the Euclidean metric $g_N = dz^2 + dw^2$ equipped with the Levi-Civita connection $\hat{\nabla}^N$ and a statistical connection $\nabla^N = \hat{\nabla}^N + K^N$, where K^N is defined as

$$K^N(\partial_z, \partial_z) = \lambda\partial_z, \quad K^N(\partial_z, \partial_w) = 0 = K^N(\partial_w, \partial_z)\,, \quad K^N(\partial_w, \partial_w) = 0.$$

Define the differentiable map $\pi : (M^2, \nabla^M, g_M) \to (N^2, \nabla^N, g_N)$ by $\pi(x, y) = (x, 0)$ with differential

$$\pi_* = \begin{pmatrix} 1 & 0 \\ 0 & 0 \end{pmatrix},$$

so that

$$\pi_*\partial_x = \partial_z, \quad \pi_*\partial_y = 0.$$

Thus, the kernel of π_* is $\pi_* = Span\{\partial_y\}$, and its orthogonal complement is $(\ker\pi_*)^\perp = Span\{\partial_x\}$.

For all $X, Y \in \Gamma((\ker\pi_*)^\perp) = Span\{\partial_x\}$, the condition holds. Specially:

1. For $X = Y = \partial_x$, we have

$$K^N(\partial_x, \pi_*\partial_x) = K^N(\partial_x, \partial_z) = \lambda\partial_z$$

and

$$\pi_*(K^M(\partial_x, \partial_x)) = \pi_*(\lambda\partial_z) = \lambda\partial_z.$$

Hence, the condition holds.

2. For $X = \partial_x$ and $Y = \partial_y$, we get

$$K^N(\partial_x, \pi_*\partial_y) = K^N(\partial_x, 0) = 0$$

and

$$\pi_*(K^M(\partial_x, \partial_y)) = \pi_*(0) = 0.$$

Thus, the condition holds.

3. For $X = \partial_y$ and $Y = \partial_x$, we find

$$K^N(\partial_y, \pi_*\partial_x) = K^N(\partial_y, \partial_z) = 0$$

and

$$\pi_*(K^M(\partial_y, \partial_x)) = \pi_*(0) = 0.$$

Thus, we conclude that for all $X, Y \in \Gamma((\ker\pi_*)^{\perp})$, the condition $K(X, \pi_*Y) = \pi_*(K(X, Y))$ is satisfied.

From now on, for simplicity, we denote by ∇^N both the linear connection of (N, ∇^N, g_N) and its pullback along π.

We now suppose that π is a statistical map; then $S_V\pi$ and $S_V^*\pi$ are defined as

$$\nabla^N_{\pi_*X}V = -S_V\pi_*X + \nabla_X^{\pi^\perp}V, \tag{3.8}$$

$$\nabla^{*N}_{\pi_*X}V = -S_V^*\pi_*X + \nabla_X^{*\pi^\perp}V \tag{3.9}$$

for any $X \in \Gamma((\ker\pi_*)^{\perp})$ and $V \in \Gamma((range\pi_*)^{\perp})$. So, from (2.1), (3.5), and (3.8), we have

$$g_N(S_V\pi_*X, \pi_*Y) = g_N(V, (\nabla^*\pi_*)(X, Y)). \tag{3.10}$$

Similarly, we find that

$$g_N(S_V^*\pi_*X, \pi_*Y) = g_N(V, (\nabla\pi_*)(X, Y)). \tag{3.11}$$

Using the concept of doubly totally geodesic submanifold from [8], we are able to define the following:

Definition 3.10 Let (M, ∇^M, g_M) and (N, ∇^N, g_N) be two statistical manifolds and $\pi : M \to N$ be a statistical map from these manifolds. If $(\nabla\pi_*)(X, Y) = (\nabla^*\pi_*)(X, Y) = 0$, for $X, Y \in \Gamma(TM)$, then the map π is called doubly totally geodesic map.

Theorem 3.11 *Let $\pi : (M^m, \nabla^M, g_M) \to (N^n, \nabla^N, g_N)$ be a statistical map which satisfies the condition $K(X, \pi_*Y) = \pi_*(K(X, Y))$; then π is doubly totally geodesic if and only if:*

1. $A_X^*Y = 0$.
2. $S_V^*\pi_*X = 0$.
3. $T_UV = 0$ *for* $X, Y \in \Gamma((\ker\pi_*)^{\perp})$ *and* $U, V \in \Gamma((range\pi_*)^{\perp})$.

Proof For $V \in \Gamma((range\pi_*)^{\perp})$, we have $(\nabla\pi_*)(X, V) \in \Gamma(range\pi_*)$. Then we get

$$g_N((\nabla\pi_*)(X, V), \pi_*Y) = 0.$$

In this case, we have

$$0 = -g_M(\nabla^M_X V, Y) = -g_M(A_X V, Y) = g_M(A^*_X Y, V).$$

Also, we get

$$0 = g_N((\nabla \pi_*)(X, Y), \pi_* V) = g_N(S^*_V \pi_* X, \pi_* Y).$$

Finally, for $U, V \in \Gamma((range\pi_*)^\perp)$,

$$0 = g_N((\nabla \pi_*)(U, V), \pi_* X) = -g_M(\nabla^M_U V, X) = -g_M(T_U V, X).$$

□

If $\pi : (M^m, \nabla^M, g_M) \to (N^n, \nabla^N, g_N)$ is a statistical map, then considering π^h_* at each $p_1 \in M$ as a linear transformation

$$\pi^h_{*p_1} : ((\ker \pi_*)^\perp(p_1), g_{Mp_1((\ker \pi_*)^\perp(p_1))}) \longrightarrow (range\pi_*(p_2), g_{Np_2(range\pi_*)(p_2))}),$$

we state the adjoint of π^h_* as $^*\pi^h_{*p_1}$. Let us assume that the adjoint of $\pi_{*p_1} : (T_{p_1}M, g_{Mp_1}) \longrightarrow (T_{p_2}N, g_{Np_2})$ is $^*\pi_{*p_1}$. Therefore the linear transformation $(^*\pi_{*p_1})^h : range\pi_*(p_2) \longrightarrow (\ker \pi_*)^\perp(p_1)$ defined as $(^*\pi_{*p_1})^h w =^* \pi_{*p_1} w$, where $w \in \Gamma(range\pi_{*p_1})$, $p_2 = \pi(p_1)$, is an isomorphism and $(\pi^h_{*p_1})^{-1} = (^*\pi_{*p_1})^h =^* (\pi^h_{*p_1})$.

Using (3.6), (3.8), and (3.9), we have

$$\begin{aligned} R^N(\pi_* X, \pi_* Y)\pi_* Z = \pi_*(R^M(X, Y)Z) &+ (\nabla^M_X(\nabla \pi_*))(Y, Z) \\ &- (\nabla^M_Y(\nabla \pi_*))(X, Z) \\ &+ S_{(\nabla \pi_*)(X,Z)}\pi_* Y - S_{(\nabla \pi_*)(Y,Z)}\pi_* X \end{aligned} \tag{3.12}$$

and

$$\begin{aligned} R^{*^N}(\pi_* X, \pi_* Y)\pi_* Z = \pi_*(R^{*^M}(X, Y)Z)&(\nabla^{*^M}_X(\nabla^* \pi_*))(Y, Z) \\ &- (\nabla^{*^M}_Y(\nabla^* \pi_*))(X, Z) \\ &+ S^*_{(\nabla^* \pi_*)(X,Z)}\pi_* Y - S^*_{(\nabla^* \pi_*)(Y,Z)}\pi_* X \end{aligned} \tag{3.13}$$

for $X, Y, Z \in \Gamma((ker\pi)^\perp)$, where R^M (respectively, R^{*^M}) and R^N (respectively, R^{*^N}) denote the curvature tensors of ∇^M (respectively, ∇^{*^M}) on M and ∇^N (respectively, ∇^{*^N}) on N. Moreover, $(\nabla^M_X(\nabla \pi_*))(Y, Z)$ and $(\nabla^{*^M}_X(\nabla^* \pi_*))(Y, Z)$ are defined by

$$(\nabla^M_X(\nabla \pi_*))(Y, Z) = \nabla^{\pi\perp}_X(\nabla^M \pi_*)(Y, Z) - (\nabla \pi_*)(\nabla^M_X Y, Z) - (\nabla \pi_*)(Y, \nabla^M_X Z)$$

and

$$(\nabla_X^{*M}(\nabla^*\pi_*))(Y,Z) = \nabla_X^{*\pi\perp}(\nabla^{*M}\pi_*)(Y,Z) - (\nabla^*\pi_*)(\nabla_X^{*M}Y,Z) \\ -(\nabla^*\pi_*)(Y,\nabla_X^{*M}Z).$$

This leads us to formulate Chen's first inequality for the statistical map π in the next section.

4 Chen's First Inequality

In this section, we establish Chen's first inequality for a statistical map π into a statistical manifold of constant curvature c, under the assumption that $\operatorname{rank}\pi = r \geq 3$.

Let $\pi : (M^m, \nabla^M, g_M) \to (N^n, \nabla^N, g_N)$ be a statistical map. Suppose that $K(X, \pi_*Y) = \pi_*K(X,Y)$. Then, using Eqs. (3.12) and (3.13), the Gauss equation for π is

$$\begin{aligned} g_N(R^N(\pi_*X,\pi_*Y)\pi_*Z,\pi_*W) &= g_M(R^M(X,Y)Z,W) \\ &+ g_N((\nabla\pi_*)(X,Z),(\nabla^*\pi_*)(Y,W)) \\ &- g_N((\nabla\pi_*)(Y,Z),(\nabla^*\pi_*)(X,W)), \end{aligned} \tag{4.1}$$

and its dual can be written as

$$\begin{aligned} g_N(R^{*N}(\pi_*X,\pi_*Y)\pi_*Z,\pi_*W) &= g_M(R^{*M}(X,Y)Z,W) \\ &+ g_N((\nabla^*\pi_*)(X,Z),(\nabla\pi_*)(Y,W)) \\ &- g_N((\nabla^*\pi_*)(Y,Z),(\nabla\pi_*)(X,W)), \end{aligned} \tag{4.2}$$

for all $X, Y, Z, W \in \Gamma((ker\pi)^\perp)$.

Given an orthonormal basis $\{e_i | i = 1, 2, 3, \cdots, r-1, r\}$ of $(ker\pi_*)^\perp$, the scalar curvature defined on $(ker\pi_*)^\perp$ is expressed as

$$\tau = \sum_{1\leq i<j\leq r} g_M(R^M(e_i,e_j)e_j,e_i),$$

and for an orthonormal basis $\{v_\alpha | \alpha = r+1, r+2, \cdots, n\}$, we put

$$h_{ij}^\alpha = g_N((\nabla\pi_*)(e_i,e_j),v_\alpha),$$
$$h_{ij}^{*\alpha} = g_N((\nabla^*\pi_*)(e_i,e_j),v_\alpha),$$

$$||h||^2 = \sum_{i,j=1}^{r} g_N((\nabla\pi_*)(e_i, e_j), (\nabla\pi_*)(e_i, e_j)),$$

$$||h^*||^2 = \sum_{i,j=1}^{r} g_N((\nabla^*\pi_*)(e_i, e_j), (\nabla^*\pi_*)(e_i, e_j)),$$

$$trace(h) = \sum_{i=1}^{r}(\nabla\pi_*)(e_i, e_i), \quad trace(h^*) = \sum_{i=1}^{r}(\nabla\pi_*)(e_i, e_i),$$

$$||trace(h)||^2 = g_N(trace(h), trace(h)),$$

$$||trace(h^*)||^2 = g_N(trace(h^*), trace(h^*)).$$

For a point $p \in M$, consider a plane section $\mathcal{L} \subset T_pM$ spanned by $\{E = e_1, F = e_2\}$. The sectional curvature $\mathcal{K}(\mathcal{L})$ is then given by

$$\mathcal{K}(\mathcal{L}) = g_M(R^M(e_1, e_2)e_2, e_1).$$

Substituting $X = T = e_1$ and $Y = Z = e_2$ into Eq. (4.1), we obtain

$$\begin{aligned}\mathcal{K}(\mathcal{L}) &= c - \sum_{\alpha=r+1}^{n}\left(2(h_{12}^{0\alpha})^2 - \frac{1}{2}((h_{12}^{\alpha})^2 + (h_{12}^{*\alpha})^2)\right)\\ &+ \sum_{\alpha=r+1}^{n}\left(2h_{11}^{0\alpha}h_{22}^{0\alpha} - \frac{1}{2}(h_{11}^{\alpha}h_{22}^{\alpha} + h_{11}^{*\alpha}h_{22}^{*\alpha})\right)\\ &= c + 2\sum_{\alpha=r+1}^{n}\left(h_{11}^{0\alpha}h_{22}^{0\alpha} - (h_{12}^{0\alpha})^2\right)\\ &+ \frac{1}{2}\sum_{\alpha=r+1}^{n}\left((h_{12}^{\alpha})^2 + (h_{12}^{*\alpha})^2 - h_{11}^{\alpha}h_{22}^{\alpha} - h_{11}^{*\alpha}h_{22}^{*\alpha}\right).\end{aligned} \tag{4.3}$$

By reformulating Eq. (4.3), we get

$$\begin{aligned}\mathcal{K}(\mathcal{L}) &= -c + 2\mathcal{K}^0(\mathcal{L})\\ &+ \frac{1}{2}\sum_{\alpha=r+1}^{n}\left((h_{12}^{\alpha})^2 + (h_{12}^{*\alpha})^2 - h_{11}^{\alpha}h_{22}^{\alpha} - h_{11}^{*\alpha}h_{22}^{*\alpha}\right),\end{aligned} \tag{4.4}$$

where $\mathcal{K}^0(\mathcal{L})$ represents the sectional curvature of the Levi-Civita connection corresponding to the plane section $\mathcal{L}$.

Alternatively, by setting $X = T = e_i$ and $Y = Z = e_j$ in Eq. (4.1), we arrive at

$$r(r-1)\frac{c}{2} = \tau + \sum_{1\le i<j\le r}\Big(g_N((\nabla\pi_*)(e_i, e_j), (\nabla^*\pi_*)(e_i, e_j))$$
$$- g_N((\nabla\pi_*)(e_j, e_j), (\nabla^*\pi_*)(e_i, e_i))\Big).$$

But we know that $2h_{ij}^{0\alpha} = h_{ij}^{\alpha} + h_{ij}^{*\alpha}$, where $h_{ij}^{0\alpha} = g_N((\hat{\nabla}\pi_*)(e_i, e_j), \upsilon_\alpha)$. Then we can write

$$r(r-1)\frac{c}{2} = \tau + 2\sum_{1\le i<j\le r} g_N((\hat{\nabla}\pi_*)(e_i, e_j), (\hat{\nabla}\pi_*)(e_i, e_j))$$
$$-\frac{1}{2}\sum_{1\le i<j\le r}\Big(g_N((\nabla\pi_*)(e_i, e_j), (\nabla\pi_*)(e_i, e_j))$$
$$+ g_N((\nabla^*\pi_*)(e_i, e_j), (\nabla^*\pi_*)(e_i, e_j))\Big)$$
$$-2\sum_{1\le i<j\le r} g_N((\hat{\nabla}\pi_*)(e_j, e_j), (\hat{\nabla}\pi_*)(e_i, e_i))$$
$$+\frac{1}{2}\sum_{1\le i<j\le r}\Big(g_N((\nabla\pi_*)(e_j, e_j), (\nabla\pi_*)(e_i, e_i))$$
$$+ g_N((\nabla^*\pi_*)(e_j, e_j), (\nabla^*\pi_*)(e_i, e_i))\Big),$$

which can be reduced to

$$r(r-1)\frac{c}{2} = \tau + 2\sum_{\alpha=r+1}^{n}\sum_{1\le i<j\le r}(h_{ij}^{0\alpha})^2$$
$$-\frac{1}{2}\sum_{\alpha=r+1}^{n}\sum_{1\le i<j\le r}\Big((h_{ij}^{\alpha})^2 + (h_{ij}^{*\alpha})^2\Big)$$
$$-2\sum_{\alpha=r+1}^{n}\sum_{1\le i<j\le r} h_{ii}^{0\alpha}h_{jj}^{0\alpha}$$
$$+\frac{1}{2}\sum_{\alpha=r+1}^{n}\sum_{1\le i<j\le r}\Big(h_{ii}^{\alpha}h_{jj}^{\alpha} + h_{ii}^{*\alpha}h_{jj}^{*\alpha}\Big). \tag{4.5}$$

Upon recalling the Gauss equation associated with the Levi-Civita connection, Eq. (4.5) becomes

$$r(r-1)\frac{c}{2} = \tau + r(r-1)c - 2\tau^0 - \frac{1}{2}\sum_{\alpha=r+1}^{n}\sum_{1\le i<j\le r}\left((h_{ij}^{\alpha})^2 + (h_{ij}^{*\alpha})^2\right)$$
$$+\frac{1}{2}\left(h_{ii}^{\alpha}h_{jj}^{\alpha} + h_{ii}^{*\alpha}h_{jj}^{*\alpha}\right), \tag{4.6}$$

where τ^0 represents the scalar curvature of Levi-Civita connection defined on $(ker\pi_*)^{\perp}$.

From the combination of Eqs. (4.4) and (4.6), it follows that

$$\tau - \mathcal{K}(\mathcal{L}) = 2(\tau^0 - \mathcal{K}^0(\mathcal{L})) - (r-2)(r+1)\frac{c}{2}$$
$$+\frac{1}{2}\sum_{\alpha=r+1}^{n}\sum_{1\le i<j\le r}\left((h_{ij}^{\alpha})^2 + (h_{ij}^{*\alpha})^2\right)$$
$$-\frac{1}{2}\left(h_{ii}^{\alpha}h_{jj}^{\alpha} + h_{ii}^{*\alpha}h_{jj}^{*\alpha}\right)$$
$$-\frac{1}{2}\sum_{\alpha=r+1}^{n}\left((h_{12}^{\alpha})^2 + (h_{12}^{*\alpha})^2 - h_{11}^{\alpha}h_{22}^{\alpha} - h_{11}^{*\alpha}h_{22}^{*\alpha}\right).$$

The terms in the above equation resemble those in the following algebraic lemma from [5]:

Lemma 4.1 *Let $s \geq 3$ be an integer and a_i, $i = 1, 2, 3, \cdots, s$, be s real numbers. Then, we have*

$$\sum_{1\le i<j\le s} a_i a_j - a_1 a_2 \le 2\left(\frac{s-2}{s-1}\right)\left(\sum_{i=1}^{s} a_i\right)^2.$$

Moreover, equality in the above inequality holds if and only if $a_1 + a_2 = a_3 = \cdots = a_s$.

Consequently, we have

$$\tau - \mathcal{K}(\mathcal{L}) \geq 2(\tau^0 - \mathcal{K}^0(\mathcal{L})) - (r-2)(r+1)\frac{c}{2}$$
$$+\frac{1}{2}\sum_{\alpha=r+1}^{n}\sum_{1\le i<j\le r}\left((h_{ij}^{\alpha})^2 + (h_{ij}^{*\alpha})^2\right)$$

$$
\begin{aligned}
& -\frac{1}{2}\sum_{\alpha=r+1}^{n}\left((h_{12}^{\alpha})^2+(h_{12}^{*\alpha})^2\right) \\
& +\left(\frac{r-2}{r-1}igg)\left(\sum_{i=1}^{r}h_{ii}^{\alpha}\right)^2+\left(\frac{r-2}{r-1}\right)\left(\sum_{i=1}^{r}h_{ii}^{*\alpha}\right)^2 \\
& \geq 2(\tau^0-\mathcal{K}^0(\mathcal{L}))-(r-2)(r+1)\frac{c}{2} \\
& +\left(\frac{r-2}{r-1}\right)\left(||trace(h)||^2+||trace(h^*)||^2\right).
\end{aligned}
$$

As a result, we derive the following theorem:

Theorem 4.2 *Let $\pi : M \to N$ be a statistical map from a statistical manifold (M^m, ∇^M, g_M) to a statistical manifold $(N^n(c), \nabla^N, g_N)$ of constant curvature $c \in \mathbb{R}$ with $r \geq 3$. Then*

$$
\begin{aligned}
\tau-\mathcal{K}(\mathcal{L})-2(\tau^0-\mathcal{K}^0(\mathcal{L})) \geq & -(r-2)(r+1)\frac{c}{2} \\
& +\left(\frac{r-2}{r-1}\right)\left(||trace(h)||^2+||trace(h^*)||^2\right).
\end{aligned}
$$

Moreover, the equality holds if and only if

$$
\begin{aligned}
& h_{11}^{\alpha}+h_{22}^{\alpha}=h_{33}^{\alpha}=\cdots=h_{rr}^{\alpha}, \\
& h_{11}^{*\alpha}+h_{22}^{*\alpha}=h_{33}^{*\alpha}=\cdots=h_{rr}^{*\alpha}, \\
& h_{ij}^{\alpha}-h_{ij}^{*\alpha}=0,
\end{aligned}
$$

$i \neq j$, $(i, j) \neq (1, 2), (2, 1)$, *and for any* $\alpha \in \{r+1, \cdots, n\}$.

Acknowledgments The authors would like to express their sincere gratitude to the anonymous reviewers for their valuable comments and constructive suggestions, which have significantly improved the quality and clarity of this manuscript.

References

1. Abe, N., Hasegawa, K.: An affine submersion with horizontal distribution and its applications. Differ. Geom. Appl. **14**, 235–250 (2001)
2. Ali, S.A., Cafaro, C., Kim, D.H., Mancini, S.: The effect of microscopic correlations on the information geometric complexity of Gaussian statistical models. Physica A **389**, 3117–3127 (2010)
3. Amari, S.: Differential-geometrical methods in statistics. In: Lecture Notes in Statistics, vol. 28, Springer, New York (1985)

4. Ay, N., Tuschmann, W.: Dually flat manifolds and global information geometry. Open Syst. Inf. Dyn. **9**, 195–200 (2002)
5. Chen, B.Y., Mihai, A., Mihai, I.: A Chen first inequality for statistical submanifolds in Hessian manifolds of constant Hessian curvature. Result. Math. **74**, 165 (2019)
6. Fischer, A.E.: Riemannian maps between Riemannian manifolds. Contemp. Math. **132**, 331–366 (1992)
7. Furuhata, H.: Hypersurfaces in statistical manifolds. Differ. Geom. Appl. **27**, 420–429 (2009)
8. Furuhata, H.: Toward differential geometry of statistical submanifolds. Inf. Geom. **7**(1), 99–108 (2024)
9. Gomez, I.S.: Notions of the ergodic hierarchy for curved statistical manifolds. Physica A **484**, 117–131 (2017)
10. Kazan, A.: Conformally-projectively flat trans-Sasakian statistical manifolds. Physica A **535**, 122441 (2019)
11. Kazan, S., Kazan, A.: Sasakian Statistical Manifolds with Semi-Symmetric Metric Connection. Universal J. Math. Appl. **1**(4), 226–232 (2018)
12. Kazan, S., Takano, K.: Anti-invariant holomorphic statistical submersions. Results Math. **78**(4), 128 (2023)
13. O'Neill, B.: The fundamental equations of a submersion. Michigan Math. J. **13**, 459–469 (1966)
14. O'Neill, B.: Semi-Riemannian Geometry with Application to Relativity. Academic Press, New York (1983)
15. O'Neill, B.: Submersions and geodesics. Duke Math. J. **34**, 363–373 (1967)
16. Opozda, B.: Bochner's technique for statistical structures. Ann. Global Anal. Geom. **48**, 357–395 (2015)
17. Opozda, B.: A sectional curvature for statistical structures. Linear Algebra Appl. **497**, 134–161 (2016)
18. Rao, C.R.: Information and accuracy attainable in the estimation of statistical parameters. Bull. Calcutta Math. Soc. **37**, 81–91 (1945)
19. Sahin, B.: Chen's first inequality for Riemannian maps. Ann. Polon. Math. **117**(3), 249–258 (2016)
20. Sahin, B.: Riemannian Submersions, Riemannian Maps in Hermitian Geometry, and their Applications. Academic Press, Elsevier, Amsterdam (2017)
21. Siddiqui, A.N., Ahmad, K.: A study of statistical submersions. Tamkang J. Math. (2023). https://doi.org/10.5556/j.tkjm.55.2024.5044
22. Siddiqui, A.N., Al-Solamy, F.R., Shahid, M.H., Mihai, I.: On CR-statistical submanifolds of holomorphic statistical manifolds. Filomat **35**(11), 3571–3584 (2021)
23. Siddiqui, A.N., Chen, B.Y., Siddiqi, M.D.: Chen inequalities for statistical submersions between statistical manifolds. Int. J. Geom. Methods Mod. Phys. **18**(4), paper no. 2150049, 17 (2021)
24. Siddiqui, A.N., Mofarreh, F., Alkhaldi, A.H., Ali, A.: Optimization methods to statistical submanifolds in statistical warped product manifolds. Filomat **38**(18) (2024)
25. Siddiqui, A.N., Uddin, S., Shahid, M.H., Chen, B.Y.: Chen's inequality for Kähler-like statistical submersions. Intern. Elec. J. Geom. **15**, 277–286 (2022)
26. Takano, K.: Examples of the statistical submersion on the statistical model. Tensor, N. S. **65**, 170–178 (2004)
27. Takano, K.: Statistical manifolds with almost complex structures and its statistical submersions. Tensor, N. S. **65**, 123–137 (2004)
28. Takano, K.: Statistical manifolds with almost contact structures and its statistical submersions. J. Geom. **85**, 171–187 (2006)
29. Vîlcu, G.E.: Almost product structures on statistical manifolds and para-Kähler-like statistical submersions. Bull. Sci. Math. **171**, paper no. 103018, 21 (2021)
30. Vîlcu, A.D., Vîlcu, G.E.: Statistical manifolds with almost quaternionic structures and quaternionic Kähler-like statistical submersions. Entropy **17**, 6213–6228 (2015)
31. Watson, B.: Almost Hermitian submersions. J. Differ. Geom. **11**(1), 147–165 (1976)

Hyperbolic Ricci-Yamabe Solitons and η-Hyperbolic Ricci-Yamabe Solitons

Mohd Danish Siddiqi

Abstract We introduced and discussed the idea of hyperbolic Ricci-Yamabe solitons associated with perfect fluid spacetime in this research note. Additionally, we examine a perfect fluid spacetime, which accommodates the hyperbolic Ricci-Yamabe solitons and the rate of change of the hyperbolic Ricci-Yamabe solitons coupled with a conformal vector field, $\varphi(\mathcal{Q})$-vector field with Ricci collineation condition. We also analyze the rate of change of the hyperbolic Ricci soliton and investigate the gradient hyperbolic Ricci-Yamabe soliton on perfect fluid spacetime with scalar concircular field. Furthermore, we investigate the energy conditions for perfect fluid spacetime in terms of gradient hyperbolic Ricci-Yamabe solitons with a scalar concircular field. In the end, we introduced a more generic notion of η-hyperbolic Ricci-Yamabe solitons and proved that a spacetime admitting the η-hyperbolic Ricci-Yamabe solitons with a conformal vector field is a perfect fluid spacetime.

Keywords Hyperbolic Ricci-Yamabe solitons · Gradient hyperbolic Ricci-Yamabe soliton · η-hyperbolic Ricci-Yamabe solitons · Conformal vector field · Scalar concircular field · Energy condition

1 Introduction

The General Theory of Relativity (GTR) is the name given to Einstein's theory of gravity. This theory states that the energy-momentum tensor $mathcalT$ is the source of the gravitational field and gravitational waves, which are represented by the spacetime curvature. GTR is the foundation of all theoretical disciplines. All current particle physics equations used in astrophysics, plasma physics, nuclear physics, etc. are based on the Einstein equations, which describe the evolution of

M. D. Siddiqi (✉)
Department of Mathematics, College of Science, Jazan University, Jazan, Saudi Arabia
e-mail: msiddiqi@jazanu.edu.sa

B.-Y. Chen, M. A. Choudhary (eds.), *Geometric Inequalities and Applications*, Infosys Science Foundation Series in Mathematical Sciences,
https://doi.org/10.1007/978-981-95-5148-4_9

spacetime curvature. The best way to understand general relativity is through the mathematical development of differential geometry and relativistic fluid models. Spacetime is best represented as a curved manifold, according to GTR's central notion [1].

The spacetime of GTR and cosmology can be simulated using a time-oriented connected four-dimensional Lorentzian manifold, a special subclass of pseudo-Riemannian manifolds with Lorentzian metric g and signature $(-,+,+,+)$. For GTR, this manifold has significant consequences. The first step in establishing the geometry of the Lorentzian manifold $\mathcal{M}^4$ is to study vector nature on it. Therefore, the ideal choice for addressing GTR [1] is the Lorentzian manifold $(\mathcal{M}^4, g)$.

One of the main elements of the matter of spacetime is the energy-momentum tensor $\mathcal{T}$. In addition to density, pressure, and other dynamical and kinematical characteristics including shear, expansion, acceleration, and velocity, matter is believed to be a fluid [46]. The universe's matter component is thought to behave like a perfect fluid in conventional cosmological models. The absence of viscosity and heat conduction characterizes a perfect fluid, which is also known as an isotropic or star-shaped fluid at rest. The dust matter fluid ($p = 0$) [29, 33] is the most basic illustration of the perfect fluid. In GTR, perfect fluids are frequently used to simulate idealized distributions. Moreover, In GTR, "stiff matter fluid" is described by the relation $p = \sigma$ [29].

Definition 1.1 ([29, 46]) A quasi-Einstein Lorentzian manifold is referred to as perfect fluid spacetime (PFS) if the Ricci tensor has the composition

$$\mathcal{S}_{Ric} = ag + b\eta \otimes \eta, \tag{1.1}$$

wherein g is the Lorentzian metric, a and b are scalars, and 1-form η is metrically equivalent to a unit time-like vector field.

Furthermore, the Lorentzian manifold is a manifold that permits a time-like vector field [27, 28].

The investigation of exact solutions to the Einstein field equations gave rise to quasi-Einstein manifolds. For instance, the quasi-Einstein manifolds [27, 28] are the *Robertson-Walker spacetime*. In GTR, they can also be viewed as a model of the perfect fluid spacetime [32].

Definition 1.2 ([29]) The energy-momentum tensor $\mathcal{T}$ in conjunction with a perfect fluid has the following shape:

$$\mathcal{T}(F_1, F_2) = \rho g(F_1, F_2) + (\sigma + p)\eta(F_1)\eta(F_2), \tag{1.2}$$

for any vector field $F_1, F_2 \in \chi(\mathcal{M}^4)$, and the isotropic pressure is denoted by p, the energy density by σ, and the Lorentzian metric by g. This means that $g(F_1, \zeta) = \eta(F_1)$, where η is 1-form, which corresponds to the fluid's time-like velocity vector ζ, and $g(\zeta, \zeta) = -1$ [32].

If $\sigma = 3p$ [29], then matter in spacetime originates from a radiation fluid. Furthermore, there are important uses for Eq. (1.2) in star structure and cosmology. The scalar field theory and electromagnetic energy-momentum tensors are two more instances of energy-momentum tensors.

The field equation governing perfect fluid motion is Einstein's gravitational equation (in short $EGFE$) [29].

$$\mathcal{S}_{Ric} = \kappa \mathcal{T} - \left(\Lambda - \frac{\mathcal{R}_{scal}}{2} \right) g, \tag{1.3}$$

where the scalar curvature of g is $\mathcal{R}_{scal}$, the cosmological constant is Λ, the gravitational constant is κ (which may be expressed as $8\pi G$, with G being the universal gravitational constant), and the Ricci tensor is $\mathcal{S}_{Ric}$.

Over the past 20 years, a number of scholars have used a range of geometric tools, such as curvature tensors [2] and, most significantly, geometric flows [26, 45], to thoroughly examine the properties of symmetries in the perfect fluid spacetime (PFS). There are many symmetries in the geometry of matter and spacetime [34, 42].

Metric symmetries are essential because they facilitate the resolution of numerous problems. They are mostly employed in GTR to classify solutions to Einstein field equations. Among these symmetries are Ricci solitons associated with the Ricci flow of spacetime. Ricci flow is important because it helps understand the concepts of entropy and energy in GTR [21]. Ricci solitons are areas where the curvatures obey a self-likeness [22].

First, spacetime symmetries were studied in connection with the Ricci soliton by Ahsan and Ali [4]. Blaga used Einstein, Ricci, and their extensions, that is, η-Einstein solitons [7] and η-Ricci solitons [15] in a PFS, respectively, to explain the geometrical axioms of a PFS in [8]. Furthermore, Venkatesha and Kumara [48] used Ricci solitons to investigate the characterization of PFS using the torse-forming vector field and Jacobi. In [38], Danish and Shah-Alam talked about the conformal Ricci solitons on PFS. Recently, Danish and Fatemah in [36, 37] looked at various properties of PFS with hyperbolic Ricci solitons. Additionally, perfect fluid spacetime, magneto-fluid spacetime [35, 44], and static spacetime [41] were studied by Siddiqi et al. utilizing Ricci-Yamabe solutions. Recent research on Ricci-Yamabe solitons on imperfect fluid-generalized Robertson-Walker spacetime [3, 40] was conducted by Alkhaldi et al. [5].

In 2010, Dai and colleagues introduced the notion of hyperbolic geometric flow. Later, Faraji, and colleagues [6, 18] introduced the concepts of gradient hyperbolic Ricci soliton and hyperbolic Ricci solitons. Recently, Blaga and Özgür have investigated the idea of hyperbolic Yamabe solitons and hyperbolic Ricci solitons in different methods (for more details, see [9, 11, 25]).

In the present study, we introduced the notion of the hyperbolic Ricci-Yamabe soliton ($HRYS$) and gradient hyperbolic Ricci-Yamabe soliton ($GHRYS$). The hyperbolic Ricci-Yamabe soliton is a generalization of the hyperbolic Ricci solitons and hyperbolic Yamabe solitons and their gradient version. We analyze the relativistic PFS in terms of a gradient hyperbolic Ricci-Yamabe soliton ($GHRYS$) and a hyperbolic Ricci-Yamabe soliton ($HRYS$) with different vector fields, drawing inspiration from previous research.

2 The Einstein Field Equation Is Satisfied by Perfect Fluid Spacetime with the Cosmic Constant $\Lambda > 0$

From Einstein's field equations by adding a cosmological constant, which creates a static universe, in accordance with his theory, in modern cosmology, it is thought to be a potential dark energy contender that could account for the Universe's faster expansion.

We find from Eqs. (1.2) and (1.3)

$$\mathcal{S}_{Ric}(F_1, F_2) = -\left(\Lambda - \frac{\mathcal{R}_{scal}}{2} + \kappa p\right) g(F_1, F_2) + \kappa(\sigma + p)\eta(F_1)\eta(F_2). \tag{2.1}$$

The values of pressure and density for PFS $(\mathcal{M}^4, g)$ are determined by comparing (10.2) and (2.1).

$$p = \frac{1}{\kappa}\left(\frac{\mathcal{R}_{scal}}{2} - \Lambda - a\right), \qquad \sigma = \frac{1}{\kappa}\left(a + b + \Lambda - \frac{\mathcal{R}_{scal}}{2}\right). \tag{2.2}$$

In addition, we gain

$$a = \frac{\mathcal{R}_{scal}}{2} - \kappa p - \Lambda, \qquad b = \kappa(\sigma + p). \tag{2.3}$$

In light of (4.7) one can articulate the following

Theorem 2.1 *If a PFS $(\mathcal{M}^4, g)$ fulfills the $EGFE$ with the cosmological constant Λ, then density σ and the pressure p are governed by (2.2).*

Let $(\mathcal{M}^4, g)$ be a PFS fulfilling (2.1). Contracting (2.1) and assumed that $g(\zeta, \zeta) = -1$, we turn up

$$\mathcal{R}_{scal} = \kappa(\sigma - 3p) + 4\Lambda. \tag{2.4}$$

Theorem 2.2 *If a PFS $(\mathcal{M}^4, g)$ with pressure p and density σ fulfils the EGFE with the cosmological constant Λ, then the scalar curvature $\mathcal{R}_{scal}$ is $4\Lambda + \kappa(\sigma - 3p)$.*

Remark 2.3 In case, $\Lambda > 0$, the cosmological constant Λ is essential to interpreting the observed accelerating expansion of the cosmos and supernova [30]. Moreover, for positive cosmological constant Λ, a spacetime is a *de-Sitter spacetime*.

Remark 2.4 ([30]) The Universe contains dark sector for the negative cosmological constant $\Lambda < 0$, that is, ΛCDM. Therefore, model cannot accelerate, and the late time expansion of Universe occurs.

If the source of matter is radiation type and stiff matter type, then from (2.4), we get

$$\Lambda = \frac{\mathcal{R}}{4}, \qquad \Lambda = \frac{\mathcal{R}}{4} + \frac{\kappa p}{2}. \tag{2.5}$$

Given (2.5) and Remark 2.3, we may now state the following findings:

Theorem 2.5 *If the source of matter is radiation type in a spacetime $(\mathcal{M}^4, g)$ with pressure p and density σ and fulfils the $EGFE$ with the cosmological constant $\Lambda > 0$, then the spacetime is an accelerating Universe.*

Theorem 2.6 *If the stiff matter is the source of spacetime $(\mathcal{M}^4, g)$ with pressure p and density σ and fulfils the $EGFE$ with the cosmological constant $\Lambda > 0$, then the spacetime is an accelerating Universe.*

Corollary 2.7 *If the source of matter is radiation type in a spacetime $(\mathcal{M}^4, g)$ with pressure p and density σ and fulfils the $EGFE$ with the cosmological constant $\Lambda > 0$, then the spacetime is a de-Sitter spacetime.*

Corollary 2.8 *If the stiff matter is the source in spacetime $(\mathcal{M}^4, g)$ with pressure p and density σ and fulfils the $EGFE$ with the cosmological constant $\Lambda > 0$, then the spacetime is a de-Sitter spacetime.*

Corollary 2.9 *If the source of matter is radiation type in a spacetime $(\mathcal{M}^4, g)$ with pressure p and density σ and fulfils the $EGFE$ with the cosmological constant $\Lambda > 0$, then the spacetime is a Supernova.*

Corollary 2.10 *If the stiff matter is the source in spacetime $(\mathcal{M}^4, g)$ with pressure p and density σ and fulfils the $EGFE$ with the cosmological constant $\Lambda > 0$, then the spacetime is a Supernova.*

3 Dust Matter Fluid Spacetime Satisfying Einstein Field Equation with the Cosmological Constant $\Lambda < 0$

For dust fluid [33], from (2.4), we turn up

$$\Lambda = \frac{\mathcal{R} - \kappa\sigma}{4}. \tag{3.1}$$

Equation (3.1) implies that if $\mathcal{R} > \kappa\sigma$, then we can obtain similar results to mentioned above for dust matter fluid with positive Λ. However, now we are interested in the situation if $\Lambda < 0$, that is, $\mathcal{R} < \kappa\sigma$.

Thus, using the aforementioned facts and Remark 2.8, we obtain the following outcomes.

Theorem 3.1 *If the dust matter is the source of spacetime* $(\mathcal{M}^4, g)$ *with pressure* p *and density* σ *and fulfils the* $EGFE$ *with the cosmological constant* $\Lambda < 0$*, then the spacetime is a non-accelerating Universe with a late time expansion rate of the Universe.*

Corollary 3.2 *If the dust matter is the source of spacetime* $(\mathcal{M}^4, g)$ *with pressure* p *and density* σ *and fulfils the* $EGFE$ *with the cosmological constant* $\Lambda < 0$*, then the non-accelerating Universe contains dark sector.*

Theorem 3.3 *If a dust fluid spacetime* $(\mathcal{M}^4, g)$ *with pressure* p *and density* σ *fulfils the* $EGFE$ *with the cosmological constant* $\Lambda < 0$*, then the dust matter is the* ΛCDM*.*

4 Development of Hyperbolic Ricci-Yamabe Solitons

The principles of Ricci flow were first presented by Hamilton [21] in 1988. It demonstrates that the limit of the Ricci flow's solutions is the soliton of Ricci. In addition, geometric flow theory, and the Ricci flow in particular, has attracted the attention of many mathematicians throughout the last 20 years.

The Ricci flow [21] occurs when the family of metrics $g(t)$ on a Riemannian manifold M evolves, if

$$\frac{\partial}{\partial t} g(t) = -2\mathcal{S}_{Ric}(t) g(t), \quad g_0 = g(0). \tag{4.1}$$

Definition 4.1 ([21]) A Ricci soliton on the Riemannian manifold $(\mathcal{M}, g)$ is a data (g, ζ, λ) that obeys

$$\frac{1}{2} L_{\zeta} g + \lambda g + \mathcal{S}_{Ric} = 0, \tag{4.2}$$

wherein $\mathcal{S}_{Ric}$ is the Ricci tensor, and for the vector field ζ, the Lie-derivative is $L_{\zeta} g$. A Ricci soliton is shrinking, expanding or stable soliton is the manifold (M, g, ζ, λ), depending on the constant λ, regardless of whether $\lambda < 0$, $\lambda > 0$, or $\lambda = 0$.

Kong and Liu, however, explored the hyperbolic Ricci flow [16]. A system of second-order nonlinear evolution partial differential equations makes up this flow. Hyperbolic Ricci flow illustrates the wave properties of metrics and manifold curvatures. The evolution equation that follows explains the hyperbolic Ricci flow, which is consequently driven by Ricci flow.

$$\frac{1}{2} \frac{\partial^2}{\partial t^2} g(t) = -\mathcal{S}_{Ric}(t) g(t), \quad g_0 = g(0), \quad \frac{\partial}{\partial t} g_{ij} = h_{ij}, \tag{4.3}$$

where h_{ij} is a symmetric 2-tensor field. Thus, a self-similar solution of hyperbolic Ricci flow is called a hyperbolic Ricci soliton (HRS) and has the following properties:

Definition 4.2 ([18]) A Riemannian manifold $(\mathcal{M}^n, g)$ is a HRS if and only if M has a vector field ζ and real scalars μ and λ such that

$$\frac{1}{2}L_\zeta L_\zeta g + \lambda L_\zeta g + \mathcal{S}_{Ric} = \mu g. \tag{4.4}$$

The types of solitons and the rate of the underlying type are indicated by λ and μ in (4.4), respectively. Additionally, μ has geometric meaning and denotes the rate of change in the solutions. Regardless of whether $\mu < 0$, $\mu > 0$, or $\mu = 0$, the rate of change of the HRS can be expanding, contracting, or roughly stable, depending on the constant μ.

Example 4.3 Let $\mathbb{H}_3$ denote the three-dimensional Heisenberg group. Since, any simply connected nilpotent Lie group is diffeomorphic to $\mathbb{R}^n$. So give $\mathbb{R}^3$ its standard coordinates (x, y, z). We consider a left-invariant Lorentzian metric g on $\mathbb{H}_3$ which is defined by

$$g = -dx^2 + dy^2 + (xdy + dz)^2.$$

Define a vector field $X = (2z + xy)F_1 + yF_2 + xF_3$, where

$$F_1 = \frac{\partial}{\partial z}, \quad F_2 = \frac{\partial}{\partial y} - x\frac{\partial}{\partial z}, \quad F_3 = \frac{\partial}{\partial x}$$

are frame fields. Then, one can easily check that

$$\mathrm{Ric}(g) - \frac{5}{2}\mathcal{L}_X g + \mathcal{L}_X(\mathcal{L}_X g) = -\frac{5}{2}g.$$

Hence, $\left(\mathbb{H}_3, g, X, -\frac{5}{2}, -\frac{5}{2}\right)$ is a hyperbolic Ricci structure.

Since $\nabla_1 X_2 - \nabla_2 X_1 = x \neq 0$, the hyperbolic Ricci soliton is not of gradient type.

Hyperbolic Yamabe flow and hyperbolic Yamabe solitons were proposed by Blaga and Özgür in [10] as an evolution equation

$$\frac{\partial^2}{\partial t^2}g(t) = -\mathcal{R}_{scal}(t)g(t), \tag{4.5}$$

where $\mathcal{R}_{scal}$ indicates the scalar curvature, and an equation-satisfying stationary solution of it

$$L_\zeta L_\zeta g + \lambda L_\zeta g = (\mu - \mathcal{R}_{scal})g \tag{4.6}$$

with two scalars λ and μ and a smooth vector field $zeta$, and the derivative of the metric g in the direction of ζ is $L_\zeta g$.

Guler and Crasmareanu [20] have presented the research of a novel geometric flow known as the Ricci-Yamabe map, which is a scalar combination of the Ricci and Yamabe flows. The Ricci-Yamabe flow of type (α, β) is another name for this. According to [20], the Ricci-Yamabe flow is the development of metrics at the Riemannian or semi-Riemannian manifold.

$$\frac{\partial}{\partial t}g(t) = -2\alpha\mathcal{S}_{Ric}(t) + \beta\mathcal{R}_{scal}(t)g(t), \quad g_0 = g(0), \quad t \in (a, b). \tag{4.7}$$

The sign of the associated scalars α and β determines whether the Ricci-Yamabe flow is Riemannian, semi-Riemannian, or singular Riemannian. Some geometrical or physical models, such as relativistic theories, can benefit from this type of multiple choices. Consequently, the Ricci-Yamabe soliton for the Ricci-Yamabe flow naturally appears as the soliton limit.

5 Hyperbolic Ricci-Yamabe Flow

The author was greatly inspired by this to present the idea of the *hyperbolic Ricci-Yamabe solitons*, which are described as the subsequent development of the hyperbolic Ricci-Yamabe flow equation in such a way that

$$\frac{\partial^2}{\partial t^2}g(t) = -2\alpha\mathcal{S}_{Ric}(t)+\beta\mathcal{R}_{scal}(t)g(t), \quad g_0 = g(0), \quad \frac{\partial}{\partial t}g(t) = h(t), \tag{5.1}$$

where h is a symmetric 2-tensor field and $g(t)$ is the solution of the hyperbolic Ricci-Yamabe flow on a Riemannian manifold $(\mathcal{M}^n, g)$ if there exist a function $f(t)$ and 1-parametric flow $\psi(t) : M \longrightarrow M$ such that the solution of (5.1) is

$$g(t) = f(t)\psi(t)^* g(0). \tag{5.2}$$

Definition 5.1 A stationary solution $g(t)$ (or self-similar solution) of (5.1) on a hyperbolic Ricci-Yamabe soliton ($HRYS$) is a Riemannian manifold $(\mathcal{M}^n, g)$ if a vector field ζ on $\mathcal{M}$ and real scalars μ and λ exist such that

$$\frac{1}{2}L_\zeta L_\zeta g + \lambda L_\zeta g + \alpha\mathcal{S}_{Ric} = (\mu - \beta\mathcal{R}_{scal})g. \tag{5.3}$$

A hyperbolic Ricci-Yamabe soliton is shrinker, expander, or stable soliton if the constant λ, regardless of whether $\lambda < 0$, $\lambda > 0$, or $\lambda = 0$. In addition the rate of change of hyperbolic Ricci-Yamabe soliton is shrinking, expanding, or stable soliton depending on the constant μ, whether $\mu < 0$, $\mu > 0$, or $\mu = 0$ [18].

Remark 5.2 A hyperbolic Ricci-Yamabe flow of type (α, β), which is precisely:

- Hyperbolic Ricci flow [16] if $\alpha = 1, \beta = 0$ (hyperbolic Ricci solitons [18])
- Hyperbolic Yamabe flow [10] if $\alpha = 0, \beta = 1$ (hyperbolic Yamabe solitons [10])
- Hyperbolic Einstein flow [13] if $\alpha = 1, \beta = -1$ (hyperbolic Einstein solitons [13])

A gradient hyperbolic Ricci-Yamabe soliton $(GHRYS)$ is called a $HRYS$ (g, λ, ζ, μ) [18] if there is a potential function f such that $\zeta = \nabla f$. Because of this, (5.3) can be translated as (4.41).

$$L_{\nabla f}(Hessf) + 2\lambda Hessf + \alpha \mathcal{S}_{Ric} = (\mu - \beta \mathcal{R}_{scal})g. \tag{5.4}$$

6 Results

Definition 6.1 ([17]) A vector field $zeta$ on a Riemannian manifold $(\mathcal{M}, g)$ is referred to as a conformal vector field if it meets the following relation:

$$L_\zeta g = 2\omega g, \tag{6.1}$$

where ω is the arbitrary nonzero smooth functions on $\mathcal{M}$. The smooth function σ is also known as a conformal coefficient. In particular, the conformal vector field with a vanishing conformal coefficient ($\omega = 0$) reduces to the Killing vector field, and with constant conformal coefficient ($\omega = cont.$) it becomes homothetic vector field.

Definition 6.2 ([12]) A Riemannian manifold $(\mathcal{M}, g)$ is considered to be admitted to Ricci collineation (RC) if

$$L_\zeta \mathcal{S}_{Ric} = 0, \tag{6.2}$$

wherein $\mathcal{S}_{Ric}$ is the Ricci tensor.

7 Main Results

Now, by the definition of hyperbolic Ricci-Yamabe solitons,

$$\begin{aligned}&(L_\zeta L_\zeta g)(F_1, F_2) + 2\lambda L_\zeta g(F_1, F_2) + 2\alpha \mathcal{S}_{Ric}(F_1, F_2)\\&= 2(\mu - \beta \mathcal{R}_{scal})g(F_1, F_2).\end{aligned} \tag{7.1}$$

Now, let ζ be a conformal vector field, and from (7.1) and (6.1) we turn up

$$L_\zeta(L_\zeta g)(F_1, F_2) + 2\lambda(L_\zeta g)(F_1, F_2) + 2\alpha\mathcal{S}_{Ric}(F_1, F_2)$$
$$= 2(\mu - \beta\mathcal{R}_{scal})g(F_1, F_2). \tag{7.2}$$

$$L_\zeta(2\omega g(F_1, F_2)) + 2\lambda(2\omega g(F_1, F_2) + 2\alpha\mathcal{S}_{Ric}(F_1, F_2)$$
$$= 2(\mu - \beta\mathcal{R}_{scal})g(F_1, F_2). \tag{7.3}$$

Once more using the conformal vector field formulation in (refdx5), we get

$$\mathcal{S}_{Ric}(F_1, F_2) = \frac{\mu - (\beta\mathcal{R}_{scal} + 2\lambda\omega + 2\omega^2)}{\alpha} g(F_1, F_2). \tag{7.4}$$

Consequently, we can draw the following conclusions:

Theorem 7.1 *If a PFS $(\mathcal{M}^4, g)$ admits the $HRYS$ $(g, \lambda, \zeta, \mu, \alpha, \beta)$ with a conformal vector field ζ, then the PFS $(\mathcal{M}^4, g)$ is Einstein, and the Einstein factor is* $\frac{\mu-(\beta\mathcal{R}_{scal}+2\lambda\omega+2\omega^2)}{\alpha}$.

In light of (7.4) and (2.1) we gain

$$-\left\{\Lambda - \frac{\mathcal{R}_{scal}}{2} + \kappa p + \frac{\mu - (\beta\mathcal{R}_{scal} + 2\lambda\omega + 2\omega^2)}{\alpha}\right\} g(F_1, F_2)$$
$$+ \kappa(\sigma + p)\eta(F_1)\eta(F_2) = 0. \tag{7.5}$$

Putting $F_1 = F_2 = \zeta$ in (7.5), we obtain

$$\lambda = \frac{\mu}{2\alpha\omega} - \left\{\frac{\kappa(\sigma + 2p)}{2\omega} + \left(1 - \frac{\beta}{\alpha}\right)\frac{\mathcal{R}_{scal}}{2\omega} + \frac{\Lambda}{2\omega} + \omega\right\}. \tag{7.6}$$

Furthermore, we discover the following corollary in the context of Definition 4.1.

Theorem 7.2 *If a PFS $(\mathcal{M}^4, g)$ admits the $HRYS$ $(g, \lambda, \zeta, \mu, \alpha, \beta)$ with a conformal vector ζ field, then the $HRYS$ is shrinking, steady, or expanding referring to as:*

1. $\frac{\mu}{2\alpha\omega} < \left\{\frac{\kappa(\sigma+2p)}{2\omega} + \left(1 - \frac{\beta}{\alpha}\right)\frac{\mathcal{R}_{scal}}{2\omega} + \frac{\Lambda}{2\omega} + \omega\right\}$
2. $\frac{\mu}{2\alpha\omega} = \left\{\frac{\kappa(\sigma+2p)}{2\omega} + \left(1 - \frac{\beta}{\alpha}\right)\frac{\mathcal{R}_{scal}}{2\omega} + \frac{\Lambda}{2\omega} + \omega\right\}$
3. $\frac{\mu}{2\alpha\omega} > \left\{\frac{\kappa(\sigma+2p)}{2\omega} + \left(1 - \frac{\beta}{\alpha}\right)\frac{\mathcal{R}_{scal}}{2\omega} + \frac{\Lambda}{2\omega} + \omega\right\}$, *respectively, provided* $\alpha, \mu \neq 0$

Depending on the constant μ, the rate of $HRYS$ is expanding, shrinking, or remaining steady, regardless of whether $\mu < 0$, $\mu > 0$, or $\mu = 0$. The hyperbolic Ricci-Yamabe flow's rate in a PFS $(\mathcal{M}^4, g)$ with a conformal vector field $zeta$ is thus determined from (7.6).

$$\mu = 2\alpha\lambda\omega + \kappa(\sigma + 2p) + (\alpha - \beta)\mathcal{R}_{scal} + \alpha(\Lambda + 2\omega^2). \tag{7.7}$$

Thus, we can articulate the following result:

Theorem 7.3 *If a PFS $(\mathcal{M}^4, g)$ admits the $HRYS$ $(g, \lambda, \zeta, \mu, \alpha, \beta)$ with a CVF ζ, then the rate of change of the $HRYS$ is expanding.*

Now, in view of Remark 5.2 and using $\alpha = 0$, $\beta = 1$ in (7.7), we can state the next corollary:

Corollary 7.4 *If a PFS $(\mathcal{M}^4, g)$ admits the hyperbolic Yamabe soliton $(g, \lambda, \zeta, \mu, \alpha = 0, \beta = 1)$ with a $CKVF$ ζ, then the rate of change of the hyperbolic Yamabe soliton is expanding, steady, or shrinking referring to as, respectively:*

1. $(\sigma + 2p) > \frac{\mathcal{R}_{scal}}{\kappa}$
2. $(\sigma + 2p) = \frac{\mathcal{R}_{scal}}{\kappa}$
3. $(\sigma + 2p) < \frac{\mathcal{R}_{scal}}{\kappa}$

In addition, inserting $\alpha = 1$ and $\beta = -1$ in (10) and using Remark 5.2, we turn up

$$\lambda = \frac{\mu}{\omega} - \left\{\frac{\kappa(\sigma + 2p)}{2\omega} + \frac{2\mathcal{R}_{scal}}{2\omega} + \frac{\Lambda}{2\omega} + \omega\right\}. \tag{7.8}$$

$$\mu = 2\lambda\omega + \left\{\frac{\kappa(\sigma + 2p)}{2\omega} + 4\mathcal{R}_{scal} + \Lambda + 2\omega^2\right\}. \tag{7.9}$$

Thus, Eqs. (7.8) and (7.9) entail the following results for hyperbolic Einstein soliton and rate of change of the hyperbolic Einstein soliton:

Corollary 7.5 *If a PFS $(\mathcal{M}^4, g)$ admits the hyperbolic Einstein soliton $(g, \lambda, \zeta, \mu, 1, -1)$ with a CVF ζ, then the hyperbolic Einstein is shrinking, steady, or expanding referring to as, respectively:*

1. $\frac{\mu}{\omega} < \left\{\frac{\kappa(\sigma+2p)}{2\omega} + \frac{2\mathcal{R}_{scal}}{2\omega} + \frac{\Lambda}{2\omega} + \omega\right\}$
2. $\frac{\mu}{\omega} = \left\{\frac{\kappa(\sigma+2p)}{2\omega} + \frac{2\mathcal{R}_{scal}}{2\omega} + \frac{\Lambda}{2\omega} + \omega\right\}$
3. $\frac{\mu}{\omega} > \left\{\frac{\kappa(\sigma+2p)}{2\omega} + \frac{2\mathcal{R}_{scal}}{2\omega} + \frac{\Lambda}{2\omega} + \omega\right\}$

In light of (7.9), we gain the following:

Corollary 7.6 *If a PFS $(\mathcal{M}^4, g)$ admits a hyperbolic Einstein soliton $(g, \lambda, \zeta, \mu, 1, -1)$ with a CVF ζ, then the rate of change of the hyperbolic Einstein soliton is expanding.*

Now, in view of (7.6) and Remark 2.3, we get

$$\Lambda = \frac{\mu}{2} - \left\{\kappa(\sigma + 2p) + \left(1 - \frac{\beta}{\alpha}\right)\mathcal{R}_{scal} + 2\omega(\lambda + 1)\right\}. \tag{7.10}$$

Theorem 7.7 *If a PFS $(\mathcal{M}^4, g)$ admits the $HRYS$ $(g, \lambda, \zeta, \mu, \alpha, \beta)$ with a CVF ζ, then the PFS $(\mathcal{M}^4, g, \lambda, \zeta, \mu, \alpha, \beta)$ is an accelerating Universe if*

$$\frac{\mu}{2} > \left\{\kappa(\sigma + 2p) + \left(1 - \frac{\beta}{\alpha}\right)\mathcal{R}_{scal} + 2\omega(\lambda + 1)\right\}.$$

Theorem 7.8 *If a PFS $(\mathcal{M}^4, g)$ admits the $HRYS$ $(g, \lambda, \zeta, \mu, \alpha, \beta)$ with a CVF ζ, then the PFS $(\mathcal{M}^4, g, \lambda, \zeta, \mu, \alpha, \beta)$ is de-Sitter spacetime if*

$$\frac{\mu}{2} > \left\{\kappa(\sigma + 2p) + \left(1 - \frac{\beta}{\alpha}\right)\mathcal{R}_{scal} + 2\omega(\lambda + 1)\right\}.$$

Theorem 7.9 *If a PFS $(\mathcal{M}^4, g)$ admits the $HRYS$ $(g, \lambda, \zeta, \mu, \alpha, \beta)$ with a CVF ζ, then the PFS $(\mathcal{M}^4, g, \lambda, \zeta, \mu, \alpha, \beta)$ is a non-accelerating Universe with a late time expansion rate of the Universe if*

$$\frac{\mu}{2} < \left\{\kappa(\sigma + 2p) + \left(1 - \frac{\beta}{\alpha}\right)\mathcal{R}_{scal} + 2\omega(\lambda + 1)\right\}.$$

Corollary 7.10 *If a PFS $(\mathcal{M}^4, g)$ admits the $HRYS$ $(g, \lambda, \zeta, \mu, \alpha, \beta)$ with a CVF ζ, then the PFS $(\mathcal{M}^4, g, \lambda, \zeta, \mu, \alpha, \beta)$ contains dark sector if*

$$\frac{\mu}{2} < \left\{\kappa(\sigma + 2p) + \left(1 - \frac{\beta}{\alpha}\right)\mathcal{R}_{scal} + 2\omega(\lambda + 1)\right\}.$$

8 Hyperbolic Ricci-Yamabe Soliton with a $\varphi(\mathcal{Q})$-Vector Field on PFS

Definition 8.1 ([24]) A vector field φ on a Riemannian manifold M, if M obeys, is considered a $\varphi(\mathcal{Q})$-vector field

$$\nabla_{\zeta}\varphi = \Omega\mathcal{Q}\zeta, \tag{8.1}$$

where $\mathcal{Q}$, Ω, and ∇ represent the Ricci operator, a constant, and the Levi-Civita connection, respectively, $g(\mathcal{Q}m, n) = \mathcal{S}_{Ric}(m, n)$. $\varphi(\mathcal{Q})$ is considered to be covariantly constant if $\Omega = 0$ in (8.1), and φ is a valid $\varphi(\mathcal{Q})$-vector field if $\Omega \neq 0$.

Using (8.1) and the Lie-derivative formulation, we arrive to

$$(L_{\varphi}g)(F_1, F_2) = 2\Omega \mathcal{S}_{Ric}(F_1, F_2) \quad (8.2)$$

for any $F_1, F_2 \in \chi(\mathcal{M}^4)$.

Taking into consideration (7.1) and (8.2), we discover

$$(2\lambda\Omega + \alpha)\mathcal{S}_{Ric}(F_1, F_2) + \Omega L_X \mathcal{S}_{Ric}(F_1, F_2) = (\mu - \beta \mathcal{R}_{scal})g(F_1, F_2). \quad (8.3)$$

Again using that ζ also holds the Ricci collineation condition, then (8.3) entails that

$$\mathcal{S}_{Ric}(F_1, F_2) = \frac{(\mu - \beta \mathcal{R}_{scal})}{(2\lambda\Omega + \alpha)} g(F_1, F_2). \quad (8.4)$$

We can therefore state the following outcome.

Theorem 8.2 *If a PFS $(\mathcal{M}^4, g)$ admits the $HRYS$ $(g, \lambda, \zeta, \mu, \alpha, \beta)$ with a proper $\varphi(\mathcal{Q})$-vector field ζ and if ζ holds RC in the PFS, then the PFS is Einstein, and Einstein's factor is $\frac{(\mu - \beta \mathcal{R}_{scal})}{(2\lambda\Omega + \alpha)}$.*

Corollary 8.3 *If a PFS $(\mathcal{M}^4, g)$ admits the $HRYS$ $(g, \lambda, \zeta, \mu, \alpha, \beta)$ with a covariantly constant $\varphi(Q)$-vector field ($\Omega = 0$) ζ and if ζ holds RC in the PFS, then the PFS is an Einstein.*

Moreover, in the light of Remark 5.2 and Theorem 8.2, we obtain the following corollary:

Corollary 8.4 *If a PFS $(\mathcal{M}^4, g)$ admits the hyperbolic Yamabe soliton $(g, \lambda, \zeta, \mu, \alpha = 0, \beta = 1)$ with a proper $\varphi(\mathcal{Q})$-vector field ζ and if ζ holds RC in the PFS, then the PFS is Einstein, and Einstein's factor is $\frac{(\mu - \mathcal{R}_{scal})}{2\lambda\Omega}$.*

Putting $m = n = \zeta$ in (8.4), we obtain

$$\mathcal{S}_{Ric}(\zeta, \zeta) = \frac{(\beta \mathcal{R}_{scal} - \mu)}{(2\lambda\Omega + \alpha)}. \quad (8.5)$$

Using Eqs. (8.5) and (2.1), we turn up

$$\lambda = \frac{(\beta \mathcal{R}_{scal} - \mu)}{2\Omega(2\Lambda - \mathcal{R}_{scal} + \kappa p)} - \frac{\alpha}{2\Omega}. \quad (8.6)$$

In case of hyperbolic Yamabe soliton $\alpha = 0$ and $\beta = 1$. Thus from (8.6), we also get

$$\lambda = \frac{(\mathcal{R}_{scal} - \mu)}{2\Omega(2\Lambda - \mathcal{R}_{scal} + \kappa p)}. \quad (8.7)$$

Now, Theorem 8.2 and Eqs. (8.6) and (8.7) entail the following results:

Theorem 8.5 *If a PFS $(\mathcal{M}^4, g)$ admits the $HRYS$ $(g, \lambda, \zeta, \mu, \alpha, \beta)$ with a proper $\varphi(\mathcal{Q})$-vector field ζ and if ζ holds RC in the PFS, then the $HRYS$ is expanding, steady, or shrinking referring to as, respectively:*

1. $\frac{(\beta\mathcal{R}_{scal}-\mu)}{2\Omega(2\Lambda-\mathcal{R}_{scal}+\kappa p)} > \frac{\alpha}{2\Omega}$
2. $\frac{(\beta\mathcal{R}_{scal}-\mu)}{2\Omega(2\Lambda-\mathcal{R}_{scal}+\kappa p)} = \frac{\alpha}{2\Omega}$
3. $\frac{(\beta\mathcal{R}_{scal}-\mu)}{2\Omega(2\Lambda-\mathcal{R}_{scal}+\kappa p)} < \frac{\alpha}{2\Omega}$

Corollary 8.6 *If a PFS $(\mathcal{M}^4, g)$ admits the hyperbolic Yamabe soliton $(g, \lambda, \zeta, \mu, \alpha = 0, \beta = 1)$ with a proper $\varphi(\mathcal{Q})$-vector field ζ and if ζ holds RC in the PFS, then the hyperbolic Yamabe soliton is expanding.*

Once again with the help of Eqs. (8.6) and (8.7), we gain

$$\mu = \beta\mathcal{R}_{scal} - \left[2\lambda\Omega(2\Lambda - \mathcal{R}_{scal} + \kappa p) + \frac{\alpha}{2\Omega}\right], \tag{8.8}$$

and

$$\mu = -2\lambda\Omega(2\Lambda - \mathcal{R}_{scal} + \kappa p), \tag{8.9}$$

Thus we can state the following outcomes:

Theorem 8.7 *If a PFS $(\mathcal{M}^4, g)$ admits the $HRYS$ $(g, \lambda, \zeta, \mu, \alpha, \beta)$ with a proper $\varphi(\mathcal{Q})$-vector field ζ and if ζ holds RC in the PFS, then the rate of change of the $HRYS$ is expanding, steady, or shrinking according as:*

1. $\beta\mathcal{R}_{scal} > [2\lambda\Omega(2\Lambda - \mathcal{R}_{scal} + \kappa p) + \frac{\alpha}{2\Omega}]$
2. $\beta\mathcal{R}_{scal} = [2\lambda\Omega(2\Lambda - \mathcal{R}_{scal} + \kappa p) + \frac{\alpha}{2\Omega}]$
3. $\beta\mathcal{R}_{scal} < [2\lambda\Omega(2\Lambda - \mathcal{R}_{scal} + \kappa p) + \frac{\alpha}{2\Omega}]$, *respectively*

Corollary 8.8 *If a PFS $(\mathcal{M}^4, g)$ admits the hyperbolic Yamabe soliton $(g, \lambda, \zeta, \mu, 0, 1)$ with a proper $\varphi(\mathcal{Q})$-vector field ζ and if ζ holds RC in the PFS, then the rate of change of the hyperbolic Yamabe soliton is shrinking.*

9 Gradient Hyperbolic Ricci-Yamabe Solitons

In this portion, we use a scalar concircular field to determine gradient hyperbolic Ricci-Yamabe solitons in PFS. Therefore, we offer the definition that follows.

Definition 9.1 ([19]) Scalar fields are defined as scalar concircular fields ($SCFs$) if they satisfy the equation $f \in C^{\infty}(\mathcal{M})$.

$$Hessf = \pi g, \tag{9.1}$$

where π is a scalar field and g is the Riemannian metric. Additionally, the equation transforms into an ordinary differential equation for an arc-length c geodesic given as

$$\frac{d^2 f}{dc^2} = \pi. \tag{9.2}$$

Now, using Eq. (5.4) with (9.1), we find

$$\begin{aligned} &L_{\nabla f}(Hessf(F_1, F_2)) + 2\lambda Hessf(F_1, F_2) + \alpha \mathcal{S}_{Ric}(F_1, F_2) \\ &= (\mu - \beta \mathcal{R}_{scal}) g(F_1, F_2). \end{aligned} \tag{9.3}$$

$$L_{\nabla f}(\pi g(F_1, F_2)) + 2\lambda\pi g(F_1, F_2) + \alpha \mathcal{S}_{Ric}(F_1, F_2) = (\mu - \beta \mathcal{R}_{scal}) g(F_1, F_2). \tag{9.4}$$

$$\pi(Hessf(F_1, F_2)) + 2\lambda\pi g(F_1, F_2) + \alpha \mathcal{S}_{Ric}(F_1, F_2) = (\mu - \beta \mathcal{R}_{scal}) g(F_1, F_2). \tag{9.5}$$

$$\mathcal{S}_{Ric}(F_1, F_2) = \frac{\left[\mu - (\pi^2 + 2\lambda\pi + \beta \mathcal{R}_{scal})\right]}{\alpha} g(F_1, F_2). \tag{9.6}$$

Therefore, we can state the following result.

Theorem 9.2 *If a PFS $(\mathcal{M}^4, g)$ admits the $GHRYS$ $(g, \lambda, \zeta = \nabla f, \mu, \alpha, \beta)$ with an SCF f, then the PFS is Einstein.*

Putting $m = n = \zeta$ in (9.6) and using (2.1), we obtain

$$\lambda = \left[\frac{\mu}{2\pi} + \frac{\alpha \mathcal{R}_{scal}}{2\pi} + \frac{\alpha\kappa(\sigma + 2p)}{2\pi}\right] - \left[\frac{\beta \mathcal{R}_{scal}}{\pi} + \frac{\Lambda}{\pi}\right]. \tag{9.7}$$

$$\mu = \left[\pi^2 + 2\lambda\pi + \beta \mathcal{R}_{scal} + \Lambda\right] - \left[\frac{\alpha \mathcal{R}_{scal}}{2} + \alpha\kappa(\sigma + 2p)\right]. \tag{9.8}$$

Hence, we articulate the next theorems and corollaries:

Theorem 9.3 *If a PFS $(\mathcal{M}^4, g)$ admits the $GHRYS$ $(g, \lambda, \zeta = \nabla f, \mu, \alpha, \beta)$ with an SCF f, then the $GHRYS$ is expanding, steady, or shrinking according as:*

1. $\left[\frac{\mu}{2\pi} + \frac{\alpha \mathcal{R}_{scal}}{2\pi} + \frac{\alpha\kappa(\sigma+2p)}{2\pi}\right] > \left[\frac{\beta \mathcal{R}_{scal}}{\pi} + \frac{\Lambda}{\pi}\right]$
2. $\left[\frac{\mu}{2\pi} + \frac{\alpha \mathcal{R}_{scal}}{2\pi} + \frac{\alpha\kappa(\sigma+2p)}{2\pi}\right] = \left[\frac{\beta \mathcal{R}_{scal}}{\pi} + \frac{\Lambda}{\pi}\right]$
3. $\left[\frac{\mu}{2\pi} + \frac{\alpha \mathcal{R}_{scal}}{2\pi} + \frac{\alpha\kappa(\sigma+2p)}{2\pi}\right] < \left[\frac{\beta \mathcal{R}_{scal}}{\pi} + \frac{\Lambda}{\pi}\right]$, *respectively*

Theorem 9.4 *If a PFS $(\mathcal{M}^4, g)$ admits the $GHRYS$ $(g, \lambda, \zeta = \nabla f, \mu, \alpha, \beta)$ with an SCF f, then the rate of change of $GHRYS$ is expanding, steady, or shrinking according as:*

1. $\left[\pi^2 + 2\lambda\pi + \beta \mathcal{R}_{scal} + \Lambda\right] > \left[\frac{\alpha \mathcal{R}_{scal}}{2} + \alpha\kappa(\sigma + 2p)\right]$

2. $\left[\pi^2 + 2\lambda\pi + \beta\mathcal{R}_{scal} + \Lambda\right] = \left[\frac{\alpha\mathcal{R}_{scal}}{2} + \alpha\kappa(\sigma + 2p)\right]$
3. $\left[\pi^2 + 2\lambda\pi + \beta\mathcal{R}_{scal} + \Lambda\right] < \left[\frac{\alpha\mathcal{R}_{scal}}{2} + \alpha\kappa(\sigma + 2p)\right]$, *respectively*

Corollary 9.5 *If a PFS $(\mathcal{M}^4, g)$ admits the gradient hyperbolic Yamabe soliton $(g, \lambda, \zeta = \nabla f, \mu, 0, 1)$ with an SCF f, then the gradient hyperbolic Yamabe soliton is expanding, steady, or shrinking according as:*

1. $\frac{\mu}{2\pi} > \frac{\mathcal{R}_{scal}}{\pi} + \frac{\Lambda}{\pi}$
2. $\frac{\mu}{2\pi} = \frac{\mathcal{R}_{scal}}{\pi} + \frac{\Lambda}{\pi}$
3. $\frac{\mu}{2\pi} < \frac{\mathcal{R}_{scal}}{\pi} + \frac{\Lambda}{\pi}$, *respectively*

Corollary 9.6 *If a PFS $(\mathcal{M}^4, g)$ admits the gradient hyperbolic Yamabe soliton $(g, \lambda, \zeta = \nabla f, \mu, \alpha, \beta)$ with an SCF f, then the rate of change of gradient hyperbolic Yamabe soliton is expanding.*

10 Energy Constraints with Gradient Hyperbolic Ricci-Yamabe Soliton in Perfect Fluid Spacetime

In this section, we know whether the Ricci tensor $\mathcal{S}_{Ric}$ in the spacetime satisfies the condition, referring to [31].

$$\mathcal{S}_{Ric}(\zeta, \zeta) > 0; \tag{10.1}$$

the *time-like convergence condition (TCC)* is Eq. (10.1) for any time-like vector fields $\zeta \in \chi(\mathcal{M}^4)$.

From (2.1) and (9.6), it gives

$$\mathcal{S}_{Ric}(\zeta, \zeta) = \left[\pi^2 + 2\lambda\pi + \beta\mathcal{R}_{scal} + \Lambda\right] - \left[\frac{\alpha\mathcal{R}_{scal}}{2} + \alpha\kappa(\sigma + 2p) + \mu\right].$$

If the PFS in question satisfies the TCC, that is, then $\mathcal{S}_{Ric}(\zeta, \zeta) > 0$.

$$\left[\pi^2 + 2\lambda\pi + \beta\mathcal{R}_{scal} + \Lambda\right] > \left[\frac{\alpha\mathcal{R}_{scal}}{2} + \alpha\kappa(\sigma + 2p) + \mu\right]. \tag{10.2}$$

$$\left[\pi^2 + 2\lambda\pi + \mathcal{R}_{scal} + \Lambda\right] > \mu. \tag{10.3}$$

The spacetime obeys the cosmological strong energy constraint (SEC) [47]. In light of the above information given and from (10.2), we can state the following:

Theorem 10.1 *If a PFS $(\mathcal{M}^4, g)$ admits a $GHRYS$ $(g, \lambda, \zeta = \nabla f, \mu, \alpha, \beta)$ with an SCF f, then the PFS $(\mathcal{M}^4, g)$ satisfies SEC, provided the rate of change of $GHRYS$ is expanding.*

We now have an intriguing observation.

Remark 10.2 1. Hawking and Ellis [23] demonstrate in 1973 that the condition of null energy is $SEC \Rightarrow NEC$.

The following theorem is obtained by combining Remark 10.2 with Theorem 10.1:

Theorem 10.3 *If a PFS $(\mathcal{M}^4, g)$ admits a $GHRYS$ $(g, \lambda, \zeta = \nabla f, \mu, \alpha, \beta)$ with an SCF f, then the PFS $(\mathcal{M}^4, g)$ satisfies NEC, if (10.2) holds, provided the rate of change of $GHRYS$ is expanding.*

In the light of (10.3), we turn up the following corollaries:

Corollary 10.4 *If a PFS $(\mathcal{M}^4, g)$ admits a gradient hyperbolic Yamabe soliton $(g, \lambda, \zeta = \nabla f, \mu, 0, 1)$ with an SCF f, then the PFS $(\mathcal{M}^4, g)$ satisfies SEC, provided the rate of change of the gradient hyperbolic Yamabe soliton is expanding.*

Corollary 10.5 *If a PFS $(\mathcal{M}^4, g)$ admits a gradient hyperbolic Yamabe soliton $(g, \lambda, \zeta = \nabla f, \mu, 0, 1)$ with an SCF f, then the PFS $(\mathcal{M}^4, g)$ satisfies NEC, if (10.2) holds, provided that the rate of change of gradient hyperbolic Yamabe soliton is expanding.*

11 η-Hyperbolic Ricci-Yamabe Solitons

Finally, in this section, we presented a more general concept of η-hyperbolic Ricci-Yamabe solitons and demonstrated that a spacetime with a conformal vector field that admits the η-hyperbolic Ricci-Yamabe solitons is a perfect fluid spacetime.

As a generalization of Ricci soliton, the η-Ricci soliton was introduced by Cho and Kimura [15] in the following form:

$$\frac{1}{2}L_{\zeta}g + \mathcal{S}_{Ric} + \lambda g + \tau\eta \otimes \eta = 0, \tag{11.1}$$

where λ and τ are real constants.

In [43] Siddiqi et al. introduced the notion of the η-Ricci-Yamabe solitons. Specifically, an η-Ricci-Yamabe soliton on the Riemannian manifold $(\mathcal{M}, g)$ is a data $(g, \lambda, \tau, \alpha, \beta)$ satisfying

$$\frac{1}{2}L_{\zeta}g + \alpha\mathcal{S}_{Ric} + \left(\lambda - \frac{\beta}{2}\mathcal{R}_{scal}\right) g + \tau\eta \otimes \eta = 0, \tag{11.2}$$

where τ is a constant.

Now, in light of (5.3) and (11.2), one can introduce the new concept similarly by amending the expression (5.3) that explains the type of soliton by a multiple of a specific $(0, 2)$-tensor field $\eta \otimes \eta$. These findings result in a significantly more comprehensive concept, termed an η-hyperbolic Ricci-Yamabe soliton (briefly an η-HRY soliton) of type (α, β) defined as

$$\frac{1}{2}L_{\zeta}L_{\zeta}g + \lambda L_{\zeta}g + \alpha\mathcal{S}_{Ric} = (\mu - \beta\mathcal{R}_{scal})g + \tau\eta\otimes\eta. \tag{11.3}$$

Next, adopting (11.3) and (6.1), we gain

$$\mathcal{S}_{Ric}(m, n) = Ag(m, n) + B\eta(m)\eta(n), \tag{11.4}$$

where $A = \frac{\mu-(\beta\mathcal{R}_{scal}+2\lambda\omega+2\omega^2)}{\alpha}$ and $B = \tau$.

Theorem 11.1 *If a spacetime $(\mathcal{M}^4, g)$ admits the η-hyperbolic Ricci-Yamabe solitons $(g, \lambda, \zeta, \mu, \tau, \alpha, \beta)$ with a conformal vector field ζ, then the spacetime $(\mathcal{M}^4, g)$ is a PFS.*

12 Open Problems

This kind of research can be expanded to include other solitons, such as the hyperbolic conformal Ricci soliton and the hyperbolic Ricci-Bourguignon soliton from conformal Ricci soliton and Ricci-Bourguignon soliton [14, 39], respectively. In reality, the author already explored certain results of hyperbolic Ricci-Bourguignon soliton in his next publication, which is under print.

References

1. Ahsan, Z.: Tensors: Mathematics of Differential Geometry and Relativity. PHI Learning Pvt. Ltd. Delhi (2017, second printing)
2. Ahsan, Z., Siddiqui, S.A.: Concircular curvature tensor and fluid spacetimes. Int. J. Theor. Phys. **48**, 3202–3212 (2009)
3. Al-Dayel, I., Deshmukh, S., Siddiqi, M.D.: A characterization of GRW spacetimes. Mathematics **9**(18), 2209 (2021)
4. Ali, M., Ahsan, Z.: Ricci solitons and symmetries of space time manifold of general relativity. J. Adv. Res. Class. Mod. Geom. **1**(2),75–84 (2014)
5. Alkhaldi, A.H., Siddiqi, M.D., Khan, M.A., Alqahtani, L.S.: Imperfect fluid generalized Robertson Walker spacetime admitting Ricci-Yamabe Metric. Adv. Math. Phys. 1–10 (2021). https://doi.org/10.1155/2021/2485804
6. Azami, S., Fasihi-Ramandi, G.: Hyperbolic Ricci soliton on warped product manifolds. Filomat **37**, 6843–6853 (2023)
7. Blaga, A.M.: On gradient η-Einstein solitons. Kragujevac J. Math. **42**(2), 229–237 (2018)
8. Blaga, A.M.: Solitons and geometrical structures in a perfect fluid spacetime. Rocky Mt. J. Math. **50**(1), 41–53 (2020)
9. Blaga, A.M., Özgür, C.: Results of hyperbolic Ricci solitons. Symmetry **15**(8), 1548 (2023)

10. Blaga, A.M., Özgür, C.: Some properties of hyperbolic Yamabe solitons. arXiv:2310.15814v1 [math.DG] 24 Oct 2023
11. Blaga, A.M., Özgür, C.: 2-killing vector fields on multiply warped product manifolds. Chaos Solitons and Fractals **180**, 114561 (2024)
12. Bokhari, A.H., Qadir, A.: Collineations of the Ricci tensor. J. Math. Phys. **34**, 3543–3552 (1993)
13. Catino, G., Mazzieri, L.: Gradient Einstein solitons. Nonlinear Anal. **132**, 66–94 (2016)
14. Chaubey, S.K., Siddiqi, M.D., Yadav, S.K.: Almost η-Ricci-Bourguignon solitons on submersions from Riemannian submersions. Balk. J. Geom. Appl. **27**, 24–38 (2022)
15. Cho, J.T., Kimura, M.: Ricci solitons and real hypersurfaces in a complex space form. Tohoku Math. J. **61**(2), 205–212 (2009)
16. Dai, W., Kong, D., Liu, K.: Hyperbolic geometric flow (I): short-time existence and nonlinear stability. Pure Appl. Math. Q. **6**(2), 331–359 (2010)
17. Deshmukh, S.: Conformal vector fields and eigenvectors of Laplace operator. Math. Phys. Anal. Geom. **15**, 163–172 (2012)
18. Faraji, H., Azami, S., Ramandi, G.F.: Three dimensional Homogeneous Hyperbolic Ricci solitons. J. Nonlinear Math. Phys. **30**, 135–155 (2023)
19. Fialkow, A.: Conformal geodesic. Trans. Am. Math. Soc. **45**, 443–473 (1989)
20. Güler, S., Crasmareanu, M.: Ricci-Yamabe maps for Riemannian flow and their volume variation and volume entropy. Turk. J. Math. **43**, 2631–2641 (2019)
21. Hamilton, R.S.: The Ricci flow on surfaces, mathematics and general relativity. Contemp. Math. Am. Math. Soc. **71**, 237–262 (1988)
22. Hamilton, R.S.: The Ricci flow on surfaces. Math. Gen. Relativ. (Santa Cruz, CA, 1986) **71**, 237–262 (1998), Contemp. Math.
23. Hawking, S.W., Ellis, G.F.R.: The Large Scale Structure of Spacetime. Cambridge University Press, Cambridge (1973)
24. Hinterleitner, I., Kiosak, V.A.: $\varphi(Ric)$-vector fields in Riemannian spaces. Arch. Math. **44**, 385–390 (2008)
25. Kaya, D.A., Özgür, C.: Hyperbolic Ricci solitons on sequential warped product manifolds. Filomat **38**(3), 1023–1032 (2024)
26. Li, Y., Siddiqi, M.D., Khan, M.A., Al-Dayel, I., Youssef, M.: Solitonic effect on relativistic string cloud spacetime attached with strange quark matter. AIMS Math. **9**, 14487–14503 (2024)
27. Mantica, C.A., Molinari, L.G.: A condition for a perfect fluid spacetime to be a generalized Robertson-Walker spacetimes, a survey. J. Math. Phys. **57**(2), 049901, 6 p. (2016)
28. Mantica, C.A., Molinari, L.G.: Generalized Robertson-Walker spacetimes, a survey. Int. J. Geom. Math. Mod. Phys. **14**(3), 102502, 6 p. (2017)
29. O'Neill, B.: Semi-Riemannian Geometry with Applications to Relativity. Academic, New York (1983)
30. Peebles, P.J., Ratra, B.: The cosmological constant and dark energy. Rev. Mod. Phys. **75**, 559 (2003)
31. Sachs, R.K., Hu, W.: General Relativity for Mathematicians. Springer, New York (1997)
32. Sańchez, M.: On the geometry of generalized Robertson-Walker spacetime geodesics. Gen. Relativ. Gravit. **27**, 71–84 (1998)
33. Siddiqi, M.D.: Ricci ρ-soliton and geometrical structure in a dust fluid and viscous fluid spacetime. Bulg. J. Phys. **46**, 163–173 (2019)
34. Siddiqi, M.D., De, U.C.: Relativistic magneto-fluid spacetimes. J. Geom. Phys. **170**, 104370 (2021)
35. Siddiqi, M.D., De, U.C.: Relativistic perfect fluid spacetimes and Ricci-Yamabe solitons. Lett. Math. Phys. **112**(1), 1 (2022)
36. Siddiqi, M.D., Mofarreh, F.: Hyperbolic Ricci soliton and gradient hyperbolic Ricci soliton on relativistic prefect fluid spacetime. AIMS Math. **9**, 21628–21640 (2024)
37. Siddiqi, M.D., Mofarreh, F.: Modified $\mathcal{F}(\mathcal{R}, \mathcal{T}^2)$-gravity coupled with perfect fluid admitting hyperbolic Ricci soliton type symmetry. Axioms **13**(10), 708 (2024). https://doi.org/10.3390/axioms13100708

38. Siddiqi, M.D., Siddqui, S.A.: Conformal Ricci soliton and Geometrical structure in a perfect fluid spacetime. Int. J. Geom. Methods Mod. Phys. **2050083**, 18 (2020)
39. Siddiqui, A.N., Siddiqi, M.D.: Almost Ricci-Bourguignon solitons and geometrical structure in a relativistic perfect fluid spacetime. Balk. J. Geom. Appl. **26**, 126–138 (2021)
40. Siddiqi, M.D., Siddqui, S.A.: k-Yamabe and quasi k-Yamabe solitons on imperfect fluid Generalized Robertson-Walker spacetime. Math. Phys. Comput. Model. Volgograd **25**(1), 21–33 (2022)
41. Siddiqi, M.D., De, U.C., Deshmukh, S.: Estimation of almost Ricci-Yamabe solitons on Static spacetimes. Filomat **36**(2), 397–407 (2022)
42. Siddiqi, M.D., Khan, M.A., Al-Dayel, I., Masood, K.: Geometrization of string cloud spacetime in general relativity. AIMS Math. **8**(12), 29042–29057 (2023)
43. Siddiqi, M.D., Mofarreh, F., Akyol, M.A., Hakami, A.H.: η-Ricci-Yamabe Solitons along Riemannian Submersions. Axioms **12**(8), 796 (2023)
44. Siddiqi, M.D., Mofarreh, F., Chaubey, S.K.: Solitonic aspect of relativistic magneto-fluid spacetime with some specific vector fields. Mathematics **11**(7), 1596 (2023)
45. Siddiqi, M.D., Mofarreh, F., Siddiqui, A.N., Siddiqui, S.A.: Geometrical structure in a relativistic thermodynamical fluid spacetime. Axioms **12**(2), 138 (2023). https://doi.org/10.3390/axioms12020138
46. Stephani, H.: General Relativity, an Introduction to the Theory of Gravitational Field. Cambridge University Press, Cambridge (1982)
47. Tipler, F.J.: Energy condition and spacetime singularities. Phys. Rev. D **17**, 2521 (1978)
48. Venkatesha, A., Kumara, H.: Ricci solitons and geometrical structure in a perfect fluid spacetime with Torse-forming vector filed. Afrika Mathematika **30**, 725–736 (2019)

A Survey on Hitchin–Thorpe Inequality and Its Extensions

Bang-Yen Chen, Majid Ali Choudhary, and Mohammed Nisar

Abstract The well-known Hitchin–Thorpe inequality is an important geometric inequality which states that if a closed, oriented 4-manifold M admits an Einstein metric, then the Euler characteristic $\chi(M)$ and the signature $\tau(M)$ of M satisfy

$$\chi(M) \geq \frac{3}{2}|\tau(M)|.$$

In 1974, N. Hitchin further established a thorough explanation of the equality case. In addition, Hitchin proved that if (M, g) is an Einstein manifold for which equality of this inequality is attained, consequently, either M is flat or its universal cover is a K3 surface, and (M, g) is Ricci-flat.

After this important work of Thorpe and Hitchin, there exist many important works related closely to the Hitchin–Thorpe inequality done by many mathematicians.

The purpose of this chapter is to delve into recent advancements of the Hitchin–Thorpe inequality and its extensions, showcasing how contemporary studies have extended their applicability.

Keywords Hitchin-Thorpe inequality · Euler characteristic · Signature · Ricci soliton · Myers' theorem

---

B.-Y. Chen (✉)
Department of Mathematics, Michigan State University, East Lansing, MI, USA
e-mail: chenb@msu.edu

M. A. Choudhary · M. Nisar
Department of Mathematics, Maulana Azad National Urdu University, Hyderabad, India

B.-Y. Chen, M. A. Choudhary (eds.), *Geometric Inequalities and Applications*,
Infosys Science Foundation Series in Mathematical Sciences,
https://doi.org/10.1007/978-981-95-5148-4_10

1 Introduction

The study of 4-manifolds is a central theme in both differential topology and differential geometry, revealing intricate relationships between the geometric structures of manifolds and their topological invariants. On the other hand, Einstein manifolds are essential Riemannian manifolds in physics and geometry, and they produce the vacuum solutions to the Einstein field equations in general relativity and are suitable options for canonical metrics on manifolds.

An important result on Einstein manifolds is the well-known Hitchin–Thorpe inequality which asserts that in the event of a closed, oriented, smooth 4-manifold M admits an Einstein metric, and then $\chi(M)$ and $\tau(M)$ of M satisfy

$$\chi(M) \geq \frac{3}{2}|\tau(M)|. \tag{1.1}$$

The Hitchin–Thorpe inequality will be referred to as the H–T inequality, χ represents the Euler characteristic, and τ represents the signature throughout this chapter.

The H–T inequality was initially mentioned by J. Thorpe in a footnote of his 1969 paper [31], which concentrated on manifolds in higher dimensions. Later in 1974, N. Hitchin [14] rediscovered this inequality and gave an in-depth review of the equality case. Moreover, it was demonstrated by Hitchin that if (M, g) represents an Einstein manifold for which equality holds, then (M, g) is Ricci-flat, and either M is flat or its universal cover is a $K3$ surface. In the same paper [14], Hitchin also laid the groundwork for these inequalities by exploring the properties of spinors and Dirac operators on manifolds, leading to a deeper understanding of how curvature conditions constrain topological properties. Independently, Thorpe's research, in his 1969 paper [31], also highlighted significant links between scalar curvature and the topology of manifolds, particularly through the Gauss–Bonnet integral. Together, their contributions have provided important developments in the field of 4-manifolds. However, C. LeBrun [20] demonstrated in 1995 that there are an unlimited number of non-homeomorphic compact, smooth, oriented 4-manifolds M that are orientated, compact, and smooth that do not contain Einstein metrics but yet satisfy the strict inequality case of (1.1). Therefore, the H–T inequality (1.1) is not a sufficient condition for a closed, oriented, 4-manifold to admit an Einstein metric. Note that M. Berger [2] observed in 1965 that every compact Einstein 4-manifold has a nonnegative Euler number.

After the works of Thorpe and Hitchin, there exist many nice works closely related with the H–T inequality obtained by many mathematicians. The aim of this chapter is to delve into recent advancements of the H–T inequality and its extensions, showcasing how contemporary studies have extended their applicability. By synthesizing classical and modern perspectives, this chapter aims to provide a detailed and accessible resource for researchers and students, fostering a deeper appreciation of the pivotal role these inequalities play in differential geometry and topology.

2 Preliminaries

The H–T inequality can be formally stated as follows: Consider a compact, smooth, oriented 4-manifold M that admits a Riemannian metric with nonnegative scalar curvature, and the inequality is given by

$$2\chi(M) \geq 3|\tau(M)|.$$

The signature, which captures the difference among the number of positive and negative eigenvalues of the intersection form on the middle dimensional cohomology, is related to the Euler characteristic, a measure of the manifold's shape or structure, by this inequality. Such a relationship underscores the constraints imposed by geometric properties (e.g., scalar curvature) on the topological invariants of the manifold.

2.1 Definitions and Background

1. **Euler Characteristic** (χ) A topological invariant that characterizes a topological space's structure or form is called the Euler characteristic. For a 4-manifold, it can be computed using the formula:

$$\chi(M) = \sum_{i=0}^{4}(-1)^i b_i,$$

where b_i are the Betti numbers, which represent the rank of the i-th homology group of the manifold.
2. **Signature** (τ) The signature of a 4-manifold is a topological invariant defined as the difference between the count of positive and negative eigenvalues of the intersection form on the middle cohomology. Formally, for a 4-manifold M:

$$\tau(M) = b_2^+ - b_2^-,$$

in the second cohomology group $H^2(M)$, the maximal positive-definite and negative-definite subspaces have dimensions b_2^+ and b_2^-.
3. **Ricci Soliton** The Ricci soliton, as defined by Hamilton [13], is a complete Riemannian manifold (M, g) that allows for the admission of a smooth vector field $X \in \mathfrak{X}(M)$, such that

$$\mathrm{Ric}_g + \frac{1}{2}\mathcal{L}_X g = \lambda g, \tag{2.1}$$

where λ is a real number. The Lie derivative in the direction of X is $\mathcal{L}_X$. Depending on whether $\lambda > 0$, $\lambda = 0$, or $\lambda < 0$, the soliton (M, g) is categorized as shrinking, steady, or expanding accordingly. In Einstein manifolds, where X is a Killing vector field, Ricci solitons are frequently seen applications. We refer to the soliton as trivial in this particular instance. As self-similar solutions and common models for singularities, Ricci solitons are essential in the study of Ricci flow [5]. (M, g) is called a gradient Ricci soliton when the vector field X can be written as the gradient ∇f of a differentiable function $f : M \to \mathbb{R}$, which is called a potential function. Equation (2.1) in this instance becomes

$$\mathrm{Ric}_g + H_f = \lambda g, \tag{2.2}$$

where H_f is the Hessian of the function f.

4. **Myers' Theorem** A complete Riemannian manifold M is compact and has a diameter of at most $\pi/\sqrt{k}$ if its Ricci curvature is bounded below by a positive constant k, that is, $\mathrm{Ric}(X, X) \geq (n-1)k$ for all tangent vectors X (Myers, 1941). This result implies that positive Ricci curvature places strong restrictions on the global geometry of the manifold, particularly ensuring finiteness in extent.

2.2 Historical Context and Development

The inequality was first suggested by Nigel Hitchin in [14] and later formalized through independent work by John A. Thorpe. The inequality connects differential geometry and topology by linking the curvature of a manifold with its topological invariants.

1. **Nigel Hitchin's Contribution** Hitchin's work focused on the properties of spinors and Dirac operators on manifolds, which led to insights into the curvature and topology of 4-manifolds. His results are pivotal in understanding the relationship between curvature conditions and topological constraints [14].
2. **John A. Thorpe's Contribution** Thorpe's research also dealt with connections between curvature and topology. His work emphasized the significance of scalar curvature in imposing constraints on the topology of manifolds [31].

2.3 Derivation and Proof

The H–T inequality can be derived using several techniques in differential geometry and topology. One common approach involves the study of the scalar curvature and the Atiyah–Singer index theorem. Here is an outline of the proof:

1. **Scalar Curvature and Gauss–Bonnet Formula [19]** The integral of the scalar curvature R over the 4-manifold M is associated with the Euler characteristic via the Gauss–Bonnet formula:

$$\int_M R\,d\text{vol} = 8\pi^2\chi(M) - \int_M \left(|W^+|^2 + |W^-|^2\right) d\text{vol},$$

where W^- and W^+ are the anti-self-dual and self-dual parts of the Weyl curvature tensor, respectively.
2. **Signature and Hirzebruch Signature Theorem [8]** The signature $\tau(M)$ can be computed using the Hirzebruch signature theorem, which relates the signature to the Pontryagin classes of the manifold:

$$\tau(M) = \frac{1}{3}\left(p_1^+(M) - p_1^-(M)\right),$$

where p_1^+ and p_1^- are the Pontryagin classes associated with the positive and negative parts of the curvature.
3. **Combining Results** By combining these results, one can derive the H–T inequality, showing that the scalar curvature constraints imply a relationship between χ and τ.

2.4 Applications of H–T Inequality

The H–T inequality has far-reaching implications in the study of 4-manifolds, providing critical constraints and insights into their geometric and topological properties. Below, we discuss several key applications of these inequalities, illustrating their utility with specific examples.

1. **Classification of 4-Manifolds** One of the primary applications of the H–T inequality is in the classification of 4-manifolds. By examining the relationship between τ and χ, mathematicians can identify whether certain manifolds can admit specific geometric structures.
Example: $K3$ Surface [8] A surface of $K3$ is a smooth, compact, simply connected 4-manifold with trivial canonical bundle. For a $K3$ surface, $\chi = 24$ and $\tau = -16$. Plugging these values into the H–T inequality,

$$2\chi = 2 \times 24 = 48$$

$$3|\tau| = 3 \times 16 = 48.$$

We see that $2\chi = 3|\tau|$, which satisfies the inequality. This confirms that $K3$ surfaces are among the special 4-manifolds that meet the strict criteria set by the H–T inequality.

2. **Existence of Metrics with Positive Scalar Curvature** The H–T inequality is particularly useful in determining the existence of Riemannian metrics with positive scalar curvature on 4-manifolds. A manifold satisfying $2\chi < 3|\tau|$ cannot support a metric of positive scalar curvature.
   Example: Complex Projective Plane $\mathbb{CP}^2$ [19] Consider the complex projective plane $\mathbb{CP}^2$ where $\chi = 3$ and $\tau = 1$. For $\mathbb{CP}^2$, we have

$$2\chi = 2 \times 3 = 6$$

$$3|\tau| = 3 \times 1 = 3.$$

   Since $2\chi > 3|\tau|$, $\mathbb{CP}^2$ can admit a metric with nonnegative scalar curvature. Indeed, $\mathbb{CP}^2$ can be endowed with the Fubini–Study metric, which has nonnegative scalar curvature.

3. **Constraints on Topology of 4-Manifolds** The H–T inequality imposes constraints on the topological structure of 4-manifolds, limiting the possible combinations of χ and τ for manifolds that can support certain curvature conditions.
   Example: $S^2 \times S^2$ [12] Consider the product of two 2-spheres, $S^2 \times S^2$ where $\chi = 4$ and $\tau = 0$. For $S^2 \times S^2$,

$$2\chi = 2 \times 4 = 8$$

$$3|\tau| = 3 \times 0 = 0.$$

   Clearly, $2\chi > 3|\tau|$, which satisfies the H–T inequality. Hence, $S^2 \times S^2$ can potentially allow a metric of nonnegative scalar curvature. This manifold indeed admits metrics of both positive and zero scalar curvatures, such as the product of standard round metrics on S^2.

4. **Generalizations to Higher Dimensions** While the original H–T inequality applies to 4-manifolds, similar ideas have been explored in higher dimensions, although they often involve more complex conditions and invariants.
   Example: 6-Manifolds [3] In higher dimensions, the relationship between curvature and topology becomes more intricate. For instance, the study of 6-manifolds often involves additional invariants like Pontryagin classes and more refined curvature conditions. While no direct analog to the H–T inequality exists in six dimensions, researchers have developed related inequalities that provide analogous constraints on the topology and geometry of higher dimensional manifolds.

3 H–T Inequality: Recent Progress on Compact Ricci Solitons

3.1 A Diameter Upper Bound for Compact Ricci Solitons with Applications to the H–T Inequality

First we mention the following.

Theorem 3.1 ([11]) *Consider* (M, g) *as a 4-dim compact, connected shrinking Ricci soliton that satisfies equation* (2.2). *If*

$$\operatorname{diam}(M, g) \leqslant \max\left\{2\sqrt{\frac{2}{C-c}}, \sqrt{\frac{2}{C-\lambda}}, \sqrt{\frac{2}{\lambda-c}}\right\},$$

in this case, the soliton satisfies the H–T inequality (1.1).

In 2018, Tadano [26] established several sufficient conditions under which 4-dim compact Ricci solitons comply with the H–T inequality and proved the following results.

Corollary 3.2 ([26]) *Consider* (M, g)*, a 4-dim connected, compact shrinking Ricci soliton that satisfies* (2.2). *If*

$$\sqrt{\frac{R_{\max} - R_{\min}}{\lambda^2}\left(16 + 6\pi^2\right)} \leqslant \operatorname{diam}(M, g),$$

then the soliton satisfies the H–T inequality (1.1).

Corollary 3.3 ([26]) *Consider* (M, g)*, a 4-dim compact, connected shrinking Ricci soliton that satisfies* (2.2). *If*

$$\frac{R_{\max}}{6\lambda} \cdot \frac{1}{\lambda}\left(\sqrt{4\left(R_{\max} - R_{\min}\right) + 3\lambda\pi^2} + 2\sqrt{R_{\max} - R_{\min}}\right) \leqslant \operatorname{diam}(M, g),$$

then the soliton satisfies the H–T inequality (1.1).

3.2 Enhanced Estimates for Oscillations and the H–T Inequality on Compact Ricci Solitons

In 2023, Tadano [30] introduced multiple new sufficient conditions under which compact 4-dim normalized shrinking Ricci solitons satisfy the H–T inequality. The following are few results from his work.

Theorem 3.4 ([30]) *Let (M, g) be a connected compact 4-dim normalized shrinking Ricci soliton with $\lambda = \frac{1}{2}$ that satisfies* (2.2). *If the diameter of (M, g) satisfies*

$$\operatorname{diam}(M, g) \geqslant \left(\frac{\sqrt{6}\pi}{\sqrt{\ln 5}} + 4 \right) \sqrt{R_{\max} - R_{\min}},$$

consequently the H–T inequality (1.1) *must be satisfied by the soliton.*

Corollary 3.5 ([30]) *Replacing $R_{\max}$, $R_{\min}$ with $f_{\max}$, $f_{\min}$, respectively, in Theorem 3.4,*

$$\operatorname{diam}(M, g) \geqslant \left(\frac{\sqrt{6}\pi}{\sqrt{\ln 5}} + 4 \right) \sqrt{f_{\max} - f_{\min}},$$

consequently the H–T inequality (1.1) *must be satisfied by the soliton.*

Theorem 3.6 ([30]) *Let (M, g) be a compact connected 4-dim nontrivial normalized shrinking Ricci soliton satisfying* (2.2) *with $\lambda = \frac{1}{2}$. If the diameter of (M, g) satisfies*

$$\operatorname{diam}(M, g) \leqslant \frac{2\pi\sqrt{2\ln 5}(\sqrt{2} - 1)}{\sqrt{R_{\max} - R_{\min}}},$$

then the soliton must satisfy the H–T inequality (1.1).

Corollary 3.7 ([30]) *Replacing $R_{\max}$, $R_{\min}$ with $f_{\max}$, $f_{\min}$, respectively, in Theorem 3.6,*

$$\operatorname{diam}(M, g) \leqslant \frac{2\pi\sqrt{2\ln 5}(\sqrt{2} - 1)}{\sqrt{f_{\max} - f_{\min}}},$$

consequently the H–T inequality (1.1) *must be satisfied by the soliton.*

3.3 Kähler Metrics and H–T Inequality for Compact Almost Ricci Soliton

In 2014, A. Brasil et al. [4] demonstrated the H–T inequality for a 4-dim compact almost Ricci soliton. They also demonstrated that a compact 4-dim almost Ricci soliton is isometric to the standard sphere under appropriate integral conditions. Moreover, they proved that a compact 4-dim Ricci soliton with a harmonic self-dual component of the Weyl tensor is Kähler–Einstein or isometric to the standard sphere S^4 under a minimal set of conditions. The following findings were proven:

Theorem 3.8 ([4]) *Consider a compact 4-dim almost Ricci soliton* $\left(M^4, g, \nabla f, \lambda\right)$ *whose scalar curvature* R *is positive.*

1. *If*

$$\int_M R^2 \,\mathrm{d}\mu \leq 6 \int_M \lambda R \,\mathrm{d}\mu,$$

then $\chi \geq \frac{3}{2}\tau$. *Specifically, assuming* λ *is constant and* $\int_M R^2 \,\mathrm{d}\mu \leq 24\lambda^2 V$, *then* $\chi \geq \frac{3}{2}|\tau|$.

2. *If* $\left(M^4, g\right)$ *is Kählerian, then*

$$3\tau + 2\chi = \frac{1}{2\pi^2} \int_M \lambda R \,\mathrm{d}\mu.$$

Theorem 3.9 ([4]) *Assume that* M^4 *is a 4-dim compact manifold whose scalar curvature* R *is positive and Ricci curvature Ric.*

1. *If Ric* $\geq \rho > 0$ *and* $R \leq 6\rho$ *or Ric* $\leq -\rho < 0$ *and* $R \geq -6\rho$, *then* $\chi \geq \frac{3|\tau|}{2}$.
2. *Assume that* M^4 *Kählerian is naturally oriented. If Ric* $\geq \rho > 0$ *and* $R \leq (2\sqrt{3}+6)\rho$ *or Ric* $\leq -\rho < 0$ *and* $R \geq -(2\sqrt{3}+6)\rho$, *then* $\chi \geq -\frac{3\tau}{2}$.

3.4 Remarks on Compact Shrinking Ricci Solitons

In 2013, L. Ma [21] gave a generalization of H–T inequality

Corollary 3.10 ([21]) *Suppose* (M, g) *is a 4-dim shrinking gradient Ricci soliton, where* $\rho > 0$ *is the shrinking constant as above. If we also suppose that*

$$\int_M s^2 \leqslant \rho^2 24 \operatorname{vol}(M), \tag{3.1}$$

then

$$2\chi(M) \pm 3\tau(M) \geqslant \frac{1}{2\pi^2} \int_M |W_\pm|^2 \geqslant 0.$$

He essentially demonstrated that the condition (3.1) is equivalent to the following:

$$\int_M \sigma_2\left(Rc - \frac{s}{6} g\right) \geqslant 0.$$

The second fundamental symmetric function of the eigenvalues of the matrix $A := Rc - \frac{s}{6}g$ is denoted by $\sigma_2(A)$ in this instance.

3.5 H–T Inequality and Euler Characteristic for Compact Ricci Solitons

Cheng et al. [6] investigated the geometry of Ricci solitons with compact gradients in four dimensions in 2023. They demonstrated that if there is an upper bound on the potential function's range, a 4-dim compact gradient Ricci soliton must meet the traditional H–T inequality.

They suppose that the gradient shrinking Ricci solitons obey the following equation, without losing generality:

$$\text{Hess} f + \text{Ric} = \frac{1}{2}g. \tag{3.2}$$

The metric may be scaled to accomplish this normalization. They first proved the subsequent outcome:

Theorem 3.11 ([6]) *Consider a compact 4-dim shrinking gradient Ricci soliton* (M^4, g, f) *that satisfies* (3.2). *Next, it asserts that*

$$8\pi^2\chi(M) \geq \int_M |W|^2 dV_g + \frac{1}{24}\text{Vol}(M)\left(5 - e^{f_{\max} - f_{\min}}\right),$$

where the Weyl tensor is W*, the volume of* M^4 *is* $\text{Vol}(M)$*, and the minimum and maximum of the potential function* f *on* M^4 *are* $f_{\min}$ *and* $f_{\max}$*, respectively. Furthermore, the equality is true only in the situation that g is an Einstein metric, in which instance* f *is constant.*

As a result of Theorem 3.11, we derive the following corollary.

Corollary 3.12 ([6]) *A compact 4-dim shrinking gradient Ricci soliton satisfying* (3.2) *is denoted by* (M^4, g, f)*. In the event that* $f_{\max} - f_{\min} \leq \log 5$*, the H–T inequality* (1.1) *holds on* M*.*

3.6 Topological Barriers to the Existence of Ricci Solitons That Are Compact and Shrinking

In 2024, the difficulty of applying the H–T inequality to shrinking gradient Ricci soliton metrics was put out by Cameron [22], who also looked at the shortcomings of previous findings in this field. He displayed the following outcomes:

Theorem 3.13 ([22]) *Assume that the compact oriented 4-dim Einstein manifold is* (M^4, g)*. In that case, the H–T inequality is satisfied by* M^4*.*

Theorem 3.14 ([22]) *Let* (M^4, g) *be a compact gradient shrinking Ricci soliton. Then the H–T inequality holds given the following sufficient condition:*

$$\int_{M^4} R^2 dV_g \leq 6\,\mathrm{Vol}\left(M^4, g\right).$$

In order to prove this theorem, he used this lemma

Lemma 3.15 ([22]) *For a compact shrinking gradient Ricci soliton* (M^4, g, f)*, the following is true:*

$$\int_M \frac{1}{2}R^2 - |Rc|^2 dV_g = \mathrm{Vol}(M, g).$$

4 H–T Inequality: Recent Progress on Einstein Manifolds

4.1 H–T Inequality for Noncompact Einstein Manifolds

An H–T inequality was established in 2007 by Dai and Wei [7] for Einstein noncompact 4-manifolds with certain asymptotic geometries at infinity. These asymptotic geometries can be described as a cusp bundle over a compact space (fibered cusps) or as a fiber bundle over a cone with a compact fiber. At infinity, these geometries are frequently found in numerous noncompact Einstein manifolds. The following are some of the outcomes of their work:

Theorem 4.1 ([7]) *Consider a noncompact complete Einstein manifold* (M^4, g) *that asymptotically approaches a fibered boundary or cusp at infinity. They additionally require* $\dim F > 0$ *in the fibered border case. Then*

$$\chi(M) \geqslant \frac{3}{2}\left|\tau(M) - \lim \eta + \frac{1}{2}a\right|.$$

Here, $a - \lim \eta$ *represents the adiabatic limit of the* η*-invariant of* $\partial \bar{M}$ *(associated with the signature operator). Furthermore, equality is achieved if and only if* (M, g) *is a complete Calabi–Yau manifold.*

Corollary 4.2 ([7]) *Let* (M^4, g) *be a complete noncompact Einstein manifold, whose fibration is given by a circle bundle over a surface, and which asymptotically approaches a fibered cusp/boundary at infinity. Then*

$$\chi(M) \geqslant \frac{3}{2}\left|\tau(M) - \frac{1}{3}e + \mathrm{sign}\, e\right|,$$

where e is the circular bundle's Euler number. Furthermore, if (M, g) is a complete Calabi–Yau manifold, then the equality holds.

Theorem 4.3 ([7]) *A cone* over $(\partial\bar{M}, g_{\partial\bar{M}})$ *is asymptotically connected to a complete Einstein 4-manifold (M^4, g). Then*

$$\chi(M) \geqslant \alpha(\partial\bar{M}) + \frac{1}{2\pi^2}\,\mathrm{vol}(\partial\bar{M}) + \frac{3}{2}\left|\tau(M) + \frac{1}{2}\eta(\partial\bar{M})\right|,$$

where the η-invariant of $(\partial\bar{M}, g_{\partial\bar{M}})$ is $\eta(\partial\bar{M})$ and the geometric invariant $\alpha(\partial\bar{M})$, which is defined by

$$\begin{aligned}\alpha(\partial\bar{M}) &= \frac{1}{8\pi^2}\int_{\partial\bar{M}} \epsilon_{abc}\omega^a \wedge \left[\Omega_c^b - \omega^b \wedge \omega^c\right] \\ &= \frac{1}{8\pi^2}\int_{\partial\bar{M}} \epsilon_{abc}\omega^a \wedge \Omega_c^b - \frac{3}{4\pi^2}\,\mathrm{vol}(\partial\bar{M}).\end{aligned}$$

In this case, Ω_c^b indicates the 2-form components of the curvature of $\partial\bar{M}$ with regard to this orthonormal basis, and ω^a denotes the dual 1-forms of an orthonormal basis for $\partial\bar{M}$. Furthermore, the equality holds if and only if M is a Calabi–Yau manifold that is asymptotically conical.

4.2 Compact Spin Gradient m-Quasi-Einstein Manifolds Satisfy H–T Inequality

In 2020, Klatt [18] demonstrated that a compact, oriented, and connected 4-dim gradient m-quasi-Einstein manifold, where $m \in [1, \infty)$, must satisfy the H–T inequality if it is also a spin manifold. Specifically, he claimed the following theorems:

Theorem 4.4 ([18]) *If $m \in [1, \infty]$ and (M^4, g) is a compact, oriented, spin 4-dim gradient m-quasi-Einstein manifold. After that, if (M^4, g) is a nontrivial gradient m-quasi-Einstein manifold, we obtain the rigorous condition $2\chi \pm 3\tau > 0$.*

He characterizes the topology more thoroughly when the hypothesis is true by only marginally expanding.

Theorem 4.5 ([18]) *If (M^4, g) is a connected, compact, oriented, spin 4-dim nontrivial gradient m-quasi-Einstein manifold with $m \in [1, \infty]$, then the universal cover $\tilde{M}^4$ satisfies $\tilde{M}^4 \approx S^4\#k(S^2 \times S^2)$ for some k.*

4.3 Einstein Structure of Squashed 4-Spheres

In 2023, Ho et al. [15] claimed a stronger version of H–T inequality by redefining topological invariants which can be expressed in terms of $SU(2)_{\pm}$connections using the decompositions. They give the following results:

$$\chi(M) = \frac{1}{2\pi^2}\int_M \left(\left(\widetilde{f}^{ij}_{(++)}\right)^2 + \left(\widetilde{f}^{ij}_{(--)}\right)^2 + \frac{R^2}{96}\right) d\mu \tag{4.1}$$

$$\tau(M) = \frac{1}{3\pi^2}\int_M \left(\left(\widetilde{f}^{ij}_{(++)}\right)^2 - \left(\widetilde{f}^{ij}_{(--)}\right)^2\right) d\mu, \tag{4.2}$$

where

$$f^{ij}_{(\pm\pm)} = \text{diag}\left(a^1_{\pm}, a^2_{\pm}, a^3_{\pm}\right).$$

Then they arrived at the following inequality:

$$\chi(M) - \frac{\lambda^2}{12\pi^2}\,\text{vol}(M) \geq \frac{3}{2}|\tau(M)|. \tag{4.3}$$

The inequality (4.1) and (4.2) are crucial in identifying an unlimited number of compact simply connected differentiable four-manifolds that meet the strict H–T inequality $\chi > \frac{3}{2}|\tau|$, even though they do not admit Einstein metrics (see [20]).

5 H–T Inequality: Recent Progress on Riemannian Manifolds

5.1 Note on a Diameter Bound for Complete Riemannian Manifolds with Positive Bakry–Émery Ricci Curvature

In 2015, Tadano [25] introduced new sufficient conditions under which 4-dim compact Ricci solitons admit the H–T inequality. The following are some of the results:

Corollary 5.1 ([25]) *Let (M, g) be a compact connected shrinking Ricci soliton in four dimensions that satisfies* (2.2). *Assume that the normalization of the soliton equals*

$$|\nabla f|^2 + R = 2\lambda f.$$

If

$$\sqrt{\frac{R_{\max}}{\lambda^2}\left(\frac{\pi^2}{2}+4\pi\right)} \leqslant \operatorname{diam}(M, g),$$

then (1.1)*, the H–T inequality, is satisfied by the soliton.*

Corollary 5.2 ([25]) *Let* (M, g) *be a compact connected shrinking Ricci soliton in four dimensions that satisfies* (2.2)*. Assume that the normalcy of the soliton equals*

$$|\nabla f|^2 + R = 2\lambda f.$$

If

$$\frac{\pi}{\sqrt{\lambda}} \cdot \frac{R_{\max}}{6\lambda}\sqrt{\frac{4R_{\max}}{\pi\lambda}} + 3 \leqslant \operatorname{diam}(M, g),$$

then (1.1)*, the H–T inequality, is satisfied by the soliton.*

5.2 *H–T-Type Inequalities for Pseudo-Riemannian Manifolds of Neutral Metric*

In 2001, Matsushita et al. [23] gave the following assertion:

Let M be a compact pseudo-Riemannian 4-manifold with neutral metric of signature $(++--)$ and structure group $SO_o(2, 2)$. Then, M can be regarded as a double almost pseudo-Hermitian 4-manifold. Provided the curvature is of the appropriate type, the H–T type inequality holds under a less restrictive condition known as the diagonal Einstein condition, rather than the standard Einstein condition. To support this, they proved the following theorem and corollary.

Theorem 5.3 ([23]) *For a double almost pseudo-Hermitian 4-manifold* (M, g, J, J')*, suppose that M satisfies the diagonal Einstein condition* $r_A = r_D = 0$.

(A) If W^+ *is not of Type* I_b*, then we have* $\chi \leqslant \frac{3}{2}\tau$.
(B) If W^- *is not of Type* I_b*, then we have* $-\chi \geqslant \frac{3}{2}\tau$.

Corollary 5.4 *Let* $M = (M, g, J, J')$ *be the same 4-manifold as considered in the above theorem. That is, it satisfies the diagonal Einstein condition* $r_A = r_D = 0$. *If the pair* (W^+, W^-) *of self-dual and anti-self-dual Weyl curvatures is neither of Type* $(\mathrm{I}_b, *)$ *nor of Type* $(*, \mathrm{I}_b)$*, then M satisfies the* $H-T$ *type inequality.*

6 H–T Inequality: Recent Progress on Ricci Flow

6.1 Harmonic Spinors in Ricci Flow

A strong parabolic H–T inequality for simply connected spin 4-manifolds was found in 2024 by Baldauf [1]. In this parabolic H–T inequality, the conditions for equality are characterized by the following theorem:

Theorem 6.1 ([1]) *If a non-singular solution $g(t)$ to the normalized Ricci flow is admitted on a closed, spin 4-manifold M satisfying $2\chi = 3|\sigma| > 0$, then $g(t)$ in this instance converges to a hyper-Kähler metric on a finite quotient of $K3$ in the smooth Cheeger–Gromov sense.*

Corollary 6.2 ([1]) *(Parabolic H–T inequality) If the normalized Ricci flow has a non-singular solution $g(t)$ on a closed, simply connected spin 4-manifold M, then*

$$2\chi \geq 3|\sigma|,$$

with equality iff $g(t)$ converges to a hyper-Kähler metric in a smooth Cheeger–Gromov manner once M is diffeomorphic to $K3$.

6.2 Normalized Ricci Flow Equation Non-singular Solutions

In 2008, Fang and colleagues [9] investigated non-singular solutions of the Ricci flow on a closed manifold with at least four dimensions. The normalized Ricci flow equation was explored on a closed smooth n-dimensional manifold M.

$$\tfrac{\partial}{\partial t} g = -2 \operatorname{Ric} + \tfrac{2r}{n} g g(0) = g_0, \tag{6.1}$$

where r represents the average scalar curvature $\frac{\int_M R dv}{\int_M dv}$.

The well-known H–T inequality for non-singular solutions to the Ricci flow on closed 4-manifolds is generalized as follows.

Theorem 6.3 ([9]) *Assume that M is a closed oriented 4-manifold and that the non-singular solution to* (6.1) *is $\{g(t)\}, t \in [0, \infty)$. Then, M satisfies one of the following conditions*

1. *M allows for a shrinking Ricci soliton.*
2. *M accepts an F-structure of positive rank.*
3. *The H–T type inequality* (1.1) *holds.*

Conjecture 6.4 ([9]) Condition 3 of Theorem 6.3 is replaced by the following H–T–Gromov–Kotschick type inequality

$$2\chi(M) - 3|\tau(M)| \geq \frac{1}{1,296\pi^2}\|M\|,$$

where $\|M\|$ is a simplicial volume of M.

6.3 H–T Inequality and Ricci Flow in 4-Manifolds

For closed 4-manifolds with a non-positive Yamabe invariant that allows long-time solutions to the normalized Ricci flow equation with a restricted scalar curvature, Y. Zhang and Z. Zhang [34] developed an H–T type inequality in 2012. They considered the normalized Ricci flow is

$$\frac{\partial}{\partial t}g(t) = -2Ric_t + \frac{2r(t)}{n}g(t), \tag{6.2}$$

where R_t is the scalar curvature of $g(t)$ and $r(t) = \frac{\int_M R_t dv_{g(t)}}{Vog_{g(t)}(M)}$ is its mean scalar curvature.

Theorem 6.5 ([34]) *Consider a 4-dim closed oriented manifold M with $\bar{\lambda}_M \leq 0$. Suppose M admits a long-time solution $g(t)$ to Eq.* (6.2), *where for every t in the interval* $[0, \infty)$, *the scalar curvature satisfies $|R_t| < C$ for some constant C that is independent of t. Then*

$$2\chi(M) - 3|\tau(M)| \geq \frac{1}{96\pi^2}\bar{\lambda}_M^2.$$

For many circumstances, the $\bar{\lambda}_M \leq 0$ hypothesis is true.

6.4 A Note on the H–T Inequality and Ricci Flows

In 2009, Y. Zhang and Z. Zhang discovered an H–T type inequality for closed oriented 4-manifolds with a zero Yamabe invariant. This inequality allows for long-time solutions to the normalized Ricci flow equation with bounded scalar curvature. They established the following results:

Theorem 6.6 ([33]) *Given $\bar{\lambda}_M = 0$, let M be an oriented closed 4-manifold. For a constant C independent of t, if M allows a long-time solution $g(t), t \in [0, \infty)$, of* (6.2) *with scalar curvature $|R_t| < C$, then H–T inequality* (1.1) *holds.*

Theorem 6.7 ([33]) *Let M be an oriented closed 4-manifold with $\bar{\lambda}_M \leq 0$, and*

$$2\chi(M) + 3\tau(M) = 0.$$

Assume that there is a long-time solution $g(t), t \in [0, \infty)$, of (6.2) *with bounded Ricci curvature $\mid Ric_t \mid < C$ for a constant C independent of t, which is non-collapsing, i.e., for any $t > 0$ and $r \leq 1$, there is an $x_t \in M$ such that*

$$\kappa r^4 \leq \mathrm{Vol}_{g(t)}\left(B_{g(t)}(x_t, r)\right),$$

for a constant $\kappa > 0$ that is independent of t. Then, there exists a sequence of times $t_k \longrightarrow \infty$ satisfying that $(M, g(t_k))$ converges to (N, g_∞) in the Gromov–Hausdorff sense, where N is a compact 4-orbifold with finite singular points $S = \{p_i\}$, and g_∞ is a Ricci-flat anti-self-dual orbifold metric, i.e., $\mathrm{Ric}(g_\infty) \equiv 0$ *and $W^+(g_\infty) \equiv 0$. Furthermore, if $\chi(M) = 0$, then a finite covering of M is a torus, and g_∞ is flat.*

6.5 Normalized Ricci Flow Non-singular Solutions on Noncompact Manifolds with Finite Volume

It was shown in 2010 by Fang et al. [10] that the Euler characteristic $\chi(M)$ must be nonnegative if $g(t)$ is a complete, non-singular solution to the normalized Ricci flow on a noncompact 4-manifold M with finite volume.

Theorem 6.8 ([10]) *The Euler characteristic number fulfills if $g(t)$ is a full non-singular solution of* (6.2) *of finite volume on a 4-manifold M.*

$$2\chi(M) \geq \left|\frac{1}{16\pi^2}\int_M \left(|W_0^+|^2 - |W_0^-|^2\right) dv_{g(0)}\right| \geq 0.$$

Theorem 6.9 ([10]) *Let M be as described in Theorem 6.8. When $(M, g(0))$ approaches a fibered cusp asymptotically, the strict H–T inequality is established.*

$$3\left|\tau(M) + \frac{1}{2}\mathrm{a}\lim \eta(\partial\bar{M})\right| < 2\chi(M),$$

where the η-invariant of the boundary is the adiabatic limit of $\mathrm{a}\lim \eta(\partial\bar{M})$.

6.6 The Normalized Ricci Flow on 4-Manifolds and Exotic Smooth Structures

In 2008, Ishida [16] explored the connection between smooth structures on closed 4-manifolds and the presence or absence of non-singular solutions to the normalized Ricci flow. In this context, non-singular solutions refer to those with uniformly bounded sectional curvature that persist for all time $t \in [0, \infty)$. The study also highlighted that several compact topological 4-manifolds possess unique or exotic smooth structures.

Definition 6.10 ([16]) If $T = \infty$ and the scalar curvature $s_{g(t)}$ of $g(t)$ satisfy the following condition, then for any $t \in [0, T)$, the maximal solution $\{g(t)\}$ to the normalized Ricci flow on X is called quasi-non-singular, $\sup_{X \times [0,T)} |s_{g(t)}| < \infty$.

The authors of [9] observed, among other things, that any closed, oriented, smooth 4-manifold X must adhere to the topological constraint (1.1), which involves the Euler characteristic $\chi(X)$ and the signature $\tau(X)$ of X:

$$2\mathcal{X}(X) \geq 3|\tau(X)|,$$

if there exists a quasi-non-singular solution to the normalized Ricci flow exists on X and if this solution also satisfies the condition:

$$\hat{s}_{g(t)} \leq -c < 0, \tag{6.3}$$

hence, for a given Riemannian metric g, the constant c is defined as $\hat{s}_g := \min_{x \in X} s_g(x)$ and is independent on t.

The author included the following proof for completeness: Let X be a Riemannian 4-manifold that is closed and orientated. According to the Hirzebruch signature formula and the Chern–Gauss–Bonnet formula, the following equations hold true for any Riemannian metric g on X:

$$\tau(X) = \frac{1}{12\pi^2} \int_X \left(\left|W_g^+\right|^2 - \left|W_g^-\right|^2 \right) d\mu_g,$$

$$\chi(X) = \frac{1}{8\pi^2} \int_X \left(\frac{s_g^2}{24} + \left|W_g^+\right|^2 + \left|W_g^-\right|^2 - \frac{\left|\overset{\circ}{r}_g\right|^2}{2} \right) d\mu_g.$$

Here, W_g^+ and W_g^- refer to the self-dual and anti-self-dual components of the Weyl curvature associated with the metric g, while $\overset{\circ}{r}_g$ denotes the trace-free portion of the Ricci curvature of g. Furthermore, s_g represents the scalar curvature of g, and $d\mu_g$ is the volume form corresponding to g. Using these expressions, we can establish the following significant equality:

$$2\chi(X) \pm 3\tau(X) = \frac{1}{4\pi^2}\int_X \left(2\left|W_g^{\pm}\right|^2 + \frac{s_g^2}{24} - \frac{\left|\overset{\circ}{r}_g\right|^2}{2}\right) d\mu_g.$$

If X possesses an Einstein metric g, then $\overset{\circ}{r}_g \equiv 0$. As a result, the formula mentioned above indicates that any 4-dim Einstein manifold must satisfy (1.1), which is simply the H–T inequality.

Theorem 6.11 ([16]) *Consider a closed, oriented, 4-dim Riemannian manifold X, and suppose that the normalized Ricci flow admits a quasi-non-singular solution as defined in Definition 6.10. Additionally, assume that this solution satisfies the uniform bound given by* (6.3)*, meaning that*

$$\hat{s}_{g(t)} \leq -c < 0.$$

The inequality is valid, where the constant c is independent of t and is defined by $\hat{s}_g := \min_{x\in X} s_g(x)$ for a given Riemannian metric g. As a result, X must satisfy the condition (1.1).

7 H–T Inequality: More Recent Progress

7.1 Myers Type Theorems and H–T Inequality

In 2019, Tadano [28] enhanced his previous results [29] on Myers type theorems and the validity of the H–T inequality for shrinking Ricci solitons. He established the following results related to the H–T inequality.

Theorem 7.1 ([21]) *Let (M, g) be a compact, shrinking Ricci soliton in four dimension that satisfies* (2.2)*. If the scalar curvature meets*

$$24\lambda^2 \operatorname{vol}(M, g) \geqslant \int_M R^2 dv,$$

thus (1.1)*, the H–T inequality, must be satisfied by the soliton.*

Theorem 7.2 ([27]) *Assume that (M, g) is a compact 4-dim nontrivial Ricci soliton that satisfies* (2.2)*. If the upper bound on the diameter of (M, g) is*

$$\operatorname{diam}(M, g) \leqslant \frac{2\pi\sqrt{2}(\sqrt{2}-1)}{\sqrt{R_{\max} - R_{\min}}},$$

thus (1.1)*, the H–T inequality, must be satisfied by the soliton.*

7.2 Topology of Toric Gravitational Instantons

In 2023, Gustav [24] utilized H–T type inequalities on Ricci-flat ALE/ALF manifolds and derived essential conditions that the rod structures of toric ALE/ALF instantons must satisfy, with the goal of furthering the classification of these spaces. Below are some of the findings from his research.

Theorem 7.3 (H–T Inequality for Ricci-Flat ALE Manifolds [24]) *Suppose (M, g) is an oriented Ricci-flat asymptotically locally Euclidean (ALE) manifold associated with the group Γ. Then,*

$$3\left|\tau(M) + \eta_S\left(S^3/\Gamma\right)\right| \leq 2\left(\chi(M) - \frac{1}{|\Gamma|}\right).$$

An equality holds if and only if the universal cover of M is hyper-Kähler.

The term $\eta_S\left(S^3/\Gamma\right)$ appearing in (7.3) refers to the η-invariant of the signature operator for the space form S^3/Γ. This is a spectral invariant associated with the space form.

Theorem 7.4 (H–T Inequality for Ricci-Flat ALF- A_k Manifolds [24]) *Suppose (M, g) is an oriented Ricci-flat manifold that is ALF-A_k for some integer k. Then,*

$$3\left|\tau(M) - \frac{e}{3} + \operatorname{sgn}(e)\right| \leq 2\chi(M),$$

in which e is the Euler number of the asymptotic circle bundle. If M has a hyper-Kähler universal cover, then equality is preserved.

7.3 H–T Inequality for Manifolds with Foliated Boundaries

In 2017, Zeroual [32] developed an H–T inequality for noncompact 4-manifolds with foliated geometry at infinity, extending the earlier work of Dai and Wei. The principal outcome of his efforts is as follows.

Theorem 7.5 *Assuming the conditions given in [32], suppose M is a 4-manifold with foliated geometry at infinity. When M permits an accurate $\mathcal{F}$- or $\mathcal{F}_c$-metric proposed by Einstein, then*

$$\chi(M) \geq \frac{3}{2}\left|\frac{1}{|\Gamma|}\left\{\sum_{a \neq Id}\sum_{z \in \operatorname{Fix}(a)} \operatorname{def}\left(a, {}^{g}\widetilde{B}\right)|z + \frac{\chi(E)}{3}\right\} + \tau(M) - \epsilon(E)\right|.$$

The universal cover of M is a complete Ricci-flat (anti-)self-dual manifold if equality holds.

7.4 Stable Cohomotopy Seiberg–Witten Invariants

In 2015, Ishida and Sasahira [17] applied the Gromov–H–T inequality to uncover new results concerning the presence of exotic differentiable structures. They presented the following conclusions.

Theorem 7.6 *Let X_m be as in Theorem A stated and proved in [17], and assume moreover that X_m is a minimal Kähler surface. Assume that N is a 4-manifold that is closed and orientated. Its Riemannian metric has nonnegative scalar curvature, and $b^+(N) = 0$. Then, the invariant of a connected sum $M := \left(\#_{m=1}^n X_m\right) \#N$ for any real number $k \geq \frac{2}{3}$, $\bar{\lambda}_k$ for $n = 2, 3$, is given by*

$$\bar{\lambda}_k(M) = -4k\pi \sqrt{2 \sum_{m=1}^{n} c_1^2 (X_m)}.$$

Here notice that minimality of X_m forces that

$$c_1^2 (X_m) = 2\chi (X_m) + 3\tau (X_m) \geq 0.$$

Theorem 7.7 ([17]) *Assume that $b^+(N) = 0$ and that N is a closed oriented smooth 4-manifold. Let X_m be a closed oriented almost complex 4-manifold for $m = 1, 2, 3$ such that $b^+ (X_m) > 1$ and such that*

$$b^+ (X_m) - b_1 (X_m) \equiv 3(\text{mod} 4).$$

Assume that $SW_{X_m} \left(\Gamma_{X_m}\right) \equiv 1(\text{mod} 2)$. Let Γ_{X_m} be a spin^c structure on X_m that is induced by the almost complex structure. Furthermore, under Definition 3 of [17], suppose that for any m, the following is true:

$$\mathfrak{S}^{ij} \left(\Gamma_{X_m}\right) \equiv 0 \bmod 2 \quad \textit{for all } i, j.$$

Then a connected sum $M := \left(\#_{m=1}^n X_m\right) \#N$, where $n = 2, 3$, cannot admit any Einstein metric if the following holds:

$$4n - (2\chi(N) + 3\tau(N)) \geq \frac{1}{3} \sum_{m=1}^{n} (2\chi (X_m) + 3\tau (X_m)).$$

Theorem 7.8 ([17]) *The following three properties can be satisfied by an infinite number of closed topological spin 4-manifolds:*

- *$\|M\| \neq 0$, that is, any four manifold M has a nontrivial simplicial volume.*

- *The strict Gromov–H–T inequality is satisfied by every four manifold M. That is to say,*

$$2\chi(M) - 3|\tau(M)| > \frac{1}{81\pi^2}\|M\|.$$

- *Each 4-manifold M admits infinitely many distinct differentiable structures for which no compatible Einstein metric exists.*

References

1. Baldauf, J.: Harmonic spinors in the Ricci flow. J. Geom. Anal. **34**(8), 235 (2024)
2. Berger, M.: Sur quelques variétés riemanniennes compactes d'Einstein. Comptes Rendus de l'Académie des Sciences. Série I. Mathématique **260**, 1554–1557 (1965)
3. Besse, A.L.: Einstein Manifolds. Springer, Berlin (2008)
4. Brasil, A., Costa, E., Ribeiro Jr., E.: Hitchin–Thorpe inequality and Kaehler metrics for compact almost Ricci soliton. Annali di Matematica Pura ed Applicata **193**, 1851–1860 (2014)
5. Cao, H.-D.: Recent Progress on Ricci Solitons. Recent Advances in Geometric Analysis. Advanced Lectures in Mathematics (ALM), vol. 11, pp. 1–38. International Press, Somerville (2010)
6. Cheng, X., Ribeiro Jr., E., Zhou, D.: On Euler characteristic and Hitchin–Thorpe inequality for 4-dimensional compact Ricci solitons. Proc. Am. Math. Soc. Ser. B **10**(3), 33–45 (2023)
7. Dai, X., Wei, G.: Hitchin–Thorpe inequality for noncompact Einstein 4-manifolds. Adv. Math. **214**(2), 551–570 (2007)
8. Donaldson, S.K., Kronheimer, P.B.: The Geometry of Four-Manifolds. Oxford University Press, Oxford (1997)
9. Fang, F., Zhang, Y., Zhang, Z.: Non-singular solutions to the normalized Ricci flow equation. Math. Ann. **340**(3), 647–674 (2008)
10. Fang, F., Zhang, Z., Zhang, Y.: Non-singular solutions of normalized Ricci flow on noncompact manifolds of finite volume. J. Geom. Anal. **20**, 592–608 (2010)
11. Fernández-López, M., García-Río, E.: Diameter bounds and Hitchin–Thorpe inequalities for compact Ricci solitons. Q. J. Math. **61**(3), 319–327 (2010)
12. Gromov, M., Lawson, H.B.: The classification of simply connected manifolds of positive scalar curvature. Ann. Math. **111**(3), 423–434 (1980)
13. Hamilton, R.S.: The Ricci flow on surfaces, mathematics and general relativity. Contemp. Math. **71**, 237–261 (1988)
14. Hitchin, N.: Compact 4-dimensional Einstein manifolds. J. Diff. Geom. **9**(3), 435–441 (1974)
15. Ho, J., Kim, K.K., Yang, H.S.: Einstein structure of squashed four-spheres (2023). ArXiv Preprint ArXiv:2309.05335
16. Ishida, M.: The normalized Ricci flow on four-manifolds and exotic smooth structures. Asian J. Math. **20**(5), 903–917 (2016)
17. Ishida, M., Sasahira, H.: Stable cohomotopy Seiberg–Witten invariants of connected sums of four-manifolds with positive first Betti number, I: non-vanishing theorem. Int. J. Math. **26**(6), 1541004 (2015)
18. Klatt, B.: Compact 4-dimensional spin gradient m-quasi-Einstein manifolds satisfy the Hitchin–Thorpe inequality (2020). ArXiv Preprint ArXiv:2012.13848
19. Lawson, H.B., Michelsohn, M.-L.: Spin Geometry. Princeton University Press, Princeton (1989)
20. LeBrun, C.: 4-manifolds without Einstein metrics. Math. Res. Lett. **3**, 133–147 (1996)

21. Ma, L.: Remarks on compact shrinking Ricci solitons of dimension four. Comptes Rendus Mathématique **351**(21–22), 817–823 (2013)
22. MacMahon, C.: Topological obstructions to the existence of compact shrinking Ricci solitons in dimension four (2024). ArXiv Preprint ArXiv:2407.21079
23. Matsushita, Y., Law, P.: Hitchin–Thorpe-type inequalities for pseudo-Riemannian 4-manifolds of metric signature $(++--)$. Geometriae Dedicata **87**(1), 65–89 (2001)
24. Nilsson, G.: Topology of toric gravitational instantons. Diff. Geom. Appl. **96**, 102171 (2024)
25. Tadano, H.: Remark on a diameter bound for complete Riemannian manifolds with positive Bakry–Émery Ricci curvature. Diff. Geom. Appl. **44**, 136–143 (2016)
26. Tadano, H.: An upper diameter bound for compact Ricci solitons with application to the Hitchin–Thorpe inequality. II. J. Math. Phys. **59**(4), 043507, 3 p. (2018)
27. Tadano, H.: Remark on a lower diameter bound for compact shrinking Ricci solitons. Diff. Geom. Appl. **66**, 231–241 (2019)
28. Tadano, H.: Some Myers type theorems and Hitchin–Thorpe inequalities for shrinking Ricci solitons, II. Rendiconti del Seminario Matematico. Università e Politecnico Torino, vol. 77, pp. 83–11 (2019)
29. Tadano, H.: Some Myers type theorems and Hitchin–Thorpe inequalities for shrinking Ricci solitons. Rendiconti del Seminario Matematico. Università e Politecnico Torino, vol. 73, issue 1, pp. 183–199 (2015)
30. Tadano, H.: Improved oscillation estimates and the Hitchin–Thorpe inequality on compact Ricci solitons. J. Math. Phys. **64**(8), 083505, 6 p. (2023)
31. Thorpe, J.A.: Some remarks on the Gauss-Bonnet integral. J. Math. Mech. **18**(8), 779–786 (1969)
32. Zerouali, A.J.: On a Hitchin–Thorpe inequality for manifolds with foliated boundaries. Annales Mathématiques du Qu'ebec **41**(1), 169–197 (2017)
33. Zhang, Y., Zhang, Z.: Hitchin–Thorpe inequality and Ricci flow on 4-manifolds (2009). ArXiv Preprint ArXiv:0903.2539
34. Zhang, Y., Zhang, Z.: A note on the Hitchin–Thorpe inequality and Ricci flow on 4-manifolds. Proc. Am. Math. Soc. **140**(5), 1777–1783 (2012)

The Principal Eigenvalue of a (p, q)-Biharmonic System Along the Ricci Flow

Shahroud Azami and Gh. Fasihi-Ramandi

Abstract In this chapter, we first provide evolution formulas for the eigenvalue of some cooperative (p, q)-biharmonic system on Riemannian manifolds under the (unnormalized) Ricci flow and normalized Ricci flow. Then, we provide some monotonic quantities under this flow. Moreover, as an application of the evolution equation, we give an example.

Keywords Laplace Operator · Reimannian manifolds

1 Introduction

We investigate the evolution for the principal eigenvalue of a (p, q)-biharmonic system on manifold (M, g) along the Ricci flow (briefly RF). Let g be a Riemannian metric on a manifold M. In coordinate $\{x^i\}$, the Laplace-Beltrami operator Δ is given as follows:

$$\Delta = g^{ij}\left(\frac{\partial^2}{\partial x^i \partial x^j} - \Gamma_{ij}^k \frac{\partial}{\partial x^k}\right), \tag{1.1}$$

where Γ_{ij}^k are Christoffel symbols of g. Also, for any function $h \in W_0^{2,p}(M)$, the p-biharmonic operator is defined by

$$\Delta_p^2 h = \Delta(|\Delta h|^{p-2}\Delta h). \tag{1.2}$$

The p-biharmonic operator is an elliptic operator. For $p = 2$, (1.2) describes the clamped plate problem.

S. Azami (✉) · Gh. Fasihi-Ramandi
Faculty of Basic Science, Department of Pure Mathematics, Imam Khomeini International University, Qazvin, Iran
e-mail: azami@sci.ikiu.ac.ir; fasihi@sci.ikiu.ac.ir

B.-Y. Chen, M. A. Choudhary (eds.), *Geometric Inequalities and Applications*, Infosys Science Foundation Series in Mathematical Sciences,
https://doi.org/10.1007/978-981-95-5148-4_11

The family $g(t)$ of Riemannian metrics on M with Ricci tensor S is said to be an unnormalized RF (or URF) if

$$\frac{d}{dt}g(t) = -2S(g(t)), \ g(0) = g_0. \tag{1.3}$$

RF was introduced by Hamilton in 1982. Let us denote by $r = \frac{\int_M R d\mu}{\int_M d\mu}$ the scalar curvature, and one often considers the normalized RF (or NRF)

$$\frac{d}{dt}g(t) = -2S(g(t)) + \frac{2r}{n}g, \ g(0) = g_0. \tag{1.4}$$

The volume of manifolds remains constant under the NRF. Short time existence and uniqueness for solutions to the RF on $[0, T)$ have been shown by Hamilton in [8] and by DeTurk in [9].

Recently, the study of geometric operators and their eigenvalues has been an important tool for determine topology and geometry of manifolds. In [13], Perelman introduced the functional $F(g(t), f(t)) = \int_M (R + |\nabla f|^2)e^{-f} d\mu$ and showed that it is nondecreasing under the RF with backward heat-type equation and by it concluded that the first eigenvalue of the operator $-4\Delta + R$ is nondecreasing along the RF. This was a new beginning for the study of eigenvalues, and then mathematicians, especially individuals who worked on geometric analysis, studied the eigenvalues of the geometrical operators under different geometric flows. For instance, in [6], Cao showed the first eigenvalue of $-\Delta + cR$ for $c \geq \frac{1}{4}$ is nondecreasing under the RF. Also, in [1] and [14] the authors have studied the monotonic of the eigenvalue of p-Laplacian under the Ricci-harmonic flow and RF, respectively. The first author [3, 4] studied the monotonicity of the first eigenvalue of the clamped plate problem under the RF and the eigenvalues of a (p, q)-Laplacian system along the mean curvature flow. Also, Abolarinwa [2] studied the eigenvalues of p-bi-Laplacian along the RF.

In the present chapter, we consider the following eigenvalue problem:

$$\begin{aligned} &\Delta_p^2 u = \lambda |u|^{p-2}u + \lambda |u|^{\alpha-1}|v|^{\beta+1}u \ \text{ in } M,\\ &\Delta_q^2 v = \lambda |v|^{q-2}v + \lambda |u|^{\alpha+1}|v|^{\beta-1}v \ \text{ in } M,\\ &u = \Delta u = v = \Delta v = 0 \ \text{ on } \partial M, \end{aligned} \tag{1.5}$$

on a closed Riemannian manifold (or CRM) (M, g) whose metric satisfies the RF, and investigate the monotonicity of principal eigenvalue of the (p, q)-biharmonic system (1.5) under the Ricci flow. In [11] and [12] the authors have investigated the principal eigenvalue of (1.5). We will prove the following theorems:

Theorem 1.1 *Suppose that $(M^n, g(t))$, $t \in [0, T)$, is a solution of the RF (1.3) on a CRM (M^n, g_0) with the positive Ricci curvature. For $p, q \geq 2$, the first eigenvalue*

$\lambda(t)$ of (1.5) satisfies $\lim_{t\to T} \lambda(t) = \infty$, where $S_{ij} - \gamma R g_{ij} > 0$ on $M \times [0, T)$ for some positive constant γ.

Theorem 1.2 *Suppose that $(M^n, g(t))$ is a solution of the URF on the smooth CRM (M^n, g_0) satisfying $R_{ij} \geq \epsilon R g_{ij}$ with $\epsilon > \max\{\frac{1}{p}, \frac{1}{q}\}$ for $t \in [0, T)$. Suppose that $\lambda(t)$ is the first nonzero eigenvalue along the URF.*

1. *If $R_{\min}(g_0)$ is a constant, then $\lambda(t)e^{-2R_{\min}(g)t}$ is monotonically nondecreasing along RF.*
2. *If $R \geq R_{\min}(g_0) \geq 0$, then $\lambda(t_0) \geq \lambda(t_1)e^{2\int_{t_1}^{t_0} R_{\min}(t)dt}$ for $[t_1, t_0] \subseteq [0, T_{\max})$.*
3. *If $R \geq R_{\min}(g_0) > 0$, then $\lambda(t)(R_{\min}^{-1}(0) - \frac{2}{n}t)^n$ is monotonically nondecreasing along Ricci flow and $\lambda(t)$ is nondecreasing along RF.*

Theorem 1.3 *If closed surface (M^2, g_0) has nonnegative scalar curvature, then for $p \geq 2$ and $q \geq 2$, the eigenvalues of (1.5) are increasing along the URF.*

1.1 Eigenvalues of (p, q)-Biharmonic System

Let (M^n, g) be a CRM. We consider a cooperative (p, q)-biharmonic system (1.5) where α, β and p, q are real constants and

$$p > 1,\ \ q > 1, \alpha > 0,\ \beta > 0, \quad \frac{\alpha+1}{p} + \frac{\beta+1}{q} = 1. \tag{1.6}$$

λ is called an eigenvalue of (1.5), if for some functions f and h,

$$\int_M f\Delta_p^2 f\, d\mu = \lambda \int_M |\nabla f|^p + \lambda \int_M |f|^{\alpha+1}|h|^{\beta+1} d\mu, \tag{1.7}$$

$$\int_M h\Delta_q^2 h\, d\mu = \lambda \int_M |\nabla h|^q + \lambda \int_M |f|^{\alpha+1}|h|^{\beta+1} d\mu, \tag{1.8}$$

where $d\mu$ is the volume element of g. The pair (f, h) is said to be eigenfunctions of (1.5). The first positive eigenvalue for (1.5) is defined by

$$\inf\{A(f, h) :\ B(f, h) = 1\}, \tag{1.9}$$

where

$$A(f, h) = \frac{\alpha+1}{p}\int_M |\Delta f|^p d\mu + \frac{\beta+1}{q}\int_M |\Delta h|^q d\mu,$$

$$B(f, h) = \int_M |f|^{\alpha+1}|h|^{\beta+1} d\mu + \frac{\alpha+1}{p}\int_M |f|^p d\mu + \frac{\beta+1}{q}\int_M |h|^q d\mu.$$

Suppose that $(M^n, g(t))$ is a solution of the RF on the smooth CRM (M^n, g_0) in the interval $[0, T)$; then

$$G(g, f, h) = \frac{\alpha+1}{p} \int_M |\nabla f|^p d\mu_t + \frac{\beta+1}{q} \int_M |\nabla h|^q d\mu_t \tag{1.10}$$

provides the evolution for eigenvalues of (1.5), along $g(t)$ where the eigenfunctions corresponding to $\lambda(t)$ are normalized, that is, $B(f, h) = 1$. We prove some results on $\lambda(t)$ under the RF.

We do not know whether the principal eigenvalue of (1.5) or its corresponding eigenfunctions are C^1-differentiable or not under the RF; then we apply methods the same as in [5, 15] and define general smooth functions with respect to t under the RF in what follows. We first assume that at time $t_0 \in [0, T)$, $(f_0, h_0) = (f(t_0), h(t_0))$ is the eigenfunctions for $\lambda(t_0)$ of (p, q)-biharmonic system (1.5). We define smooth functions $h(t)$ and $f(t)$ along the Ricci flow, as follows:

$$u(t) = u_0 \left[\frac{\det[g_{ij}(t)]}{\det[g_{ij}(t_0)]}\right]^{\frac{1}{p}}, \qquad v(t) = v_0 \left[\frac{\det[g_{ij}(t)]}{\det[g_{ij}(t_0)]}\right]^{\frac{1}{q}}. \tag{1.11}$$

Let

$$f(t) = \frac{u(t)}{K^{\frac{1}{p}}}, \qquad h(t) = \frac{v(t)}{K^{\frac{1}{q}}}, \tag{1.12}$$

where

$$K = \int_M |u(t)|^{\alpha+1} |v(t)|^{\beta+1} d\mu + \frac{\alpha+1}{p} \int_M |u(t)|^p d\mu + \frac{\beta+1}{q} \int_M |v(t)|^q d\mu.$$

Then $f(t)$, $h(t)$ are smooth functions along the RF and

$$\int_M |f|^{\alpha+1} |h|^{\beta+1} d\mu + \frac{\alpha+1}{p} \int_M |f|^p d\mu + \frac{\beta+1}{q} \int_M |h|^q d\mu = 1. \tag{1.13}$$

Also at time t_0, $(f(t_0), h(t_0))$ are the eigenfunctions for $\lambda(t_0)$ of (1.5); that is, $\lambda(t_0) = G(g(t_0), f(t_0), h(t_0))$ provide a new smooth eigenvalue function. Let M^n be a CRM and $g(t)$ be a smooth solution of RF. Set

$$\lambda(f, h, t) := \frac{\alpha+1}{p} \int_M |\Delta f|^p d\mu + \frac{\beta+1}{q} \int_M |\Delta h|^q d\mu, \tag{1.14}$$

where f, h are smooth functions such that

$$\int_M |f|^{\alpha+1}|h|^{\beta+1}d\mu + \frac{\alpha+1}{p}\int_M |f|^p d\mu + \frac{\beta+1}{q}\int_M |h|^q d\mu = 1. \tag{1.15}$$

If (f, h) are the eigenfunctions for $\lambda(t)$ at t_0, then $\lambda(f, h, t_0) = \lambda(t_0)$.

2 Variation of the Eigenvalue $\lambda(t)$

In the following, we will prove some evolution formulas for $\lambda(t)$ along the RF. From [8] we have the following lemma:

Lemma 2.1 *Under the RF, we get:*

1. $\frac{\partial}{\partial t}g^{ij} = 2S^{ij}$
2. $\frac{\partial}{\partial t}(d\mu) = -Rd\mu$
3. $\frac{\partial}{\partial t}(\Gamma_{ij}^k) = -\nabla_j S_i^k - \nabla_i S_j^k + \nabla^k S_{ij}$
4. $\frac{\partial}{\partial t}R = \Delta R + 2|S|^2$
5. $\frac{\partial}{\partial t}(\Delta u) = 2S^{ij}\nabla_i\nabla_j u + \Delta u_t$

and along the UNF (1.4), we have

1. $\frac{\partial}{\partial t}g^{ij} = 2S^{ij} - \frac{2}{n}rg^{ij}$
2. $\frac{\partial}{\partial t}(d\mu) = (r - R)d\mu$
3. $\frac{\partial}{\partial t}(\Gamma_{ij}^k) = -\nabla_j S_i^k - \nabla_i S_j^k + \nabla^k S_{ij}$
4. $\frac{\partial}{\partial t}(\Delta u) = 2S^{ij}\nabla_i\nabla_j u - \frac{2}{n}r\Delta u + \Delta u_t$

where R denotes the scalar curvature.

Proposition 2.2 *Suppose that $(M^n, g(t))$ is a solution to the URF on the smooth CRM (M^n, g_0). The first eigenvalue of (1.5) along the URF satisfies*

$$\begin{aligned}
\frac{d}{dt}\lambda(f, h, t)|_{t=t_0} = {} & 2(\alpha+1)\int_M (\Delta f)(S^{ij}\nabla_i\nabla_j f)|\Delta f|^{p-2}d\mu \\
& -\frac{\alpha+1}{p}\int_M |\Delta f|^p R d\mu \\
& +2(\beta+1)\int_M (\Delta h)(S^{ij}\nabla_i\nabla_j h)|\Delta h|^{q-2}d\mu \\
& -\frac{\beta+1}{q}\int_M |\Delta h|^q R d\mu \\
& +\lambda(t_0)\int_M R|f|^{\alpha+1}|h|^{\beta+1}d\mu + \lambda(t_0)\frac{\alpha+1}{p}\int_M |f|^p R d\mu
\end{aligned}$$

$$+\lambda(t_0)\frac{\beta+1}{q}\int_M |h|^q R d\mu. \tag{2.1}$$

Proof $\lambda(f,h,t)$ is a smooth function along the RF, and by derivative of (1.14) with respect to t, we get

$$\begin{aligned}\frac{d}{dt}\lambda(f,h,t)|_{t=t_0} &= \frac{\alpha+1}{p}\left(\int_M \frac{\partial}{\partial t}(|\Delta f|^p)d\mu_t + \int_M |\Delta f|^p \frac{\partial}{\partial t}(d\mu_t)\right) \\ &+\frac{\beta+1}{q}\left(\int_M \frac{\partial}{\partial t}(|\Delta h|^q)d\mu_t + \int_M |\Delta h|^q \frac{\partial}{\partial t}(d\mu_t)\right).\end{aligned} \tag{2.2}$$

On the manifold $(M, g(t))$ along the RF, we arrive at

$$\frac{\partial}{\partial t}(d\mu_t) = -Rd\mu, \tag{2.3}$$

and

$$\frac{\partial}{\partial t}|\Delta f|^p = p(\Delta f)(2S^{ij}\nabla_i\nabla_j f + \Delta f_t)|\Delta f|^{p-2}. \tag{2.4}$$

Therefore we obtain

$$\begin{aligned}\frac{d}{dt}\lambda(f,h,t)|_{t=t_0} &= (\alpha+1)\int_M \left\{(\Delta f)(2S^{ij}\nabla_i\nabla_j f + \Delta f_t)|\Delta f|^{p-2}\right\}d\mu \\ &+ (\beta+1)\int_M \left\{(\Delta h)(2S^{ij}\nabla_i\nabla_j h + \Delta h_t)|\Delta h|^{q-2}\right\}d\mu \\ &- \frac{\alpha+1}{p}\int_M |\Delta f|^p R d\mu - \frac{\beta+1}{q}\int_M |\Delta h|^q R d\mu.\end{aligned} \tag{2.5}$$

Differentiating both sides of $B(f,h) = 1$ gives

$$\begin{aligned}&(\alpha+1)\int_M |f|^{\alpha-1}|h|^{\beta+1} f_t u d\mu + (\beta+1)\int_M |f|^{\alpha+1}|h|^{\beta-1} v v_t d\mu \\ &+(\alpha+1)\int_M |f|^{p-2} u u_t d\mu + (\beta+1)\int_M |h|^{q-2} v v_t d\mu \\ &= \int_M R|f|^{\alpha+1}|h|^{\beta+1} d\mu + \frac{\alpha+1}{p}\int_M |f|^p R d\mu + \frac{\beta+1}{q}\int_M |h|^q R d\mu.\end{aligned} \tag{2.6}$$

Also, integrating by parts shows

$$(\alpha+1)\int_M (\Delta f)(\Delta f_t)|\Delta f|^{p-2}d\mu + (\beta+1)\int_M (\Delta h)(\Delta h_t)|\Delta h|^{q-2}d\mu \quad (2.7)$$
$$= (\alpha+1)\int_M f_t\Delta_p^2 f\, d\mu + (\beta+1)\int_M h_t\Delta_q^2 h\, d\mu$$
$$= \lambda(\alpha+1)\int_M |f|^{p-2}uu_t d\mu + \lambda(\alpha+1)\int_M |f|^{\alpha-1}|h|^{\beta+1} f_t u d\mu$$
$$+\lambda(\beta+1)\int_M |h|^{q-2}vv_t d\mu + \lambda(\beta+1)\int_M |f|^{\alpha+1}|h|^{\beta-1}vv_t d\mu$$
$$= \lambda\int_M R|f|^{\alpha+1}|h|^{\beta+1}d\mu + \lambda\frac{\alpha+1}{p}\int_M |f|^p R d\mu + \lambda\frac{\beta+1}{q}\int_M |h|^q R d\mu.$$

The last equality is obtained of (2.6). Replacing (2.7) in (2.5) completes the proof of the proposition.

We state the variation of $\lambda(t)$ along the NRF.

Proposition 2.3 *Suppose that $(M^n, g(t))$ is a solution to the NRF on the smooth CRM (M^n, g_0). The first eigenvalue of (1.5) along the NRF satisfies*

$$\frac{d}{dt}\lambda(f,h,t)|_{t=t_0} = 2(\alpha+1)\int_M (\Delta f)(S^{ij}\nabla_i\nabla_j f)|\Delta f|^{p-2}d\mu$$
$$-\frac{\alpha+1}{p}\int_M |\Delta f|^p R d\mu$$
$$+2(\beta+1)\int_M (\Delta h)(S^{ij}\nabla_i\nabla_j h)|\Delta h|^{q-2}d\mu$$
$$-\frac{\beta+1}{q}\int_M |\Delta h|^q R d\mu$$
$$+\lambda(t_0)\int_M R|f|^{\alpha+1}|h|^{\beta+1}d\mu$$
$$+\lambda(t_0)\frac{\alpha+1}{p}\int_M |f|^p R d\mu \quad (2.8)$$
$$+\lambda(t_0)\frac{\beta+1}{q}\int_M |h|^q R d\mu - \frac{2r}{n}(\alpha+1)\int_M |\Delta f|^p d\mu$$
$$-\frac{2r}{n}(\beta+1)\int_M |\Delta h|^q d\mu.$$

Proof Along the NRF, we have

$$\frac{\partial}{\partial t}(d\mu_t) = (r-R)d\mu, \quad (2.9)$$

and

$$\frac{\partial}{\partial t}|\Delta f|^p = p(\Delta f)(2S^{ij}\nabla_i\nabla_j f - \frac{2}{n}r\Delta f + \Delta f_t)|\Delta f|^{p-2}. \tag{2.10}$$

Equality $B(f,h) = 1$ implies that

$$\begin{aligned}(\alpha+1)\int_M (\Delta f)(\Delta f_t)|\Delta f|^{p-2}d\mu + (\beta+1)\int_M (\Delta h)(\Delta h_t)|\Delta h|^{q-2}d\mu \\ = \lambda\int_M R|f|^{\alpha+1}|h|^{\beta+1}d\mu + \lambda\frac{\alpha+1}{p}\int_M |f|^p R d\mu + \lambda\frac{\beta+1}{q}\int_M |h|^q R d\mu - r\lambda.\end{aligned} \tag{2.11}$$

We can then write

$$\begin{aligned}\frac{d}{dt}\lambda(f,h,t)|_{t=t_0} &= \frac{\alpha+1}{p}\left(\int_M \frac{\partial}{\partial t}(|\Delta f|^p)d\mu_t + \int_M |\Delta f|^p\frac{\partial}{\partial t}(d\mu_t)\right) \\ &\quad+\frac{\beta+1}{q}\left(\int_M \frac{\partial}{\partial t}(|\Delta h|^q)d\mu_t + \int_M |\Delta h|^q\frac{\partial}{\partial t}(d\mu_t)\right) \\ &= (\alpha+1)\int_M \left\{(\Delta f)(2R^{ij}\nabla_i\nabla_j f - \frac{2}{n}r\Delta f + \Delta f_t)|\Delta f|^{p-2}\right\}d\mu \\ &\quad+(\beta+1)\int_M \left\{(\Delta h)(2R^{ij}\nabla_i\nabla_j h - \frac{2}{n}r\Delta h + \Delta h_t)|\Delta h|^{q-2}\right\}d\mu \\ &\quad-\frac{\alpha+1}{p}\int_M |\Delta f|^p R d\mu - \frac{\beta+1}{q}\int_M |\Delta h|^q R d\mu + r\lambda(t_0).\end{aligned} \tag{2.12}$$

Therefore the proposition is obtained by replacing (2.11) in (2.12).

Proof (Proof of Theorem 1.1) For any $u \in C^\infty(M^n)$, the Bochner formula is

$$\frac{1}{2}\Delta|\nabla u|^2 = |\nabla\nabla u|^2 + \langle\nabla u, \nabla\Delta u\rangle + S(\nabla u, \nabla u). \tag{2.13}$$

Now, by the Cauchy-Schwartz inequality $|\nabla\nabla u|^2 \geq \frac{1}{n}|\Delta u|^2$, we find

$$\frac{1}{2}\Delta|\nabla u|^2 \geq \frac{1}{n}|\Delta u|^2 + \langle\nabla u, \nabla\Delta u\rangle + S(\nabla u, \nabla u). \tag{2.14}$$

Taking integration on (2.14) on CRM M, we get

$$\frac{n-1}{n}\int_M (\Delta u)^2 d\mu \geq \int_M S(\nabla u, \nabla u)d\mu. \tag{2.15}$$

The inequalities $S_{ij} - \gamma R g_{ij} > 0$ and $R \geq R_{\min}(t)$ along the RF imply that

$$\int_M (\Delta u)^2 d\mu \geq \frac{n\gamma}{n-1}R_{\min}(t)\int_M |\nabla u|^2 d\mu. \tag{2.16}$$

For $s \geq 2$, applying the Hölder inequality to above relation gives

$$\frac{n\gamma}{n-1} R_{\min}(t) \int_M |\nabla u|^2 d\mu \leq \left(\int_M |\Delta u|^s d\mu\right)^{\frac{2}{s}} \left(\int_M d\mu\right)^{\frac{s-2}{s}}$$
$$= \left(\mathrm{vol}(M)\right)^{\frac{s-2}{s}} \left(\int_M |\Delta u|^s d\mu\right)^{\frac{2}{s}},$$

which yields

$$\int_M |\Delta u|^s d\mu \geq \left(\mathrm{vol}(M)\right)^{-\frac{s-2}{2}} \left(\frac{n\gamma}{n-1} R_{\min}(t)\right)^{\frac{s}{2}} \left(\int_M |\nabla u|^2 d\mu\right)^{\frac{s}{2}}. \tag{2.17}$$

Therefore, for $p, q \in [2, +\infty)$ we have

$$\lambda(t) \geq \frac{\alpha+1}{p} \left[\left(\mathrm{vol}(M)\right)^{-\frac{p-2}{2}} \left(\frac{n\gamma}{n-1} R_{\min}(t)\right)^{\frac{p}{2}} \left(\int_M |\nabla f|^2 d\mu\right)^{\frac{p}{2}}\right]$$
$$+\frac{\beta+1}{q} \left[\left(\mathrm{vol}(M)\right)^{-\frac{q-2}{2}} \left(\frac{n\gamma}{n-1} R_{\min}(t)\right)^{\frac{q}{2}} \left(\int_M |\nabla h|^2 d\mu\right)^{\frac{q}{2}}\right].$$

From [7, 10] we have $\lim_{t \to T} R_{\min}(t) = \infty$; then $\lim_{t \to T} \lambda(t) = \infty$.

***Proof** (Proof of Theorem 1.2)* From (2.1) and $S_{ij} \geq \epsilon R g_{ij}$, we can write

$$\begin{aligned}\frac{d}{dt}\lambda(f, h, t)|_{t=t_0} &\geq (\alpha+1)\Big(2\epsilon - \frac{1}{p}\Big) \int_M R|\Delta f|^p d\mu \\ &+ (\beta+1)\Big(2\epsilon - \frac{1}{q}\Big) \int_M R|\Delta h|^q d\mu \\ &+ \lambda(t_0) \int_M R|f|^{\alpha+1}|h|^{\beta+1} d\mu + \lambda(t_0)\frac{\alpha+1}{p} \int_M |f|^p R d\mu \\ &+ \lambda(t_0)\frac{\beta+1}{q} \int_M |h|^q R d\mu.\end{aligned} \tag{2.18}$$

1. The positivity of scalar curvature remains unchanged under the URF, and we have

$$\frac{d}{dt}\lambda(f, h, t)|_{t=t_0} \geq R_{\min}(g_0)(\alpha+1)\Big(2\epsilon - \frac{1}{p}\Big) \int_M |\Delta f|^p d\mu$$
$$+R_{\min}(g_0)(\beta+1)\Big(2\epsilon - \frac{1}{q}\Big) \int_M |\Delta h|^q d\mu$$

$$+R_{\min}(g_0)\lambda(t_0)\int_M |f|^{\alpha+1}|h|^{\beta+1}d\mu$$
$$+R_{\min}(g_0)\lambda(t_0)\frac{\alpha+1}{p}\int_M |f|^p d\mu$$
$$+R_{\min}(g_0)\lambda(t_0)\frac{\beta+1}{q}\int_M |h|^q d\mu. \tag{2.19}$$

Using the condition $B(f,h)=1$, we deduce

$$\frac{d}{dt}\lambda(f,h,t)|_{t=t_0} \geq 2R_{\min}(g_0)\lambda(t_0). \tag{2.20}$$

Function $\lambda(f,h,t)$ is smooth with respect to t, and then on arbitrary sufficiently small neighborhood of t_0 as I, we get $\frac{d}{dt}\lambda(f,h,t) \geq 2R_{\min}(g_0)\lambda(f,g,t)$. Taking integration on $[t_1,t_0)$ with respect to time t for t_1 sufficiently close to t_0 yields

$$\ln\frac{\lambda(f(t_0),h(t_0),t_0)}{\lambda(f(t_1),h(t_1),t_1)} \geq 2R_{\min}(g_0)(t_0-t_1). \tag{2.21}$$

Notice $\lambda(f(t_0),h(t_0),t_0) = \lambda(t_0)$ and $\lambda(f(t_1),h(t_1),t_1) \geq \lambda(t_1)$. These imply $\ln\frac{\lambda(t_0)}{\lambda(t_1)} \geq 2R_{\min}(g_0)(t_0-t_1)$, and then $\lambda(t_0)e^{-2R_{\min}(g_0)t_0} \geq \lambda(t_1)e^{-2R_{\min}(g_1)t_1}$. Since $t_0 \in [0,T_{\max})$ is arbitrary, thus $\lambda(t)e^{-2R_{\min}(g)t}$ is monotonically nondecreasing under the RF.

2. Suppose $R \geq R_{\min}(t) \geq 0$. From (1.3), we conclude

$$\frac{d}{dt}\lambda(f,h,t)|_{t=t_0} \geq 2R_{\min}(t)\lambda(t_0), \tag{2.22}$$

which implies $\lambda(t_0) \geq \lambda(t_1)e^{2\int_{t_1}^{t_0}R_{\min}(t)dt}$ on $[t_1,t_0] \subseteq [0,T_{\max})$.
3. Suppose $R \geq R_{\min}(0) \neq 0$. Using (4) of Lemma (2.1) and $|S|^2 \geq \frac{2}{n}R^2$, we obtain

$$\frac{\partial R}{\partial t} \geq \Delta R + \frac{2}{n}R^2. \tag{2.23}$$

The maximum principle leads to $R(t) \geq \frac{1}{R_{\min}^{-1}(0)-\frac{2}{n}t}$. By (1.3) for any $[t_1,t_0] \subseteq [0,T_{\max})$, we have

$$\lambda(t_0) \geq \lambda(t_1)e^{2\int_{t_1}^{t_0}\frac{dt}{R_{\min}^{-1}(0)-\frac{2}{n}t}} = \lambda(t_1)\left(\frac{R_{\min}^{-1}(0)-\frac{2}{n}t_1}{R_{\min}^{-1}(0)-\frac{2}{n}t_0}\right)^n; \tag{2.24}$$

then we get $\lambda(t_0)(R_{\min}^{-1}(0) - \frac{2}{n}t_0)^n \geq \lambda(t_1)(R_{\min}^{-1}(0) - \frac{2}{n}t_1)^n$ which yields $\lambda(t)(R_{\min}^{-1}(0) - \frac{2}{n}t)^n$ increasing under the RF. Since $(R_{\min}^{-1}(0) - \frac{2}{n}t)^n$ is decreasing, then $\lambda(t)$ is nondecreasing under the RF.

2.1 *Variation of* $\lambda(t)$ *on Surfaces*

Now, we write Propositions 2.2 and 2.3 in a particular case.

Corollary 2.4 *Suppose that* $(M^2, g(t))$ *is a solution of the URF on a closed surface* (M^2, g_0). *The first nonzero eigenvalue* $\lambda(t)$ *of (1.5) under URF satisfies*

$$\begin{aligned}\frac{d}{dt}\lambda(f,h,t)|_{t=t_0} &= (\alpha+1)\frac{p-1}{p}\int_M |\Delta f|^p R d\mu + (\beta+1)\frac{q-1}{q}\int_M |\Delta h|^q R d\mu\\ &+\lambda(t_0)\int_M R|u|^{\alpha+1}|h|^{\beta+1}d\mu + \lambda(t_0)\frac{\alpha+1}{p}\int_M |f|^p R d\mu\\ &+\lambda(t_0)\frac{\beta+1}{q}\int_M |h|^q R d\mu. \end{aligned}\tag{2.25}$$

Proof For $n = 2$, we get $S = \frac{1}{2}Rg$; then (2.1) leads to (2.25).

Corollary 2.5 *Suppose that* $(M^2, g(t))$ *is a solution of the NRF on a closed surface* (M^2, g_0). *The first nonzero eigenvalue* $\lambda(t)$ *of (1.5) under NRF satisfies*

$$\begin{aligned}\frac{d}{dt}\lambda(f,h,t)|_{t=t_0} &= (\alpha+1)\frac{p-1}{p}\int_M |\Delta f|^p R d\mu\\ &+(\beta+1)\frac{q-1}{q}\int_M |\Delta h|^q R d\mu \qquad (2.26)\\ &+\lambda(t_0)\int_M R|f|^{\alpha+1}|h|^{\beta+1}d\mu + \lambda(t_0)\frac{\alpha+1}{p}\int_M |f|^p R d\mu\\ &+\lambda(t_0)\frac{\beta+1}{q}\int_M |h|^q R d\mu - r(\alpha+1)\int_M |\Delta f|^p d\mu\\ &-r(\beta+1)\int_M |\Delta h|^q d\mu. \end{aligned}$$

Proof For $n = 2$, we have $S = \frac{1}{2}Rg$; then (2.8) yields (2.26).

Remark 2.6 Suppose that $(M^2, g(t))$ is a solution of the NRF on a compact surface. From [8], for a constant k depending only on g_0, we get:

(i) If $r < 0$, then

$$r - ke^{rt} \leq R \leq r + ke^{rt}. \tag{2.27}$$

Therefore, from (2.26) in every small enough neighborhood of t_0, for $p \geq q$ we obtain

$$-qke^{rt} \leq \frac{1}{\lambda(f,h,t)}\frac{d}{dt}\lambda(f,h,t) \leq pke^{rt}. \tag{2.28}$$

For every t_1 sufficiently close to t_0, on $[t_1, t_0]$, we get

$$\ln\frac{\lambda(t_0)}{\lambda(t_1)} \geq \ln\frac{\lambda(f(t_0), h(t_0), t_0)}{\lambda(f(t_1), h(t_1), t_1)} \geq -\frac{qk}{r}(e^{rt_0} - e^{rt_1}), \tag{2.29}$$

equivalently

$$\ln\lambda(t_0) + \frac{qk}{r}e^{rt_0} \geq \ln\lambda(t_1) + \frac{qk}{r}e^{rt_1}. \tag{2.30}$$

Thus, $\ln\lambda(t)+\frac{qc}{r}e^{rt}$ is increasing under the NRF. Also, for any t_2 close enough to t_0, on $[t_0, t_2]$, it follows

$$\ln\frac{\lambda(t_2)}{\lambda(t_0)} \leq \ln\frac{\lambda(f(t_2), h(t_2), t_2)}{\lambda(f(t_0), h(t_0), t_0)} \leq \frac{pk}{r}(e^{rt_2} - e^{rt_0}); \tag{2.31}$$

hence $\ln\lambda(t) - \frac{pk}{r}e^{rt}$ is decreasing along the NRF.

(ii) If $r = 0$, then

$$-\frac{k}{1+kt} \leq R \leq k. \tag{2.32}$$

Similar to the above, in this case, quantities $\lambda(t)(1+kt)^q$ and $\lambda(t)e^{-pkt}$ are increasing and decreasing along the NRF, respectively.

(iii) If $r > 0$, then

$$-ke^{rt} \leq R \leq r + ke^{rt}. \tag{2.33}$$

By use of this inequality, quantities $\ln\lambda(t)+qke^{rt}+rpt$ and $\ln\lambda(t)-pke^{rt}-(p-q)rt$ are increasing and decreasing along the NRF, respectively.

Proof ***(Proof of Theorem 1.3)*** From [8], under the URF on a surface, we infer

$$\frac{\partial}{\partial t}R = \Delta R + R^2.$$

Applying the scalar maximum principle, the nonnegativity of the scalar curvature is preserved under the RF, (2.25). It leads to $\frac{d}{dt}\lambda(f, h, t) > 0$; therefore $\lambda(t)$ is increasing.

2.2 Variation of $\lambda(t)$ on Homogeneous Manifolds

In this section, we study the behavior of $\lambda(t)$ when an initial metric is homogeneous.

Proposition 2.7 *Suppose that $(M^n, g(t))$ is a solution of the URF on the smooth CRM (M^n, g_0) in which g_0 is homogeneous. The first nonzero eigenvalue $\lambda(t)$ of (1.5) under the URF satisfies*

$$\begin{aligned}\frac{d}{dt}\lambda(f, h, t)|_{t=t_0} &= 2(\alpha+1)\int_M (\Delta f)(S^{ij}\nabla_i\nabla_j f)|\Delta f|^{p-2}d\mu \\ &+2(\beta+1)\int_M (\Delta h)(S^{ij}\nabla_i\nabla_j h)|\Delta h|^{q-2}d\mu. \qquad (2.34)\end{aligned}$$

Proof The homogeneous metric remains homogeneous under RF, and the scalar curvature of a homogeneous manifold is constant. Therefore (2.1) leads to

$$\begin{aligned}\frac{d}{dt}\lambda(f, h, t)|_{t=t_0} &= 2(\alpha+1)\int_M (\Delta f)(S^{ij}\nabla_i\nabla_j f)|\Delta f|^{p-2}d\mu \\ &-R\frac{\alpha+1}{p}\int_M |\Delta f|^p d\mu \\ &+2(\beta+1)\int_M (\Delta h)(S^{ij}\nabla_i\nabla_j h)|\Delta h|^{q-2}d\mu \\ &-R\frac{\beta+1}{q}\int_M |\Delta h|^q d\mu \\ &+R\lambda(t_0)\int_M |f|^{\alpha+1}|h|^{\beta+1}d\mu + R\lambda(t_0)\frac{\alpha+1}{p}\int_M |f|^p d\mu \\ &+R\lambda(t_0)\frac{\beta+1}{q}\int_M |h|^q d\mu\end{aligned}$$

$$= 2(\alpha+1)\int_M (\Delta f)(S^{ij}\nabla_i\nabla_j f)|\Delta f|^{p-2}d\mu$$
$$+2(\beta+1)\int_M (\Delta h)(S^{ij}\nabla_i\nabla_j h)|\Delta h|^{q-2}d\mu.$$

Note 2.8 In Proposition 2.7, if we suppose that $(M^n, g(t))$ is a solution of the NRF on the smooth CRM (M^n, g_0) in which g_0 is homogeneous, then (2.8) yields

$$\frac{d}{dt}\lambda(f,h,t)|_{t=t_0} = 2(\alpha+1)\int_M (\Delta f)(S^{ij}\nabla_i\nabla_j f)|\Delta f|^{p-2}d\mu$$
$$+2(\beta+1)\int_M (\Delta h)(S^{ij}\nabla_i\nabla_j h)|\Delta h|^{q-2}d\mu$$
$$-\frac{2r}{n}(\alpha+1)\int_M |\Delta f|^p d\mu - \frac{2r}{n}(\beta+1)\int_M |\Delta h|^q d\mu.$$

2.3 Variation of $\lambda(t)$ on Three-Dimensional Manifolds

Now, we investigate the behavior of $\lambda(t)$ on three-dimensional manifolds.

Proposition 2.9 *Suppose that $(M^3, g(t))$, $t\in[0,T)$ is a solution of the URF on a CRM (M^3, g_0) with a positive Ricci curvature. If $S_{ij}\geq \epsilon R g_{ij}$ for some constant ϵ with $\max\{\frac{1}{p},\frac{1}{q}\}\leq\epsilon\leq\frac{1}{3}$ at time $t=0$, then the principal eigenvalue of (1.5) is increasing along the RF.*

Proof For any solution of the RF [10] on a CRM (M^3, g_0) with a positive curvature, the inequality $S_{ij}\geq\epsilon R g_{ij}$ remains by the RF on $[0,T)$. Hence, Theorem 1.2 results that $\lambda(t)$ is increasing.

Proposition 2.10 *Let $(M^3, g(t))$ be a solution to the URF on a CRM (M^3, g_0) with homogeneous metric g_0 and nonnegative Ricci curvature; then the principal eigenvalue of (1.5) is increasing.*

Proof The nonnegativity of the Ricci tensor remains under the RF [10] in dimension 3. From (3.1), $\lambda(t)$ is increasing.

3 Example

We give an example of variation of $\lambda(t)$ on some CRMs.

Example 3.1 Suppose that (M^n, g_0) is an Einstein manifold, i.e., $Ric(g_0) = cg_0$ for some constant c. Let

$$g(t) = u(t)g_0, \quad u(0) = 1, \tag{3.1}$$

be a solution to the RF for some positive function $u(t)$. By direct computation, we have

$$\frac{\partial g}{\partial t} = u'(t)g_0.$$

Hence,

$$u'(t)g_0 = -2S(g(t)) = -2S(u(t)g_0) = -2S(g_0) = -2cg_0. \tag{3.2}$$

This shows that $u(t) = -2ct + 1$. Thus,

$$g(t) = (1 - 2ct)g_0,$$

that is, $g(t)$ is an Einstein metric. Also,

$$S(g(t)) = S(g_0) = cg_0 = \frac{c}{1-2ct}g(t),\ R(g(t)) = \frac{1}{1-2ct}R(g_0) = \frac{cn}{1-2ct}, \tag{3.3}$$

and

$$d\mu_{g(t)} = (1 - 2ct)^{\frac{n}{2}} d\mu_{g_0}. \tag{3.4}$$

Using equation (2.1), for $q \geq p$, we get

$$\frac{d}{dt}\lambda(f, h, t)|_{t=t_0} \geq \frac{2pc}{1-2ct_0}\lambda(f, h, t)|_{t=t_0}. \tag{3.5}$$

Hence $\lambda(t)(1 - 2ct)^p$ is increasing along the RF.

Declarations

Funding This work does not receive any funding.

Conflict of Interests We declare that we do not have any commercial or associative interest that represents a conflict of interest in connection with the work submitted.

Availability of Data and Material All data generated or analyzed during this study are included in this published article.

Author's Contributions All authors contributed equally in the preparation of this manuscript.

References

1. Abolarinwa, A.: Evolution and monotonicity of the first eigenvalue of p-Laplacian under the Ricci-harmonic flow. J. Appl. Anal. **21**, 147–160 (2015)
2. Abolarinwa, A.: Some monotonic quantities involving the eigenvalues of p-bi-Laplacian along the Ricci flow. Iran. J. Sci. Technol. Trans. A: Sci. **46**, 219–228 (2022)
3. Azami, S.: Evolution of the first eigenvalue of the clamped plate on manifold along the Ricci flow. J. Complex Var. Elliptic Equations (2019). https://doi.org/10.1080/17476933.2019.1624730
4. Azami, S.: Eigenvalues of a (p, q)-Laplacian system under the mean curvature flow. Iran. J. Sci. Technol. Trans. A Sci. **43**(4), 1881–1888 (2019)
5. Cao, X.-D.: Eigenvalues of $(-\Delta+\frac{R}{2})$ on manifolds with nonnegative curvature operator. Math. Ann. **337**(2), 435–441 (2007)
6. Cao, X.-D.: First eigenvalues of geometric operators under the Ricci flow. Proc. AMS **136**(1), 4075–4078 (2008)
7. Cerbo, L.F.D.: Eigenvalues of the Laplacian under the Ricci flow. Rendiconti di Mathematica, Serie VII, Roma **27**, 183–195 (2007)
8. Chow, B., Knopf, D.: The Ricci Flow: An Introduction. Mathematical Surveys and Monographs, vol. 110. AMS, New York (2004)
9. DeTurck, D.M.: Deforming metrics in the direction of their Ricci tensors. J. Diff. Geom. **18**, 157–162 (1983)
10. Hamilton, R.S.: Three-manifolds with positive Ricci curvature. J. Diff. Geom. **17**, 255–306 (1982)
11. Leadi, L.A., Toyou, R.L.: Principal eigenvalue for cooperative (p, q)- biharmonic system. J. Part. Diff. Eq. **32**(1), 33–51 (2019)
12. Li, L., Tang, C.-L.: Existence of three solutions for (p, q)-biharmonic systems. Nonlinear Anal. **73**, 796–805 (2010)
13. Perelman, G.: The entropy formula for the Ricci flow and its geometric applications (2002). ArXiv:math. Dc/0211159
14. Wu, J.Y., Wang, E.M., Zheng, Y.: First eigenvalue of the p-Laplace operator along the Ricci flow. Ann. Glob. Ana. Geom. **38**, 27–55 (2010)
15. Wu, J.Y.: First eigenvalue monotonicity for the p-Laplace operator under the Ricci flow. Acta Mathematica Sinica **27**(8), 1591–1598 (2011)

The Jacobi Geometry of Plane Parametrized Curves and Associated Inequalities

Mircea Crasmareanu

Abstract Firstly, we introduce a new frame and a new curvature function for a fixed parametrization r of a plane curve C. This new frame is called *Jacobi* since it involves the rotation with the first two Jacobi elliptic functions of the usual Frenet frame. The Jacobi-curvature involves only the third Jacobi elliptic function w and is computed for some remarkable examples; the inequalities satisfied by w imply inequalities for the Jacobi-curvature. Secondly, we introduce a whole family of new parametrizations r_ρ for C with $r = r_{\rho=0}$. The expression of r_ρ involves an integral containing the curvature function k of r, and all r_ρ have the same curvature.

Keywords Plane parametrized curve · Jacobi elliptic functions · Inequalities · Jacobi-curvature · Jacobi mate

1 Introduction

The delightful note [1] of Bishop proposes a new frame, as alternative to the classical frame of Frenet, for the study of curves. Following this path, in an almost half of a century, some new frames are considered, especially for space curves; see, for example, the papers [9] and [10]. In order to consider the case of a plane curve C, we introduce in the paper [3] a deformation (following the Masur terminology from [8]) called *flow-frame* since it is the rotated version of the Frenet frame, the rotation angle being exactly the time t of a current point of C. It follows naturally a new curvature, called *flow-curvature*. We point out that some other curvature functions are defined in the paper [7].

In the present work we firstly generalize this construction by defining the Jacobi-frame of C using the well-known Jacobi elliptic functions u, v, w. These functions

M. Crasmareanu (✉)
Faculty of Mathematics, University "Al. I. Cuza", Iasi, Romania
e-mail: mcrasm@uaic.ro
http://www.math.uaic.ro/~mcrasm

B.-Y. Chen, M. A. Choudhary (eds.), *Geometric Inequalities and Applications*, Infosys Science Foundation Series in Mathematical Sciences,
https://doi.org/10.1007/978-981-95-5148-4_12

are defined through a modulus ρ, and the vanishing of ρ implies $u = \cos$ and $v = \sin$. Correspondingly, this new frame defines a curvature, called *Jacobi* by us from natural reasons and denoted k_J. Hence, the main theoretical result of this note is the computation of k_J and a comparison with the usual curvature k, as well as a relationship with the flow-curvature k_f. As is usually, we focus then on examples, with a special view toward periodicity induced by the periodicity of the third Jacobi function w.

Secondly, we generalize a given arc-length parametrization $r = r(s)$ of C. The main tool of this new approach is an integral involving the ratio of k and w, and it is worth to remark that all new parametrizations are also arc-length and with the same curvature k. In fact, the initial r corresponds to $\rho = 0$. The difficulties in working with general Jacobi elliptic functions force us to restrict the examples to the circles with center in the origin of $\mathbb{R}^2$.

The contents are as follows. The next section reviews the flow-frame and the flow-curvature as starting point for our generalization. The main section, namely 2, concerns with the new frame and the new curvature, computed in Proposition 2.2 and calculated for some examples. Also, we point out an extension of the notion of *Jacobi-frame* to space curves. In the last section we construct the family of new parametrizations r_ρ preserving the natural parameter s and the curvature k.

We finish this introduction by pointing out that the efficiency of the Jacobi generalization is already proved by our study [5]. As potential applications we consider that the Computer Design and the Machine Learning can benefit from such new tools.

2 The Flow-Curvature of a Plane Parametrized Curve

Fix an open interval $I \subseteq \mathbb{R}$, and consider $C \subset \mathbb{R}^2$ a regular parametrized curve of equation:

$$C : r(t) = (x(t), y(t)) = x(t)\bar{i} + y(t)\bar{j}, \quad \|r'(t)\| > 0, \quad t \in I. \tag{2.1}$$

The ambient setting $\mathbb{R}^2$ is a Euclidean vector space with respect to the canonical inner product:

$$\langle u, v\rangle = x^1y^1 + x^2y^2, \quad u = (x^1, x^2) \in \mathbb{R}^2, \quad v = (y^1, y^2) \in \mathbb{R}^2,$$
$$0 \leq \|u\|^2 = \langle u, u\rangle. \tag{2.2}$$

The infinitesimal generator of the rotations in $\mathbb{R}^2 = \mathbb{C}$ is the linear vector field, called *angular*:

$$\xi(u) := -x^2\frac{\partial}{\partial x^1} + x^1\frac{\partial}{\partial x^2}, \quad \xi(u) = i \cdot u = i \cdot (x^1 + ix^2), \quad i = \sqrt{-1}. \tag{2.3}$$

It is a complete vector field with integral curves the circles $\mathcal{C}(O, r)$:

$$\begin{cases} \gamma^{\xi}_{u_0}(t) = (u^1_0 \cos t - u^2_0 \sin t, u^1_0 \sin t + u^2_0 \cos t) = R(t) \cdot \begin{pmatrix} u^1_0 \\ u^2_0 \end{pmatrix}, \quad t \in \mathbb{R}, \\ r = \|u_0\| = \|(u^1_0, u^2_0)\|, \quad R(t) := \begin{pmatrix} \cos t & -\sin t \\ \sin t & \cos t \end{pmatrix} \in SO(2) = S^1, \end{cases} \tag{2.4}$$

and since the rotations $R(t)$ are isometries of the Riemannian metric $g_{can} = dx^2 + dy^2 = |dz|$, it follows that ξ is a Killing vector field of the Riemannian manifold $(\mathbb{R}^2, g_{can})$. The first integrals of ξ are the Gaussian functions, i.e., multiples of the square norm: $f_\alpha(x, y) = \alpha(x^2 + y^2)$, $\alpha \in \mathbb{R}$.

The Frenet apparatus of the curve C is provided by the Frenet frame $\{T, N\}$ and its curvature function k:

$$\begin{cases} T(t) = \frac{r'(t)}{\|r'(t)\|} \in S^1, \quad N(t) = i \cdot T(t) = \frac{1}{\|r'(t)\|}(-y'(t), x'(t)) \in S^1, \\ k(t) = \frac{1}{\|r'(t)\|}\langle T'(t), N(t)\rangle = \frac{1}{\|r'(t)\|^3}\langle r''(t), ir'(t)\rangle \\ = \frac{1}{\|r'(t)\|^3}[x'(t)y''(t) - y'(t)x''(t)]. \end{cases} \tag{2.5}$$

The starting point in defining a new frame is the identity:

$$\frac{d}{dt}R(t) = R\left(t + \frac{\pi}{2}\right) = R(t)R\left(\frac{\pi}{2}\right) = R\left(\frac{\pi}{2}\right)R(t), \tag{2.6}$$

and remark that the Frenet equations can be unified by means of the column matrix $\mathcal{F}(t) = \begin{pmatrix} T \\ N \end{pmatrix}(t)$ as

$$\frac{d}{dt}\mathcal{F}(t) = \|r'(t)\|k(t)R\left(-\frac{\pi}{2}\right)\mathcal{F}(t). \tag{2.7}$$

In a previous paper, namely [3], we have defined a new frame and correspondingly a new curvature function for C:

Definition 2.1 The *flow-frame* of C consists in the pair of unit vectors $(E^f_1(t), E^f_2(t)) \in T^2 := S^1 \times S^1$ given by

$$\mathcal{E}(t) := \begin{pmatrix} E^f_1 \\ E^f_2 \end{pmatrix}(t) = R(t)\mathcal{F}(t) = \begin{pmatrix} \cos t T(t) - \sin t N(t) \\ \sin t T(t) + \cos t N(t) \end{pmatrix} \tag{2.8}$$

the letter f being the initial of the word "flow." The *flow-curvature* of C is the smooth function $k_f : I \to \mathbb{R}$ given by *the flow-equations*:

$$\frac{d}{dt}\mathcal{E}(t) = \|r'(t)\| k_f(t) R\left(-\frac{\pi}{2}\right)\mathcal{E}(t). \tag{2.9}$$

Hence, the main result of the cited work is the following:

Proposition 2.2 *The expression of the flow-curvature is*

$$k_f(t) = k(t) - \frac{1}{\|r'(t)\|} < k(t). \tag{2.10}$$

Proof We have directly in the flow-frame

$$\|r'(t)\| k_f(t) R\left(-\frac{\pi}{2}\right) = R\left(t + \frac{\pi}{2}\right) R(-t) + \|r'(t)\| k(t) R(t) R\left(-\frac{\pi}{2}\right) R(-t), \tag{2.11}$$

and the conclusion follows. □

3 The Jacobi-Curvature of a Plane Parametrized Curve

Fix now the real number $\rho \in (-1, 1)$ as *the modulus* for the differential system [6, p. 130]:

$$\begin{cases} \frac{du}{dt} = -wv, & u(0) = 1, \\ \frac{dv}{dt} = wu, & v(0) = 0, \\ \frac{dw}{dt} = -\rho^2 uv, & w(0) = 1. \end{cases} \tag{3.1}$$

Recall that its solutions are called *Jacobi elliptic functions* and there are usually denoted $cn(\cdot, \rho)$, $sn(\cdot, \rho)$, respectively, $dn(\cdot, \rho)$; we prefer the simple notation used above. As solutions of the ODE system (3.1) these functions satisfy two remarkable identities:

$$u^2 + v^2 = 1, \quad \rho^2 v^2 + w^2 = 1. \tag{3.2}$$

Also, both functions $u(\cdot)$ and $v(\cdot)$ are periodic with $L = 4\tilde{L}$ for [6, p. 131]:

$$\tilde{L} = \tilde{L}(\rho) := \int_0^1 \frac{ds}{\sqrt{(1 - s^2)(1 - \rho^2 s^2)}}, \tag{3.3}$$

while w is periodic of period $2\tilde{L}$. In particular, $\tilde{L}(0) = \arcsin s|_0^1 = \frac{\pi}{2}$ for the usual trigonometrical functions $cn(\cdot, 0) = \cos(\cdot)$ and $sn(\cdot, 0) = \sin(\cdot)$. The *complementary modulus* is $\rho' := \sqrt{1-\rho^2} \in (0, 1]$, and the third Jacobi function is bounded by

$$0 < \rho' \leq w(t) \leq 1. \tag{3.4}$$

The *self-complementary case* $\rho' = \rho$ is provided by $\rho = \frac{1}{\sqrt{2}}$ and being in the interval (0, 1) is the eccentricity of an ellipse, called *self-complementary* and studied in [2]. The picture of the function $w = w(\cdot, \rho = \frac{1}{\sqrt{2}})$ is below with the half-period:

$$\tilde{L}\left(\rho = \frac{1}{\sqrt{2}}\right) = \frac{\sqrt{2}\Gamma(1/4)\Gamma(1/2)}{4\Gamma(3/4)} \simeq 1.85407.$$

Following the path of the first section we introduce a new frame and a new curvature function for the given curve (Fig. 1):

Definition 3.1 The *Jacobi-frame* of C consists in the pair of unit vectors $(E_1^J(t), E_2^J(t)) \in T^2$ given by

$$\begin{cases} \mathcal{E}^J(t) := \begin{pmatrix} E_1^J \\ E_2^J \end{pmatrix}(t) = R^J(t)\mathcal{F}(t) = \begin{pmatrix} u(t)T(t) - v(t)N(t) \\ v(t)T(t) + u(t)N(t) \end{pmatrix}, \\ R^J(t) := \begin{pmatrix} u(t) & -v(t) \\ v(t) & u(t) \end{pmatrix} \in SO(2) = S^1. \end{cases} \tag{3.5}$$

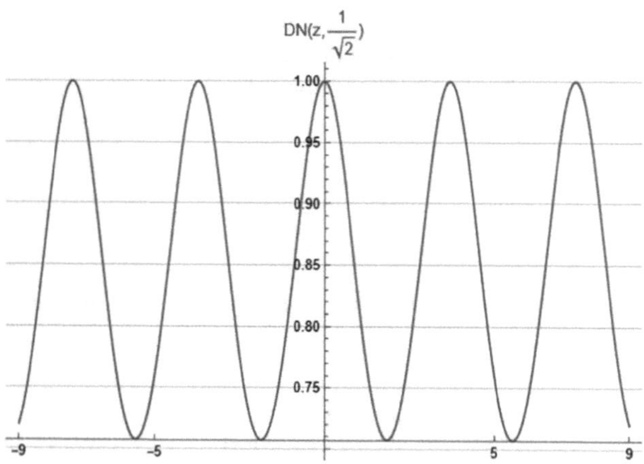

Fig. 1 The Jacobi function$w = w(t, \rho = 1/\sqrt{2}), \quad t \in (-9, 9)$

The *Jacobi-curvature* of C is the smooth function $k_J : I \to \mathbb{R}$ given by *the Jacobi-frame equations*:

$$\frac{d}{dt}\mathcal{E}^J(t) = \|r'(t)\|k_J(t)R\left(-\frac{\pi}{2}\right)\mathcal{E}^J(t). \tag{3.6}$$

It follows now the main result, with a similar proof as above:

Proposition 3.2 *The expression of the Jacobi-curvature is*

$$k_J(t) = k(t) - \frac{w(t)}{\|r'(t)\|} \in \left[k_f(t), k(t) - \frac{\rho'}{\|r'(t)\|} < k(t)\right]. \tag{3.7}$$

Remark 3.3

(i) If we use Eq. (2.8) with R replaced by $R \circ \Omega$ to define the notion of Ω-frame for the plane curve C, then the corresponding Ω-curvature of the plane curve C is

$$k_\Omega(t) = k(t) - \frac{\Omega'(t)}{\|r'(t)\|}, \tag{3.8}$$

and the curves in polar coordinates with vanishing Ω-curvature are provided by

$$\rho(t) = Re^{\int_{t_0}^t \cot[\Omega(u)-u+C]du}, \quad R > 0, \quad C \in \mathbb{R}. \tag{3.9}$$

The flow-curvature corresponds to the identity map $\Omega = 1_\mathbb{R}$, while the Jacobi-curvature corresponds to the function $\Omega = W := \int w$. This last function is usually called *amplitude*, and we supposed to be *strictly positive*.

(ii) It is well known the identity:

$$t = \int_0^{W(t)} \frac{d\xi}{\sqrt{1-\rho^2\sin^2\xi}},$$

and then we have the function $W \to t(W)$. The first two Jacobi differential equations become

$$\frac{du}{dW} = -v \circ t(W), \quad \frac{dv}{dW} = u \circ t(W),$$

which are similar to the differential equations satisfied by the trigonometrical functions cos and sin.

(iii) Let $s \in (0, L(C) > 0)$ be a natural parameter for the curve C, i.e., $\|r'(s)\| = 1$ for all s. Here, $L(C)$ is the length of the curve. Let also $K = K(s)$ be *the structural angle* of C, i.e., $k = \frac{dK}{ds}$. Then k_J is a derivative:

$$k_J(s) = (K - W)'(s).$$

(iv) Suppose now that the curve C is in the space $\mathbb{R}^3$ and is bi-regular, i.e., $\|r'(t) \times r''(t)\| > 0$ for all $t \in I$; hence it has the Frenet frame (T, N, B) and the pair (curvature, torsion) = $(k > 0, \tau)$. We define its *Jacobi-frame* as

$$\begin{pmatrix} T \\ E_2^J \\ E_3^J \end{pmatrix}(t) := \begin{pmatrix} 1 & 0_2(h) \\ 0_2(v) & R^J(t) \end{pmatrix} \begin{pmatrix} T \\ N \\ B \end{pmatrix}, \quad 0_2(h) := (0, 0), \quad 0_2(v) := \begin{pmatrix} 0 \\ 0 \end{pmatrix}, \tag{3.10}$$

and then, its matrix moving equation is

$$\frac{d}{dt}\begin{pmatrix} T \\ E_2^J \\ E_3^J \end{pmatrix}(t) = \|r'(t)\| \begin{pmatrix} 0 & k_J^2(t) & k_J^3(t) \\ -k_J^2(t) & 0 & \tau_J(t) \\ -k_J^3(t) & -\tau_J(t) & 0 \end{pmatrix} \begin{pmatrix} T \\ E_2^J \\ E_3^J \end{pmatrix}(t). \tag{3.11}$$

A similar computation yields

$$k_J^2(t) = k(t)u(t), \quad k_J^3(t) = k(t)v(t), \quad \tau_f(t) = \tau(t) - \frac{w(t)}{\|r'(t)\|} < \tau(t). \tag{3.12}$$

□

From now on we focus on computing some relevant examples:

Example 3.4

(i) If C is the line $r_0 + tu, t \in \mathbb{R}$, with the vector $u \neq \bar{0} = (0, 0)$, then k_J is periodic with the period $2\tilde{L}$ since

$$k_J(t) = -\frac{w(t)}{\|u\|} \in \left[k_f(t) = -\frac{1}{\|u\|}, -\frac{\rho'}{\|u\|}\right]. \tag{3.13}$$

In particular, if u is a unit vector, then $k_J(t) = -w(t) \in [-1, -\rho' < 0]$.

(ii) The circle $C(O, R)$ with the usual parametrization $r(t) = Re^{it}$ has

$$k_J(t) = \frac{1 - w(t)}{R} \in \left[k_f = 0, \frac{1 - \rho'}{R}\right] \tag{3.14}$$

again a $2\tilde{L}$-periodic k_J curvature. Also, it follows a geometrical interpretation of the third Jacobi function: w is the function $1 - k_J$ of the unit circle S^1.

□

Example 3.5 The involute of the unit circle S^1 is

$$C : r(t) = (\cos t + t \sin t, \sin t - t \cos t) = (1 - it)e^{it}, \quad t \in (0, +\infty). \tag{3.15}$$

A direct computation gives

$$r'(t) = (t \cos t, t \sin t) = te^{it}, \quad \|r'(t)\| = t, \quad k(t) = \frac{1}{t} > 0,$$
$$k^J(t) = \frac{1 - w(t)}{t}. \tag{3.16}$$

The parametrization (3.15) suggests the following generalization; we call *Jacobi-involute of* S^1 the curve:

$$C_\rho : r_\rho(t) = (u(t) + W(t)v(t), v(t) - W(t)u(t)) \tag{3.17}$$

with

$$r'_\rho(t) = W(t)w(t)(u(t), v(t)), \quad \|r'_\rho(t)\| = W(t)w(t). \tag{3.18}$$

Finally, we obtain its curvatures:

$$k_\rho(t) = \frac{1}{W(t)}, \quad k_{\rho J} \equiv 0 \tag{3.19}$$

by recalling the hypothesis from Remark 3.3 (i) that $W(t) > 0$; hence C_ρ is a Jacobi-flat curve. The length of the curve $C_\rho|_{(0,L)}$ is

$$Length(C_\rho|_{(0,L)}) = \int_0^L W(t)W'(t)dt = \frac{W^2(L)}{2} = \frac{\pi^2}{8}. \tag{3.20}$$

□

Example 3.6 Recall that for $R > 0$ the cycloid of radius R has the equation

$$C : r(t) = R(t - \sin t, 1 - \cos t) = R[(t, 1) - e^{i(\frac{\pi}{2} - t)}], \quad t \in \mathbb{R}. \tag{3.21}$$

We have immediately

$$r'(t) = R(1 - \cos t, \sin t) = R[(1,0) - e^{it}], \|r'(t)\| = 2R|\sin\frac{t}{2}|, k(t) = -\frac{1}{4R|\sin\frac{t}{2}|}, \tag{3.22}$$

and then we restrict our definition domain to $(0, \pi)$. It follows

$$k_f(t) = -\frac{3}{4R\sin\frac{t}{2}} < 0, \quad k_J(t) = -\frac{1 + 2w(t)}{4R\sin\frac{t}{2}}. \tag{3.23}$$

□

The expression of k_J suggests to define *the Jacobi-cycloid* as being the regular curve C defined on $(0, 2\tilde{L})$ whose Frenet curvature k is

$$k(t) = -\frac{1 + 2w(t)}{4Rv(\frac{t}{2})} \tag{3.24}$$

since the fundamental theorem of plane curves assures the existence of such a curve.

□

4 The Jacobi Mates of an Arc-Length Parametrization

Suppose again that the given parametrization is an arc-length one: $r'(s) \in S^1$. Here, $L(C)$ is the length of the curve. Recall, after Remark 3.3(iii), the function $K : (0, L(C)) \to \mathbb{R}$ as the antiderivative of the curvature function $k = k(s)$. Then the fundamental theorem of plane curves states that the velocity vector field is given by

$$r'(s) = (-\sin K(s), \cos K(s)) = (\cos\theta(s), \sin\theta(s)), \quad \theta(s) := \frac{\pi}{2} + K(s). \tag{4.1}$$

Sometimes, the function θ is called *the structural angle* of the curve C since $\theta' = k$.

An immediate application of the above relation is the fact that the defining functions x, y of the parametrization r of C satisfy the third-order differential equation [4]:

$$kU''' - k'U'' + k^3U' = 0. \tag{4.2}$$

Hence, the aim of this section is to define a generalization of (4.1). Following the path of the previous section, we introduce a new antiderivative for the given curve:

$$K^{\rho}(s) := \int_0^s \frac{k(t)}{w(t)} dt. \tag{4.3}$$

This integral can be considered as the antiderivative of k with *the weight* $\frac{1}{w} > 0$, and the relationship with K is

$$K^{\rho}(s) = \frac{K}{w}\Big|_0^s - \rho^2 \int_0^s \frac{K(t)u(t)v(t)}{w^2(t)} dt = \frac{K(s)}{w(s)} - \rho^2 \int_0^s \frac{K(t)u(t)v(t)}{w^2(t)} dt. \tag{4.4}$$

Also, if k is strictly positive (e.g., C is a convex curve), then we have the inequalities:

$$K \leq K^{\rho} \leq \frac{K}{\rho'}. \tag{4.5}$$

Definition 4.1 *The Jacobi mate* of the arc-length parametrization r is the function $r_{\rho} : (0, L(C)) \to \mathbb{R}^2$ with the derivative

$$r'_{\rho}(s) := (-v(K^{\rho}(s)), u(K^{\rho}(s))). \tag{4.6}$$

Remark 4.2

(i) The curvature of r_{ρ} is also k since the acceleration function for r_{ρ} is

$$r''_{\rho}(s) := (-k(s)u(K^{\rho}(s)), -k(s)v(K^{\rho}(s))). \tag{4.7}$$

Hence, r_{ρ} is a re-parametrization of the same curve C, and its components x^{ρ}, y^{ρ} satisfy the same ODE (3.2).

(ii) The function $r_{\frac{1}{\sqrt{2}}}$ can be called *the self-complementary parametrization* of C.

□

Due to the complexity of computations we restrict to a single relevant example:

Example 4.3 The circle $C(O, R)$ has the usual arc-length parametrization $r(s) = Re^{i\frac{s}{R}}$ for $s \in (0, 2\pi R)$. We need the function

$$\tilde{W}(s) := \int_0^s \frac{dt}{w(t)} = \frac{1}{i\rho'} \ln \frac{u(s) + i\rho' v(s)}{w(s)}, \tag{4.8}$$

and hence, $K^{\rho} = \frac{\tilde{W}}{R}$. Then the Jacobi mate is r_{ρ} with the derivative

$$r'_{\rho}(s) := \left(-v\left(\frac{\tilde{W}(s)}{R}\right), u\left(\frac{\tilde{W}(s)}{R}\right)\right). \tag{4.9}$$

Indeed, the case $\rho = 0$ recasts the usual r since then $\rho' = 1$, and the Euler formula gives $\tilde{W}(s) = s$. □

5 Conclusions

The Jacobi elliptic functions permit us to move beyond classical confines and provide us with a framework in which we generalize some usual notions of the differential geometry of plane curves.

6 Chapter Information

Acknowledgments The author would like to express his sincere thanks to one of the Editors, namely Prof. Dr. Majid Ali Choudhary, and to two anonymous reviewers for their helpful comments and suggestions.

Conflict of Interest Disclosure No potential conflict of interest was declared by the author.

Supporting/Supporting Organizations No grants were received from any public, private or non-profit organizations for this research.

Ethical Approval and Participant Consent It is declared that during the preparation process of this study, scientific and ethical principles were followed and all the studies benefited from are stated in the bibliography.

Funding This research received no external funding.

References

1. Bishop, R.L.: There is more than one way to frame a curve. Am. Math. Mon. **82**, 246–251 (1975). Zbl 0298.53001
2. Crasmareanu, M.: Magic conics, their integer points and complementary ellipses. An. Ştiinţ. Univ. Al. I. Cuza Iaşi Mat. **67**(1), 129–148 (2021). Zbl 1513.11093
3. Crasmareanu, M.: The flow-curvature of plane parametrized curves. Commun. Fac. Sci. Univ. Ankara Ser. A1 Math. Stat. **72**(2), 417–428 (2023). MR4615669
4. Crasmareanu, M.: The adjoint map of Euclidean plane curves and curvature problems. Tamkang J. Math. **55**(4), 331–337 (2024). Zbl 7973538

5. Crasmareanu, M.: The Jacobi mate of an oval. Eur. J. Math. Anal. **4**, 18 (2024)
6. Cushman, R.H., Bates, L.M.: Global Aspects of Classical Integrable Systems, 2nd edn. Birkhäuser/Springer, Basel (2015). Zbl 1321.70001
7. Góźdź, S.: Curvature type functions for plane curves. An. Ştiinţ. Univ. Al. I. Cuza Iaşi Mat. **39**, 295–303 (1993). Zbl 0851.53001
8. Mazur, B.: Perturbations, deformations, and variations (and "near-misses") in geometry, physics, and number theory. Bull. Am. Math. Soc. **41**(3), 307–336 (2004). Zbl 1057.11033
9. Özen, K.E., Tosun, M.: A new moving frame for trajectories with non-vanishing angular momentum. J. Math. Sci. Model. **4**(1), 7–18 (2021).
10. Soliman, M.A., Nassar, H.A.-A., Hussien, R.A., Youssef, T.: Evolutions of the ruled surfaces via the evolution of their directrix using quasi frame along a space curve. J. Appl. Math. Phys. **6**, 1748–1756 (2018)

B.-Y. Chen Inequalities for Submanifolds of Conformally Flat Manifolds

Cihan Özgür

Abstract We obtain B.-Y. Chen inequalities for submanifolds in m-dimensional, $m > 4$, conformally flat manifolds and conformally flat manifolds endowed with a semi-symmetric metric connection. The equality cases are also considered.

Keywords Conformally flat manifold · B.-Y. Chen inequality · k-Ricci curvature · Semi-symmetric metric connection

1 Introduction

In recent years, the problem of establishing simple fundamental relations between the intrinsic (mainly the scalar curvature, the sectional curvature, and the Ricci curvature) and extrinsic invariants (mainly the squared mean curvature) of a submanifold has become one of the most fundamental problems in the theory of the submanifolds. As the first result, in 1993, B.-Y. Chen obtained a basic inequality involving the sectional curvature and the squared mean curvature of submanifolds in a real space form [2]. For submanifolds of real space forms, the inequalities between the Ricci curvature, k-Ricci curvature, and squared mean curvature were given in [4]. Now, these inequalities are known as B.-Y. Chen inequalities. The inequalities obtained by B.-Y. Chen have attracted great attention, and similar inequalities for submanifolds in various space forms have been obtained by many authors. For the collections of the results in these directions, see [5–8, 11], and the references therein.

Motivated by the above studies, in the present study, we find B.-Y. Chen inequalities for submanifolds in m-dimensional, $m > 4$, conformally flat manifolds and conformally flat manifolds endowed with a semi-symmetric metric connection. We obtain relations between the mean curvature, scalar and sectional curvatures, and k-Ricci curvatures. Our results generalize some of the results obtained for

C. Özgür (✉)
Department of Mathematics, Izmir Democracy University, Izmir, Türkiye
e-mail: cihan.ozgur@idu.edu.tr

B.-Y. Chen, M. A. Choudhary (eds.), *Geometric Inequalities and Applications*,
Infosys Science Foundation Series in Mathematical Sciences,
https://doi.org/10.1007/978-981-95-5148-4_13

submanifolds in real space forms (see [2–4] and [12]) and in a Riemannian manifold of quasi-constant curvature (see [14] and [16]).

2 Preliminaries

The Weyl conformal curvature tensor of an m-dimensional Riemannian manifold $(\widetilde{M}, \widetilde{g})$ is defined by

$$\widetilde{C}(X_1, X_2, X_3, X_4) = \widetilde{R}(X_1, X_2, X_3, X_4)$$

$$-\frac{1}{m-2}\left\{\widetilde{Ric}(X_2, X_3)\widetilde{g}(X_1, X_4) - \widetilde{Ric}(X_1, X_3)\widetilde{g}(X_2, X_4)\right.$$

$$\left.+\widetilde{Ric}(X_1, X_4)\widetilde{g}(X_2, X_3) - \widetilde{Ric}(X_2, X_4)\widetilde{g}(X_1, X_3)\right\}$$

$$+\frac{2\widetilde{\tau}}{(m-1)(m-2)}\left[\widetilde{g}(X_2, X_3)\widetilde{g}(X_1 X_4) - \widetilde{g}(X_1, X_3)\widetilde{g}(X_2, X_4)\right], \tag{2.1}$$

where $\widetilde{Ric}$ and $\widetilde{\tau}$ denote the Ricci tensor and scalar curvature of $(\widetilde{M}, \widetilde{g})$, respectively. It is known that for $n \geq 4$, the manifold is conformally flat if and only if $\widetilde{C} = 0$ [1].

Using (2.1), the Gauss equation for a submanifold M^n of a conformally flat manifold N^m is

$$R(X_1, X_2, X_3, X_4)$$

$$=\frac{1}{m-2}\left\{Ric^N(X_2, X_3)g(X_1, X_4) - Ric^N(X_1, X_3)g(X_2, X_4)\right.$$

$$\left.+Ric^N(X_1, X_4)g(X_2, X_3) - Ric^N(X_2, X_4)g(X_1, X_3)\right\}$$

$$-\frac{2\tau^N}{(m-1)(m-2)}\left[g(X_2, X_3)g(X_1, X_4) - g(X_1, X_3)g(X_2, X_4)\right],$$

$$+g\left(h(X_1, X_4), h(X_2, X_3)\right) - g\left(h(X_1, X_3), h(X_2, X_4)\right), \tag{2.2}$$

where h denotes the second fundamental form of M^n in N^m and Ric^N and τ^N are the Ricci tensor and scalar curvature of N^m, respectively.

Let $\{e_1, ..., e_m\}$ be an orthonormal basis of the tangent space $T_x M^n$. Then the scalar curvature τ at $x \in M^n$ is defined by

$$\tau(x) = \sum_{1 \leq i < j \leq n} K(e_i \wedge e_j),$$

where $K(e_i \wedge e_j) = K_{ij}$ denotes the sectional curvature of a 2-plane section spanned by e_i, e_j.

Let M^n be an n-dimensional Riemannian manifold, Π a k-plane section of $T_x M^n$, $x \in M^n$, and X a unit vector in Π.

Let $\{e_i\}$, $1 \leq i \leq k$, be an orthonormal basis of Π such that $e_1 = X$.

The *Ricci curvature* (or *k-Ricci curvature*) [4] of Π at X is defined by

$$Ric_{\Pi}(X) = \sum_{i=2}^{k} K_{1i}.$$

For each integer k, $2 \leq k \leq n$, the Riemannian invariant Θ_k on M^n is defined by

$$\Theta_k(x) = \frac{1}{k-1} \inf_{\Pi, X} Ric_{\Pi}(X), \ x \in M^n,$$

where Π runs over all k-plane sections in $T_x M^n$ and X runs over all unit vectors in Π [4].

3 Submanifolds of Conformally Flat Manifolds

Let H be the mean curvature vector of M^n at a point x.

Now let $M^n, n \geq 3$, be an n-dimensional submanifold of an m-dimensional conformally flat manifold N^m, $m > 4$, and $x \in M^n$, $\{e_i\}$, $1 \leq i \leq n$, and $\{e_j\}$, $n+1 \leq j \leq m$ be orthonormal bases of $T_x M^n$ and $T_x^{\perp} M^n$, respectively. The square norm $\|h\|^2$ of the second fundamental (see [2]) is defined by

$$\|h\|^2 = \sum_{i,j=1}^{n} g(h(e_i, e_j), h(e_i, e_j)).$$

Let $x \in M^n, \pi \subset T_x M^n, \pi = sp\,\{e_1, e_2\}$.

For a submanifold of a conformally flat manifold, we prove the following first Chen inequality:

Theorem 3.1 *Let $M^n, n \geq 3$, be an n-dimensional submanifold of an m-dimensional conformally flat manifold N^m, $m > 4$. Then*

$$\tau - K(\pi) \leq -\frac{1}{m-2}\left\{Ric^N(e_1, e_1) + Ric^N(e_2, e_2)\right\} + \frac{n^2(n-2)}{2(n-1)}\|H\|^2$$

$$+\frac{n-1}{m-2}\left[\sum_{j=1}^{n} Ric^N(e_j, e_j)\right] - \frac{\tau^N[n(n-1)-2]}{(m-1)(m-2)}, \tag{3.1}$$

where π is a 2-plane section of T_xM^n, $x \in M^n$. The equality case of inequality (3.1) holds at a point $x \in M^n$ if and only if there exist an orthonormal basis $\{e_i\}$, $1 \leq i \leq n$, of T_xM^n and an orthonormal basis $\{e_p\}$, $n+1 \leq p \leq m$, of $T_x^{\perp}M^n$ such that the shape operators of M^n in N^m at x have the following forms:

$$A_{e_{n+1}} = \begin{pmatrix} \lambda & 0 & 0 & \cdots & 0 \\ 0 & \mu & 0 & \cdots & 0 \\ 0 & 0 & \lambda+\mu & \cdots & 0 \\ \vdots & \vdots & \vdots & \ddots & \vdots \\ 0 & 0 & 0 & \cdots & \lambda+\mu \end{pmatrix}, \tag{3.2}$$

$$A_{e_s} = \begin{pmatrix} h^s_{11} & h^s_{12} & 0 \cdots 0 \\ h^s_{12} & -h^s_{11} & 0 \cdots 0 \\ 0 & 0 & 0 \cdots 0 \\ \vdots & \vdots & \vdots \cdots \vdots \\ 0 & 0 & 0 \cdots 0 \end{pmatrix}, \quad n+2 \leq s \leq m, \tag{3.3}$$

where we denote $h^s_{ij} = g(h(e_i, e_j), e_s)$, $1 \leq i, j \leq n$ and $n+2 \leq s \leq m$, and Ric^N, τ^N are the Ricci tensor and scalar curvature of N^m, respectively.

Proof For $X_1 = X_4 = e_i$, $X_2 = X_3 = e_j$, $i \neq j$, from the Gauss equation (2.2), by summation after $1 \leq i, j \leq n$, we have

$$2\tau = \frac{2(n-1)}{m-2}\left[\sum_{j=1}^{n} Ric^N(e_j, e_j)\right] - \frac{2\tau^N}{(m-1)(m-2)}n(n-1)$$
$$+ n^2 \|H\|^2 - \|h\|^2 . \tag{3.4}$$

When we take

$$\rho = 2\tau - \frac{n^2(n-2)}{n-1}\|H\|^2 + \frac{2\tau^N}{(m-1)(m-2)}n(n-1)$$
$$- \frac{2(n-1)}{m-2}\left[\sum_{j=1}^{n} Ric^N(e_j, e_j)\right], \tag{3.5}$$

from (3.4) and (3.5), we get

$$n^2 \|H\|^2 = (n-1)\left(\|h\|^2 + \rho\right). \tag{3.6}$$

Let the normal vector e_{n+1} be a unit vector in the direction of the mean curvature vector H at x. Then using (3.6), we obtain

$$\left(\sum_{i=1}^{n} h_{ii}^{n+1}\right)^2 = (n-1)\left(\sum_{i,j=1}^{n}\sum_{s=n+1}^{m}(h_{ij}^s)^2 + \rho\right),$$

or equivalently,

$$\left(\sum_{i=1}^{n} h_{ii}^{n+1}\right)^2 = (n-1)\left\{\sum_{i=1}^{n}(h_{ii}^{n+1})^2 + \sum_{i\neq j}(h_{ij}^{n+1})^2 + \sum_{i,j=1}^{n}\sum_{s=n+2}^{m}(h_{ij}^s)^2 + \rho\right\}. \tag{3.7}$$

By using Lemma 3.1 in [2], Eq. (3.7) gives us

$$2h_{11}^{n+1}h_{22}^{n+1} \geq \sum_{i\neq j}(h_{ij}^{n+1})^2 + \sum_{i,j=1}^{n}\sum_{s=n+2}^{m}(h_{ij}^s)^2 + \rho. \tag{3.8}$$

When we use the Gauss equation for $X_1 = X_4 = e_1$, $X_2 = X_3 = e_2$, we have

$$\begin{aligned}
K(\pi) = R(e_1, e_2, e_2, e_1) &= \frac{1}{m-2}\left\{Ric^N(e_1, e_1) + Ric^N(e_2, e_2)\right\} \\
&\quad - \frac{2\tau^N}{(m-1)(m-2)} + \sum_{s=n+1}^{m}[h_{11}^s h_{22}^s - (h_{12}^s)^2] \\
&\geq \frac{1}{m-2}\left\{Ric^N(e_1, e_1) + Ric^N(e_2, e_2)\right\} - \frac{2\tau^N}{(m-1)(m-2)} \\
&\quad + \frac{1}{2}\left[\sum_{i\neq j}(h_{ij}^{n+1})^2 + \sum_{i,j=1}^{n}\sum_{s=n+2}^{m}(h_{ij}^s)^2 + \rho\right] \\
&\quad + \sum_{s=n+2}^{m} h_{11}^s h_{22}^s - \sum_{s=n+1}^{m}(h_{12}^s)^2 = \frac{1}{m-2}\left\{Ric^N(e_1, e_1) + Ric^N(e_2, e_2)\right\} \\
&\quad - \frac{2\tau^N}{(m-1)(m-2)} + \frac{1}{2}\sum_{i\neq j}(h_{ij}^{n+1})^2 + \frac{1}{2}\sum_{i,j=1}^{n}\sum_{s=n+2}^{m}(h_{ij}^s)^2 \\
&\quad + \frac{1}{2}\rho + \sum_{s=n+2}^{m} h_{11}^s h_{22}^s - \sum_{s=n+1}^{m}(h_{12}^s)^2
\end{aligned}$$

$$= \frac{1}{m-2}\left\{Ric^N(e_1,e_1)+Ric^N(e_2,e_2)\right\}-\frac{2\tau^N}{(m-1)(m-2)}$$

$$+\frac{1}{2}\sum_{i\neq j}(h_{ij}^{n+1})^2+\frac{1}{2}\sum_{s=n+2}^{m}\sum_{i,j>2}(h_{ij}^s)^2+$$

$$+\frac{1}{2}\sum_{s=n+2}^{m}(h_{11}^s+h_{22}^s)^2+\sum_{j>2}\left[(h_{1j}^{n+1})^2+(h_{2j}^{n+1})^2\right]+\frac{1}{2}\rho$$

$$\geq \frac{1}{m-2}\left\{Ric^N(e_1,e_1)+Ric^N(e_2,e_2)\right\}-\frac{2\tau^N}{(m-1)(m-2)}+\frac{\rho}{2},$$

which implies

$$K(\pi)\geq \frac{1}{m-2}\left\{Ric^N(e_1,e_1)+Ric^N(e_2,e_2)\right\}-\frac{2\tau^N}{(m-1)(m-2)}+\frac{\rho}{2}. \tag{3.9}$$

Using (3.5), it follows from (3.9) that

$$K(\pi)\geq \tau+\frac{1}{m-2}\left\{Ric^N(e_1,e_1)+Ric^N(e_2,e_2)\right\}-\frac{n^2(n-2)}{2(n-1)}\|H\|^2$$

$$-\frac{n-1}{m-2}\left[\sum_{j=1}^{n}Ric^N(e_j,e_j)\right]+\frac{\tau^N[n(n-1)-2]}{(m-1)(m-2)},$$

which proves the inequality (3.1).

Similar to the proof of Lemma 3.2 in [2], it can be easily seen that the equality case holds at a point $x\in M^n$ if and only if with respect to a suitable frame field, the shape operators of M^n are of the forms (3.2) and (3.3). □

Now using Theorem 3.1, we prove the following relationship between the k-Ricci curvature and the squared mean curvature $\|H\|^2$:

Theorem 3.2 *Let $M^n, n \geq 3$, be an n-dimensional submanifold of an m-dimensional conformally flat manifold N^m, $m > 4$. Then*

$$\|H\|^2\geq \frac{2\tau}{n(n-1)}-\frac{2}{n(m-2)}\left[\sum_{j=1}^{n}Ric^N(e_j,e_j)\right]$$

$$+\frac{2\tau^N}{(m-1)(m-2)}. \tag{3.10}$$

Proof Equation (3.4) can be written as

$$2\tau + \|h\|^2 - n^2\|H\|^2 = \frac{2(n-1)}{m-2}\left[\sum_{j=1}^{n} Ric^N(e_j, e_j)\right] - \frac{2\tau^N}{(m-1)(m-2)} n(n-1). \tag{3.11}$$

Let $\{e_i\}$, $1 \leq i \leq m$, be an orthonormal basis of T_xN^m at x such that the normal vector e_{n+1} is parallel to the mean curvature vector H and $\{e_i\}$, $1 \leq i \leq n$, diagonalize the shape operator $A_{e_{n+1}}$. Then $A_{e_{n+1}} = [a_1, ..., a_n]_{n\times n}$ and $A_{e_s} = (h^s_{ij})$, $i, j = 1, ..., n$; $s = n+2, ..., m$, trace $A_r = 0$. So in view of (3.11), we get

$$n^2\|H\|^2 = 2\tau(x) + \sum_{i=1}^{n} a_i^2 + \sum_{r=n+2}^{m}\sum_{i,j=1}^{n}(h^r_{ij})^2$$

$$-\frac{2(n-1)}{m-2}\left[\sum_{j=1}^{n} Ric^N(e_j, e_j)\right] + \frac{2\tau^N(x)}{(m-1)(m-2)} n(n-1). \tag{3.12}$$

On the other hand, it is trivial that

$$\sum_{i=1}^{n} a_i^2 \geq n\|H\|^2.$$

From (3.12), we have

$$n^2\|H\|^2 \geq 2\tau + n\|H\|^2 - \frac{2(n-1)}{m-2}\left[\sum_{j=1}^{n} Ric^N(e_j, e_j)\right] + \frac{2\tau^N}{(m-1)(m-2)} n(n-1), \tag{3.13}$$

or, equivalently,

$$\|H\|^2 \geq \frac{2\tau}{n(n-1)} - \frac{2}{n(m-2)}\left[\sum_{j=1}^{n} Ric^N(e_j, e_j)\right] + \frac{2\tau^N}{(m-1)(m-2)},$$

and this proves the theorem. □

Using Theorem 3.2, we obtain the following:

Theorem 3.3 *Let $M^n, n \geq 3$, be an n-dimensional submanifold of an m-dimensional conformally flat manifold N^m, $m > 4$. Then, for any integer k, $2 \leq k \leq n$, and any point $x \in M^n$,*

$$\|H\|^2 \geq \Theta_k(x) - \frac{2}{n\,(m-2)}\left[\sum_{j=1}^{n} Ric^N(e_j, e_j)\right] + \frac{2\tau^N}{(m-1)(m-2)}. \tag{3.14}$$

Proof Let $\{e_i\}$, $1 \leq i \leq n$, be an orthonormal basis of $T_x M^n$. Denote by $\Pi_{\alpha_1 \ldots \alpha_k}$ the k-plane section spanned by $\{e_{\alpha_i}\}$, $1 \leq i \leq k$. By definition, we have (see [4])

$$\tau(x) \geq \frac{n(n-1)}{2}\Theta_k(x),$$

which implies (3.14). □

4 Submanifolds of Conformally Flat Manifolds Endowed with a Semi-symmetric Metric Connection

Let N^m be an m-dimensional Riemannian manifold and $\widetilde{\nabla}$ a linear connection on N^m. If the torsion tensor $\widetilde{T}$ of $\widetilde{\nabla}$ satisfies

$$\widetilde{T}\left(\widetilde{X}_1, \widetilde{X}_2\right) = \eta(\widetilde{X}_2)\widetilde{X}_1 - \eta(\widetilde{X}_1)\widetilde{X}_2$$

for a 1-form η, then the connection $\widetilde{\nabla}$ is called a *semi-symmetric connection*. Let $\widetilde{g}$ be a Riemannian metric on N^m. If $\widetilde{\nabla}\widetilde{g} = 0$, then $\widetilde{\nabla}$ is called a *semi-symmetric metric connection* on N^m.

By Yano [15], a semi-symmetric metric connection $\widetilde{\nabla}$ on N^m is given by

$$\widetilde{\nabla}_{\widetilde{X}_1}\widetilde{X}_2 = \widetilde{D}_{\widetilde{X}_1}\widetilde{X}_2 + \eta(\widetilde{X}_2)\widetilde{X}_1 - \widetilde{g}(\widetilde{X}_1, \widetilde{X}_2)E,$$

for any vector fields $\widetilde{X}_1$ and $\widetilde{X}_2$ on N^m, where $\widetilde{D}$ denotes the Levi-Civita connection with respect to the Riemannian metric $\widetilde{g}$ and E is a vector field defined by $\widetilde{g}(E, \widetilde{X}) = \eta(\widetilde{X})$ for any vector field $\widetilde{X}$.

Let M^n be an n-dimensional submanifold of an m-dimensional Riemannian manifold N^m. Let ∇ and D denote the induced semi-symmetric metric connection and the induced Levi-Civita connection, respectively.

Let $\widetilde{R}$, $\widetilde{R}^{\widetilde{D}}$, R, and R^D denote the curvature tensors of $\widetilde{\nabla}$, $\widetilde{D}$, ∇, and D, respectively.

The Gauss formulas with respect to $\widetilde{\nabla}$, respectively, $\widetilde{D}$ are given by

$$\widetilde{\nabla}_{X_1}X_2 = \nabla_{X_1}X_2 + \sigma(X_1, X_2), \quad X_1, X_2 \in \chi(M),$$

$$\widetilde{D}_{X_1}X_2 = D_{X_1}X_2 + h(X_1, X_2), \quad X_1, X_2 \in \chi(M),$$

where h is the second fundamental form of M^n in N^m and σ is a (0, 2)-tensor on M^n (see [1] and [13]). According to the formula (7) from [13], σ is also symmetric and $\sigma = h$ when E is tangent to M^n.

Let N^m be a conformally flat manifold, $m \geq 4$, endowed with a semi-symmetric metric connection $\widetilde{\nabla}$.

The curvature tensor $\widetilde{R}$ with respect to the semi-symmetric metric connection $\widetilde{\nabla}$ on N^m can be written as (see [10])

$$\begin{aligned}\widetilde{R}(X_1, X_2, X_3, X_4) = \widetilde{R}^{\widetilde{D}}&(X_1, X_2, X_3, X_4) - \beta(X_2, X_3)\widetilde{g}(X_1, X_4) \\ &+\beta(X_1, X_3)\widetilde{g}(X_2, X_4) - \beta(X_1, X_4)\widetilde{g}(X_2, X_3) \\ &+\beta(X_2, X_4)\widetilde{g}(X_1, X_3),\end{aligned} \tag{4.1}$$

where $X_i \in \chi(M^n)$, $1 \leq i \leq 4$, and β is a (0, 2)-tensor field defined by

$$\beta(X_1, X_2) = \left(D_{X_1}\eta\right) X_2 - \eta(X_1)\eta(X_2) + \frac{1}{2}\eta(E)\widetilde{g}(X_1, X_2).$$

From (2.1) and (4.1), it follows that the curvature tensor $\widetilde{R}$ can be expressed as

$$\begin{aligned}&\widetilde{R}(X_1, X_2, X_3, X_4) \\ &= \frac{1}{m-2}\Big\{Ric^N(X_2, X_3)g(X_1, X_4) - Ric^N(X_1, X_3)g(X_2, X_4) \\ &\quad +Ric^N(X_1, X_4)g(X_2, X_3) - Ric^N(X_2, X_4)g(X_1, X_3)\Big\} \\ &\quad - \frac{2\tau^N}{(m-1)(m-2)}\left[g(X_2, X_3)g(X_1, X_4) - g(X_1, X_3)g(X_2, X_4)\right] \\ &\quad -\beta(X_2, X_3)g(X_1, X_4) + \beta(X_1, X_3)g(X_2, X_4) \\ &\quad -\beta(X_1, X_4)g(X_2, X_3) + \beta(X_2, X_4)g(X_1, X_3).\end{aligned} \tag{4.2}$$

Let $\lambda = trace\,(\beta)$.

The Gauss equation for a submanifold M^n into a conformally flat manifold N^m, $m > 4$, is

$$\widetilde{R}^{\widetilde{D}}(X_1, X_2, X_3, X_4) = R^D(X_1, X_2, X_3, X_4) + g(h(X_1, X_3), h(X_2, X_4)) - g(h(X_1, X_4), h(X_2, X_3)), \tag{4.3}$$

and from [13], the Gauss equation with respect to the semi-symmetric metric connection is

$$\widetilde{R}(X_1, X_2, X_3, X_4) = R(X_1, X_2, X_3, X_4) + g(\sigma(X_1, X_3), \sigma(X_2, X_4)) - g(\sigma(X_2, X_3), \sigma(X_1, X_4)). \tag{4.4}$$

We denote the mean curvature $\mathcal{H}$ with respect to the semi-symmetric metric connection by

$$\mathcal{H} = \frac{1}{n}\text{trace}\,(\sigma)$$

and

$$\beta(e_1, e_1) + \beta(e_2, e_2) = \lambda - \text{trace}\left(\beta_{|_{\pi^\perp}}\right).$$

From [9], we know that $\mathcal{H} = H$ if and only if the vector field E is tangent to M^n.

Now let $M^n, n \geq 3$, be an n-dimensional submanifold of an m-dimensional conformally flat manifold N^m, $m > 4$, endowed with a semi-symmetric metric connection $\widetilde{\nabla}$ and $x \in M^n$, $\{e_i\}$, $1 \leq i \leq n$, and $\{e_p\}$, $n + 1 \leq p \leq m$ be orthonormal bases of $T_x M^n$, and $T_x^\perp M^n$, respectively. Let $x \in M^n$, $\pi \subset T_x M^n$, $\pi = sp\,\{e_1, e_2\}$.

We prove the following first Chen inequality:

Theorem 4.1 *Let $M^n, n \geq 3$, be an n-dimensional submanifold of an m-dimensional conformally flat manifold N^m, $m > 4$, endowed with a semi-symmetric metric connection $\widetilde{\nabla}$. Then*

$$\tau^\nabla - K^\nabla(\pi) \leq (n-2)\left[\frac{n^2}{2(n-1)}\,\|\mathcal{H}\|^2 - \lambda\right] - trace\left(\beta_{|_{\pi^\perp}}\right) - \frac{1}{m-2}\left\{Ric^N(e_1, e_1) + Ric^N(e_2, e_2)\right\} + \frac{n-1}{m-2}\left[\sum_{j=1}^{n} Ric^N(e_j, e_j)\right] - \frac{\tau^N\,[n(n-1)-2]}{(m-1)(m-2)}, \tag{4.5}$$

where π is a 2-plane section of T_xM^n, $x \in M^n$, and τ^∇, K^∇ are the scalar curvature and the sectional curvature of the induced semi-symmetric metric connection ∇, respectively, and Ric^N and τ^N are the Ricci tensor and the scalar curvature of N^m with respect to the Levi-Civita connection, respectively.

Proof For $X_1 = X_4 = e_i$, $X_2 = X_3 = e_j$, $i \neq j$, from Eqs. (4.2) and (4.4), by summation after $1 \leq i, j \leq n$, it follows that

$$2\tau^\nabla + \|\sigma\|^2 - n^2\|\mathcal{H}\|^2 = -2(n-1)\lambda$$

$$+\frac{2(n-1)}{m-2}\left[\sum_{j=1}^{n} Ric^N(e_j, e_j)\right] - \frac{2\tau^N}{(m-1)(m-2)}n(n-1), \tag{4.6}$$

where

$$\|\sigma\|^2 = \sum_{i,j=1}^{n} g(\sigma(e_i, e_j), \sigma(e_i, e_j)).$$

When we take

$$\rho = 2\tau^\nabla - \frac{n^2(n-2)}{n-1}\|\mathcal{H}\|^2 + 2(n-1)\lambda$$

$$-\frac{2(n-1)}{m-2}\left[\sum_{j=1}^{n} Ric^N(e_j, e_j)\right] + \frac{2\tau^N}{(m-1)(m-2)}n(n-1), \tag{4.7}$$

we have

$$n^2\|\mathcal{H}\|^2 = (n-1)\left(\|\sigma\|^2 + \rho\right).$$

Let the normal vector e_{n+1} be a unit vector in the direction of the mean curvature vector $\mathcal{H}$ at x. Similar to the proof of Theorem 3.1, we have

$$2\sigma_{11}^{n+1}\sigma_{22}^{n+1} \geq \sum_{i \neq j}(\sigma_{ij}^{n+1})^2 + \sum_{i,j=1}^{n}\sum_{s=n+2}^{m}(\sigma_{ij}^{s})^2 + \rho.$$

For $X_1 = X_4 = e_1, X_2 = X_3 = e_2$, the Gauss equation with respect to the semi-symmetric metric connection gives

$$K^{\nabla}(\pi) = R(e_1, e_2, e_2, e_1) = \frac{1}{m-2}\left\{Ric^N(e_1, e_1) + Ric^N(e_2, e_2)\right\}$$

$$- \frac{2\tau^N}{(m-1)(m-2)} - \beta(e_1, e_1) - \beta(e_2, e_2) + \sum_{s=n+1}^{m} [\sigma_{11}^s \sigma_{22}^s - (\sigma_{12}^s)^2]$$

$$\geq \frac{1}{m-2}\left\{Ric^N(e_1, e_1) + Ric^N(e_2, e_2)\right\} - \frac{2\tau^N}{(m-1)(m-2)}$$

$$-\beta(e_1, e_1) - \beta(e_2, e_2) + \frac{1}{2}\left[\sum_{i \neq j}(\sigma_{ij}^{n+1})^2 + \sum_{i,j=1}^{n} \sum_{s=n+2}^{m} (\sigma_{ij}^s)^2 + \rho\right]$$

$$+ \sum_{s=n+2}^{m} \sigma_{11}^s \sigma_{22}^s - \sum_{s=n+1}^{m} (\sigma_{12}^s)^2 = \frac{1}{m-2}\left\{Ric^N(e_1, e_1) + Ric^N(e_2, e_2)\right\}$$

$$- \frac{2\tau^N}{(m-1)(m-2)} - \beta(e_1, e_1) - \beta(e_2, e_2)$$

$$+ \frac{1}{2}\sum_{i \neq j}(\sigma_{ij}^{n+1})^2 + \frac{1}{2}\sum_{i,j=1}^{n} \sum_{s=n+2}^{m} (\sigma_{ij}^s)^2 + \frac{1}{2}\rho + \sum_{s=n+2}^{m} \sigma_{11}^s \sigma_{22}^s - \sum_{s=n+1}^{m} (\sigma_{12}^s)^2$$

$$= \frac{1}{m-2}\left\{Ric^N(e_1, e_1) + Ric^N(e_2, e_2)\right\} - \frac{2\tau^N}{(m-1)(m-2)}$$

$$- \beta(e_1, e_1) - \beta(e_2, e_2) + \frac{1}{2}\sum_{i \neq j}(\sigma_{ij}^{n+1})^2 + \frac{1}{2}\sum_{s=n+2}^{m} \sum_{i,j>2} (\sigma_{ij}^s)^2$$

$$+ \frac{1}{2}\sum_{s=n+2}^{m} (\sigma_{11}^s + \sigma_{22}^s)^2 + \sum_{j>2}[(\sigma_{1j}^{n+1})^2 + (\sigma_{2j}^{n+1})^2] + \frac{1}{2}\rho$$

$$\geq \frac{1}{m-2}\left\{Ric^N(e_1, e_1) + Ric^N(e_2, e_2)\right\} - \frac{2\tau^N}{(m-1)(m-2)}$$

$$- \beta(e_1, e_1) - \beta(e_2, e_2) + \frac{\rho}{2},$$

which implies

$$K^{\nabla}(\pi) \geq \frac{1}{m-2}\left\{Ric^N(e_1,e_1)+Ric^N(e_2,e_2)\right\} - \frac{2\tau^N}{(m-1)(m-2)}$$
$$-\beta(e_1,e_1)-\beta(e_2,e_2)+\frac{\rho}{2}.$$

Using (4.7), we get

$$K^{\nabla}(\pi) \geq \tau^{\nabla} + (n-2)\left[-\frac{n^2}{2(n-1)}\,\|\mathcal{H}\|^2+\lambda\right] + \operatorname{trace}\left(\beta_{|_{\pi^\perp}}\right)$$
$$+\frac{1}{m-2}\left\{Ric^N(e_1,e_1)+Ric^N(e_2,e_2)\right\}$$
$$-\frac{n-1}{m-2}\left[\sum_{j=1}^{n} Ric^N(e_j,e_j)\right] + \frac{\tau^N\,[n(n-1)-2]}{(m-1)(m-2)},$$

which proves the inequality. □

When the vector field E is tangent to M^n, for the equality case of the inequality (4.5), we have the following theorem:

Theorem 4.2 *If the vector field E is tangent to M^n, the equality case of the inequality (4.5) holds at a point $x \in M^n$ if and only if for suitable chosen orthonormal bases of T_xM^n and $T_x^{\perp}M^n$, the shape operators of M^n in N^m at x have the forms (3.2) and (3.3).*

Similar to Theorem 3.2, using Theorem 4.1, we give the following relationship between the k-Ricci curvature and the squared mean curvature $\|H\|^2$:

Theorem 4.3 *Let $M^n, n \geq 3$, be an n-dimensional submanifold of an m-dimensional conformally flat manifold N^m, $m > 4$, endowed with a semi-symmetric metric connection $\widetilde{\nabla}$ such that the vector field E is tangent to M^n. Then*

$$\|H\|^2 \geq \frac{2\tau^{\nabla}}{n(n-1)} - \frac{2}{n\,(m-2)}\left[\sum_{j=1}^{n} Ric^N(e_j,e_j)\right]$$
$$+\frac{2\tau^N}{(m-1)(m-2)}+\frac{2}{n}\lambda.$$

Proof The relation (4.6) is equivalent to

$$n^2 \|H\|^2 = 2\tau^{\nabla} + \|h\|^2 + 2(n-1)\lambda$$

$$-\frac{2(n-1)}{m-2}\left[\sum_{j=1}^{n} Ric^N(e_j, e_j)\right] + \frac{2\tau^N}{(m-1)(m-2)} n(n-1). \tag{4.8}$$

Similar to the proof of Theorem 3.2, from (4.8), we get

$$n^2 \|H\|^2 = 2\tau^{\nabla} + \sum_{i=1}^{n} a_i^2 + \sum_{r=n+2}^{m} \sum_{i,j=1}^{n} (h_{ij}^r)^2$$

$$+2(n-1)\lambda - \frac{2(n-1)}{m-2}\left[\sum_{j=1}^{n} Ric^N(e_j, e_j)\right] + \frac{2\tau^N}{(m-1)(m-2)} n(n-1).$$

Hence, we have

$$n^2 \|H\|^2 \geq 2\tau^{\nabla} + n\|H\|^2 + 2(n-1)\lambda \tag{4.9}$$

$$-\frac{2(n-1)}{m-2}\left[\sum_{j=1}^{n} Ric^N(e_j, e_j)\right] + \frac{2\tau^N}{(m-1)(m-2)} n(n-1),$$

or, equivalently,

$$\|H\|^2 \geq \frac{2\tau^{\nabla}}{n(n-1)} - \frac{2}{n(m-2)}\left[\sum_{j=1}^{n} Ric^N(e_j, e_j)\right]$$

$$+\frac{2\tau^N}{(m-1)(m-2)} + \frac{2}{n}\lambda.$$

□

Similar to Theorem 3.3, using Theorem 4.3, we can state the following theorem:

Theorem 4.4 *Let $M^n, n \geq 3$, be an n-dimensional submanifold of an m-dimensional conformally flat manifold N^m, $m > 4$, endowed with a semi-symmetric metric connection $\widetilde{\nabla}$ such that the vector field E is tangent to M^n. Then, for any integer k, $2 \leq k \leq n$, and any point $x \in M^n$, we have*

$$\|H\|^2 \geq \Theta_k(x) + \frac{2}{n}\lambda$$

$$-\frac{2}{n\,(m-2)}\left[\sum_{j=1}^{n} Ric^{N}(e_j, e_j)\right]+\frac{2\tau^{N}}{(m-1)(m-2)}.$$

Proof The proof is obtained by a similar way given in the proof of Theorem 3.3. □

References

1. Chen, B.Y.: Geometry of Submanifolds and its Applications. Science University of Tokyo, Tokyo (1981)
2. Chen, B.Y.: Some pinching and classification theorems for minimal submanifolds. Arch. Math. (Basel) **60**, 568–578 (1993)
3. Chen, B.Y.: Mean curvature and shape operator of isometric immersions in real space form. Glasgow Math. J. **38**, 87–97 (1996)
4. Chen, B.Y.: Relations between Ricci curvature and shape operator for submanifolds with arbitrary codimensions. Glasg. Math. J. **41**, 33–41 (1999)
5. Chen, B.Y.: δ-invariants, inequalities of submanifolds and their applications. In: Mihai, A., Mihai, I., Miron, R. (eds.) Topics in Differential Geometry, pp. 29–156. Editura Academiei Romane, Bucuresti (2008)
6. Chen, B.Y.: Pseudo-Riemannian geometry, δ-invariants and applications, World Scientific Publishing Co. Pte. Ltd., Hackensack, NJ (2011)
7. Chen, B.Y., Blaga, A.M.: Recent developments on Chen-Ricci inequalities in differential geometry. In: Geometry of Submanifolds and Applications. Infosys Science Foundation Series in Mathematical Sciences, pp. 1–61. Springer, Singapore (2024)
8. Chen, B.Y., Vîlcu, G.E.: Recent developments on the first Chen inequality in differential geometry. Mathematics **11**(19), 4186 (2023)
9. Imai, T.: Hypersurfaces of a Riemannian manifold with semi-symmetric metric connection. Tensor (N.S.) **23**, 300–306 (1972)
10. Imai, T.: Notes on semi-symmetric metric connections. Tensor (N.S.) **24**, 293–296 (1972)
11. Mihai, A.: Modern Topics in Submanifold Theory. Editura Universitatii Bucuresti, Bucharest (2006)
12. Mihai, A., Özgür, C.: Chen inequalities for submanifolds of real space forms with a semi-symmetric metric connection. Taiwanese J. Math. **14**, 1465–1477 (2010)
13. Nakao, Z.: Submanifolds of a Riemannian manifold with semisymmetric metric connections. Proc. Am. Math. Soc. **54**, 261–266 (1976)
14. Özgür, C.: B. Y. Chen inequalities for submanifolds of a Riemannian manifold of quasi-constant curvature. Turkish J. Math. **35**, 501–509 (2011)
15. Yano, K.: On semi-symmetric metric connection. Rev. Roumaine Math. Pures Appl. **15**, 1579–1586 (1970)
16. Zhang, P., Zhang, L., Song, W.: Chen's inequalities for submanifolds of a Riemannian manifold of quasi-constant curvature with a semi-symmetric metric connection. Taiwanese J. Math. **18**, 1841–1862 (2014)

General Chen Inequalities for Statistical Submanifolds in Kenmotsu Statistical Manifolds of Constant ϕ-Sectional Curvature

Simona Decu and Gabriel-Eduard Vîlcu

Abstract In this chapter, we prove general Chen inequalities in statistical submanifolds of Kenmotsu statistical manifolds of constant ϕ-sectional curvature. Furthermore, we investigate the equality case of these inequalities. Finally, we point out a representative example.

Keywords Kenmotsu statistical manifold · Statistical submanifold · Chen inequalities · Chen invariants

1 Introduction

The concept of curvature is one of the central notions of differential geometry, distinguishing the geometrical core of the subject from those aspects that are analytic, algebraic, or topological [35]. The curvature invariants play a key role in physics, chemistry, geology, biology, art, technology, etc. Honeycombs and shells, crystals and galaxies, DNA-molecules, red blood cells, flowers, stems, tissues, pollen grains of plants, the relativistic space-time universe itself, etc. all do assume shapes in accordance with similar natural curvature conditions [24].

S. Decu (✉)
Department of Applied Mathematics, Bucharest University of Economic Studies, Bucharest, Romania

Romanian Academy, Center of Mountain Economy (CE-MONT), "Costin C. Kiriţescu" National Institute of Economic Research, Bucharest, Romania
e-mail: simona.marinescu@csie.ase.ro

G.-E. Vîlcu
Faculty of Applied Sciences, Department of Mathematics and Informatics, National University of Science and Technology Politehnica Bucharest, Bucharest, Romania

Romanian Academy, "Gheorghe Mihoc-Caius Iacob" Institute of Mathematical Statistics and Applied Mathematics, Bucharest, Romania
e-mail: gabriel.vilcu@upb.ro

B.-Y. Chen, M. A. Choudhary (eds.), *Geometric Inequalities and Applications*, Infosys Science Foundation Series in Mathematical Sciences,
https://doi.org/10.1007/978-981-95-5148-4_14

Motivated by the challenges of applying the famous Nash's embedding theorem [32] to submanifold theory, B.-Y. Chen formulated the following fundamental problem [6].

Problem 1.1 Establish simple relationships between main extrinsic invariants and main intrinsic invariants of a submanifold.

Solutions of this problem are focused on some geometrical inequalities involving intrinsic invariants and extrinsic invariants of submanifolds. On the one hand, there are intrinsic curvatures like sectional curvature, scalar curvature, and Chen invariants, which give together like the DNA-structure of the Riemannian manifolds involved [7]. On the other hand, there are extrinsic curvatures like mean curvature, shape operator, and Casorati curvature, which fundamentally relate to the shape that these submanifolds assume in their ambient space.

Fascinating solutions of this basic problem are revealed by B.-Y. Chen who defined in the 1990s new types of intrinsic invariants called *δ-invariants* or *Chen invariants*, involved in optimal inequalities for submanifolds in real space forms [5]. The theory of δ-invariants turns out to be a very fruitful branch of the differential geometry (see, e.g., [14, 16, 25, 31, 33, 37, 38]). Very ample surveys on this research field can be explored in [9, 12, 13, 15, 29, 30].

Moreover, new answers of this basic problem refer to other types of geometric inequalities, like Casorati inequalities and Wintgen inequalities (see, e.g., [1, 3, 4, 9–11, 13, 17, 18]).

Kenmotsu geometry was first studied in 1972 by K. Kenmotsu [26] as a field of contact geometry, leading to a broad spectrum of applications in physics and control theory [36]. On the other hand, S. Amari defined the concept of *statistical manifold* in 1985 in a study of information geometry [2]. Moreover, H. Furuhata introduced the notion of *Kenmotsu statistical manifold*, which is locally constructed as a warped product of a holomorphic statistical manifold and a real line [22]. In the context of the above basic problem, new solutions are achieved considering a Kenmotsu statistical manifold. Thus, Casorati inequalities for submanifolds in Kenmotsu statistical manifolds of constant ϕ-sectional curvature are demonstrated in [18]. In addition, other Casorati inequalities are proved in [19] for submanifolds in the latter ambient space endowed with semi-symmetric metric connection. Recently, the authors of the present work established in [20] the first Chen inequality for statistical submanifolds in Kenmotsu statistical manifolds of constant ϕ-sectional curvature. The statistical manifolds endowed with almost product structures are investigated in [39].

In this chapter, we obtain general Chen inequalities in terms of the general Chen invariants (intrinsic invariants) and the mean curvature (extrinsic invariant) of statistical submanifolds in Kenmotsu statistical manifolds having a constant ϕ-sectional curvature. Furthermore, the equality case of these inequalities is investigated. Finally, a representative example is emphasized.

2 Preliminaries

We consider $(\bar{M}, \bar{g})$ a Riemannian manifold, $\bar{g}$ a Riemannian metric on $\bar{M}$, and $\bar{\nabla}$ an affine connection on $\bar{M}$. The triplet $(\bar{M}, \bar{g}, \bar{\nabla})$ is named a *statistical manifold* if the torsion tensor field of $\bar{\nabla}$ vanishes and $\bar{\nabla}g$ as a $(0, 3)$-tensor is symmetric [21]. The pair $(\bar{g}, \bar{\nabla})$ is called a *statistical structure* on $\bar{M}$ [34]. For any connection $\bar{\nabla}$ on $(\bar{M}, \bar{g})$, one defines its *dual connection* $\bar{\nabla}^*$ with respect to $\bar{g}$ as follows:

$$\bar{g}(\bar{\nabla}_X Y, Z) + \bar{g}(Y, \bar{\nabla}^*_X Z) = X\, \bar{g}(Y, Z),$$

for any vector fields X, Y, Z on $\bar{M}$. Furthermore, denote the Levi-Civita connection on $\bar{M}$ by $\bar{\nabla}^0$, defined by [40]

$$\bar{\nabla}^0 = \frac{\bar{\nabla} + \bar{\nabla}^*}{2}.$$

If $(\bar{M}, \bar{g}, \bar{\nabla})$ is a statistical manifold, then so is $(\bar{M}, \bar{g}, \bar{\nabla}^*)$.

For a statistical structure $(\bar{g}, \bar{\nabla})$, we set a tensor field $\bar{K} \in \Gamma(T\bar{M}^{(1,2)})$ by

$$\bar{K}_X Y = \bar{K}(X, Y) = \bar{\nabla}_X Y - \bar{\nabla}^0_X Y. \tag{2.1}$$

Moreover, $\bar{K}$ is given by

$$\bar{K}(X, Y) = \bar{\nabla}^0_X Y - \bar{\nabla}^*_X Y = \frac{1}{2}(\bar{\nabla}_X Y - \bar{\nabla}^*_X Y).$$

It is clear that $\bar{K}$ satisfies the properties:

$$\bar{K}(X, Y) = \bar{K}(Y, X),$$

$$\bar{g}(\bar{K}(X, Y), Z) = \bar{g}(Y, \bar{K}(X, Z)).$$

Let M be a submanifold of a statistical manifold $(\bar{M}, \bar{g}, \bar{\nabla})$ with g the induced metric on M and ∇ the induced connection on M. Then it is known that (M, g, ∇) is a statistical manifold as well.

The Gauss formulas are given by Furuhata and Hasegawa [21]

$$\bar{\nabla}_X Y = \nabla_X Y + h(X, Y),$$

$$\bar{\nabla}^*_X Y = \nabla^*_X Y + h^*(X, Y),$$

for any $X, Y \in \Gamma(TM)$, where we denoted h and h^* the imbedding curvature tensor of M in $\bar{M}$ for $\bar{\nabla}$ and $\bar{\nabla}^*$, respectively.

Next, denote by R, $\bar{R}$, R^*, and $\bar{R}^*$ the $(0, 4)$-*curvature tensors* for the connections ∇, $\bar{\nabla}$, ∇^*, and $\bar{\nabla}^*$, respectively. We define the *statistical curvature tensor field* [21] on M and $\bar{M}$, denoted by S and $\bar{S}$, respectively,

$$S(X,Y)Z = \frac{1}{2}\{R(X,Y)Z + R^*(X,Y)Z\}, \tag{2.2}$$

for any $X, Y, Z \in \Gamma(TM)$, and

$$\bar{S}(X,Y)Z = \frac{1}{2}\{\bar{R}(X,Y)Z + \bar{R}^*(X,Y)Z\}, \tag{2.3}$$

for any $X, Y, Z \in \Gamma(T\bar{M})$.

For any $X, Y \in \Gamma(T\bar{M})$, one defines $\bar{Q}$ the $(1,3)$-tensor field on $\bar{M}$ given by [34]

$$\bar{Q}(X,Y) = [\bar{K}_X, \bar{K}_Y].$$

Recall that $\bar{Q}$ is called the Hessian curvature tensor for the connection $\bar{\nabla}$.

It is known that $\bar{Q}(X,Y)$ satisfies the relation [34]:

$$\bar{R}(X,Y) + \bar{R}^*(X,Y) = 2\bar{R}^0(X,Y) + 2\bar{Q}(X,Y).$$

Afterward, let $(\bar{M}, \bar{g}, \phi, \xi)$ be a $(2n+1)$-dimensional Kenmotsu manifold defined as an almost contact metric manifold $\bar{M}$ which satisfies for any $X, Y \in \Gamma(T\bar{M})$ the relations:

$$(\bar{\nabla}^0_X \phi)(Y) = \bar{g}(\phi X, Y)\xi - \eta(Y)\phi X,$$

$$\bar{\nabla}^0_X \xi = X - \eta(X)\xi,$$

where $\phi \in \Gamma(T\bar{M}^{(1,1)})$, $\xi \in \Gamma(T\bar{M})$, and η is a 1-form on $\bar{M}$ with

$$\eta(X) = \bar{g}(X,\xi).$$

A Kenmotsu manifold $\bar{M}$ equipped with a statistical structure $(\bar{g}, \bar{\nabla})$ is called a Kenmotsu statistical manifold [22] if the following expression holds for any $X, Y \in \Gamma(T\bar{M})$:

$$\bar{K}(X, \phi Y) = -\phi \bar{K}(X,Y),$$

where $\bar{K}$ is the tensor field defined in (2.1).

A Kenmotsu statistical manifold $(\bar{M}, \bar{\nabla}, \bar{g}, \phi, \xi)$ is said to be of constant ϕ-sectional curvature c if the statistical curvature tensor field $\bar{S}$ is given by Furuhata and Hasegawa [22]

$$\bar{S}(X,Y)Z = \frac{c-3}{4}\{\bar{g}(Y,Z)X - \bar{g}(X,Z)Y)\} \tag{2.4}$$

$$+\frac{c+1}{4}\{\bar{g}(\phi Y, Z)\phi X - \bar{g}(\phi X, Z)\phi Y - 2\bar{g}(\phi X, Y)\phi Z$$
$$- \bar{g}(Y, \xi)\bar{g}(Z, \xi)X + \bar{g}(X, \xi)\bar{g}(Z, \xi)Y$$
$$+ \bar{g}(Y, \xi)\bar{g}(Z, X)\xi - \bar{g}(X, \xi)\bar{g}(Z, Y)\xi\},$$

for any $X, Y, Z \in \Gamma(T\bar{M})$.

Let M be an $(m+1)$-dimensional submanifold of a Kenmotsu statistical manifold $\bar{M}$ of dimension $2n+1$. Then the *Gauss equations* are the following [23]:

$$\bar{g}(\bar{R}(X, Y)Z, W) = g(R(X, Y)Z, W) - \bar{g}(h(Y, Z), h^*(X, W)) + \bar{g}(h(X, Z), h^*(Y, W)), \tag{2.5}$$

$$2\bar{g}(\bar{S}(X, Y)Z, W) = 2g(S(X, Y)Z, W) - \bar{g}(h(Y, Z), h^*(X, W)) + \bar{g}(h(X, Z), h^*(Y, W)) - \bar{g}(h^*(Y, Z), h(X, W)) + \bar{g}(h^*(X, Z), h(Y, W)), \tag{2.6}$$

$$4\bar{g}(\bar{R}^0(X, Y)Z, W) = 4g(R^0(X, Y)Z, W) - \bar{g}(h(Y, Z) + h^*(Y, Z), h(X, W) + h^*(X, W)) + \bar{g}(h(X, Z) + h^*(X, Z), h(Y, W) + h^*(Y, W)), \tag{2.7}$$

where h and h^* are the *imbedding curvature tensor* of M in $\bar{M}$ with respect to the dual connections $\bar{\nabla}$ and $\bar{\nabla}^*$.

The *mean curvature* vector fields of M are defined by, respectively,

$$H = \frac{1}{m+1}\sum_{i=1}^{m+1} h(e_i, e_i), \quad H^* = \frac{1}{m+1}\sum_{i=1}^{m+1} h^*(e_i, e_i).$$

These latter notions imply

$$2h^0 = h + h^*$$

and

$$2H^0 = H + H^*,$$

where h^0 and H^0 are the second fundamental form and the mean curvature field of M, respectively, with respect to the Levi-Civita connection ∇^0 on M.

Then, the *squared mean curvatures* of the submanifold M in $\bar{M}$ are given by

$$\|H\|^2 = \frac{1}{(m+1)^2}\sum_{\alpha=m+2}^{2n+1}\left(\sum_{i=1}^{m+1} h_{ii}^{\alpha}\right)^2, \quad \|H^*\|^2 = \frac{1}{(m+1)^2}\sum_{\alpha=m+2}^{2n+1}\left(\sum_{i=1}^{m+1} h_{ii}^{*\alpha}\right)^2,$$

where $h^{\alpha}_{ij} = g(h(e_i, e_j), e_\alpha)$ and $h^{*\alpha}_{ij} = g(h^*(e_i, e_j), e_\alpha)$, for $i, j \in \{1, \ldots, m+1\}$, $\alpha \in \{m+2, \ldots, 2n+1\}$.

If $p \in M$ and $\pi \subset T_pM$ is a nondegenerate 2-plane, then the *sectional curvature* is defined as [21]

$$\mathcal{K}(\pi) = \frac{g(S(X, Y)Y, X)}{g(X, X)g(Y, Y) - g^2(X, Y)}, \tag{2.8}$$

where $\{X, Y\}$ is a basis of π.

The *scalar curvature* τ of (M, ∇, g) at a point $p \in M$ is defined by

$$\tau(p) = \sum_{1 \leq i < j \leq m+1} g(S(e_i, e_j)e_j, e_i), \tag{2.9}$$

where $\{e_1, \ldots, e_{m+1}\}$ is an orthonormal basis at p.

If $L \subset T_pM$ is an r-dimensional subspace, then the scalar curvature of L is defined by

$$\tau(L) = \sum_{1 \leq \alpha < \beta \leq r} \mathcal{K}(e_\alpha \wedge e_\beta), \tag{2.10}$$

where $\{e_1, \ldots, e_r\} \subset L$ is an orthonormal basis.

Let $k \in \mathbb{N}^*$ and $n_1, \ldots, n_k \geq 2$ be integers such that $n_1 < m+1$ and $n_1 + \ldots + n_k \leq m+1$. For each k-tuple $(n_1, \ldots, n_k)$ and any $p \in M$, the Chen invariant $\delta(n_1, \ldots, n_k)$ is defined by

$$\delta(n_1, \ldots, n_k)(p) = \tau(p) - inf\{\tau(L_1) + \ldots + \tau(L_k)\},$$

where $L_1, \ldots, L_k$ are mutually orthogonal subspaces of T_pM such that $dim\ L_i = n_i, \forall\, i = 1, \ldots, k$.

In particular, $\delta(2) = \tau - inf\mathcal{K}$ is the Chen first invariant.

B.-Y. Chen proved in [8] a fundamental inequality involving the general Chen invariants (intrinsic invariants) and the squared mean curvature (extrinsic invariant) for submanifolds M^n in Riemannian space forms $\bar{M}^m(c)$:

$$\delta(n_1, \ldots, n_k) \leq \frac{n^2(n + k - \sum_{i=1}^k n_i - 1)}{2(n + k - \sum_{i=1}^k n_i)} \|H\|^2 + \frac{1}{2}[n(n-1) - \sum_{i=1}^k n_i(n_i - 1)]c.$$

These inequalities are known as Chen inequalities. General Chen inequalities for statistical submanifolds in Hessian manifolds of constant Hessian curvature are established in [31].

Next, we consider the following result.

Lemma 2.1 ([31]) *Let $m \geq 2$, $k \geq 1$ be two integers, and let $n_1, n_2, \ldots, n_k \geq 2$ be integers such that $n_1 < m+1$, $n_1 + \ldots + n_k \leq m+1$. Denote $N_0 = 0$, $N_i = n_1 + \ldots + n_i$ for $i = 1, \ldots, k$. Then, for any real numbers $a_1, \ldots, a_{m+1}$, we have*

$$\sum_{1 \leq i < j \leq m+1} a_i a_j - \sum_{i=1}^{k} \sum_{N_{i-1}+1 \leq \alpha_i < \beta_i \leq N_i} a_{\alpha_i} a_{\beta_i} \leq \frac{m+k-\sum_{i=1}^{k} n_i}{2(m+1+k-\sum_{i=1}^{k} n_i)} \left(\sum_{j=1}^{m+1} a_j \right)^2 . \tag{2.11}$$

Moreover, the equality holds if and only if

$$\sum_{\alpha_i = N_{i-1}+1}^{N_i} a_{\alpha_i} = a_{N_k+1} = \ldots = a_{m+1}, \ \forall\, i = 1, \ldots, k.$$

3 Main Inequality

Let $(\bar{M}, \bar{\nabla}, \bar{g}, \phi, \xi)$ be a $(2n+1)$-dimensional Kenmotsu statistical manifold of constant ϕ-sectional curvature c, and let M be an $(m+1)$-dimensional statistical submanifold of $(\bar{M}, \bar{\nabla}, \bar{g}, \phi, \xi)$.

Let Q be the Hessian curvature tensor for the connection ∇, that is,

$$Q(X, Y) = [K_X, K_Y],$$

for any $X, Y \in \Gamma(TM)$. Then it is clear that we have (see also [31])

$$R(X, Y) + R^*(X, Y) = 2R^0(X, Y) + 2Q(X, Y).$$

Let π be a plane in T_pM, for $p \in M$. Take an orthonormal basis $\{X, Y\}$ of π, and define the sectional Q-curvature $\mathcal{K}^Q(\pi)$ of the plane section π by [34]

$$\begin{aligned} \mathcal{K}^Q(\pi) &= g(Q(X, Y)Y, X) \\ &= \frac{1}{2}\{g(R(X, Y)Y, X) + g(R^*(X, Y)Y, X) - 2g(R^0(X, Y)Y, X)\}. \end{aligned} \tag{3.1}$$

Then $\mathcal{K}^Q(\pi)$ can be written also

$$\mathcal{K}^Q(\pi) = g(S(X, Y)Y, X) - g(R^0(X, Y)Y, X). \tag{3.2}$$

We denote by

$$\mathcal{K}_0 = \mathcal{K}_0(X, Y) = g(R^0(X, Y)Y, X)$$

the sectional curvature of the Levi-Civita connection ∇^0 on M and by h^0 the second fundamental form of M.

In this chapter, we assume that the structure vector field ξ is tangent to the submanifold M. Then, we consider $\{e_1, \ldots, e_m, e_{m+1} = \xi\}$ and $\{e_{m+2}, \ldots, e_{2n+1}\}$ orthonormal bases of T_pM and $T_p^{\perp}M$, respectively, for any $p \in M$.

The Q-scalar curvature of M, denoted by τ^Q, corresponding to the sectional Q-curvature of M is defined by

$$\begin{aligned}
\tau^Q &= \sum_{1\le i<j\le m+1} \mathcal{K}^Q(e_i \wedge e_j)\\
&= \frac{1}{2} \sum_{1\le i<j\le m+1} [g(R(e_i, e_j)e_j, e_i) + g(R^*(e_i, e_j)e_j, e_i)\\
&- 2g(R^0(e_i, e_j)e_j, e_i)]\\
&= \sum_{1\le i<j\le m+1} g(S(e_i, e_j)e_j, e_i) - \sum_{1\le i<j\le m+1} g(R^0(e_i, e_j)e_j, e_i).
\end{aligned}$$

Then, τ^Q becomes

$$\tau^Q = \tau - \tau_0, \tag{3.3}$$

where τ is the statistical scalar curvature of ∇ on M and τ_0 is the scalar curvature of the Levi-Civita connection ∇^0 on M.

Next, we denote $\phi X = PX + FX$, where PX is the tangent component of ϕX and FX is the normal component of ϕX.

Equations (2.4) and (2.6) imply

$$\begin{aligned}
\tau &= \frac{m(m+1)(c-3)}{8} + \frac{3(c+1)}{8}\|P\|^2 - \frac{(c+1)m}{4}\\
&+ \frac{1}{2}\sum_{\alpha=m+2}^{2n+1} \sum_{1\le i<j\le m+1} (h_{ii}^{*\alpha} h_{jj}^{\alpha} + h_{ii}^{\alpha} h_{jj}^{*\alpha} - 2h_{ij}^{\alpha} h_{ij}^{*\alpha}),
\end{aligned} \tag{3.4}$$

where $\|P\|^2$ denotes the squared norm of P, given by (see [27, 28])

$$\|P\|^2 = \sum_{1\le i,j\le m+1} \bar{g}^2(e_i, Pe_j).$$

Furthermore, summing over $1 \le i < j \le m+1$ for $X = W = e_i$ and $Y = Z = e_j$ in the Gauss equation (2.7), we get

$$
\begin{aligned}
2\tau_0 = 2\bar{\tau}_0 &+ \frac{1}{2}\sum_{\alpha=m+2}^{2n+1}\sum_{1\le i<j\le m+1}[h_{ii}^{\alpha}h_{jj}^{\alpha} - (h_{ij}^{\alpha})^2] \\
&+ \frac{1}{2}\sum_{\alpha=m+2}^{2n+1}\sum_{1\le i<j\le m+1}[h_{ii}^{*\alpha}h_{jj}^{*\alpha} - (h_{ij}^{*\alpha})^2] \\
&+ \frac{1}{2}\sum_{\alpha=m+2}^{2n+1}\sum_{1\le i<j\le m+1}(h_{jj}^{\alpha}h_{ii}^{*\alpha} + h_{ii}^{\alpha}h_{jj}^{*\alpha} - 2h_{ij}^{\alpha}h_{ij}^{*\alpha}).
\end{aligned} \tag{3.5}
$$

Then, using Eqs. (3.4) and (3.5) in Eq. (3.3), we obtain

$$
\begin{aligned}
\tau^{Q} &= (\tau - 2\tau_0) + \tau_0 \\
&= \frac{m(m+1)(c-3)}{8} + \frac{3(c+1)}{8}\|P\|^2 - \frac{(c+1)m}{4} \\
&+ \tau_0 - 2\bar{\tau}_0 - \frac{1}{2}\sum_{\alpha=m+2}^{2n+1}\sum_{1\le i<j\le m+1}[h_{ii}^{\alpha}h_{jj}^{\alpha} - (h_{ij}^{\alpha})^2] \\
&- \frac{1}{2}\sum_{\alpha=m+2}^{2n+1}\sum_{1\le i<j\le m+1}[h_{ii}^{*\alpha}h_{jj}^{*\alpha} - (h_{ij}^{*\alpha})^2].
\end{aligned} \tag{3.6}
$$

On the other hand, let $L_1, \ldots, L_k$ be k mutually orthogonal subspaces of T_pM, $dim\ L_i = n_i$, for any $i = 1, \ldots, k$, defined by

$$
\begin{aligned}
L_1 &= sp\{e_1, \ldots, e_{n_1}\}, \\
L_2 &= sp\{e_{n_1+1}, \ldots, e_{n_1+n_2}\}, \\
&\ldots\ldots\ldots\ldots \\
L_k &= sp\{e_{n_1+\ldots+n_{k-1}+1}, \ldots, e_{n_1+\ldots+n_{k-1}+n_k}\}.
\end{aligned}
$$

Moreover, we denote $N_0 = 0$, $N_i = n_1 + \ldots + n_i$ for $i = 1, \ldots, k$.
From the Gauss equations (2.6) and (2.7), we obtain

$$
\begin{aligned}
\tau^{Q}(L_i) = \frac{1}{2} &\sum_{N_{i-1}+1\le\alpha_i<\beta_i\le N_i}[g(R(e_{\alpha_i}, e_{\beta_i})e_{\beta_i}, e_{\alpha_i}) + g(R^*(e_{\alpha_i}, e_{\beta_i})e_{\beta_i}, e_{\alpha_i})] \\
&- \sum_{N_{i-1}+1\le\alpha_i<\beta_i\le N_i} g(R^0(e_{\alpha_i}, e_{\beta_i})e_{\beta_i}, e_{\alpha_i})
\end{aligned}
$$

$$
\begin{aligned}
&= \sum_{N_{i-1}+1\le\alpha_i<\beta_i\le N_i} g(S(e_{\alpha_i}, e_{\beta_i})e_{\beta_i}, e_{\alpha_i})\\
&- \sum_{N_{i-1}+1\le\alpha_i<\beta_i\le N_i} g(R^0(e_{\alpha_i}, e_{\beta_i})e_{\beta_i}, e_{\alpha_i})\\
&= \tau(L_i) - \tau_0(L_i)\\
&= \tau_0(L_i) - 2\bar{\tau}_0(L_i) + \frac{n_i(n_i-1)(c-3)}{8} + \frac{3(c+1)}{4}\Psi(L_i)\\
&- \frac{(c+1)(n_i-1)}{4}\\
&-\frac{1}{2}\sum_{r=m+2}^{2n+1}\sum_{N_{i-1}+1\le\alpha_i<\beta_i\le N_i}[h^r_{\alpha_i\alpha_i}h^r_{\beta_i\beta_i} - (h^r_{\alpha_i\beta_i})^2]\\
&-\frac{1}{2}\sum_{r=m+2}^{2n+1}\sum_{N_{i-1}+1\le\alpha_i<\beta_i\le N_i}[h^{*r}_{\alpha_i\alpha_i}h^{*r}_{\beta_i\beta_i} - (h^{*r}_{\alpha_i\beta_i})^2],
\end{aligned}
$$

where we denote $\Psi(L_i) = \sum_{\alpha_i<\beta_i}\bar{g}^2(Pe_{\alpha_i}, e_{\beta_i})$. By summing over $i = 1, \ldots, k$ the latest relation, we get

$$
\begin{aligned}
\sum_{i=1}^{k}\tau^Q(L_i) = &\sum_{i=1}^{k}[\tau_0(L_i) - 2\bar{\tau}_0(L_i) + \frac{n_i(n_i-1)(c-3)}{8} + \frac{3(c+1)}{4}\Psi(L_i)\\
&- \frac{(c+1)(n_i-1)}{4}]\\
&-\frac{1}{2}\sum_{r=m+2}^{2n+1}\sum_{i=1}^{k}\sum_{N_{i-1}+1\le\alpha_i<\beta_i\le N_i}[h^r_{\alpha_i\alpha_i}h^r_{\beta_i\beta_i} - (h^r_{\alpha_i\beta_i})^2] \qquad (3.7)\\
&-\frac{1}{2}\sum_{r=m+2}^{2n+1}\sum_{i=1}^{k}\sum_{N_{i-1}+1\le\alpha_i<\beta_i\le N_i}[h^{*r}_{\alpha_i\alpha_i}h^{*r}_{\beta_i\beta_i} - (h^{*r}_{\alpha_i\beta_i})^2].
\end{aligned}
$$

By subtracting Eq. (3.7) from Eq. (3.6), we obtain

$$
\begin{aligned}
[\tau^Q - \sum_{i=1}^{k}\tau^Q(L_i)] - [\tau_0 - \sum_{i=1}^{k}\tau_0(L_i)] = &\frac{m(m+1)(c-3)}{8} + \frac{3(c+1)}{8}\|P\|^2\\
&- \frac{(c+1)m}{4}
\end{aligned}
$$

$$
\begin{aligned}
&-\frac{1}{2}\sum_{r=m+2}^{2n+1}\sum_{1\le i<j\le m+1}\left\{[h_{ii}^{r}h_{jj}^{r}-(h_{ij}^{r})^{2}]+[h_{ii}^{*r}h_{jj}^{*r}-(h_{ij}^{*r})^{2}]\right\}\\
&+\frac{1}{2}\sum_{r=m+2}^{2n+1}\sum_{i=1}^{k}\sum_{N_{i-1}+1\le\alpha_i<\beta_i\le N_i}\{[h_{\alpha_i\alpha_i}^{r}h_{\beta_i\beta_i}^{r}-(h_{\alpha_i\beta_i}^{r})^{2}]\\
&+[h_{\alpha_i\alpha_i}^{*r}h_{\beta_i\beta_i}^{*r}-(h_{\alpha_i\beta_i}^{*r})^{2}]\}\\
&-2\bar{\tau}_0+2\sum_{i=1}^{k}\bar{\tau}_0(L_i)-\frac{c-3}{8}\sum_{i=1}^{k}n_i(n_i-1)-\frac{3(c+1)}{4}\sum_{i=1}^{k}\Psi(L_i)\\
&+\frac{(c+1)}{4}\sum_{i=1}^{k}(n_i-1).
\end{aligned}
$$

Next, by using Lemma 2.1, we have

$$
\begin{aligned}
&\sum_{1\le i<j\le m+1}h_{ii}^{r}h_{jj}^{r}-\sum_{i=1}^{k}\sum_{N_{i-1}+1\le\alpha_i<\beta_i\le N_i}h_{\alpha_i\alpha_i}^{r}h_{\beta_i\beta_i}^{r}\\
&\qquad\le\frac{m+k-\sum_{i=1}^{k}n_i}{2\left(m+1+k-\sum_{i=1}^{k}n_i\right)}\left(\sum_{i=1}^{m+1}h_{ii}^{r}\right)^{2}
\end{aligned}
\tag{3.8}
$$

and

$$
\begin{aligned}
&\sum_{1\le i<j\le m+1}h_{ii}^{*r}h_{jj}^{*r}-\sum_{i=1}^{k}\sum_{N_{i-1}+1\le\alpha_i<\beta_i\le N_i}h_{\alpha_i\alpha_i}^{*r}h_{\beta_i\beta_i}^{*r}\\
&\qquad\le\frac{m+k-\sum_{i=1}^{k}n_i}{2\left(m+1+k-\sum_{i=1}^{k}n_i\right)}\left(\sum_{i=1}^{m+1}h_{ii}^{*r}\right)^{2}.
\end{aligned}
\tag{3.9}
$$

Applying the expressions of (3.8) and (3.9) in the latest equation, we establish the general Chen inequalities for arbitrary statistical submanifolds of a Kenmotsu statistical manifold of constant ϕ-sectional curvature c, as follows.

Theorem 3.1 *Let $(\bar{M},\bar{\nabla},\bar{g},\phi,\xi)$ be a $(2n+1)$-dimensional Kenmotsu statistical manifold of constant ϕ-sectional curvature c, and M is an $(m+1)$-dimensional* *statistical submanifold of* $(\bar{M}, \bar{\nabla}, \bar{g}, \phi, \xi)$*. Then, for any integers* $n_1, \ldots, n_k \geq 2$*,* $k \in \mathbb{N}^*$*, such that* $n_1 < m+1$*,* $n_1 + \ldots + n_k \leq m+1$*, the following inequalities hold:*

$$(\tau^Q - \tau_0) - \left[\sum_{i=1}^{k} \tau^Q(L_i) - \sum_{i=1}^{k} \tau_0(L_i)\right] \geq \frac{m(m+1)(c-3)}{8} + \frac{3(c+1)}{8}\|P\|^2$$
$$- \frac{(c+1)m}{4} - \frac{(m+1)^2(m+k-\sum_{i=1}^{k} n_i)(\|H\|^2 + \|H^*\|^2)}{4(m+1+k-\sum_{i=1}^{k} n_i)}$$
$$- 2[\bar{\tau}_0 - \sum_{i=1}^{k} \bar{\tau}_0(L_i)]$$
$$- \frac{c-3}{8}\sum_{i=1}^{k} n_i(n_i-1) - \frac{3(c+1)}{4}\sum_{i=1}^{k} \Psi(L_i) + \frac{(c+1)}{4}\sum_{i=1}^{k}(n_i-1), \tag{3.10}$$

where $L_1, \ldots, L_k$ *are mutually orthogonal subspaces of* T_pM*, with* $dim\ L_i = n_i$*,* $\forall\, i = 1, \ldots, k$*.*

Moreover, the equality case holds in (3.10) *at a point* $p \in M$ *if and only if there exist* $\{e_1, \ldots, e_{m+1}\}$ *and* $\{e_{m+2}, \ldots, e_{2n+1}\}$ *orthonormal bases of* T_pM *and* $T_p^{\perp}M$*, respectively, such that the shape operators have the expressions:*

$$A_{e_r} = \begin{pmatrix} \begin{matrix} A_1^r & \ldots & 0 \\ \vdots & \ddots & \vdots \\ 0 & \ldots & A_k^r \end{matrix} & 0 \\ 0 & \mu_r I \end{pmatrix}, \quad A_{e_r}^* = \begin{pmatrix} \begin{matrix} A_1^{*r} & \ldots & 0 \\ \vdots & \ddots & \vdots \\ 0 & \ldots & A_k^{*r} \end{matrix} & 0 \\ 0 & \mu_r^* I \end{pmatrix},$$

for $r = m+2, \ldots, 2n+1$*, where* I *is the identity matrix, and* A_i^r *and* A_i^{*r} *are symmetric* $n_i \times n_i$ *submatrices with* $trace\ A_i^r = \mu_r$ *and* $trace\ A_i^{*r} = \mu_r^*$*, for all* $i = 1, \ldots, k$*.*

We remark that the equality case of (3.10) follows immediately from the equality case of Lemma 2.1. We would like to note that the particular case of the above theorem for $k = 1$ and $n_1 = 2$ was established in [20].

We notice also that if $(\bar{M} = N \times \mathbb{R}, \bar{\nabla} = \bar{\nabla}^0 + \bar{K}, \bar{g}, \phi, \xi)$ is a Kenmotsu statistical manifold of constant ϕ-sectional curvature c, where $(N, \tilde{\nabla}, \tilde{g}, J)$ is a holomorphic statistical manifold, then $c = -1$ [22]. In this respect, we have the following result.

Corollary 3.2 *Let* $(\bar{M} = N \times \mathbb{R}, \bar{\nabla} = \bar{\nabla}^0 + \bar{K}, \bar{g}, \phi, \xi)$ *be a* $(2n+1)$*-dimensional Kenmotsu statistical manifold of constant* ϕ*-sectional curvature, where* $(N, \tilde{\nabla}, \tilde{g}, J)$ *is a holomorphic statistical manifold, and let* M *be an* $(m+1)$*-dimensional statistical submanifold of* $\bar{M}$*. Then, for any integers* $k \in \mathbb{N}^*$ *and* $n_1, \ldots, n_k \geq 2$

such that $n_1 < m+1$, $n_1 + \ldots + n_k \leq m+1$, *the following inequalities hold:*

$$(\tau^Q - \tau_0) - \left[\sum_{i=1}^{k} \tau^Q(L_i) - \sum_{i=1}^{k} \tau_0(L_i) \right] \geq -\frac{m(m+1)}{2} \tag{3.11}$$

$$- \frac{(m+1)^2(m+k-\sum_{i=1}^{k} n_i)(\|H\|^2 + \|H^*\|^2)}{4\left(m+1+k-\sum_{i=1}^{k} n_i\right)}$$

$$- 2\left[\bar{\tau}_0 - \sum_{i=1}^{k} \bar{\tau}_0(L_i) \right] + \frac{1}{2}\sum_{i=1}^{k} n_i(n_i - 1),$$

where $L_1, \ldots, L_k$ *are mutually orthogonal subspaces of* T_pM, *with* $dim\ L_i = n_i$, $\forall\, i = 1, \ldots, k$.

4 An Example

Let $(H^{2n+1}, \bar{\nabla} = \bar{\nabla}^0 + \bar{K}, \bar{g}, \phi, \xi)$ be the $(2n+1)$-dimensional Kenmotsu statistical manifold studied in [22, Examples 3.3 and 3.10]. We have

$$H^{2n+1} = \{(x_1, \ldots, x_n, y_1, \ldots, y_n, z) \in \mathbb{R}^{2n+1} | z > 0\}.$$

The structure tensors (g, ϕ, ξ) are defined by

$$\bar{g} = \frac{1}{z^2}\{(dx_1)^2 + \ldots + (dx_n)^2 + (dy_1)^2 + \ldots + (dy_n)^2 + (dz)^2\},$$

$$\phi\frac{\partial}{\partial x_\alpha} = \frac{\partial}{\partial y_\alpha},\ \phi\frac{\partial}{\partial y_\alpha} = -\frac{\partial}{\partial x_\alpha},\ \phi\frac{\partial}{\partial z} = 0,\ for\ \alpha \in \{1, \ldots, n\},$$

and

$$\xi = -z\frac{\partial}{\partial z}.$$

For any $X, Y \in \Gamma(TH^{2n+1})$, we set the $(1, 2)$-tensor field $\bar{K}$ as

$$\bar{K}(X, Y) = \nu\, \bar{\eta}(X)\, \bar{\eta}(Y)\, \xi,$$

where $\nu \in C^\infty(H^{2n+1})$ and $\bar{\eta}$ is the 1-form on H^{2n+1} dual to ξ, that is, $\bar{\eta}(X) = g(X, \xi)$.

Then $(H^{2n+1}, \bar{\nabla} = \bar{\nabla}^0 + \bar{K}, g, \phi, \xi)$ is a Kenmotsu statistical manifold with constant ϕ-sectional curvature $c = -1$. Moreover, $M = H^{2k+1}$ (where $0 < k < n$)

is a statistical submanifold of the Kenmotsu statistical manifold H^{2n+1} for which the inequality (3.10) holds with equality case.

5 Conclusions

In this chapter, we established general Chen inequalities for statistical submanifolds in Kenmotsu statistical manifolds of constant ϕ-sectional curvature. Moreover, the equality case of these inequalities is examined, and an example is revealed to highlight our results.

References

1. Alodan, H., Chen, B.Y., Deshmukh, S., Vîlcu, G.E.: A generalized Wintgen inequality for quaternionic CR-submanifolds. Rev. Real Acad. Cienc. Exactas Fis. Nat. Ser. A-Mat. **114**, 129 (2020)
2. Amari, S.: Differential-geometrical methods in statistics. In: Lecture Notes in Statistics, vol. 28. Springer, Berlin (1985)
3. Aquib, M., Lone, M.S., Neacşu, C., Vîlcu, G.E.: On δ-Casorati curvature invariants of Lagrangian submanifolds in quaternionic Kähler manifolds of constant q-sectional curvature. Rev. Real Acad. Cienc. Exactas Fis. Nat. Ser. A-Mat. **117**, 107 (2023)
4. Brubaker, N.D., Suceavă, B.D.: A geometric interpretation of Cauchy-Schwarz inequality in terms of Casorati curvature. Int. Electron. J. Geom. **11**(1), 48–51 (2018)
5. Chen, B.Y.: Some pinching and classification theorems for minimal submanifolds. Arch. Math. **60**, 568–578 (1993)
6. Chen, B.Y.: Relations between Ricci curvature and shape operator for submanifolds with arbitrary codimension. Glasgow Math. J. **41**, 33–41 (1999)
7. Chen, B.Y.: Riemannian DNA, inequalities and their applications. Tamkang J. Sci. Eng. **3**(3), 123–130 (2000)
8. Chen, B.Y.: Some new obstructions to minimal and Lagrangian isometric immersions. Japan. J. Math. **26**, 105–127 (2000)
9. Chen, B.Y.: Pseudo-Riemannian Geometry, δ-Invariants and Applications. World Scientific, Hackensack (2011)
10. Chen, B.Y.: Recent developments in δ-Casorati curvature invariants. Turk. J. Math. **45**, 1–46 (2021)
11. Chen, B.Y.: Recent developments in Wintgen inequality and Wintgen ideal submanifolds. Int. Electron. J. Geom. **20**, 6–45 (2021)
12. Chen, B.Y., Blaga, A.M.: Recent developments on Chen–Ricci inequalities in differential geometry. In: Geometry of Submanifolds and Applications. Springer, Berlin (2023/2024)
13. Chen, B.Y., Blaga, A.M., Vîlcu, G.E.: Differential geometry of submanifolds in complex space forms involving δ-invariants. Mathematics **10**, 591 (2022)
14. Chen, B.Y., Mihai, A., Mihai, I.: A Chen first inequality for statistical submanifolds in Hessian manifolds of constant Hessian curvature. Results Math. **74**, 165 (2019)
15. Chen, B.Y., Vîlcu, G.E.: Recent developments on the first chen inequality in differential geometry. Mathematics **11**, 4186 (2023)
16. Decu, S.: Optimal inequalities for submanifolds in quaternion-space-forms with semi-symmetric metric connection. Bull. Transilv. Univ. Braşov **2**, 175–184 (2009)

17. Decu, S., Haesen, S.: Chen inequalities for spacelike submanifolds in statistical manifolds of type para-Kähler space forms. Mathematics **10**, 330 (2022)
18. Decu, S., Haesen, S., Verstraelen, L., Vîlcu, G.E.: Curvature invariants of statistical submanifolds in Kenmotsu statistical manifolds of constant ϕ-sectional curvature. Entropy **20**, 529 (2018)
19. Decu, S., Vîlcu, G.E.: Casorati inequalities for statistical submanifolds in Kenmotsu statistical manifolds of constant ϕ-sectional curvature with semi-symmetric metric connection. Entropy **24**, 800 (2022)
20. Decu, S., Vîlcu, G.E.: Chen inequality for statistical submanifolds in Kenmotsu statistical manifolds of constant ϕ-sectional curvature. In: Recent Advances in Differential Geometry and Related Areas. Contemporary Mathematics **821**, 67–76 (2025)
21. Furuhata, H., Hasegawa, I.: Submanifold theory in holomorphic statistical manifolds. In: Geometry of Cauchy-Riemann Submanifolds, pp. 179–214. Springer, Singapore (2016)
22. Furuhata, H., Hasegawa, I., Okuyama, Y., Sato, K.: Kenmotsu statistical manifolds and warped product. J. Geom. **108**, 1175–1191 (2017)
23. Furuhata, H., Hasegawa, I., Satoh, N.: Chen invariants and statistical submanifolds. Commun. Korean Math. Soc. **37**(3), 851–864 (2022)
24. Gielis, J., Haesen, S., Verstraelen, L.: Universal natural shapes. Krag. J. Math. **28**, 57–68 (2005)
25. Haesen, S., Verstraelen, L.: Ideally embedded space-times. J. Math. Phys. **45**, 1497–1510 (2003)
26. Kenmotsu, K.: A class of almost contact Riemannian manifolds. Tohoku Math. J. **2**, 93–103 (1972)
27. Lee, C.W., Lee, J.W., Vîlcu, G.E.: Optimal inequalities for the normalized δ-Casorati curvatures of submanifolds in Kenmotsu space forms. Adv. Geom. **17**(3), 355–362 (2017)
28. Lone, M.A., Shahid, M.H., Vîlcu, G.E.: On Casorati curvatures of submanifolds in pointwise Kenmotsu space forms. Math. Phys. Anal. Geom. **22**, 2 (2019)
29. Mihai, A.: Modern Topics in Submanifold Theory. Bucharest University Press, Bucharest (2006)
30. Mihai, I.: Statistical manifolds and their submanifolds. Results on Chen-like invariants. Geometry of Submanifolds: AMS Special Session in Honor of Bang-Yen Chen's 75th Birthday, 20–21 October 2018, University of Michigan, Ann Arbor, Michigan, vol. 756, Contemporary Mathematics, Singapore (2020)
31. Mihai, I., Mihai, R.I.: General Chen inequalities for statistical submanifolds in Hessian manifolds of constant Hessian curvature. Mathematics **10**, 3061 (2022)
32. Nash, J.F.: The imbedding problem for Riemannian manifolds. Ann. Math. **63**, 20–63 (1956)
33. Neacşu, C.D.: On some optimal inequalities for statistical submanifolds of statistical space forms. Politehn. Univ. Bucharest Sci. Bull. Ser. A Appl. Math. Phys. **85**(1), 107–118 (2023)
34. Opozda, B.: A sectional curvature for statistical structures. Linear Algebra Appl. **497**, 134–161 (2016)
35. Osserman, R.: Curvature in the eighties. Am. Math. Monthly **97**, 731–754 (1990)
36. Pitiş, G.: Contact forms in geometry and topology. In: Topics in Modern Differential Geometry, Atlantis Press, Paris (2017)
37. Suceavă, B.D.: Remarks on B.Y. Chen's inequality involving classical invariants. An. Ştiinţ. Univ. Al. I. Cuza Iaşi. Mat. (N.S.) **45**, 405–412 (1999)
38. Suceavă, B.D., Vajiac, M.B.: Estimates of B.Y. Chen's δ-invariant in terms of Casorati curvature and mean curvature for strictly convex Euclidean hypersurfaces. Intern. Electron. J. Geom. **12**(1), 26–31 (2019)
39. Vîlcu, G.E.: Almost product structures on statistical manifolds and para-Kähler-like statistical submersions. Bull. Sci. Math. **171**, 103018 (2021)
40. Vos, P.: Fundamental equations for statistical submanifolds with applications to the Bartlett correction. Ann. Inst. Statist. Math. **41**, 429–450 (1989)

B. Y. Chen Inequalities for Pointwise Quasi Hemi-slant Submanifolds of a Kaehler Manifold

Nergiz (Önen) Poyraz, Mehmet Akif Akyol, and Erol Yaşar

Abstract In this study, we establish Chen-type inequalities for pointwise quasi hemi-slant submanifolds within Kaehler manifolds, giving explicit relations among the mean curvature, scalar curvature, sectional curvature, Ricci curvature, and the ambient space's sectional curvature. Moreover, we characterize the conditions under which these inequalities become equalities.

Keywords Pointwise quasi hemi-slant submanifold · Chen inequality · Kaehler manifold · Complex space form · Second fundamental form

1 Introduction

The theory of submanifolds is a central topic in differential geometry, with significant applications in diverse areas such as mathematical physics, image processing, economic modeling, and computer-aided design. Within complex geometry, one particularly rich research direction concerns slant submanifolds, introduced by Chen in [5] as a generalization of holomorphic and totally real submanifolds, and further summarized in [6]. Over the years, many extensions of this concept have been proposed, including semi-slant, hemi-slant, bi-slant, and quasi bi-slant

---

N. (Önen) Poyraz
Faculty of Arts and Sciences, Department of Mathematics, Çukurova University, Adana, Turkey
e-mail: nonen@cu.edu.tr

M. A. Akyol (✉)
Faculty of Engineering and Natural Sciences, Department of Mathematics, Uşak University, Uşak, Turkey
e-mail: mehmet.akyol@usak.edu.tr

E. Yaşar
Faculty of Science and Arts, Department of Mathematics, Mersin University, Mersin, Turkey
e-mail: yerol@mersin.edu.tr

B.-Y. Chen, M. A. Choudhary (eds.), *Geometric Inequalities and Applications*,
Infosys Science Foundation Series in Mathematical Sciences,
https://doi.org/10.1007/978-981-95-5148-4_15

submanifolds, each studied extensively (see [1, 3, 4, 9, 11, 13, 14, 18–21, 23, 24, 26–28, 30, 33]).

In 2002, Chen and Garay [16], inspired by Etayo's definition of quasi-slant submanifolds [17], introduced the notion of pointwise slant submanifolds in Hermitian geometry. Later, Şahin [29] defined pointwise semi-slant submanifolds. More recently, Akyol and Beyendi [2] proposed the concept of pointwise quasi hemi-slant submanifolds as a natural generalization encompassing slant, semi-slant, hemi-slant, bi-slant, and quasi bi-slant submanifolds.

In Riemannian geometry, one of the most notable curvature invariants for a Riemannian manifold (M_1, g_1) is the Chen invariant, introduced by Chen [8] as

$$\delta_{M_1} = \tau(p) - \inf(K)(p),$$

where $\tau(p)$ is the scalar curvature of M_1 and

$$\inf(K)(p) = \inf\{K(\Pi) : \Pi \text{ is a plane section of } T_pM_1\}$$

is the infimum of the sectional curvatures at the point $p \in M_1$.

A fundamental problem in submanifold theory is to discover simple yet meaningful relationships between the intrinsic and extrinsic invariants of a submanifold. In this regard, Chen established a series of influential inequalities—now known as Chen inequalities—in [7, 10]. In recent years, Chen-like inequalities have been investigated for various classes of Riemannian submanifolds (see [15, 22, 25, 31, 32, 35]).

The present work focuses on pointwise quasi hemi-slant submanifolds of Kähler manifolds. We establish several inequalities involving the mean curvature, scalar curvature, sectional curvature, Ricci curvature, and the sectional curvature of the ambient space. In addition, we analyze the conditions under which the equality cases occur.

The structure of the chapter is as follows: Sect. 2 reviews the necessary preliminaries, including basic definitions and fundamental formulas. Section 3 presents the main results, deriving inequalities for pointwise quasi hemi-slant submanifolds of Kaehler manifolds and examining the equality cases.

2 Preliminaries

Let $\tilde{M}_1$ be a smooth manifold of dimension $2m_1$. Then, $\tilde{M}_1$ is said to be an almost Hermitian manifold if it admits a tensor field J_1 of type (1, 1) and a Riemannian metric $\tilde{h}$ on $\tilde{M}_1$ satisfying

$$J_1^2 = -I, \quad \tilde{h}(J_1E_1, J_1E_2) = \tilde{h}(E_1, E_2) \tag{2.1}$$

for any vector fields E_1, E_2 on $T\tilde{M}_1$, where I denotes the identity transformation. The fundamental 2-form Ω on $\tilde{M}_1$ is defined by $\Omega(E_1, E_2) = \tilde{h}(E_1, J_1E_2)$, $\forall E_1, E_2 \in \Gamma(T\tilde{M}_1)$, with $\Gamma(T\tilde{M}_1)$ being the section of tangent bundle $T\tilde{M}_1$ of $\tilde{M}_1$. An almost Hermitian manifold $\tilde{M}_1$ is called a Kaehler manifold [34] if

$$(\tilde{\nabla}_{E_1} J_1)E_2 = 0, \tag{2.2}$$

where $\tilde{\nabla}$ is the Levi-Civita connection on $\tilde{M}_1$ with respect to h. Let M_2 be a Riemannian manifold isometrically immersed in $\tilde{M}_1$, and the induced Riemannian metric on M_2 is denoted by the same symbol $\tilde{h}$ throughout this chapter. Let $\mathcal{A}$ and h denote the shape operator and second fundamental form, respectively, of immersion of M_2 into $\tilde{M}_1$. The Gauss and Weingarten formulas of M_2 into $\tilde{M}_1$ are given by [12]

$$\tilde{\nabla}_{E_1} E_2 = \nabla_{E_1} E_2 + h(E_1, E_2) \tag{2.3}$$

and

$$\tilde{\nabla}_{E_1} F_2 = -A_{F_2}E_1 + \nabla^{\perp}_{E_1} F_2, \tag{2.4}$$

for any vector fields $E_1, E_2 \in \Gamma(TM_2)$ and $F_2 \in \Gamma(T^{\perp}M_2)$, where ∇ is the induced connection on M_2, $\nabla^{\perp}$ represents the connection on the normal bundle $T^{\perp}M_2$ of M_2, and A_{F_2} is the shape operator of M_2 with respect to normal vector $F_2 \in \Gamma(T^{\perp}M_2)$. Moreover, $\mathcal{A}_{F_2}$ and h are related by

$$\tilde{h}(h(E_1, E_2), F_2) = \tilde{h}(A_{F_2}E_1, E_2) \tag{2.5}$$

for any vector fields $E_1, E_2 \in \Gamma(TM_2)$ and $F_2 \in \Gamma(T^{\perp}M_2)$. The mean curvature vector H is given by $H = \frac{1}{n}\,\mathrm{trace}(h)$. The submanifold M_2 is totally geodesic if $h = 0$ and minimal if $H = 0$. The Gauss equation is given by

$$\begin{aligned}\tilde{R}(E_1, E_2, E_3, E_4) = R(E_1, E_2, E_3, E_4) &- \tilde{h}(h(E_1, E_4), h(E_2, E_3)) \\ &+ \tilde{h}(h(E_1, E_3), h(E_2, E_4))\end{aligned} \tag{2.6}$$

for all $E_1, E_2, E_3, E_4 \in \Gamma(TM_2)$, where R is the curvature tensor of M_2. Let $\Pi = Span\{e_i, e_j\}$ be 2-dimensional nondegenerate plane of the tangent space T_pM_2 at $p \in M_2$. Then the number

$$K_{ij} = \frac{\tilde{h}(R(e_j, e_i)e_i, e_j)}{\tilde{h}(e_i, e_i)\tilde{h}(e_j, e_j) - \tilde{h}(e_i, e_j)^2} \tag{2.7}$$

is called the sectional curvature of section Π at $p \in M_2$. Let M_2 be an m_2-dimensional Riemannian manifold. We denote by $K(\pi)$ the sectional curvature of

M_2 associated with a plane section $\pi \subset T_pM_2$, $p \in M_2$. If $\{e_1, \ldots, e_{m_2}\}$ is an orthonormal basis of the tangent space TpM_2, then the scalar curvature τ at p is defined by

$$\tau(p) = \sum_{1 \leq i < j \leq m} K_{ij}.$$

Let M_2 be an m_2-dimensional Riemannian manifold, L be a k-plane section of T_pM_2, $p \in M_2$, and E be a unit vector in L. We choose an orthonormal basis $\{e_1, \ldots, e_k\}$ of L such that $e_1 = E$. Ricci curvature (or k-Ricci curvature) of L at E is defined by

$$\mathrm{Ric}_L(E) = K_{12} + K_{13} + \ldots + K_{1k},$$

where K_{ij} denotes, as usual, the sectional curvature of the 2-plane section spanned by e_i, e_j. For each integer k, $2 \leq k \leq m_2$, the Riemannian invariant θ_k on M_2 is defined by

$$\theta_k(p) = \frac{1}{k-1} \inf_{L,X} \mathrm{Ric}_L(E), \quad p \in M, \tag{2.8}$$

where L runs over all k-plane sections in T_pM_2 and E runs over all unit vectors in L. A Kaehler manifold $\tilde{M}_1$ is named a complex space form if it has a fixed holomorphic sectional curvature represented by $\tilde{M}_1(c)$. The curvature tensor of the complex space form $\tilde{M}_1(c)$ is dedicated by

$$\begin{aligned} \tilde{R}(E_1, E_2)F_1 = \frac{c}{4}\{&\tilde{h}(E_2, F_1)E_1 - \tilde{h}(E_1, F_1)E_2 + \tilde{h}(J_1E_2, F_1)J_1E_1 \\ &- \tilde{h}(J_1E_1, F_1)J_1F_2 + 2\tilde{h}(E_1, J_1E_2)J_1F_1\} \end{aligned} \tag{2.9}$$

for any $E_1, E_2, F_1 \in \Gamma(T\tilde{M}_1)$.

Definition 2.1 A submanifold M_2 of an almost Hermitian manifold $\tilde{M}_1$ is called pointwise slant if, at each point $p \in M_2$, the Wirtinger angle $\theta(E_1)$ is independent of the choice of nonzero vector $E_1 \in T_p^*M_2$, where $T_p^*M_2$ is the tangent space of nonzero vectors. In this case, θ is called the slant function of M_2 [16].

Definition 2.2 ([2]) Let M_2 be an isometrically immersed submanifold in a Kaehler manifold $\tilde{M}_1$. Then we say that M_2 is a pointwise quasi hemi-slant submanifold if it is furnished with three orthogonal distributions $(\mathfrak{D}, \mathfrak{D}_\theta, \mathfrak{D}^\perp)$ satisfying the conditions:

(i) $TM_2 = \mathfrak{D} \oplus \mathfrak{D}_\theta \oplus \mathfrak{D}^\perp$.
(ii) The distribution $\mathfrak{D}$ is invariant, i.e., $J\mathfrak{D} = \mathfrak{D}$.

(iii) For any nonzero vector field $E_1 \in (\mathfrak{D}_\theta)_p$, $p \in M_2$, the angle θ between $J_1 E_1$ and $(\mathfrak{D}_\theta)_p$ is slant function and is independent of the choice of the point p and E_1 in $(\mathfrak{D}_\theta)_p$,
(iv) The distribution $\mathfrak{D}^\perp$ is anti-invariant, i.e., $J_1\mathfrak{D}^\perp \subseteq \mathcal{T}^\perp M$.

We call the angle θ a pointwise quasi hemi-slant angle of M_2. A pointwise quasi hemi-slant submanifold M_2 is called proper if its pointwise slant function satisfies $\theta \neq 0, \frac{\pi}{2}$, and θ is not constant on M_2.

If we represent by $\tilde{d}_1$, $\tilde{d}_2$, and $\tilde{d}_3$ the dimension of $\mathfrak{D}$, $\mathfrak{D}_\theta$, and $\mathfrak{D}^\perp$, respectively, then from our generalized definition of pointwise quasi hemi-slant submanifold M_2, we can easily see the following particular cases:

(i) If $\tilde{d}_1 = 0$, then M_2 is pointwise hemi-slant submanifold.
(ii) If $\tilde{d}_2 = 0$, then M_2 is semi-invariant submanifold
(iii) If $\tilde{d}_3 = 0$, then M_2 is pointwise semi-slant submanifold.

Let M_2 be a pointwise quasi hemi-slant submanifold of a Kaehler manifold $\tilde{M}_1$. Then, for any $\xi \in \Gamma(TM_2)$, we have

$$\xi = Q\xi + R\xi + S\xi, \tag{2.10}$$

where Q, R, and S denote the projections on the distributions $\mathfrak{D}$, $\mathfrak{D}_\theta$, and $\mathfrak{D}^\perp$, respectively.

$$J_1\xi = P\xi + F\xi, \tag{2.11}$$

where $P\xi$ and $F\xi$ are tangential and normal components on M_2. By using (2.10) and (2.11), we get immediately

$$J_1\xi = PQ\xi + FQ\xi + PR\xi + FR\xi + PS\xi + FS\xi; \tag{2.12}$$

here since $J\mathfrak{D} = \mathfrak{D}$, we have $FQ\xi = 0$. Thus we get

$$J_1(TM_2) = \mathfrak{D} \oplus T\mathfrak{D}_\theta \oplus F\mathfrak{D}_\theta \oplus J_1\mathfrak{D}^\perp \tag{2.13}$$

and

$$T^\perp M_2 = F\mathfrak{D}_\theta \oplus J_1\mathfrak{D}^\perp \oplus \mu, \tag{2.14}$$

where μ is the orthogonal complement of $F\mathfrak{D}_\theta \oplus J_1\mathfrak{D}^\perp$ in $T^\perp M_2$ and $J_1\mu = \mu$. Also, for any $\eta \in T^\perp M_2$, we have

$$J_1\eta = B\eta + C\eta, \tag{2.15}$$

where $B\eta \in \Gamma(TM_2)$ and $C\eta \in \Gamma(T^\perp M_2)$.

3 Main Results

Let M_2 be m_2-dimensional pointwise quasi hemi-slant submanifold of a complex space form $\tilde{M}_1(c)$. Then, from (2.6) and (2.9) we get following equations:

$$\begin{aligned}R(E_1, E_2, E_3, E_4) = &\tilde{R}(E_1, E_2, E_3, E_4) + \tilde{h}(h(E_1, E_4), h(E_2, E_3)) \\ &- \tilde{h}(h(E_1, E_3), h(E_2, E_4)) \\ = &\frac{c}{4}\{\tilde{h}(E_2, E_3)\tilde{h}(E_1, E_4) - \tilde{h}(E_1, E_3)\tilde{h}(E_2, E_4) + \tilde{h}(J_1E_2, E_3)\tilde{h}(J_1E_1, E_4) \\ &- \tilde{h}(J_1E_1, E_3)\tilde{h}(J_1E_2, E_4) + 2\tilde{h}(E_1, J_1E_2)\tilde{h}(J_1E_3, E_4)\} \\ &+ \tilde{h}(h(E_1, E_4), h(E_2, E_3)) - \tilde{h}(h(E_1, E_3), h(E_2, E_4)). \end{aligned} \tag{3.1}$$

We choose

$$e_1, e_2 = J_1e_1, \ldots, e_{2r_1-1}, e_{2d_1} = J_1e_{2d_1-1}, e_{2d_1+1}, e_{2d_1+2} = \sec\sigma\, Pe_{2d_1+1}, \ldots,$$
$$e_{2d_1+2d_2-1}, e_{2d_1+2d_2} = \sec\sigma\, Pe_{2d_1+2d_2-1}, e_{2d_1+2d_2+1}, \ldots, e_{2d_1+2d_2+d_3}$$

an orthonormal basis of T_pM_2, where $m_2 = 2d_1 + 2d_2 + d_3$. Then we get

$$\tilde{h}^2(Je_d, e_{d+1}) = \begin{cases} 1 & , \quad d \in \{1, 2, \ldots, 2d_1 - 1\} \\ \cos^2\sigma & , \quad d \in \{2d_1 + 1, \ldots, 2d_1 + 2d_2 - 1\} \\ 0 & , \quad d \in \{2d_1 + 2d_2 + 1, \ldots, 2d_1 + 2d_2 + d_3 - 1\} \end{cases} \tag{3.2}$$

and

$$\sum_{i,j=1}^{m_2} \tilde{h}^2(Je_i, e_j) = 2(d_1 + d_2\cos^2\sigma). \tag{3.3}$$

Lemma 3.1 *If $r \geq 2$ and $p_1, \ldots, p_r, p$ are real numbers such that*

$$\left(\sum_{i=1}^{r} p_i\right)^2 = (r-1)\left(\sum_{i=1}^{r} p_i^2 + p\right),$$

then $2p_1p_2 \geq p$ with equality holding if and only if

$$p_1 + p_2 = p_3 = \ldots = p_n.$$

Theorem 3.2 *Let M_2 be m_2-dimensional pointwise quasi hemi-slant submanifold of a complex space form $\tilde{M}_1(c)$. Then, the following statements are true:*

(i) For any plane section π invariant by P and tangent to $\mathfrak{D}$,

$$\tau(p) - K(\pi) \leq \frac{m_2^2(m_2-2)}{2(m_2-1)} \|H\|^2 + \frac{c}{8}\left((m_2^2 - m_2 - 2) + 6(d_1 - 1 + d_2 \cos^2\sigma)\right). \tag{3.4}$$

(ii) For any plane section π invariant by P and tangent to $\mathfrak{D}^\theta$,

$$\tau(p) - K(\pi) \leq \frac{m_2^2(m_2-2)}{2(m_2-1)} \|H\|^2 + \frac{c}{8}\left((m_2^2 - m_2 - 2) + 6(d_1 + (d_2 - 1)\cos^2\sigma)\right). \tag{3.5}$$

(iii) For any plane section π invariant by P and tangent to $\mathfrak{D}^\perp$,

$$\tau(p) - K(\pi) \leq \frac{m_2^2(m_2-2)}{2(m_2-1)} \|H\|^2 + \frac{c}{8}\left((m_2^2 - m_2 - 2) + 6(d_1 + d_2 \cos^2\sigma)\right). \tag{3.6}$$

Equality case of (3.4), (3.5), and (3.6) at a point $p \in M_2$ if and only if

$$A_{m_2+1} = \begin{bmatrix} a\ 0\ 0\ .. & 0 \\ 0\ b\ 0\ .. & 0 \\ 0\ 0\ \ ... & \lambda I_{m_2-2} \end{bmatrix} \tag{3.7}$$

and

$$A_r = \begin{bmatrix} h_{11}^r & h_{12}^r & 0\ .. & 0 \\ h_{12}^r & h_{22}^r & 0\ .. & 0 \\ 0 & 0 & ... & 0_{n-2} \end{bmatrix}, \quad r = m_2+2, \ldots, n, \tag{3.8}$$

where $\lambda = a+b$, $\{e_1, e_2, \ldots, e_{m_2}\}$ basis of T_pM_2 and $\{e_{m_2+1}, e_{m_2+2}, \ldots, e_n\}$ basis of $T_pM_2^\perp$.

Proof Let $p \in M_2$, $\{e_1, e_2, \ldots, e_{m_2}\}$ be the basis of T_pM_2, and $\{e_{m_2+1}, e_{m_2+2}, \ldots, e_n\}$ be the basis of $T_pM_2^{\perp}$. If we put $E_1 = E_4 = e_i$ and $E_2 = E_3 = e_j$ in Eq. (3.1), then

$$\sum_{i,j=1}^{m_2} R(e_i, e_j, e_j, e_i) = \sum_{i,j=1}^{m_2} \tilde{h}(h(e_i, e_i), h(e_j, e_j)) - \tilde{h}(h(e_i, e_j), h(e_j, e_i))$$

$$+ \frac{c}{4} \sum_{i,j=1}^{m_2} \Big\{ \tilde{h}(e_j, e_j)\tilde{h}(e_i, e_i) - \tilde{h}(e_i, e_j)\tilde{h}(e_j, e_i)$$

$$+ \tilde{h}(J_1 e_j, e_j)\tilde{h}(J_1 e_i, e_i)$$

$$- \tilde{h}(J_1 e_i, e_j)\tilde{h}(J_1 e_j, e_i) + 2\tilde{h}(e_i, J_1 e_j), \tilde{h}(J_1 e_j, e_i) \Big\} \tag{3.9}$$

holds. Hence we derive

$$2\tau(p) = m_2^2 \|H\|^2 - \|h\|^2 + \frac{c}{4} \left\{ (m_2^2 - m_2) + 6 \sum_{i,j=1}^{m_2} \tilde{h}^2(J_1 e_i, e_j) \right\}, \tag{3.10}$$

where

$$\|h\|^2 = \sum_{i,j=1}^{n} \tilde{h}(h(e_i, e_j), h(e_i, e_j)).$$

If we use (3.3) in (3.10), we get

$$2\tau(p) = m_2^2 \|H\|^2 - \|h\|^2 + \frac{c}{4} \left\{ (m_2^2 - m_2) + 6(d_1 + d_2 \cos^2 \sigma) \right\}. \tag{3.11}$$

If we denote

$$\varepsilon = 2\tau(p) - \frac{m_2^2(m_2 - 2)}{m_2 - 1} \|H\|^2 - \frac{c}{4} \left\{ (m_2^2 - m_2) + 6(d_1 + d_2 \cos^2 \sigma) \right\}, \tag{3.12}$$

then

$$m_2^2 \|H\|^2 = (m_2 - 1)(\varepsilon + \|h\|^2) \tag{3.13}$$

holds.

Let $p \in M_2$, $\pi \subset T_pM_2$, $dim\pi = 2$, and π invariant by P.

We consider three cases:

Case 1 The plane section π is tangent to $\mathfrak{D}$. We may assume that $\pi = sp\{e_1, e_2\}$ and $e_{m_2+1} = \frac{H}{\|H\|}$, and Eq. (3.13) can be rewritten by

$$\left(\sum_{i=1}^{m_2} h_{ii}^{m_2+1}\right)^2 = (m_2 - 1)\left(\sum_{i,j=1}^{m_2}\sum_{r=m_2+1}^{n}\left(h_{ij}^r\right)^2 + \varepsilon\right) \tag{3.14}$$

or

$$\left(\sum_{i=1}^{m_2} h_{ii}^{m_2+1}\right)^2 = (m_2 + 1)\left(\sum_{i=1}^{m_2}\left(h_{ii}^{m_2+1}\right)^2 + \sum_{i\neq j}^{m_2}\left(h_{ij}^{m_2+1}\right)^2 + \sum_{i,j=1}^{m_2}\sum_{r=m_2+2}^{n}\left(h_{ij}^r\right)^2 + \varepsilon\right), \tag{3.15}$$

where

$$h_{ij}^r = \tilde{h}(h(e_i, e_j), e_r).$$

If we apply Lemma 3.1 to Eq. (3.15), we get

$$2h_{11}^{m_2+1}h_{22}^{m_2+1} > \sum_{i\neq j}^{m_2}\left(h_{ij}^{m_2+1}\right)^2 + \sum_{i,j=1}^{m_2}\sum_{r=m_2+2}^{n}\left(h_{ij}^r\right)^2 + \varepsilon. \tag{3.16}$$

If we put $E_1 = E_3 = e_1$ and $E_2 = E_4 = e_2$ in Eq. (3.1), then we get

$$\begin{aligned} K(\pi) &= \frac{c}{4}\{\tilde{h}(e_2, e_2)\tilde{h}(e_1, e_1) - (\tilde{h}(e_1, e_2))^2 + \tilde{h}(J_1 e_2, e_2)\tilde{h}(J_1 e_1, e_1) \\ &\quad - \tilde{h}(J_1 e_1, e_2)\tilde{h}(J_1 e_2, e_1)\} + 2\tilde{h}(e_1, J_1 e_2)\tilde{h}(J_1 e_2, e_1)\} \\ &\quad + \sum_{r=m_2+1}^{n}\left(h_{11}^r h_{22}^r - (h_{12}^r)^2\right) \\ &= \frac{c}{4}\left\{1 + 3g^2(Je_1, e_2)\right\} + \sum_{r=m_2+1}^{n}\left(h_{11}^r h_{22}^r - (h_{12}^r)^2\right). \end{aligned} \tag{3.17}$$

By using (3.16) in (3.17)

$$K(\pi) \geq c + \frac{1}{2}\left(\sum_{i\neq j}^{m_2}\left(h_{ij}^{m_2+1}\right)^2 + \sum_{r=m_2+2}^{n}\sum_{i,j>2}^{m_2}\left(h_{ij}^r\right)^2\right.$$

$$+ \sum_{r=m_2+1}^{n} \left(h_{11}^r + h_{22}^r\right)^2 + \varepsilon \Bigg)$$

$$+ \sum_{j>2}^{m_2} \left(\left(h_{1j}^{m_2+1}\right)^2 + \left(h_{2j}^{m_2+1}\right)^2 \right).$$

Finally, we can write

$$K(\pi) \geq c + \frac{\varepsilon}{2}. \tag{3.18}$$

By virtue of (3.12) and (3.18), we have

$$\begin{aligned} \tau(p) - K(\pi) \quad &\leq \frac{m_2^2(m_2-2)}{2(m_2-1)} \|H\|^2 \\ &+ \frac{c}{8}\left((m_2^2 - m_2 - 2) + 6(d_1 - 1 + d_2\cos^2\sigma)\right). \end{aligned} \tag{3.19}$$

Hence we get (3.4).

If the equality case of (3.4) holds, then the inequalities given by (3.16) and (3.19) become equalities, and we have the equality in Lemma 3.1

$$h_{ij}^{m_2+1} = 0, \forall i \neq j,\ i, j > 2,$$

$$h_{ij}^r = 0, \forall i \neq j,\ i, j > 2, r = m_2 + 1, \ldots, n,$$

$$h_{11}^r + h_{22}^r = 0, \forall r = m_2 + 2, \ldots, n,$$

$$h_{1j}^{m_2+1} = h_{2j}^{m_2+1} = 0, \forall j > 2,$$

$$h_{11}^{m_2+1} + h_{22}^{m_2+1} = h_{33}^{m_2+1} = \ldots = h_{m_2m_2}^{m_2+1}.$$

We may choose $\{e_1, e_2\}$ such that $h_{12}^{m_2+1} = 0$, and we denote $a = h_{11}^{m_2+1}$, $b = h_{22}^{m_2+1}$, $\lambda = h_{33}^{m_2+1} = \ldots = h_{m_2m_2}^{m_2+1}$. Thus, the shape operator of M_2 takes the form given by (3.7) and (3.8). The converse is easy to follow. Similar to the proof of Case (i), one can obtain Case (ii) and Case (iii). □

From the last theorem, we have the following corollary:

Corollary 3.3 *Let M_2 be an m_2-dimensional pointwise quasi hemi-slant submanifold of a complex space form $\tilde{M}_1(c)$. Then, the following statements are true:*

$$\delta_{M_2} \leq \frac{m_2^2(m_2-2)}{2(m_2-1)} \|H\|^2 + \frac{c}{8}\left((m_2^2 - m_2 - 2) + 6(d_1 - 1 + d_2\cos^2\sigma)\right), \tag{3.20}$$

$$\delta_{M_2} \le \frac{m_2^2(m_2-2)}{2(m_2-1)}\|H\|^2 + \frac{c}{8}\Big((m_2^2 - m_2 - 2) + 6(d_1 + (d_2-1)\cos^2\sigma\Big), \tag{3.21}$$

$$\delta_{M_2} \le \frac{m_2^2(m_2-2)}{2(m_2-1)}\|H\|^2 + \frac{c}{8}\Big((m_2^2 - m_2 - 2) + 6(d_1 + d_2\cos^2\sigma)\Big). \tag{3.22}$$

Equality case of (3.20), (3.21), and (3.22) if and only if, for the basis $\{e_1, e_2, \ldots, e_{m_2}\}$ *of* T_pM_2 *and basis* $\{e_{m_2+1}, e_{m_2+2}, \ldots, e_n\}$ *of* $T_pM_2^{\perp}$, *equations (3.7) and (3.8) are satisfied.*

Theorem 3.4 *Let* M_2 *be an* m_2*-dimensional pointwise quasi hemi-slant submanifold of a complex space form* $\tilde{M}_1(c)$. *Then, the following statements are true.*

(*i*) *For each unit vector* $E \in \Gamma(\mathfrak{D})$ *we have*

$$Ric(E) \le \frac{1}{4}m_2^2\|H\|^2 + \frac{c}{4}(m_2+2). \tag{3.23}$$

(*ii*) *For each unit vector* $E \in \Gamma(\mathfrak{D}^\theta)$ *we have*

$$Ric(E) \le \frac{1}{4}m_2^2\|H\|^2 + \frac{c}{4}(m_2 - 1 + 3\cos^2\sigma). \tag{3.24}$$

(*iii*) *For each unit vector* $E \in \Gamma(\mathfrak{D}^\perp)$ *we have*

$$Ric(E) \le \frac{1}{4}m_2^2\|H\|^2 + \frac{c}{4}(m_2 - 1). \tag{3.25}$$

Also, the equality cases of (3.23)–(3.25) hold if and only if there exist an orthonormal basis $\{e_1 = E, e_2, \ldots, e_{m_2}\}$ *of* T_pM_2 *and* $\{e_{m_2+1}, e_{m_2+2}, \ldots, e_n\}$ *of* $T_pM_2^\perp$ *such that*

$$h_{12}^r = h_{13}^r = \ldots = h_{1m_2}^r = 0 \text{ and } h_{11}^r = h_{22}^r + \ldots + h_{m_2m_2}^r,\ r \in \{m_2+1, \ldots, n\}.$$

Proof Let M_2 be an m_2-dimensional hemi-slant submanifold of a complex space form $\tilde{M}_1(c)$. From (3.11), we can write

$$2\tau(p) = m_2^2\|H\|^2 - \|h\|^2 + \frac{c}{4}\Big\{(m_2^2 - m_2) + 6\big(d_1 + d_2\cos^2\sigma\big)\Big\}. \tag{3.26}$$

From (3.26), we have

$$\begin{aligned} \frac{1}{4}m_2^2\|H\|^2 &= \tau(p) - \frac{c}{8}\Big\{\big(m_2^2 - m_2\big) + 6\big(d_1 + d_2\cos^2\sigma\big)\Big\} \\ &\quad + \frac{1}{4}\sum_{r=m_2+1}^{n}\big(h_{11}^r - h_{22}^r - \ldots - h_{nn}^r\big)^2 \end{aligned} \tag{3.27}$$

$$+\sum_{r=m_2+1}^{n}\sum_{j=2}^{m_2}(h_{1j}^r)^2-\sum_{r=m_2+1}^{n}\sum_{2\le i<j\le m_2}\big(h_{ii}^r h_{jj}^r-(h_{ij}^r)^2\big).$$

Using (2.6), we also have

$$\sum_{r=m_2+1}^{n}\sum_{2\le i<j\le m_2}(h_{ii}^r h_{jj}^r-(h_{ij}^r)^2)=\sum_{2<i<j\le m_2}K_{ij}-\sum_{2\le i<j\le m_2}\widetilde{K}_{ij}. \tag{3.28}$$

On the other hand, since $\tilde{M}_1(c)$ is a complex space form, its curvature tensor $\tilde{R}$ satisfies (2.9), and we get

$$\sum_{2\le i<j\le m_2}\widetilde{K}_{ij}=\frac{c}{4}\left\{\frac{(m_2-2)(m_2-1)}{2}+3\sum_{2\le i<j\le m_2}\tilde{h}^2(J_1e_i,e_j)\right\}. \tag{3.29}$$

As $e_1\in\Gamma(D)$, we get immediately

$$\sum_{2\le i<j\le m_2}\widetilde{K}_{ij}=\frac{c}{8}\left\{\big(m_2^2-3m_2-2\big)+6\big(d_1-1+d_2\cos^2\sigma\big)\right\}. \tag{3.30}$$

By virtue of (3.27), (3.28), and (3.30), we have

$$\begin{aligned}\frac{1}{4}m_2^2\,\|H\|^2&=\tau(p)-\frac{c}{4}\left\{(m^2-m)+6\big(d_1+d_2\cos^2\sigma\big)\right\}\\&\quad+\frac{1}{4}\sum_{r=m_2+1}^{n}\big(h_{11}^r-h_{22}^r-\ldots-h_{nn}^r\big)^2\\&\quad+\sum_{r=m_2+1}^{n}\sum_{j=2}^{m}(h_{1j}^r)^2-\sum_{2<i<j\le m_2}K_{ij}\\&\quad+\frac{c}{8}\left\{(m_2^2-3m_2-2)+6(d_1-1+d_2\cos^2\sigma)\right\}\end{aligned} \tag{3.31}$$

holds. Hence, we get

$$\begin{aligned}\mathrm{Ric}(e_1)&=\frac{1}{4}m_2^2\,\|H\|^2+\frac{c}{4}(m_2-1+3)-\frac{1}{4}\sum_{r=m_2+1}^{n}\big(h_{11}^r-h_{22}^r-\ldots-h_{nn}^r\big)^2\\&\quad-\sum_{r=m_2+1}^{n}\sum_{j=2}^{m_2}(h_{1j}^r)^2.\end{aligned} \tag{3.32}$$

If we choose $e_1 = E$ as any unit vector of T_pM_2 in the above equation, we obtain (3.23). Now, we remark that the equality case of (3.23) holds if and only if the equality is attained in (3.32). However, this happens if and only if

$$h^r_{12} = h^r_{13} = \ldots = h^r_{1m_2} = 0 \text{ and } h^r_{11} = h^r_{22} + \ldots + h^r_{m_2m_2},\ r \in \{m_2+1, \ldots, n\}.$$

The proof of the converse part is straightforward. Thus we obtain Case (i). Similar to the proof of Case (i), one can get Case (ii) and Case (iii). □

Theorem 3.5 *Let M_2 be an m_2-dimensional pointwise quasi hemi-slant submanifold of a complex space form $\tilde{M}_1(c)$. Then we get*

$$\tau(p) \leq \frac{(m_2^2 - m_2)}{2} \|H\|^2 + \frac{c}{8}\left\{(m_2^2 - m_2) + 6(d_1 + d_2\cos^2\sigma)\right\}. \tag{3.33}$$

The equality case of (3.33) holds at $p \in M_2$ if and only if p is a totally umbilical point.

Proof Let $p \in M_2$ and $\{e_1, \ldots, e_{m_2}\}$ be an orthonormal basis of T_pM_2. The relation (3.10) is equivalent to

$$m_2^2\|H\|^2 = 2\tau(p) + \|h\|^2 + \frac{c}{4}\left\{(m_2^2 - m_2) + 6(d_1 + d_2\cos^2\sigma)\right\}. \tag{3.34}$$

We choose an orthonormal basis $\{e_1, \ldots, e_{m_2}, e_{m_2+1}, \ldots, e_n\}$ at p such that e_{m_2+1} is parallel to the mean curvature vector $H(p)$ and $e_1, \ldots, e_{m_2}$ diagonalize the shape operator $A_{e_{m_2+1}}$. Then the shape operators take the forms

$$A_{e_{n+1}} = \begin{bmatrix} a_1 & 0 & \ldots & 0 \\ 0 & a_2 & \ldots & 0 \\ \cdot & \cdot & \cdot & \cdot \\ \cdot & \cdot & \cdot & \cdot \\ \cdot & \cdot & \cdot & \cdot \\ 0 & 0 & \ldots & a_{m_2} \end{bmatrix}, \tag{3.35}$$

$$A_{e_r} = \left(h^r_{ij}\right), \quad i, j = 1, \ldots, m_2; \quad r = m_2+2, \ldots, n, \quad \text{trace} A_{e_r} = 0. \tag{3.36}$$

By (3.34), we have

$$\begin{aligned} m_2^2\|H\|^2 &= 2\tau(p) + \sum_{i=1}^{n} a_i^2 + \sum_{r=m_2+2}^{n}\sum_{i,j=1}^{m_2}(h^r_{ij})^2 \\ &+ \frac{c}{4}\left\{(m_2^2 - m_2) + 6(d_1 + d_2\cos^2\sigma)\right\}. \end{aligned} \tag{3.37}$$

Since

$$0 \leq \sum_{i<j} (p_i - p_j)^2 = (m_2 - 1) \sum_i p_i^2 - 2 \sum_{i<j} p_i p_j, \tag{3.38}$$

we find

$$m_2^2 \|H\|^2 = \left(\sum_{i=1}^{m_2} p_i \right)^2 = \sum_{i=1}^{m_2} p_i^2 + 2 \sum_{i<j} p_i p_j \leq m_2 \sum_{i=1}^{m_2} p_i^2, \tag{3.39}$$

which gives

$$\sum_{i=1}^{m_2} p_i^2 \geq m_2 \|H\|^2 . \tag{3.40}$$

Hence by using (3.37) and (3.40), we get

$$\begin{aligned} m_2^2 \|H\|^2 \geq 2\tau(p) + m_2 \|H\|^2 + \sum_{r=m_2+2}^{n} \sum_{i,j=1}^{m_2} (h_{ij}^r)^2 \\ + \frac{c}{4} \left\{ (m_2^2 - m_2) + 6(d_1 + d_2 \cos^2 \sigma) \right\}. \end{aligned} \tag{3.41}$$

If the equality case of (3.33) holds, then from (3.38) and (3.41) it follows that

$$p_1 = p_2 = \ldots = p_{m_2} \text{ and } A_{e_r} = 0, \; r = m_2 + 2, \ldots, n. \tag{3.42}$$

Hence, p is a totally umbilical point. The converse is straightforward. □

Theorem 3.6 *Let M_2 be an m_2-dimensional pointwise quasi hemi-slant submanifold of a complex space form $\tilde{M}_1(c)$. Then we have*

$$\theta_k(p) \leq \|H\|^2 + \frac{c}{4} \left\{ 1 + \frac{6}{m_2^2 - m_2} (d_1 + d_2 \cos^2 \sigma) \right\}. \tag{3.43}$$

Proof Let $\{e_1, \ldots, e_{m_2}\}$ be an orthonormal basis of $T_p M_2$. Denote by $L_{i_1 \ldots i_k}$ the k-plane section spanned by $\{e_{i_1}, \ldots, e_{i_k}\}$. If we consider the definitions of the Ricci and scalar curvatures, we have

$$\tau\left(L_{i_1 \ldots i_k}\right) = \frac{1}{2} \sum_{i \in \{i_1, \ldots, i_k\}} \mathrm{Ric}_{L_{i_1 \ldots i_k}}(e_i), \tag{3.44}$$

$$\tau(p) = \frac{1}{C_{m_2-2}^{k-2}} \sum_{1 \leq i_1 < \ldots \leq i_k \leq m_2} \tau\left(L_{i_1 \ldots i_k}\right). \tag{3.45}$$

By virtue of (2.8), (3.44), and (3.45), we get

$$\tau(p) \geq \frac{m_2(m_2-1)}{2}\theta_k(p). \tag{3.46}$$

Taking into account of (3.33) and (3.46), we obtain (3.43). □

References

1. Akyol, M.A., Beyendi, S.: A note on quasi bi-slant submanifolds of cosymplectic manifolds. Commun. Fac. Sci. Univ. Ank. Ser. A1 Math. Stat. **69**(2), 1508–1521 (2020)
2. Beyendi, S., Akyol, M.A., Stankovic, M.S.: Pointwise quasi hemi-slant submanifolds. Filomat **37**(1), 127–138 (2023)
3. Cabrerizo, J.L., Carriazo, A., Fernandez, L.M., Fernandez, M.: Semi-slant submanifolds of a Sasakian manifold. Geom. Dedic. **78**(2), 183–199 (1999)
4. Carriazo, A.: Bi-slant immersions. In: Proceeding of the ICRAMS 2000, Kharagpur, pp. 88–97 (2000)
5. Chen, B.Y.: Slant immersions. Bull. Aust. Math. Soc. **41**, 135–147 (1990)
6. Chen, B.Y.; Geometry of Slant Submanifolds, Katholieke Universiteit Leuven, Leuven, Belgium. View at Zentralblatt Math. (1990)
7. Chen, B.Y.: Some pinching and classification theorems for minimal submanifolds. Arch. Math. **60**(6), 568–578 (1993)
8. Chen, B.Y.: A Riemannian invariant for submanifolds in space forms and its applications. In: Geometry and Topology of Submanifolds VI (Leuven, 1993/Brussels, 1993), pp. 568–578. Word Scientific Publishing, River Edge (1994)
9. Chen, B.Y.: Mean curvature and shape operator of isometric immersion in real space forms. Glasgow Math. J. **38**, 87–97 (1996)
10. Chen, B.Y.: Strings of Riemannian invariants, inequalities, ideal immersions and their applications. In: The Third Pacific Rim Geometry Conference (Seoul, 1996). Geometry and Topology Monographs, vol. 25, pp. 7–60. International Press, Cambridge (1998)
11. Chen, B.Y.: Relation between Ricci curvature and shape operator for submanifolds with arbitrary codimension. Glasgow Math. J. **41**, 33–41 (1999)
12. Chen, B.-Y.: Riemannian submanifolds. In: Handbook of Differential Geometry, vol. I, pp. 187–418. North-Holland, Amsterdam (2000)
13. Chen, B.Y.: A general optimal inequality for a Riemannian submanifold. J. Inequal. Pure Appl. Math. **6**(3), 1–11 (2005)
14. Chen, B.Y.: Pseudo-Riemannian Geometry, δ-Invariants and Applications. World Scientific Publishing Co. Pte. Ltd., Hackensack (2011)
15. Chen, B.Y., Blaga, A.M.: Recent developments on Chen–Ricci inequalities in differential geometry. In: Geometry of Submanifolds and Applications, pp. 1–61. Springer Nature, Singapore (2024)
16. Chen, B.Y., Garay, O.J.: Pointwise slant submanifolds in almost Hermitian manifolds. Turk. J. Math. **36**, 630–640 (2012)
17. Etayo, F.: On quasi-slant submanifolds of an almost Hermitian manifold. Publ. Math. Debr. **53**, 217–223 (1998)
18. Hong, S., Tripathi, M.M.: On Ricci curvature of submanifolds. Int. J. Pure Appl. Math. Sci. **2**, 227–245 (2005)
19. Liu, X., Zhou, J.: On Ricci curvature of certain submanifolds in cosymplectic space form. Sarajeva J. Math. **2**, 95–106 (2006)

20. Lone, M.A., Lone, M.S., Shahid, M.H.: Hemi-slant submanifolds of cosymplectic manifolds. Cogent Math. **3**(1), 1204143 (2016)
21. Matsumoto, K., Mihai, I., Oiaga, A.: Shape operator for slant submanifolds in complex space forms. Bull. Yamagata Univ. Natur. Sci. **14**(4), 169–177 (2000)
22. Mihai, A., Özgür, C.: Chen inequalities for submanifolds of real space forms with a semi-symmetric metric connection. Taiwan. J. Math. **14**(4), 1465–1477 (2010)
23. Oiaga, A., Mihai, I., B. Y. Chen inequalities for slant submanifolds in complex space forms. Demonstratio Math. **32**(4), 835–846 (1999)
24. O'Neill, B.: Semi-Riemannian Geometry with Applications to Relativity. Academic, London (1983)
25. Özgür, C.: B. Y. Chen inequalities for submanifolds of a Riemannian manifold of quasi-constant curvature. Turk. J. Math. **35**(3), 501–509 (2011)
26. Papaghuic, N.: Semi-slant submanifold of Kaehlerian manifold. An. St. Univ. AI. I. Cuza. Iasi. Math.(N.S.) **9**, 55–61 (1994)
27. Prasad, R., Shukla, S.S., Haseeb, A., Kumar, S.: Quasi hemi-slant submanifolds of Kaehler manifolds. Honam Math. J. **42**(4), 795–809 (2020)
28. Şahin, B.: Non-existence of warped product semi-slant submanifolds of Kaehler manifolds. Geom. Dedicata **117**, 195–202 (2006)
29. Şahin, B.: Warped product pointwise semi-slant submanifolds of Kaehler manifolds. Port. Math. **70**, 252–268 (2013)
30. Taştan, H.M., Ozdemir, F.: The geometry of hemi-slant submanifolds of a locally product Riemannian manifold. Turk. J. Math. **39**, 268–284 (2015)
31. Tripathi, M.M.: Improved Chen-Ricci inequality for curvature-like tensor and its applications. Differ. Geom. Appl. **29**, 685–698 (2011)
32. Tripathi, M.M.: Chen-Ricci inequality for curvature like tensor and its applications. Differ. Geom. Appl. **29**(5), 685–692 (2011)
33. Uddin, S., Chen, B.Y., Al-Solamy, F.R.: Warped product bi-slant immersions in Kaehler manifolds. Mediterr. J. Math. **14**(2), 14–95 (2017)
34. Yano, K., Kon, M.: Structures on Manifolds. Series in Pure Mathematics. World Scientific Publishing Co., Singapore (1984)
35. Zhang, P., Zhang, L., Song, W.: Chen's inequalities for submanifolds of a Riemannian manifold of quasi-constant curvature with a semi-symmetric metric connection. Taiwan. J. Math. **18**, 1841–1862 (2014)

Correction to: Statistical Maps and Chen's First Inequality for These Maps

Sema Kazan and Aliya Naaz Siddiqui

Correction to:
Chapter 8 in: B.-Y. Chen, M. A. Choudhary (eds.), *Geometric Inequalities and Applications*, Infosys Science Foundation Series in Mathematical Sciences, https://doi.org/10.1007/978-981-95-5148-4_8

Owing to an unfortunate oversight during the production process, the surname of the author Aliya Naaz Suddiqui was incorrectly presented in author byline of the Chapter 8 in the initially published version. The chapter has been updated with the correction as Aliya Naaz Siddiqui.

The updated version of this chapter can be found at
https://doi.org/10.1007/978-981-95-5148-4_8

B.-Y. Chen, M. A. Choudhary (eds.), *Geometric Inequalities and Applications*,
Infosys Science Foundation Series in Mathematical Sciences,
https://doi.org/10.1007/978-981-95-5148-4_16

MIX
Papier aus verantwortungsvollen Quellen
Paper from responsible sources
FSC® C105338

If you have any concerns about our products,
you can contact us on
ProductSafety@springernature.com

In case Publisher is established outside the EU,
the EU authorized representative is:
Springer Nature Customer Service Center GmbH
Europaplatz 3, 69115 Heidelberg, Germany

Printed by Libri Plureos GmbH
in Hamburg, Germany